jOOQ Masterclass

A practical guide for Java developers to write SQL queries for complex database interactions

Anghel Leonard

BIRMINGHAM—MUMBAI

jOOQ Masterclass

Copyright © 2022 Packt Publishing

All rights reserved. No part of this book may be reproduced, stored in a retrieval system, or transmitted in any form or by any means, without the prior written permission of the publisher, except in the case of brief quotations embedded in critical articles or reviews.

Every effort has been made in the preparation of this book to ensure the accuracy of the information presented. However, the information contained in this book is sold without warranty, either express or implied. Neither the author, nor Packt Publishing or its dealers and distributors, will be held liable for any damages caused or alleged to have been caused directly or indirectly by this book.

Packt Publishing has endeavored to provide trademark information about all of the companies and products mentioned in this book by the appropriate use of capitals. However, Packt Publishing cannot guarantee the accuracy of this information.

Assistant Group Product Manager: Alok Dhuri
Senior Editor: Kinnari Chohan
Technical Editor: Maran Fernandes
Copy Editor: Safis Editing
Project Coordinator: Manisha Singh
Proofreader: Safis Editing
Indexer: Tejal Daruwale Soni
Production Designer: Aparna Bhagat
Marketing Coordinator: Sonakshi Bubbar

First published: August 2022

Production reference: 2110822

Published by Packt Publishing Ltd.
Livery Place
35 Livery Street
Birmingham
B3 2PB, UK.

ISBN 978-1-80056-689-7

www.packt.com

Foreword

It has been a great pleasure reviewing Anghel's book over the past year.

Anghel is very knowledgeable about databases in general, as well as SQL specifically and the various popular persistence technologies in the Java ecosystem. This shows in his examples, which are suitable both for beginners and for SQL gurus, who will use at least two window functions in every query.

jOOQ Masterclass is very well structured, both for jOOQ rookies who wish to learn about how to put jOOQ to best use and advanced jOOQ users who wish to have a reference for almost any use case that jOOQ supports. Anghel has a vast number of examples, which are both intuitive and powerful.

With this book at your disposal, your next jOOQ application will be a breeze.

– Lukas Eder, Founder and CEO of Data Geekery, the company behind jOOQ

Contributors

About the author

Anghel Leonard is a chief technology strategist and independent consultant with 20+ years of experience in the Java ecosystem. In his daily work, he is focused on architecting and developing Java-distributed applications that empower robust architectures, clean code, and high performance. He is also passionate about coaching, mentoring, and technical leadership. He is the author of several books, videos, and dozens of articles related to Java technologies.

I want to thank Lukas Eder who has guided me while writing this book and who has always been quick to answer my questions, suggestions, and so on.

About the reviewers

Matthew Cachia has a bachelor's degree in computer science and has been working in the payments space for more than 14 years. He is currently the technical architect of `Weavr.io`.

He is into static-type languages – most notably, Java, Scala, and Kotlin. He has become increasingly fascinated by compilers, transpilers, and languages overall – a field he regrets not picking up when reading his degree.

In his free time, he likes playing role-playing games and spending time with his family – Georgiana, Thomas, and Alex.

Lukas Eder is the founder and CEO of Data Geekery, the company behind jOOQ.

Table of Contents

Part 2: jOOQ and Queries

3
jOOQ Core Concepts 57

4
Building a DAO Layer (Evolving the Generated DAO Layer) 113

5
Tackling Different Kinds of SELECT, INSERT, UPDATE, DELETE, and MERGE 125

6
Tackling Different Kinds of JOINs 167

7
Types, Converters, and Bindings 195

8
Fetching and Mapping 235

Part 3: jOOQ and More Queries

9

CRUD, Transactions, and Locking 315

10

Exporting, Batching, Bulking, and Loading 375

11
jOOQ Keys 419

12
Pagination and Dynamic Queries 455

Part 4: jOOQ and Advanced SQL

13
Exploiting SQL Functions 493

14
Derived Tables, CTEs, and Views 553

15
Calling and Creating Stored Functions and Procedures 593

16
Tackling Aliases and SQL Templating 631

17
Multitenancy in jOOQ 659

Part 5: Fine-tuning jOOQ, Logging, and Testing

18
jOOQ SPI (Providers and Listeners) 673

19
Logging and Testing 701

Preface

The last decade has constantly changed how we think and write applications including the persistence layer, which must face new challenges such as working in microservice architectures and cloud environments. Flexibility, versatility, dialect agnostic, rock-solid SQL support, small learning curve, and high performance are just a few of the attributes that makes jOOQ the most appealing persistence technology for modern applications.

Being part of the modern technology stack, jOOQ is the new persistence trend that respects all standards of a mature, robust, and well-documented technology. This book covers jOOQ in detail, so it prepares you to become a jOOQ power-user and an upgraded version of yourself ready to handle the future of the persistence layer. Don't think of jOOQ as just another piece of technology; think of it as part of your mindset, your straightforward road to exploiting SQL instead of abstracting it, and your approach to doing things right in your organization.

Who this book is for

This book is for Java developers who write applications that interact with databases via SQL. No prior experience with jOOQ is assumed.

What this book covers

Chapter 1, *Starting jOOQ and Spring Boot*, shows how to create a kickoff application that involves jOOQ and Spring Boot in Java/Kotlin under Maven/Gradle.

Chapter 2, *Customizing the jOOQ Level of Involvement*, covers the configurations (declarative and programmatic) needed for employing jOOQ as a type-safe query builder and executor. Moreover, we set up jOOQ to generate POJOs and DAOs on our behalf. We use Java/Kotlin under Maven/Gradle.

Chapter 3, *jOOQ Core Concepts*, discusses jOOQ core concepts such as fluent API, SQL syntax correctness, emulating missing syntax/logic, jOOQ result set, jOOQ records, type safety, CRUD binding, and inlining parameters.

Chapter 4, Building a DAO Layer (Evolving the Generated DAO Layer), shows how implementing a DAO layer can be done in several ways/templates. We tackle an approach for evolving the DAO layer generated by jOOQ.

Chapter 5, Tackling Different Kinds of SELECT, INSERT, UPDATE, DELETE, and MERGE Statements, covers different kinds of `SELECT`, `INSERT`, `UPDATE`, `DELETE`, and `MERGE` queries. For example, we cover nested `SELECT`, `INSERT...DEFAULT VALUES`, `INSERT...SET` queries, and so on.

Chapter 6, Tackling Different Kinds of JOIN Statements, tackles different kinds of `JOIN`. jOOQ excels on standard and non-standard `JOIN`. We cover `INNER`, `LEFT`, `RIGHT`, `...`, `CROSS`, `NATURAL`, and `LATERAL JOIN`.

Chapter 7, Types, Converters, and Bindings, covers the custom data types, conversion, and binding.

Chapter 8, Fetching and Mapping, being one of the most comprehensive chapters, covers a wide range of jOOQ fetching and mapping techniques, including JSON/SQL, XML/SQL, and MULTISET features.

Chapter 9, CRUD, Transactions, and Locking, covers jOOQ CRUD support next to Spring/jOOQ transactions and optimistic/pessimistic locking.

Chapter 10, Exporting, Batching, Bulking, and Loading, covers batching, bulking, and loading files into the database via jOOQ. We will do single-thread and multi-thread batching.

Chapter 11, jOOQ Keys, tackles the different kinds of identifiers (auto-generated, natural IDs, and composite IDs) from the jOOQ perspective.

Chapter 12, Pagination and Dynamic Queries, covers pagination and building dynamic queries. Mainly, all jOOQ queries are dynamic, but in this chapter, we will highlight this, and we will write several filters by gluing and reusing different jOOQ artifacts.

Chapter 13, Exploiting SQL Functions, covers window functions (probably the most powerful SQL feature) in the jOOQ context.

Chapter 14, Derived Tables, CTEs, and Views, covers derived tables and recursive **Common Table Expressions** (**CTEs**) in the jOOQ context.

Chapter 15, Calling and Creating Stored Functions and Procedures, covers stored procedures and functions in the jOOQ context. This is one of the most powerful and popular jOOQ features.

Chapter 16, *Tackling Aliases and SQL Templating*, covers aliases and SQL templating. As you'll see, this chapter contains a must-have set of knowledge that will help you to avoid common related pitfalls.

Chapter 17, *Multitenancy in jOOQ*, covers different aspects of multi-tenancy/partitioning.

Chapter 18, *jOOQ SPI (Providers and Listeners)*, covers jOOQ providers and listeners. Using these kinds of artifacts, we can interfere with the default behavior of jOOQ.

Chapter 19, *Logging and Testing*, covers jOOQ logging and testing.

To get the most out of this book

To get the most out of this book, you'll need to know the Java language and be familiar with one of the following database technologies:

Software/hardware covered in the book	Operating system requirements
MySQL (preferably 8+)	Windows, Linux
PostgreSQL (preferably 13+)	
SQL Server (preferably 2017+)	
Oracle (preferably 18c)	

Refer to this link for additional installation instructions and information you need to set things up: `https://github.com/PacktPublishing/jOOQ-Masterclass/tree/master/db`.

If you are using the digital version of this book, we advise you to type the code yourself or access the code from the book's GitHub repository (a link is available in the next section). Doing so will help you avoid any potential errors related to the copying and pasting of code.

Download the example code files

You can download the example code files for this book from GitHub at `https://github.com/PacktPublishing/jOOQ-Masterclass`. If there's an update to the code, it will be updated in the GitHub repository.

We also have other code bundles from our rich catalog of books and videos available at `https://github.com/PacktPublishing/`. Check them out!

Download the color images

We also provide a PDF file that has color images of the screenshots and diagrams used in this book. You can download it here: `https://packt.link/a1q9L`.

Conventions used

There are a number of text conventions used throughout this book.

`Code in text`: Indicates code words in text, database table names, folder names, filenames, file extensions, pathnames, dummy URLs, user input, and Twitter handles. Here is an example: "For instance, the next snippet of code relies on the `fetchInto()` flavor."

A block of code is set as follows:

```
// 'query' is the ResultQuery object
List<Office> result = query.fetchInto(Office.class);
```

When we wish to draw your attention to a particular part of a code block, the relevant lines or items are set in bold:

```
public List<Office> findOfficesInTerritory(
                                    String territory) {

  List<Office> result = ctx.selectFrom(table("office"))
    .where(field("territory").eq(territory))
    .fetchInto(Office.class);

  return result;
}
```

Any command-line input or output is written as follows:

```
<result>
 <record>
  <value field="product_line">Vintage Cars</value>
  <value field="product_id">80</value>
  <value field="product_name">1936 Mercedes Benz ...</value>
 </record>
 ...
</result>
```

> **Tips or Important Notes**
> Appear like this.

Get in touch

Feedback from our readers is always welcome.

General feedback: If you have questions about any aspect of this book, email us at `customercare@packtpub.com` and mention the book title in the subject of your message.

Errata: Although we have taken every care to ensure the accuracy of our content, mistakes do happen. If you have found a mistake in this book, we would be grateful if you would report this to us. Please visit `www.packtpub.com/support/errata` and fill in the form.

Piracy: If you come across any illegal copies of our works in any form on the internet, we would be grateful if you would provide us with the location address or website name. Please contact us at `copyright@packt.com` with a link to the material.

If you are interested in becoming an author: If there is a topic that you have expertise in and you are interested in either writing or contributing to a book, please visit `authors.packtpub.com`.

Share Your Thoughts

Once you've read *jOOQ Masterclass*, we'd love to hear your thoughts! Scan the QR code below to go straight to the Amazon review page for this book and share your feedback.

https://packt.link/r/1800566891

Your review is important to us and the tech community and will help us make sure we're delivering excellent quality content.

Part 1: jOOQ as a Query Builder, SQL Executor, and Code Generator

By the end of this part, you will know how to take advantage of the aforementioned three terms in the title in different kickoff applications. You will see how jOOQ can be used as a companion or as a total replacement for your current persistence technology (most probably, an ORM).

This part contains the following chapters:

1
Starting jOOQ and Spring Boot

This chapter is a practical guide to start working with jOOQ (open source and free trial commercial) in Spring Boot applications. For convenience, let's assume that we have a Spring Boot stub application and plan to implement the persistence layer via jOOQ.

The goal of this chapter is to highlight the fact that setting the environment for generating and executing SQL queries via jOOQ in a Spring Boot application is a job that can be accomplished almost instantly in any of the Java/Kotlin and Maven/Gradle combinations. Besides that, this is a good opportunity to have your first taste of the jOOQ DSL-fluent API and to get your first impressions.

The topics of this chapter include the following:

- Starting jOOQ and Spring Boot instantly
- Using the jOOQ query DSL API to generate a valid SQL statement
- Executing the generated SQL and mapping the result set to a POJO

Let's get started!

Technical requirements

The code for this chapter can be found on GitHub at `https://github.com/PacktPublishing/jOOQ-Masterclass/tree/master/Chapter01`.

Starting jOOQ and Spring Boot instantly

Spring Boot provides support for jOOQ, and this aspect is introduced in the Spring Boot official documentation under the *Using jOOQ* section. Having built-in support for jOOQ makes our mission easier, since, among other things, Spring Boot is capable of dealing with aspects that involve useful default configurations and settings.

Consider having a Spring Boot stub application that will run against MySQL and Oracle, and let's try to add jOOQ to this context. The goal is to use jOOQ as a SQL builder for constructing valid SQL statements and as a SQL executor that maps the result set to a POJO.

Adding the jOOQ open source edition

Adding the jOOQ open source edition into a Spring Boot application is quite straightforward.

Adding the jOOQ open source edition via Maven

From the Maven perspective, adding the jOOQ open source edition into a Spring Boot application starts from the `pom.xml` file. The jOOQ open source edition dependency is available at Maven Central (`https://mvnrepository.com/artifact/org.jooq/jooq`) and can be added like this:

```
<dependency>
  <groupId>org.jooq</groupId>
  <artifactId>jooq</artifactId>
  <version>...</version> <!-- optional -->
</dependency>
```

Alternatively, if you prefer a Spring Boot starter, then rely on this one:

```
<dependency>
  <groupId>org.springframework.boot</groupId>
  <artifactId>spring-boot-starter-jooq</artifactId>
</dependency>
```

If you are a fan of Spring Initializr (`https://start.spring.io/`), then just select the jOOQ dependency from the corresponding list of dependencies.

That's all! Note that `<version>` is optional. If `<version>` is omitted, then Spring Boot will properly choose the jOOQ version compatible with the Spring Boot version used by the application. Nevertheless, whenever you want to try a different jOOQ version, you can simply add `<version>` explicitly. At this point, the jOOQ open source edition is ready to be used to start developing the persistence layer of an application.

Adding the jOOQ open source edition via Gradle

From the Gradle perspective, adding the jOOQ open source edition into a Spring Boot application can be accomplished via a plugin named `gradle-jooq-plugin` (`https://github.com/etiennestuder/gradle-jooq-plugin/`). This can be added to your `build.gradle`, as follows:

```
plugins {
  id 'nu.studer.jooq' version ...
}
```

Of course, if you rely on Spring Initializr (`https://start.spring.io/`), then just select a Gradle project, add the jOOQ dependency from the corresponding list of dependencies, and once the project is generated, add the `gradle-jooq-plugin` plugin. As you'll see in the next chapter, using `gradle-jooq-plugin` is quite convenient for configuring the jOOQ Code Generator.

Adding a jOOQ free trial (commercial edition)

Adding a free trial commercial edition of jOOQ (jOOQ Express, Professional, and Enterprise editions) to a Spring Boot project (overall, in any other type of project) requires a few preliminary steps. Mainly, these steps are needed because the jOOQ free trial commercial distributions are not available on Maven Central, so you have to manually download the one that you need from the jOOQ download page (`https://www.jooq.org/download/`). For instance, you can choose the most popular one, the jOOQ Professional distribution, which comes packaged as a ZIP archive. Once you have unzipped it, you can install it locally via the `maven-install` command. You can find these steps exemplified in a short movie in the bundled code (*Install_jOOQ_Trial.mp4*).

For Maven applications, we use the jOOQ free trial identified as `org.jooq.trial` (for Java 17) or `org.jooq.trial-java-{version}`. When this book was written, the `version` placeholder could be 8 or 11, but don't hesitate to check for the latest updates. We prefer the former, so in `pom.xml`, we have the following:

```
<dependency>
  <groupId>org.jooq.trial-java-8</groupId>
  <artifactId>jooq</artifactId>
  <version>...</version>
</dependency>
```

For Java/Gradle, you can do it, as shown in the following example, via `gradle-jooq-plugin`:

```
jooq {
  version = '...'
  edition = nu.studer.gradle.jooq.JooqEdition.TRIAL_JAVA_8
}
```

For Kotlin/Gradle, you can do it like this:

```
jooq {
  version.set(...)
  edition.set(nu.studer.gradle.jooq.JooqEdition.TRIAL_JAVA_8)
}
```

In this book, we will use jOOQ open source in applications that involve MySQL and PostgreSQL, and jOOQ free trial in applications that involve SQL Server and Oracle. These two database vendors are not supported in jOOQ open source.

If you're interested in adding jOOQ in a Quarkus project then consider this resource: `https://github.com/quarkiverse/quarkus-jooq`

Injecting DSLContext into Spring Boot repositories

One of the most important interfaces of jOOQ is `org.jooq.DSLContext`. This interface represents the starting point of using jOOQ, and its main goal is to configure the behavior of jOOQ when executing queries. The default implementation of this interface is named `DefaultDSLContext`. Among the approaches, `DSLContext` can be created via an `org.jooq.Configuration` object, directly from a JDBC connection (`java.sql.Connection`), a data source (`javax.sql.DataSource`), and a dialect needed for translating the Java API query representation, written via jOOQ into a database-specific SQL query (`org.jooq.SQLDialect`).

> **Important Note**
>
> For `java.sql.Connection`, jOOQ will give you full control of the connection life cycle (for example, you are responsible for closing this connection). On the other hand, connections acquired via `javax.sql.DataSource` will be automatically closed after query execution by jOOQ. Spring Boot loves data sources, therefore the connection management is already handled (acquire and return connection from/to the connection pool, transaction begin/commit/rollback, and so on).

All jOOQ objects, including `DSLContext`, are created from `org.jooq.impl.DSL`. For creating a `DSLContext`, the `DSL` class exposes a `static` method named `using()`, which comes in several flavors. Of these, the most notable are listed next:

```
// Create DSLContext from a pre-existing configuration
DSLContext ctx = DSL.using(configuration);

// Create DSLContext from ad-hoc arguments
DSLContext ctx = DSL.using(connection, dialect);
```

For example, connecting to the MySQL `classicmodels` database can be done as follows:

```
try (Connection conn = DriverManager.getConnection(
    "jdbc:mysql://localhost:3306/classicmodels",
    "root", "root")) {

  DSLContext ctx =
    DSL.using(conn, SQLDialect.MYSQL);
  ...
} catch (Exception e) {
  ...
}
```

Alternatively, you can connect via a data source:

```
DSLContext ctx = DSL.using(dataSource, dialect);
```

For example, connecting to the MySQL `classicmodels` database via a data source can be done as follows:

```
DSLContext getContext() {
  MysqlDataSource dataSource = new MysqlDataSource();

  dataSource.setServerName("localhost");
  dataSource.setDatabaseName("classicmodels");
  dataSource.setPortNumber("3306");
  dataSource.setUser(props.getProperty("root");
  dataSource.setPassword(props.getProperty("root");

  return DSL.using(dataSource, SQLDialect.MYSQL);
}
```

But Spring Boot is capable of automatically preparing a ready-to-inject `DSLContext` based on our database settings. For example, Spring Boot can prepare `DSLContext` based on the MySQL database settings specified in `application.properties`:

```
spring.datasource.driverClassName=com.mysql.cj.jdbc.Driver
spring.datasource.url=jdbc:mysql://localhost:3306/
                 classicmodels?createDatabaseIfNotExist=true
spring.datasource.username=root
spring.datasource.password=root

spring.jooq.sql-dialect=MYSQL
```

Once Spring Boot detects the jOOQ presence, it uses the preceding settings to create `org.jooq.Configuration`, which is used to prepare a ready-to-inject `DSLContext`.

> **Important Note**
>
> While `DSLContext` has a high degree of configurability and flexibility, Spring Boot performs only the minimum effort to serve a default `DSLContext` that can be injected and used immediately. As you'll see in this book (but especially in the official jOOQ manual – `https://www.jooq.org/doc/latest/manual/`), `DSLContext` has tons of configurations and settings that allow taking control of almost anything that happens with our SQL statements.

The `DSLContext` object provided by Spring Boot can be easily injected into our persistence repositories. For instance, the next snippet of code serves such a `DSLContext` object directly into `ClassicModelsRepository`:

```
@Repository
public class ClassicModelsRepository {

  private final DSLContext ctx;

  public ClassicModelsRepository(DSLContext ctx) {
    this.ctx = ctx;
  }
  ...
}
```

Don't conclude here that the application needs to keep a reference to `DSLContext`. That can still be used directly in a local variable, as you saw earlier (which means that you can have as many `DSLContext` objects as you want). It only means that, in a Spring Boot application, for most common scenarios, it is more convenient to simply inject it as shown previously.

Internally, jOOQ can use `java.sql.Statement` or `PreparedStatement`. By default, and for very good and strong reasons, jOOQ uses `PreparedStatement`.

Typically, the `DSLContext` object is labeled as `ctx` (used in this book) or `dsl`. But, other names such as `dslContext`, `jooq`, and `sql` are also good choices. Basically, you name it.

Okay, so far, so good! At this point, we have access to `DSLContext` provided out of the box by Spring Boot, based on our settings from `application.properties`. Next, let's see `DSLContext` at work via jOOQ's query DSL API.

Using the jOOQ query DSL API to generate valid SQL

Using the jOOQ query DSL API to generate valid SQL is a good start for exploring the jOOQ world. Let's take a simple SQL statement, and let's express it via jOOQ. In other words, let's use the jOOQ query DSL API to express a given SQL string query into the jOOQ object-oriented style. Consider the next SQL `SELECT` written in the MySQL dialect:

```
SELECT * FROM `office` WHERE `territory` = ?
```

The SQL, `SELECT * FROM `office` WHERE `territory` = ?`, is written as a plain string. This query can be generated by jOOQ if it is written via the DSL API, as follows (the value of the `territory` binding variable is supplied by the user):

```
ResultQuery<?> query = ctx.selectFrom(table("office"))
  .where(field("territory").eq(territory));
```

Alternatively, if we want to have the `FROM` clause closer to SQL look, then we can write it as follows:

```
ResultQuery<?> query = ctx.select()
  .from(table("office"))
  .where(field("territory").eq(territory));
```

Most schemas are case-insensitive, but there are databases such as MySQL and PostgreSQL that prefer mostly lowercase, while others such as Oracle prefer mostly uppercase. So, writing the preceding query in Oracle style can be done as follows:

```
ResultQuery<?> query = ctx.selectFrom(table("OFFICE"))
  .where(field("TERRITORY").eq(territory));
```

Alternatively, you can write it via an explicit call of `from()`:

```
ResultQuery<?> query = ctx.select()
  .from(table("OFFICE"))
  .where(field("TERRITORY").eq(territory));
```

The jOOQ fluent API is a piece of art that looks like fluent English and, therefore, is quite intuitive to read and write.

Reading the preceding queries is pure English: *select all offices from the OFFICE table where the TERRITORY column is equal to the given value.*

Pretty soon, you'll be amazed at how fast you can write these queries in jOOQ.

> **Important Note**
>
> As you'll see in the next chapter, jOOQ can generate a Java-based schema that mirrors the one in the database via a feature named the jOOQ Code Generator. Once this feature is enabled, writing these queries becomes even simpler and cleaner because there will be no need to reference the database schema explicitly, such as the table name or the table columns. Instead, we will reference the Java-based schema.
>
> And, thanks to the Code Generator feature, jOOQ makes the right choices for us upfront almost everywhere. We no longer need to take care of queries' type-safety and case-sensitivity, or identifiers' quotation and qualification.
>
> The jOOQ Code Generator atomically boosts the jOOQ capabilities and increases developer productivity. This is why using the jOOQ Code Generator is the recommended way to exploit jOOQ. We will tackle the jOOQ Code Generator in the next chapter.

Next, the jOOQ query (`org.jooq.ResultQuery`) must be executed against the database, and the result set will be mapped to a user-defined simple POJO.

Executing the generated SQL and mapping the result set

Executing the generated SQL and mapping the result set to a POJO via jOOQ can be done via the fetching methods available in the jOOQ API. For instance, the next snippet of code relies on the `fetchInto()` flavor:

```
public List<Office> findOfficesInTerritory(String territory) {

  List<Office> result = ctx.selectFrom(table("office"))
    .where(field("territory").eq(territory))
    .fetchInto(Office.class);

  return result;
}
```

What happened there?! Where did `ResultQuery` go? Is this black magic? Obviously not! It's just that jOOQ has immediately fetched results after constructing the query and mapped them to the `Office` POJO. Yes, the jOOQ's `fetchInto(Office.class)` or `fetch().into(Office.class)` would work just fine out of the box. Mainly, jOOQ executes the query and maps the result set to the `Office` POJO by wrapping and abstracting the JDBC complexity in a more object-oriented way. If we don't want to immediately fetch the results after constructing the query, then we can use the `ResultQuery` object like this:

```
// 'query' is the ResultQuery object
List<Office> result = query.fetchInto(Office.class);
```

The `Office` POJO is available in the code bundled with this book.

> **Important Note**
>
> jOOQ has a comprehensive API for fetching and mapping a result set into collections, arrays, maps, and so on. We will detail these aspects later on in *Chapter 8, Fetching and Mapping*.

The complete application is named *DSLBuildExecuteSQL*. Since this can be used as a stub application, you can find it available for Java/Kotlin in combination with Maven/Gradle. These applications (along with, in fact, all the applications in this book) use Flyway for schema migration. As you'll see later, Flyway and jOOQ make a great team.

So, let's quickly summarize this chapter before moving on to exploit the astonishing jOOQ Code Generator feature.

Summary

Note that we barely scratched the surface of jOOQ's capabilities by using it only for generating and executing a simple SQL statement. Nevertheless, we've already highlighted that jOOQ can generate valid SQL against different dialects and can execute and map a result set in a straightforward manner.

In the next chapter, we learn how to trust jOOQ more by increasing its level of involvement. jOOQ will generate type-safe queries, POJOs, and DAOs on our behalf.

2 Customizing the jOOQ Level of Involvement

In the previous chapter, we introduced jOOQ in a Spring Boot application and used it for generating and executing a valid non-type-safe SQL statement. In this chapter, we will continue this journey and increase the jOOQ level of involvement via an astonishing feature – the so-called jOOQ *Code Generator*. In other words, jOOQ will be in control of the persistence layer via a straightforward flow that begins with type-safe queries, continues by generating **Plain Old Java Objects** (**POJOs**) used to map the query results as objects, and ends with generating DAOs used to shortcut the most common queries in object-oriented style.

By the end of this chapter, you'll know how to write type-safe queries, and how to instruct jOOQ to generate POJOs and DAOs that have custom names in Java and Kotlin applications, using Maven and Gradle. We will cover these topics declaratively (for instance, in XML files) and programmatically.

The following topics will be covered in this chapter:

- Understanding what type-safe queries are
- Generating a jOOQ Java-based schema
- Writing queries using a Java-based schema
- Configuring jOOQ to generate POJOs
- Configuring jOOQ to generate DAOs
- Configuring jOOQ to generate interfaces
- Tackling programmatic configuration
- Introducing jOOQ settings

Let's start with a brief discussion about type-safe queries.

Technical requirements

The code files used in this chapter can be found on GitHub:

`https://github.com/PacktPublishing/jOOQ-Masterclass/tree/master/Chapter02`

Understanding what type-safe queries are

Generally speaking, what actually is a type-safe API? In short, an API is type-safe if it relies on the type system of a programming language aiming to prevent and report type errors. Specifically, jOOQ enables the compiler to do that via the Code Generator features.

Working with type-safe SQL is preferable because there is no need to validate every SQL statement via dedicated tests, and it is faster to fix things during coding than while running the application. For example, you can significantly reduce the number of unit tests dedicated to SQL validation and focus on integration tests, which is always a good thing. So, SQL type safety really matters!

Declaring SQL statements as Java `String` statements (for example, in JPQL style, which is verified at execution time) doesn't take advantage of type safety. In other words, the compiler cannot guarantee that a SQL statement is valid. This happens in each of the following examples that use different choices for the persistence layer. All these examples compile but fail at runtime.

Let's see a `JdbcTemplate` non-type-safe SQL example (with the wrong order of binding values):

```
public Manager findManager(Long id, String name) {
  String sql = "SELECT * FROM MANAGER
    WHERE MANAGER_ID=? AND MANAGER_NAME=?";

  Manager result = jdbcTemplate
    .queryForObject(sql, Manager.class, name, id);
}
```

Here, we have a Spring Data example (name should be `String`, not `int`):

```
@Query(value = "SELECT c.phone, p.cachingDate FROM Customer c
       INNER JOIN c.payments p WHERE c.customer_name = ?1")
CustomerPojo fetchCustomerWithCachingDateByName(int name);
```

Here is a Spring Data derived query method example (name should be `String`, not `int`):

```
Customer findByName(int name);
```

The following is a jOOQ query builder without the Code Generator example (instead of `v`, it should be `v.getOwnerName()`):

```
public Customer findCustomer(Voucher v) {

  ctx.select().from(table("CUSTOMER"))
     .where(field("CUSTOMER.CUSTOMER_NAME").eq(v))...;
}
```

Here's another jOOQ query builder without the Code Generator example (in our schema, there is no `OFFICES` table and no `CAPACITY` column):

```
ctx.select()
   .from(table("OFFICES"))
   .where(field("OFFICE.CAPACITY").gt(50));
```

These are just some simple cases that are easy to spot and fix. Imagine a non-type-safe complex query with a significant number of bindings.

But, if the jOOQ Code Generator is enabled, then jOOQ will compile the SQL statements against an actual Java-based schema that mirrors a database. This way, jOOQ ensures *at least* the following:

- The classes and fields that occur in SQL exist, have the expected type, and are mapped to a database.
- There are no type mismatches between the operators and operands.
- The generated query is syntactically valid.

> **Important Note**
> I said *at least* because, besides type safety, jOOQ takes care of many other aspects, such as quotations, qualification, and case sensitivity of identifiers. These aspects are not easy to handle across SQL dialects, and thanks to the Code Generator feature, jOOQ makes the right choices for us upfront almost everywhere. As Lukas Eder said: *"Using jOOQ with the Code Generator is just a little additional setup, but it will help jOOQ to make the right, carefully chosen default choices for so many silly edge cases that are so annoying to handle later on. I can't recommend it enough! :)"*

Back to type safety, let's assume that the jOOQ Code Generator has produced the needed artifacts (a suite of classes that mirrors the database tables, columns, routines, views, and so on). In this context, the previous jOOQ examples can be rewritten in a type-safe manner, as follows. Note that none of the following snippets will compile:

```
import static jooq.generated.tables.Customer.CUSTOMER;
...
public Customer findCustomer(Voucher v) {

  ctx.select().from(CUSTOMER)
     .where(CUSTOMER.CUSTOMER_NAME.eq(v))...;
}
```

Besides being less verbose than the original example, this query is type-safe as well. This time, `CUSTOMER` (which replaced `table("CUSTOMER")`) is a `static` instance (shortcut) of the `Customer` class, representing the `customer` table. Moreover, `CUSTOMER_NAME` (which replaced `field("CUSTOMER.CUSTOMER_NAME")`) is also a `static` field in the `Customer` class, representing the `customer_name` column of the `customer` table. These Java objects have been generated by the jOOQ Code Generator as part of the Java-based schema. Note how this `static` instance was nominally imported here – if you find the technique of importing each `static` artifact cumbersome, then you can simply rely on the neat trick of importing the entire schema as `import static jooq.generated.Tables.*`.

The second jOOQ example can be rewritten in a type-safe manner, as follows:

```
import static jooq.generated.tables.Office.OFFICE;
...
ctx.select().from(OFFICES).where(OFFICE.CAPACITY.gt(50));
```

The following figure is a screenshot from the IDE, showing that the compiler complains about the type safety of this SQL:

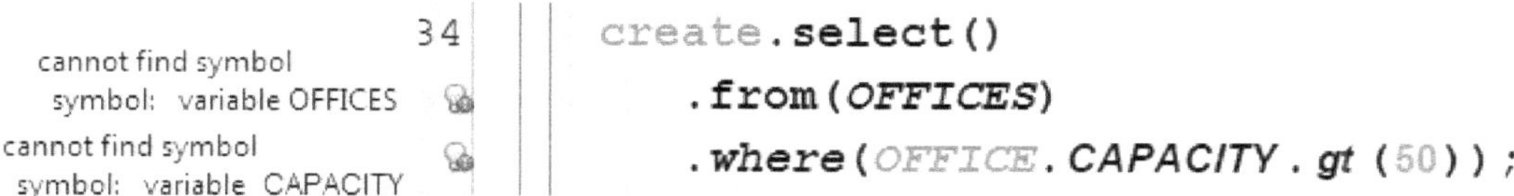

Figure 2.1 – The compiler reports a type safety error

> **Important Note**
>
> Lukas Eder said this: *"As you probably know, the IDEs help writing SQL and JPQL strings, which is nice. But IDEs doesn't fail the build when a column name changes."* Well, having type-safe queries covers this aspect, and the IDE can fail the build. So, thanks to jOOQ's fluency and expressiveness, the IDE can provide code completion and refactoring support. Moreover, with jOOQ, the bind variables are part of a non-dynamic **Abstract Syntax Tree** (**AST**); therefore, it is not possible to expose SQL injection vulnerabilities this way.

OK, but how do we obtain this Java-based schema?

Generating a jOOQ Java-based schema

All the previous queries were referencing the database schema explicitly by placing the table or column name between quotes and passing them as arguments to the jOOQ built-in `table()` and `field()` methods respectively.

But, using the jOOQ Code Generator allows the SQL statements expressed via jOOQ's query DSL API to take advantage of a Java-based schema that mirrors the one from the database. The code generation part is the job of the jOOQ generation tool (its starting point is the `org.jooq.codegen.GenerationTool` class).

Having a Java-based schema is quite useful. The SQL statements can be expressed via the Java data access layer and executed against the underlying database schema. Besides being type-safe, these SQL statements are not prone to typos, are easy to refactor (for example, to rename a column), and are less verbose than referencing the database schema explicitly.

jOOQ comes with several solutions for generating the Java-based schema via the jOOQ Code Generator. Mainly, jOOQ can generate the Java-based schema by applying the technique of reverse engineering to the database directly, the DDL files, JPA entities, or XML files containing the schema. Next, we will tackle the first three approaches, starting with the first approach, which generates the Java-based schema directly from the database. Mainly, we will use Flyway to migrate the database (Liquibase is supported as well), which is subsequently reverse engineered by jOOQ to obtain the Java-based schema.

Code generation from a database directly

The following figure represents the jOOQ Java-based schema generation flow:

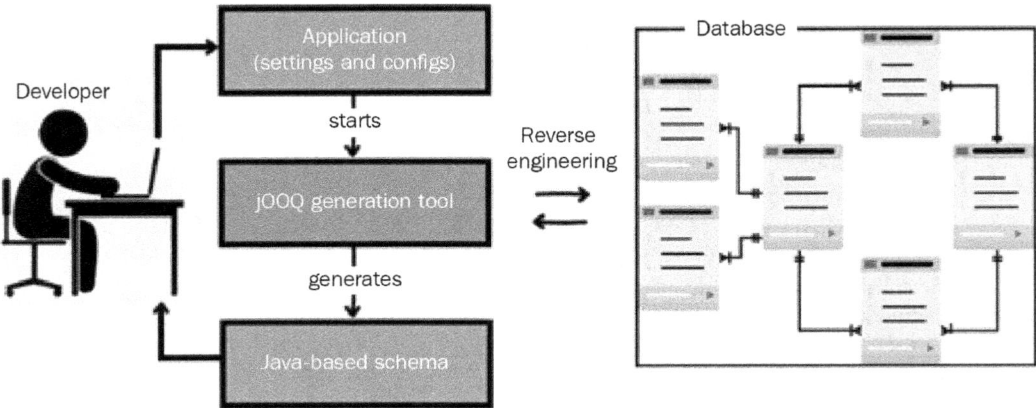

Figure 2.2 – Java-based schema generation

So far, jOOQ will regenerate the Java-based schema every time the application starts (runs).

In other words, even if the database schema has not changed, jOOQ will regenerate the Java-based schema at each run. Obviously, this is preferable to regenerating the Java-based schema only when the underlying database schema is missing or has changed (for instance, a new column has been added to a table); otherwise, this is just a waste of time.

Conscious schema change management is a good thing, and having a tool for this is great! Most probably, you'll choose between Flyway and Liquibase. While we will only cover the Flyway approach in the next section, Liquibase is very well represented in the jOOQ manual (`https://www.jooq.org/doc/latest/manual/code-generation/codegen-liquibase/`).

Adding Flyway with Maven

Flyway is a great tool for database migration (https://flywaydb.org/). Mainly, Flyway keeps track of database schema modifications via a table named flyway_schema_history (or schema_version in Flyway prior to version 5). This table is automatically added to the database and is maintained by Flyway itself.

Typically, in Spring Boot, Flyway reads and executes all the database migration scripts located in the indicated path (the default path is src/main/resources/db/migration). For instance, in this book, we use an explicit path that points to a location outside the applications in the root folder (${root}/db/migration). We do this because we want to avoid multiplying the migrations scripts in every single application. To quickly start with Flyway, simply add to pom.xml the following dependency:

```
<dependency>
  <groupId>org.flywaydb</groupId>
  <artifactId>flyway-core</artifactId>
</dependency>
```

The Flyway default Maven phase for a migrate operation is pre-integration-test (right after package). On the other hand, jOOQ needs the migrations to take place in the generate-sources phase (right after validate), therefore much earlier.

Mainly, jOOQ triggers a SELECT query against the flyway_schema_history table to check the schema version. This means that jOOQ needs to wait for migrations to take place and the schema version to be updated. If the version is updated, then jOOQ regenerates the Java-based schema; otherwise, you'll see a message like this: Existing version 1.1 is up to date with 1.1 for schema classicmodels. Ignoring schema.

Scheduling migrations in the generate-sources phase can be done via the Flyway Maven plugin, as follows:

```
<phase>generate-sources</phase>
```

Let's try using Gradle.

Adding Flyway with Gradle

If you prefer to use Gradle, then you'll need `build.gradle` in the following code:

```
plugins {
  id 'org.flywaydb.flyway' version '...'
}

dependencies {
  implementation 'org.flywaydb:flyway-core'
}

flyway {
  driver = ...
  url = ...
  ...
}
```

Next, let's add the SQL scripts following the Flyway naming conventions.

Adding SQL scripts for Flyway

In the applications developed in this book, the scripts read and executed by Flyway are named `V1.1__Create.sql` (this file contains the DDLs of the database schema) and `afterMigrate.sql` (this file contains the DMLs to populate the database) and are placed externally to the applications in the `${root}/db/migration` folder. Adding a new file that respects the Flyway naming convention (for example, `V1.2__AddColumn.sql`) will instruct Flyway to update the database schema and jOOQ to regenerate the Java-based schema. As long as no migrations happen and the jOOQ-generated classes exist, jOOQ doesn't regenerate the Java-based schema.

The following figure represents the flow, which is particularly interesting for most use cases that contain DDL changes:

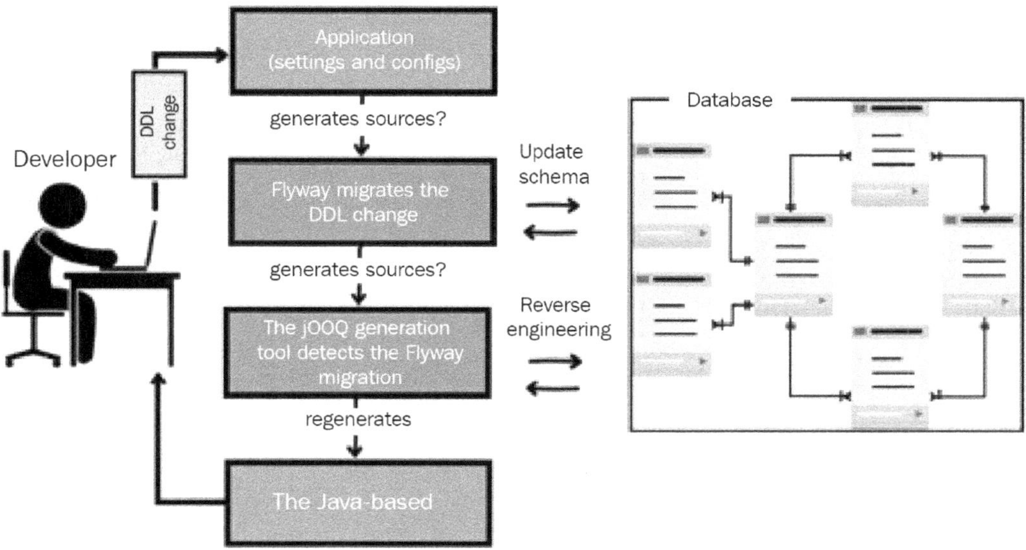

Figure 2.3 – Flyway migrations and the jOOQ Java-based schema generation

Note how Flyway migrations take place before jOOQ code generation. Finally, it's time to enable the jOOQ Code Generator.

From a developer perspective, enabling the jOOQ Code Generator is a setup task that gets materialized in a snippet of code, written in standalone migration scripts or `pom.xml` if there is a Maven-based project, or `build.gradle` if there is a Gradle-based project. jOOQ reads this information and uses it to configure and automatically execute the `org.jooq.codegen.GenerationTool` generator accordingly.

Running the Code Generator with Maven

Mainly, the jOOQ Code Generator can run in standalone mode or with Maven/Gradle. While there are no big differences between these two approaches, we prefer to go further with the Maven plugin, `jooq-codegen-maven`. Nevertheless, for a quick example of running the Code Generator from the command line in standalone mode, you have everything you need (including a `README` file) packed in a ZIP archive named `standalone-codegen-jooq.zip`. This is available for MySQL, PostgreSQL, SQL Server, and Oracle.

Now, configuring jOOQ's Code Generator requires some information that can be packed in an XML file. The climax of this file is the `<configuration>` tag used to shape an `org.jooq.meta.jaxb.Configuration` instance. Consider reading carefully each comment of the following jOOQ Code Generator configuration stub, since each comment provides important details about the tag that precedes it (in the bundled code, you'll see an expanded version of these comments, containing extra details):

```
<plugin>
  <groupId>...</groupId>
  <artifactId>jooq-codegen-maven</artifactId>

  <executions>
    <execution>

      <id>...</id>

      <phase>generate-sources</phase>

      <goals>
        <goal>generate</goal>
      </goals>

      <configuration xmlns = "...">

        <!-- Configure the database connection here -->
        <jdbc>...</jdbc>
```

Next, the `<generator/>` tag contains all the information needed for customizing the jOOQ generator:

```
        <generator>
          <!-- The Code Generator:
          org.jooq.codegen.{Java/Kotlin/Scala}Generator
          Defaults to org.jooq.codegen.JavaGenerator -->
          <name>...</name>

          <database>
            <!-- The database type. The format here is:
            org.jooq.meta.[database].[database]Database -->
            <name>...</name>
```

```
          <!-- The database schema-->
          <inputSchema>...</inputSchema>

          <!-- What should be included by the generator -->
          <includes>...</includes>

          <!-- What should be excluded by the generator -->
          <excludes>...</excludes>

          <!-- Schema version provider -->
          <schemaVersionProvider>...</schemaVersionProvider>

          <!-- Set generator queries timeout(default 5s) -->
          <logSlowQueriesAfterSeconds>
            ...
          </logSlowQueriesAfterSeconds>
        </database>

        <target>
          <!-- The output package of generated classes -->
          <packageName>...</packageName>

          <!—The output directory of generated classes -->
          <directory>...</directory>
        </target>
      </generator>
    </configuration>
  </execution>
 </executions>
</plugin>
```

Based on this stub and the comments, let's try to fill up the missing parts for configuring the jOOQ Code Generator against the `classicmodels` database in MySQL:

```
<plugin>
  <groupId>org.jooq</groupId>
  <artifactId>jooq-codegen-maven</artifactId>

  <executions>
    <execution>
```

```
<id>generate-for-mysql</id>

<phase>generate-sources</phase>

<goals>
  <goal>generate</goal>
</goals>

<configuration xmlns = "...">

  <jdbc>
    <driver>${spring.datasource.driverClassName}</driver>
    <url>${spring.datasource.url}</url>
    <user>${spring.datasource.username}</user>
    <password>${spring.datasource.password}</password>
  </jdbc>

  <generator>
    <name>org.jooq.codegen.JavaGenerator</name>

    <database>
      <name>org.jooq.meta.mysql.MySQLDatabase</name>

      <inputSchema>classicmodels</inputSchema>

      <includes>.*</includes>
      <excludes>
        flyway_schema_history | sequences
        | customer_pgs | refresh_top3_product
        | sale_.* | set_.* | get_.* | .*_master
      </excludes>

      <schemaVersionProvider>
       SELECT MAX(`version`) FROM `flyway_schema_history`
      </schemaVersionProvider>

      <logSlowQueriesAfterSeconds>
        20
```

```
          </logSlowQueriesAfterSeconds>
        </database>

        <target>
          <packageName>jooq.generated</packageName>
          <directory>target/generated-sources</directory>
        </target>
      </generator>
    </configuration>
  </execution>
 </executions>
</plugin>
```

For brevity, the alternatives for PostgreSQL, SQL Server, and Oracle are not listed here, but you can find them in the code bundled with this book in the application named *WriteTypesafeSQL*.

Additionally, the Maven plugin supports the following flags in `<configuration>`:

- Disabling the plugin via a Boolean property/constant:

  ```
  <skip>false</skip>
  ```

- Specifying an external XML configuration instead of an inline configuration:

  ```
  <configurationFile>${externalfile}</configurationFile>
  ```

- Alternatively, specifying several external configuration files, merged by using Maven's `combine.children="append"` policy:

  ```
  <configurationFiles>
    <configurationFile>${file1}</configurationFile>
    <configurationFile>...</configurationFile>
  </configurationFiles>
  ```

Next, let's run the jOOQ generator via Gradle.

Running the Code Generator with Gradle

Running the Code Generator via Gradle can be accomplished via `gradle-jooq-plugin` (`https://github.com/etiennestuder/gradle-jooq-plugin/`). The next snippet of code represents the climax of configuration for Oracle:

```
dependencies {
  jooqGenerator 'com.oracle.database.jdbc:ojdbc8'
  jooqGenerator 'com.oracle.database.jdbc:ucp'
}

jooq {
  version = '...'
  edition = nu.studer.gradle.jooq.JooqEdition.TRIAL_JAVA_8

  configurations {

    main {
      generateSchemaSourceOnCompilation = true  // default

      generationTool {

        logging = org.jooq.meta.jaxb.Logging.WARN

        jdbc {
          driver = project.properties['driverClassName']
          url = project.properties['url']
          user = project.properties['username']
          password = project.properties['password']
        }

        generator {

          name = 'org.jooq.codegen.JavaGenerator'

          database {
            name = 'org.jooq.meta.oracle.OracleDatabase'

            inputSchema = 'CLASSICMODELS'
```

```
            includes = '.*'
            schemaVersionProvider = 'SELECT MAX("version")
              FROM "flyway_schema_history"'

            excludes = '''\
              flyway_schema_history | DEPARTMENT_PKG | GET_.*
              | CARD_COMMISSION | PRODUCT_OF_PRODUCT_LINE
              ...
            '''

            logSlowQueriesAfterSeconds = 20
          }

          target {
            packageName = 'jooq.generated'
            directory = 'target/generated-sources'
          }

          strategy.name =
            "org.jooq.codegen.DefaultGeneratorStrategy"
        }
    ...
}
```

In addition, we have to bind the jOOQ generator to the Flyway migration tool to execute it only when it is really needed:

```
tasks.named('generateJooq').configure {

  // ensure database schema has been prepared by
  // Flyway before generating the jOOQ sources
  dependsOn tasks.named('flywayMigrate')

  // declare Flyway migration scripts as inputs on this task
  inputs.files(fileTree('...'))
        .withPropertyName('migrations')
        .withPathSensitivity(PathSensitivity.RELATIVE)
```

```
    // make jOOQ task participate in
    // incremental builds and build caching
    allInputsDeclared = true
    outputs.cacheIf { true }
}
```

In the bundled code, you can find the complete application (*WriteTypesafeSQL*) for MySQL, PostgreSQL, SQL Server, and Oracle, written for Java/Kotlin and Maven/Gradle combos.

Alternatively, if you prefer `Ant`, then read this: `https://www.jooq.org/doc/latest/manual/code-generation/codegen-ant/`. Next, let's tackle another approach to generating the Java-based schema.

Code generation from SQL files (DDL)

Another jOOQ approach for obtaining the Java-based schema relies on the DDL Database API, which is capable of accomplishing this task from SQL scripts (a single file or incremental files) containing the database schema. Mainly, the jOOQ SQL parser materializes our SQL scripts into an in-memory H2 database (available out of the box in Spring Boot), and the generation tool will reverse-engineer it to output the Java-based schema. The following figure depicts this flow:

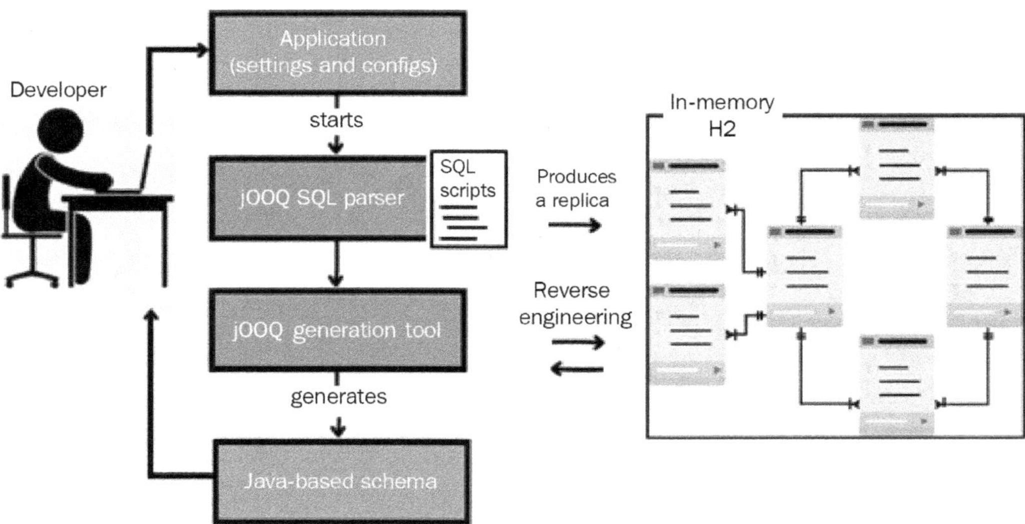

Figure 2.4 – The jOOQ Java-based schema generation via the DDL Database API

The climax of the DDL Database API configuration relies on the jOOQ Meta Extensions, represented by `org.jooq.meta.extensions.ddl.DDLDatabase`.

Running the Code Generator with Maven

In this context, running the Code Generator via Maven relies on the following XML stub. Read each comment, since they contain valuable information (in the bundled code, you'll see an expanded version of these comments):

```
<configuration xmlns = "...">
  <generator>

    <name>...</name>

    <database>
      <name>org.jooq.meta.extensions.ddl.DDLDatabase</name>

      <properties>

        <!-- Specify the location of your SQL script -->
        <property>
          <key>scripts</key>
          <value>...</value>
        </property>

        <!-- The sort order of scripts in a directory
        (semantic, alphanumeric, flyway, none) -->
        <property>
          <key>sort</key>
          <value>...</value>
        </property>

        <!-- The default schema for unqualified objects
        (public, none) -->
        <property>
          <key>unqualifiedSchema</key>
          <value>...</value>
        </property>

        <!-- The default name case for unquoted objects
        (as_is, upper, lower) -->
```

```
      <property>
        <key>defaultNameCase</key>
        <value>...</value>
      </property>
    </properties>

    <inputSchema>PUBLIC</inputSchema>

    <includes>...</includes>
    <excludes>...</excludes>

    <schemaVersionProvider>...</schemaVersionProvider>

    <logSlowQueriesAfterSeconds>
    ...
    </logSlowQueriesAfterSeconds>
  </database>

  <target>
    <packageName>...</packageName>
    <directory>...</directory>
  </target>
 </generator>
</configuration>
```

In this context, jOOQ generates the Java-based schema without connecting to the real database. It uses the DDL files to produce an in-memory H2 database that is subsequently reverse-engineered into Java classes. The `<schemaVersionProvider>` tag can be bound to a Maven constant that you have to maintain in order to avoid running the Code Generator when nothing has changed.

Besides this stub, we need the following dependency:

```
<dependency>
  <groupId>org.jooq{.trial-java-8}</groupId>
  <artifactId>jooq-meta-extensions</artifactId>
  <version>${jooq.version}</version>
</dependency>
```

Based on this stub and the explanations from the comments, let's try to fill up the missing parts to configure the jOOQ Code Generator against the `classicmodels` database in PostgreSQL:

```
<configuration xmlns = "...">
  <generator>

    <name>org.jooq.codegen.JavaGenerator</name>

    <database>
      <name>org.jooq.meta.extensions.ddl.DDLDatabase</name>

      <properties>
        <property>
          <key>scripts</key>
          <value>...db/migration/ddl/postgresql/sql</value>
        </property>
        <property>
          <key>sort</key>
          <value>flyway</value>
        </property>
        <property>
          <key>unqualifiedSchema</key>
          <value>none</value>
        </property>
        <property>
          <key>defaultNameCase</key>
          <value>lower</value>
        </property>
      </properties>

      <inputSchema>PUBLIC</inputSchema>

      <includes>.*</includes>
      <excludes>
            flyway_schema_history | akeys | avals | defined
          | delete.* | department_topic_arr | dup
          |  ...
      </excludes>
```

```
        <schemaVersionProvider>
          ${schema.version} <!-- this is a Maven constant -->
        </schemaVersionProvider>

        <logSlowQueriesAfterSeconds>
          20
        </logSlowQueriesAfterSeconds>
      </database>

    <target>
      <packageName>jooq.generated</packageName>
      <directory>target/generated-sources</directory>
    </target>
  </generator>
</configuration>
```

The Gradle alternative is available in the bundled code.

Preparing the SQL files

Currently, it is impossible to use some vendor-specific stuff; therefore, our SQL files may contain parts that the jOOQ SQL parser may not understand. In such cases, we have to prepare our SQL files by delimiting these parts with the jOOQ default conventions from the following example:

```
-- [jooq ignore start]
IF OBJECT_ID('payment', 'U') IS NOT NULL
  DROP TABLE payment;
-- [jooq ignore stop]
```

The code between `-- [jooq ignore start]` and `-- [jooq ignore stop]` is ignored by the jOOQ SQL parser. Turning on/off ignoring content between such tokens can be done via the `parseIgnoreComments` Boolean property, while customizing these tokens can be done via the `parseIgnoreCommentStart` and `parseIgnoreCommentStop` properties. For more details, refer to `https://www.jooq.org/doc/latest/manual/code-generation/codegen-ddl/`.

In the bundled code, you can see an implementation of this stub for MySQL, PostgreSQL, SQL Server, and Oracle via the Java/Kotlin and Maven/Gradle combos, under the name *DeclarativeDDLDatabase*.

Going forward, while the jOOQ SQL parser will become more powerful, this will be the recommended approach for using the jOOQ Code Generator. The goal is to delegate jOOQ to do more migration work out of the box.

Code generation from entities (JPA)

Let's assume that you have a JPA application that relies on a schema shaped as an entity model (JPA-annotated entities) and you want to obtain the jOOQ Java-based schema. If you cannot isolate the JPA entity model in a separate module of the application, then you can configure jOOQ to generate the Java-based schema directly from the real database (supposing that you have access to the real database schema during the development stage) or from the DDL files (assuming that you have such files). But, if you can easily place the entities in a separate module of the application, then you can rely on jOOQ's JPA Database API (`org.jooq.meta.extensions.jpa.JPADatabase`), which is capable of generating the Java-based schema from the JPA model. The JPA Database API requires entities in a separate module because it has to look them up from the classpath via Spring.

The following figure depicts the flow of the JPA Database API:

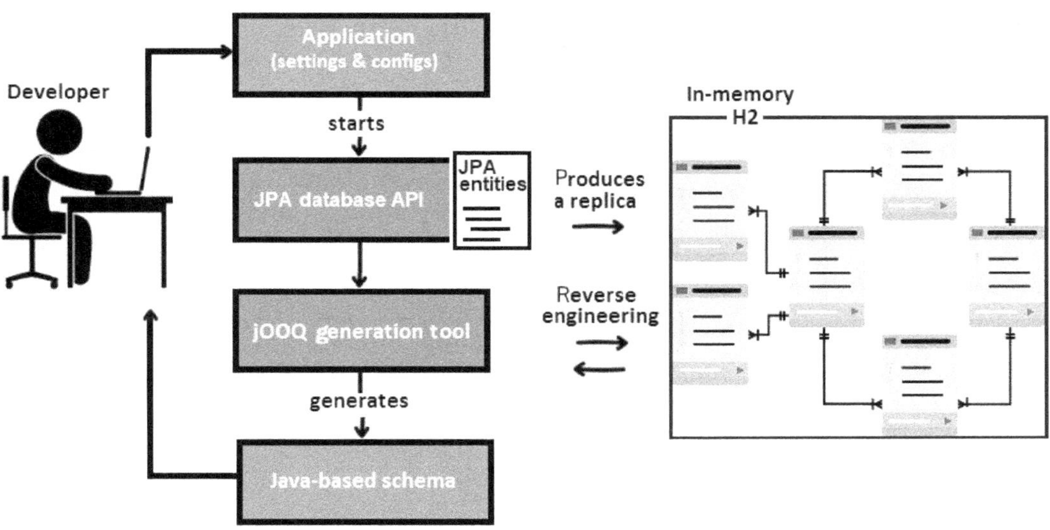

Figure 2.5 – The jOOQ Java-based schema generation via the JPA Database API

The flow of the JPA Database API uses Hibernate internally for generating an in-memory H2 database from the JPA model (entities). Subsequently, jOOQ reverse-engineers this H2 database into jOOQ classes (the Java-based schema).

Running the Code Generator with Maven

In this context, running the Code Generator via Maven relies on the following XML stub. Read each comment, since they contain valuable information (in the bundled code, you can find an expanded version of these comments):

```
<configuration xmlns="...">

  <!-- JDBC connection to the H2 in-memory database -->
  <jdbc>...</jdbc>

  <generator>
    <database>
      <name>org.jooq.meta.extensions.jpa.JPADatabase</name>

      <properties>
        <!-- The properties prefixed with hibernate... or
        javax.persistence... will be passed to Hibernate -->
        <property>
          <key>...</key>
          <value>...</value>
        </property>

        <!-- Java packages (comma separated) that
        contains your entities -->
        <property>
          <key>packages</key>
          <value>...</value>
        </property>

        <!-- Whether JPA 2.1 AttributeConverters should
        be auto-mapped to jOOQ Converters (default true) -->
        <property>
          <key>useAttributeConverters</key>
          <value>...</value>
        </property>

        <!-- The default schema for unqualified objects
        (public, none) -->
```

```
        <property>
          <key>unqualifiedSchema</key>
          <value>...</value>
        </property>
      </properties>

      <includes>...</includes>
      <excludes>...</excludes>

      <schemaVersionProvider>...</schemaVersionProvider>

      <logSlowQueriesAfterSeconds>
      ...
      </logSlowQueriesAfterSeconds>
    </database>

    <target>
      <packageName>...</packageName>
      <directory>...</directory>
    </target>
  </generator>
</configuration>
```

Based on this stub and the comments, here is an example containing the popular settings (this snippet was extracted from a JPA application that uses MySQL as the real database):

```
<configuration xmlns="...">

  <jdbc>
    <driver>org.h2.Driver</driver>
    <url>jdbc:h2:~/classicmodels</url>
  </jdbc>

  <generator>
    <database>
      <name>org.jooq.meta.extensions.jpa.JPADatabase</name>
```

```
    <properties>
      <property>
        <key>hibernate.physical_naming_strategy</key>
        <value>
          org.springframework.boot.orm.jpa
              .hibernate.SpringPhysicalNamingStrategy
        </value>
      </property>
      <property>
        <key>packages</key>
        <value>com.classicmodels.entity</value>
      </property>
      <property>
        <key>useAttributeConverters</key>
        <value>true</value>
      </property>
      <property>
        <key>unqualifiedSchema</key>
        <value>none</value>
      </property>
    </properties>

    <includes>.*</includes>
     <excludes>
           flyway_schema_history | sequences
           | customer_pgs | refresh_top3_product
           | sale_.* | set_.* | get_.* | .*_master
     </excludes>

     <schemaVersionProvider>
       ${schema.version}
     </schemaVersionProvider>

     <logSlowQueriesAfterSeconds>
       20
     </logSlowQueriesAfterSeconds>
   </database>
```

```
      <target>
        <packageName>jooq.generated</packageName>
        <directory>target/generated-sources</directory>
      </target>
    </generator>
  </configuration>
```

Besides this stub, we need the following dependency:

```
<dependency>
  <groupId>org.jooq{.trial-java-8}</groupId>
  <!-- before jOOQ 3.14.x, jooq-meta-extensions -->
  <artifactId>jooq-meta-extensions-hibernate</artifactId>
  <version>${jooq.meta.extensions.hibernate.version}
  </version>
</dependency>
```

This approach and the Gradle alternative are available in the bundled code for Java and Kotlin under the name `DeclarativeJPADatabase`.

Another approach that you'll find interesting is generating the Java-based schema from XML files: `https://www.jooq.org/doc/latest/manual/code-generation/codegen-xml/`. This is exemplified in `DeclarativeXMLDatabase` and `ProgrammaticXMLGenerator`.

Generally speaking, it is highly recommended to read the *Code generation* section of the jOOQ manual: `https://www.jooq.org/doc/latest/manual/code-generation/`. This section contains tons of settings and configurations that influence the generated artifacts.

If you need to manage multiple databases, schemas, catalogs, a shared-schema multitenancy, and so on, then refer to *Chapter 17, Multitenancy in jOOQ*.

Writing queries using a Java-based schema

Once jOOQ's Code Generator has done its job, we have access to the generated artifacts. Among these artifacts, we have the `jooq.generated.tables` folder, which contains the database tables mirrored as Java code. The generated artifacts are placed in the specified `/target` folder (in our case, `target/generated-sources`) under the specified package name (in our case, `jooq.generated`).

> **Important Note**
>
> Typically, you'll instruct the jOOQ Code Generator to store generated code under the `/target` folder (Maven), `/build` folder (Gradle), or `/src` folder. Basically, if you choose the `/target` or `/build` folder, then jOOQ regenerates the code at each build; therefore, you are sure that sources are always up to date. Nevertheless, to decide which path fits best to your strategic case, consider reading Lukas Eder's answer from Stack Overflow: `https://stackoverflow.com/questions/25576538/why-does-jooq-suggest-to-put-generated-code-under-target-and-not-under-src`. It is also recommended to check out the *Code generation and version control* section from the jOOQ manual, available at `https://www.jooq.org/doc/latest/manual/code-generation/codegen-version-control/`.

Remember that, in the previous chapter (*Chapter 1*, *Starting jOOQ and Spring Boot*), we already used the jOOQ DSL API to write the following query:

```
ResultQuery<?> query = ctx.selectFrom(table("office"))
  .where(field("territory").eq(territory));
```

This query references the database schema (table and columns). Rewriting this query referencing the Java-based schema produces the following code (jOOQ `Record` such as `OfficeRecord` are introduced in the next chapter; for now, think of it as the result set wrapped in a Java object):

```
import static jooq.generated.tables.Office.OFFICE;
import jooq.generated.tables.records.OfficeRecord;
...
ResultQuery<OfficeRecord> query = ctx.selectFrom(OFFICE)
  .where(OFFICE.TERRITORY.eq(territory));
```

Alternatively, generating and executing the query immediately can be done as follows (`Office` is a POJO):

```
public List<Office> findOfficesInTerritory(String territory) {

  List<Office> result = ctx.selectFrom(OFFICE)
    .where(OFFICE.TERRITORY.eq(territory))
    .fetchInto(Office.class);

  return result;
}
```

Depending on the database vendor, the generated SQL looks as follows with MySQL (note that jOOQ has correctly generated backticks specific to MySQL queries):

```
SELECT
  `classicmodels`.`office`.`office_code`,
  `classicmodels`.`office`.`city`,
  ...
  `classicmodels`.`office`.`territory`
FROM `classicmodels`.`office`
WHERE `classicmodels`.`office`.`territory` = ?
```

The generated SQL looks as follows with PostgreSQL (note that jOOQ has used the qualification containing the PostgreSQL schema):

```
SELECT
  "public"."office"."office_code",
  "public"."office"."city",
  ...
  "public"."office"."territory"
FROM "public"."office"
WHERE "public"."office"."territory" = ?
```

The generated SQL looks as follows with Oracle (note that jOOQ has made the identifiers uppercase, exactly as Oracle prefers):

```
SELECT
  "CLASSICMODELS"."OFFICE"."OFFICE_CODE",
  "CLASSICMODELS"."OFFICE"."CITY",
  ...
  "CLASSICMODELS"."OFFICE"."TERRITORY"
FROM "CLASSICMODELS"."OFFICE"
WHERE "CLASSICMODELS"."OFFICE"."TERRITORY" = ?
```

The generated SQL looks as follows with SQL Server (note that jOOQ has used `[]`, specific to SQL Server):

```
SELECT
  [classicmodels].[dbo].[office].[office_code],
  [classicmodels].[dbo].[office].[city],
  ...
  [classicmodels].[dbo].[office].[territory]
FROM [classicmodels].[dbo].[office]
WHERE [classicmodels].[dbo].[office].[territory] = ?
```

So, depending on the dialect, jOOQ has produced the expected query.

> **Important Note**
>
> Note that `selectFrom(table("OFFICE"))` has been rendered as `*`, while `selectFrom(OFFICE)` has been rendered as a list of column names. In the first case, jOOQ cannot infer the columns from the argument table; therefore, it projects `*`. In the second case, thanks to the Java-based schema, jOOQ projects the known columns from the table, which avoids the usage of the controversial `*`. Of course, `*` per se isn't controversial – just the fact that the columns aren't listed explicitly, as this article explains: `https://tanelpoder.com/posts/reasons-why-select-star-is-bad-for-sql-performance/`.

Let's try another example that queries the `ORDER` table. Since `ORDER` is a reserved word in most dialects, let's see how jOOQ will handle it. Note that our query doesn't do anything special to instruct jOOQ about this aspect:

```
ResultQuery<OrderRecord> query = ctx.selectFrom(ORDER)
  .where(ORDER.REQUIRED_DATE.between(startDate, endDate));
```

Or, generating and executing it immediately (`Order` is a POJO):

```
public List<Order> findOrdersByRequiredDate(
      LocalDate startDate, LocalDate endDate) {

  List<Order> result = ctx.selectFrom(ORDER)
    .where(ORDER.REQUIRED_DATE.between(startDate, endDate))
    .fetchInto(Order.class);

  return result;
}
```

Let's see the valid SQL generated for MySQL:

```
SELECT
  `classicmodels`.`order`.`order_id`,
  ...
  `classicmodels`.`order`.`customer_number`
FROM `classicmodels`.`order`
WHERE `classicmodels`.`order`.`required_date`
  BETWEEN ? AND ?
```

For brevity, we'll skip the generated SQL for PostgreSQL, Oracle, and SQL Server. Mainly, since jOOQ quotes everything by default, we can use reserved and unreserved names exactly in the same way and get back valid SQL statements.

Let's tackle one more example:

```
ResultQuery<Record2<String, LocalDate>> query = ctx.select(
         CUSTOMER.CUSTOMER_NAME, ORDER.ORDER_DATE)
  .from(ORDER)
  .innerJoin(CUSTOMER).using(CUSTOMER.CUSTOMER_NUMBER)
  .orderBy(ORDER.ORDER_DATE.desc());
```

Or, generating and executing it immediately (`CustomerAndOrder` is a POJO):

```
public List<CustomerAndOrder> findCustomersAndOrders() {

  List<CustomerAndOrder> result
    = ctx.select(CUSTOMER.CUSTOMER_NAME, ORDER.ORDER_DATE)
         .from(ORDER)
         .innerJoin(CUSTOMER).using(CUSTOMER.CUSTOMER_NUMBER)
         .orderBy(ORDER.ORDER_DATE.desc())
         .fetchInto(CustomerAndOrder.class);

  return result;
}
```

This query uses the `JOIN...USING` syntax. Basically, instead of a condition via the `ON` clause, you supply a set of fields that have an important particularity – their names are common to both tables to the left and right of the join operator. However, some dialects (for example, Oracle) don't allow us to use qualified names in `USING`. Having qualified names leads to an error such as `ORA-25154: column part of USING clause cannot have qualifier`.

jOOQ is aware of this aspect and takes action. Following the Oracle dialect, jOOQ renders `CUSTOMER.CUSTOMER_NUMBER` as `"CUSTOMER_NUMBER"`, not qualified as `"CLASSICMODELS"."CUSTOMER"."CUSTOMER_NUMBER"`. Check this here:

```
SELECT
  "CLASSICMODELS"."CUSTOMER"."CUSTOMER_NAME",
  "CLASSICMODELS"."ORDER"."ORDER_DATE"
FROM
```

```
  "CLASSICMODELS"."ORDER"
  JOIN "CLASSICMODELS"."CUSTOMER" USING ("CUSTOMER_NUMBER")
ORDER BY
  "CLASSICMODELS"."ORDER"."ORDER_DATE" DESC
```

This was just an example of how jOOQ takes care of the generated SQL by emulating the correct syntax, depending on the dialect used! Thanks to jOOQ code generation, we benefit from default choices for so many silly edge cases that are so annoying to handle later on.

Let's summarize a handful of advantages brought by jOOQ code generation:

- Type-safe SQL queries. Did I mention type-safe SQL queries?!
- No need to worry about the identifier's case sensitivity, quotation, and qualification.
- Using generated code makes for much leaner expressions. There's less wrapping noise such as `field("X", "Y")`, `field(name("X", "Y"))`, or `field(name("X", "Y"), DATA_TYPE)`. Via jOOQ code generation, this would just be `X.Y`.
- The IDE can provide code completion and refactoring support.
- We can use the IDE to find uses of tables and columns because they're Java objects.
- The code will no longer compile when the columns are renamed, rather than having to run the query for it to fail.
- Avoidance of issues caused by edge cases with vendor-specific data types.
- Since jOOQ quotes everything by default, users don't have to think of quoting reserved names such as `table(name("ORDER"))`. It's just `ORDER`, and jOOQ will produce `` `ORDER` ``, `"ORDER"`, `[ORDER]`, or whatever is specific to the used dialect.

> **Important Note**
>
> As a rule of thumb, always consider jOOQ code generation as the default way to exploit jOOQ. Of course, there are edge cases when code generation cannot be fully exploited (for instance, in the case of schemas that are created/modified dynamically at runtime), but this is a different story.

The application developed in this section is named *WriteTypesafeSQL*.

jOOQ versus JPA Criteria versus QueryDSL

All these three, jOOQ, JPA Criteria (or the Spring Data JPA Specifications API built on top of the Criteria API), and QueryDSL, can provide type-safe SQL.

If you come from a JPA background, then you know that JPA defines a Metamodel API for Criteria queries. So, the Criteria API and the Metamodel API can provide type safety for SQL as well. But, the Criteria API is quite complicated compared to QueryDSL. You don't have to take my word for it – try it! However, the Criteria API is something that you need to learn in addition to JPQL and all the JPA stuff. Also, it is not intuitive, it is poorly documented, and developers describe it as quite slow. Moreover, having 100% type safety means having to write all SQL statements that are prone to type errors via the Criteria API.

QueryDSL supports SQL type safety as well. Having support in Spring Boot, QueryDSL is well covered in this article at `https://dzone.com/articles/querydsl-vs-jooq-feature`, which contains a non-exhaustive list of jOOQ support beyond QueryDSL's "feature completeness." Nevertheless, that article is quite old and may be out of date. Meanwhile, jOOQ has even more advantages that you can find yourself by a quick search on `reddit.com`.

Next, let's go one step further and give more control to jOOQ.

Configuring jOOQ to generate POJOs

So far, we have used our own **POJOs** as our primary **Data Transfer Objects** (**DTOs**). This is a common approach in layered applications such as Spring Boot applications.

The `Office` and `Order` POJOs are Java mirrors of the `OFFICE` and `ORDER` tables, since our queries fetch all the columns from these tables. On the other hand, the `CustomerAndOrder` POJO maps columns from two different tables, `CUSTOMER` and `ORDER`. More precisely, it maps `CUSTOMER_NAME` from `CUSTOMER` and `ORDER_DATE` from `ORDER`.

Optionally, jOOQ can generate POJOs on our behalf via the jOOQ Code Generator. In Maven, this feature can be enabled via the following configuration into the `<generator>` tag:

```
<generator>
  ...
  <generate>
    <pojos>true</pojos>
  </generate>
  ...
</generator>
```

Additionally, jOOQ can add to the generated POJOs a set of Bean Validation API annotations to convey type information. More precisely, they include two well-known validation annotations – `@NotNull` (`javax/jakarta.validation.constraints.NotNull`) and `@Size` (`javax/jakarta.validation.constraints.Size`). To enable these annotations, the configuration should be as follows:

```
<generate>
  <pojos>true</pojos>
  <validationAnnotations>true</validationAnnotations>
</generate>
```

Also, you should add the dependency for `validation-api` as in the bundled code.

By default, the names of the generated POJOs are the same as the names of the tables in *Pascal* case (for instance, the table named `office_has_manager` becomes `OfficeHasManager`). Altering the default behavior can be achieved via so-called *generator strategies* – basically, in Maven, a piece of XML delimited by the `<strategy>` tag that relies on regular expressions for producing custom (user-defined) output. For example, if the POJOs are prefixed with the `Jooq` text, then the generator strategy will be the following:

```
<strategy>
  <matchers>
    <tables>
      <table>
        <pojoClass>
          <expression>JOOQ_$0</expression>
          <transform>PASCAL</transform>
        </pojoClass>
      ...
</strategy>
```

This time, the table named `office_has_manager` results in a POJO source named `JooqOfficeHasManager`. More details about the generator strategies (including the programmatic approach) are available in *Chapter 18, jOOQ SPI (Providers and Listeners)*. Also, it is recommended to read `https://www.jooq.org/doc/latest/manual/code-generation/codegen-matcherstrategy/`.

The Gradle alternative is available in the bundled code.

By default, jOOQ generates a POJO for each table in the database. Therefore, by default, jOOQ can generate a POJO as `Office` and `Order` (or `JooqOffice` and `JooqOrder`, conforming to the preceding strategy), but its purpose is not to generate more complex POJOs, such as composite POJOs or ones containing arbitrary objects (such as `CustomerAndOrder`). The following is the source code of `JooqOffice`, generated by jOOQ:

```
public class JooqOffice implements Serializable {

  private static final long serialVersionUID = 1821407394;

  private String officeCode;
  private String city;
  ...
  private String territory;

  public JooqOffice() {}

  public JooqOffice(JooqOffice value) {
    this.officeCode = value.officeCode;
    this.city = value.city;
    ...
    this.territory = value.territory;
  }

  public JooqOffice(String officeCode,
         String city, ... String territory) {
    this.officeCode = officeCode;
    this.city = city;
    ...
    this.territory = territory;
  }

  @NotNull
  @Size(max = 10)
  public String getOfficeCode() {
    return this.officeCode;
  }
```

```
  public void setOfficeCode(String officeCode) {
    this.officeCode = officeCode;
  }

  // getters and setters and toString() omitted for brevity
}
```

Similar POJOs are generated for each table of the `classicmodels` database. This means that we can still use our `CustomerAndOrder` POJO, but there is no need to write our own POJOs for `Office` and `Order` because we can use those generated by jOOQ. The following code was cut out from `ClassicModelsRepository` and uses the generated `JooqOffice` and `JooqOrder` (note the imports – jOOQ placed the POJOs in the `jooq.generated.tables.pojos` package):

```
import jooq.generated.tables.pojos.JooqOffice;
import jooq.generated.tables.pojos.JooqOrder;
...
public List<JooqOffice> findOfficesInTerritory(
                               String territory) {

  List<JooqOffice> result = ctx.selectFrom(OFFICE)
    .where(OFFICE.TERRITORY.eq(territory))
    .fetchInto(JooqOffice.class);

  return result;
}

public List<JooqOrder> findOrdersByRequiredDate(
           LocalDatestartDate, LocalDateendDate) {

  List<JooqOrder> result = ctx.selectFrom(ORDER)
    .where(ORDER.REQUIRED_DATE.between(startDate, endDate))
      .fetchInto(JooqOrder.class);

      return result;
}
```

Done! So, jOOQ-generated POJOs can be used as any regular POJOs. For instance, they can be returned from a `REST` controller, and Spring Boot will serialize them as JSON. We'll detail more types of supported POJOs later on when we tackle the mapping result set to POJOs.

The application developed in this section is available as *GeneratePojos*. Next, let's see how jOOQ can generate DAOs.

Configuring jOOQ to generate DAOs

If you are familiar with Spring Data JPA/JDBC, then you're already used to relying on a DAO layer that wraps the queries. Both Spring Data JDBC and JPA provide a built-in DAO that exposes a set of CRUD operations and can be extended via user-defined repositories.

jOOQ code generation can produce similar DAOs. Basically, for each table of the database, jOOQ can generate an `org.jooq.DAO` implementation that exposes methods such as `findById()`, `delete()`, `findAll()`, `insert()`, and `update()`.

In Maven, this feature can be enabled via the following configuration in the `<generator>` tag:

```
<generator>
  ...
  <generate>
    <daos>true</daos>
  </generate>
  ...
</generator>
```

jOOQ DAOs make use of POJOs; therefore, jOOQ will implicitly generate POJOs as well. Since we are in Spring Boot, it will be nice to have the generated DAOs annotated with `@Repository` as the built-in `SimpleJpaRepository`. To achieve this, we use the `<springAnnotations/>` flag, as follows:

```
<generate>
  <daos>true</daos>
  <springAnnotations>true</springAnnotations>
</generate>
```

By default, the names of the generated DAOs are the same as the names of the tables in *Pascal* case and suffixed with the word `Dao` (for instance, the table named `office_has_manager` becomes `OfficeHasManagerDao`). Altering the default behavior can be achieved via so-called *generator strategies*. For instance, following the Spring style, we prefer `OfficeHasManagerRepository` instead of `OfficeHasManagerDao`. This can be achieved as follows:

```
<strategy>
  <matchers>
    <tables>
      <table>
        <daoClass>
          <expression>$0_Repository</expression>
          <transform>PASCAL</transform>
        </daoClass>
      ...
</strategy>
```

The Gradle alternative is available in the bundled code. For instance, the generated `OfficeRepository` looks as follows:

```
@Repository
public class OfficeRepository
       extends DAOImpl<OfficeRecord, JooqOffice, String> {

  public OfficeRepository() {
    super(Office.OFFICE, JooqOffice.class);
  }

  @Autowired
  public OfficeRepository(Configuration configuration) {
    super(Office.OFFICE, JooqOffice.class, configuration);
  }

  @Override
  public String getId(JooqOffice object) {
    return object.getOfficeCode();
  }
```

```
  public List<JooqOffice> fetchRangeOfOfficeCode(
       String lowerInclusive, String upperInclusive) {
    return fetchRange(Office.OFFICE.OFFICE_CODE,
         lowerInclusive, upperInclusive);
  }

  // more DAO-methods omitted for brevity
}
```

Each generated DAO extends the common base implementation named `DAOImpl`. This implementation supplies common methods such as `insert()`, `update()`, `delete()`, and `findById()`.

So far, our `ClassicModelsRepository` contains three query methods, represented by `findOfficesInTerritory()`, `findOrdersByRequiredDate()`, and `findCustomersAndOrders()`.

However, let's check the query from `findOfficesInTerritory()`:

```
List<JooqOffice> result = ctx.selectFrom(OFFICE)
  .where(OFFICE.TERRITORY.eq(territory))
  .fetchInto(JooqOffice.class);
```

Here, we notice that the generated `OfficeRepository` already covers this query via the `fetchByTerritory(String territory)` method; therefore, we can use this built-in DAO method directly in our service, `ClassicModelsService`, as follows:

```
@Transactional(readOnly = true)
public List<JooqOffice> fetchOfficesInTerritory(
                              String territory) {

  return officeRepository.fetchByTerritory(territory);
}
```

Going further, check out the query from `findOrdersByRequiredDate()`:

```
List<JooqOrder> result = ctx.selectFrom(ORDER)
  .where(ORDER.REQUIRED_DATE.between(startDate, endDate))
  .fetchInto(JooqOrder.class);
```

This time, the previous query is covered in `OrderRepository` by the built-in DAO method, `fetchRangeOfRequiredDate(LocalDate li, LocalDate ui)`. So, we can drop the previous query and rely on `ClassicModelsService` on the built-in one, as follows:

```
@Transactional(readOnly = true)
public List<JooqOrder> fetchOrdersByRequiredDate(
      LocalDate startDate, LocalDate endDate) {

  return orderRepository.fetchRangeOfRequiredDate(
      startDate, endDate);
}
```

At this point, the only query method left in `ClassicModelsRepository` is `findCustomersAndOrders()`. This query method doesn't have an alternative in the default generated DAOs; therefore, we still need it.

For now, you can check the application named *GenerateDaos*. Later on, we'll discuss extending and customizing the jOOQ-generated DAO.

Configuring jOOQ to generate interfaces

Besides POJOs and DAOs, jOOQ can generate an interface for each table. Each column is associated with a getter and a setter. In Maven, this can be done as shown here:

```
<generate>
  <interfaces>true</interfaces>
  <immutableInterfaces>true</immutableInterfaces>
</generate>
```

Basically, jOOQ generates interfaces that look like Spring Data's so-called *interfaces-based closed projections*. We can use these interfaces for mapping results sets exactly as we do with *closed projections*.

Nevertheless, note that at the time of writing, this feature has been proposed to be removed. You can track the deprecation here: `https://github.com/jOOQ/jOOQ/issues/10509`.

Next, let's continue with the programmatic configuration of the jOOQ Code Generator.

Tackling programmatic configuration

If you prefer programmatic configurations, then jOOQ exposes a fluent API (`org.jooq.meta.jaxb.*`) that can be used for configuring code generation in programmatic fashion. First, for Maven, add the following dependency in `pom.xml`:

```
<dependency>
  <groupId>org.jooq{.trial-java-8}</groupId>
  <artifactId>jooq-codegen</artifactId>
</dependency>
```

Alternatively, in Gradle, add `implementation 'org.jooq{.trial-java-8}:jooq-codegen'`.

Note that `Configuration` refers to `org.jooq.meta.jaxb.Configuration`, not `org.jooq.Configuration`, which is used for creating `DSLContext` and other jOOQ contexts.

This programmatic API mirrors the declarative approach and, therefore, is very intuitive. For instance, here it is the programmatic alternative of the declarative approach presented in the *Configuring jOOQ to generate DAOs* section for the MySQL `classicmodels` schema:

```
Configuration configuration = new Configuration()
  .withJdbc(new Jdbc()
    .withDriver("com.mysql.cj.jdbc.Driver")
    .withUrl("jdbc:mysql://localhost:3306/classicmodels")
    .withUser("root")
    .withPassword("root"))
  .withGenerator(new Generator()
    .withName("org.jooq.codegen.JavaGenerator")
    .withDatabase(new Database()
      .withName("org.jooq.meta.mysql.MySQLDatabase")
      .withInputSchema("classicmodels")
      .withIncludes(".*")
      .withExcludes("flyway_schema_history | sequences"
        + " | customer_pgs | refresh_top3_product"
        + " | sale_.* | set_.* | get_.* | .*_master")
      .withSchemaVersionProvider("SELECT MAX(`version`)
          FROM `flyway_schema_history`")
      .withLogSlowQueriesAfterSeconds(20))
  .withGenerate(new Generate()
    .withDaos(true)
```

```
    .withValidationAnnotations(Boolean.TRUE)
    .withSpringAnnotations(Boolean.TRUE))
  .withStrategy(new Strategy()
    .withMatchers(new Matchers()
      .withTables(new MatchersTableType()
        .withPojoClass(new MatcherRule()
         .withExpression("Jooq_$0")
         .withTransform(MatcherTransformType.PASCAL))
        .withDaoClass(new MatcherRule()
          .withExpression("$0_Repository")
          .withTransform(MatcherTransformType.PASCAL)))))
  .withTarget(new Target()
    .withPackageName("jooq.generated")
    .withDirectory(System.getProperty("user.dir")
    .endsWith("webapp") ? "target/generated-sources"
              : "webapp/target/generated-sources")));

GenerationTool.generate(configuration);
```

The jOOQ Code Generator must generate the classes before the application's classes are compiled; therefore, the programmatic Code Generator should be placed in a separate module of your application and invoked at the proper moment before the compilation phase. As you'll see in the bundled code (*ProgrammaticGenerator*), this can be achieved via `exec-maven-plugin` for Maven or `JavaExec` for Gradle.

If you prefer the DDL Database API, then you'll love the programmatic approach from `ProgrammaticDDLDatabase`. If you prefer the JPA Database API, then check out the programmatic approach as well, *ProgrammaticJPADatabase*.

All the applications from this chapter are available for Java/Kotlin and Maven/Gradle combos.

Introducing jOOQ settings

jOOQ supports a bunch of optional settings (`org.jooq.conf.Settings`) that are mostly used to customize rendered SQL. While all these settings rely on defaults that have been carefully chosen for a wide range of cases, there are still situations when we have to alter them.

If you prefer the declarative approach, then you can alter these settings via an XML file, named `jooq-settings.xml`, placed in the application classpath. For instance, if the rendered SQL doesn't contain the name of the catalog/schema, then `jooq-settings.xml` will be as follows:

```
<?xml version="1.0" encoding="UTF-8"?>

<settings>
  <renderCatalog>false</renderCatalog>
  <renderSchema>false</renderSchema>
</settings>
```

Without these settings, jOOQ renders the name of the catalog/schema for each generated SQL. Here is an example in SQL Server:

- Without these settings, jOOQ renders `[classicmodels].[dbo].[customer].[customer_name]`.
- With these settings, jOOQ doesn't render the schema and catalog names – `[customer].[customer_name]`.

As you can see in the corresponding XSD (`https://www.jooq.org/xsd/jooq-runtime-3.x.x.xsd`), jOOQ supports a lot of settings, and most of them are for advanced users and serve only certain scenarios. Nevertheless, some of them are more popular than others, and you'll see them mentioned in the proper context throughout this book.

Moreover, jOOQ `Settings` can be programmatically shaped via `@Bean`, as follows:

```
@Bean
public Settings jooqSettings() {
  return new Settings()
   .withRenderCatalog(Boolean.FALSE)
   .withRenderSchema(Boolean.FALSE);
}
```

Via `@Bean`, we customize jOOQ settings globally (at the application level), but we can override them locally at the `DSLContext` level via the `DSLContext` constructor (`DSL.using()`), as shown in this example:

```
DataSource ds = ...;

DSLContext ctx = DSL.using(ds, SQLDialect.MYSQL,
```

```
new Settings()
   .withRenderCatalog(Boolean.FALSE)
   .withRenderSchema(Boolean.FALSE));
```

Alternatively, we can locally define `DSLContext`, derived from the current `DSLContext` (denoted as `ctx`) and having altered `Settings`:

```
ctx.configuration().derive(
  new Settings()
    .withRenderCatalog(Boolean.FALSE)
    .withRenderSchema(Boolean.FALSE))).dsl()
    ... // some query
```

During this book, you'll have plenty of occasions to see `Settings` at work, so there is no need to bother too much for the moment.

It's time to summarize this chapter!

Summary

In this chapter, we have reached several targets, but the most important was the introduction of the jOOQ Code Generator using configurative and programmatic approaches. More specifically, you saw how to write type-safe queries and how to generate and use POJOs and DAOs. These are fundamental skills in jOOQ that we'll develop during the entire book.

From this point forward, we'll focus on other topics that will help you to become a jOOQ power user.

In the next chapter, we will start diving into the jOOQ core concepts.

Part 2: jOOQ and Queries

This part covers jOOQ type-safe queries, the fluent API, converters, bindings, inlining parameters, mappers, and associations.

This part contains the following chapters:

- *Chapter 3, jOOQ Core Concepts*
- *Chapter 4, Building a DAO layer (Evolving the Generated DAO Layer)*
- *Chapter 5, Tackling Different Kinds of SELECT, INSERT, UPDATE, DELETE, and MERGE Statements.*
- *Chapter 6, Tackling Different Kinds of JOIN Statements*
- *Chapter 7, Types, Converters, and Bindings*
- *Chapter 8, Fetching and Mapping*

3

jOOQ Core Concepts

Before exploring more awesome features of jOOQ, we have to cover the core (fundamental) concepts that jOOQ relies on. Having a decent insight into jOOQ core concepts helps us to make the right decisions and to understand how jOOQ works under the hood. Don't worry, our aim is not to enter the jOOQ bowels! We aim to bring you close to the jOOQ paradigm and start thinking about your persistent layer in the jOOQ context.

The goal of this chapter is to briefly introduce the following topics:

- Hooking jOOQ results (`Result`) and records (`Record`)
- Exploring jOOQ query types
- Understanding the jOOQ fluent API
- Highlighting how jOOQ emphasizes SQL syntax correctness
- Casting, coercing, and collating
- Binding values (parameters)

By the end of this chapter, you'll be familiar with the jOOQ core concepts that will help you to easily follow the upcoming chapters.

Let's get started!

Technical requirements

The code for this chapter can be found on GitHub at `https://github.com/PacktPublishing/jOOQ-Masterclass/tree/master/Chapter03`.

Hooking jOOQ results (Result) and records (Record)

In the previous chapters, we've mapped the JDBC result set of our queries to POJOs via the jOOQ `fetchInto()` method. But, in jOOQ, between the JDBC result set and a well-known `List<POJO>` (or other data structure such as an array, map, and set), there is another fundamental layer referenced as `Result<Record>` represented from the following two interfaces:

- `org.jooq.Record`: When we trigger a `SELECT` query, we get back a result set that contains a list of columns and the corresponding list of values. Typically, we refer to the content of the result set as *records*. jOOQ maps each such *record* to its `Record` interface. Think of `Record` as the jOOQ internal representation of *records*.
- `org.jooq.Result`: The jOOQ `Result` interface is a `java.util.List` of `org.jooq.Record`. In other words, jOOQ maps each *record* of the result set to a `Record` and collects this record in `Result`. Once `Result<Record>` is complete (the whole result set was processed), it can be mapped into an array, a set/list of POJOs, or a map, or it can be returned as it is.

The following figure represents this straightforward path: JDBC result set | jOOQ `Result<Record>` | array/list/set/map:

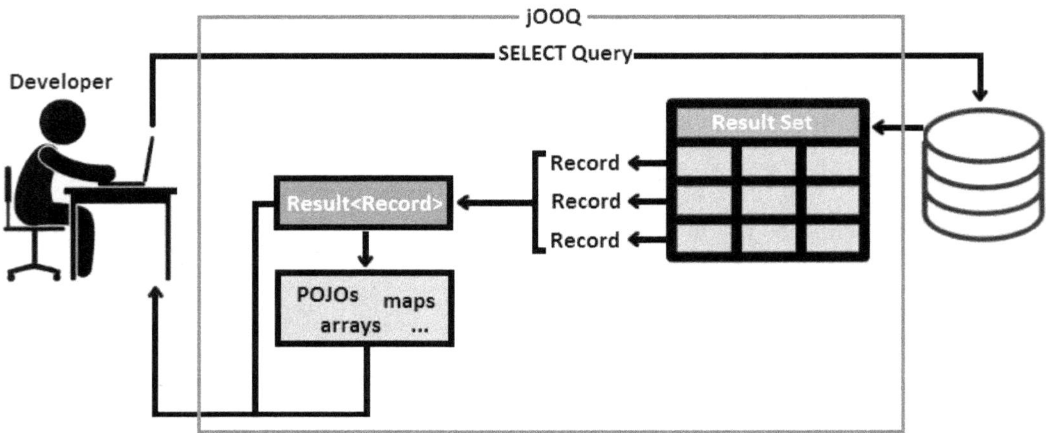

Figure 3.1 – Processing of the JDBC ResultSet

As you can see from this figure, we can fetch the result set as type-specific to the application's needs (for instance, List<POJO>), but we can fetch the result set directly as Result<Record> as well. If you come from the JPA area, then you may think that the jOOQ Record is somehow similar to JPA entities, but this is not true. In jOOQ, there is no equivalent of persistence context (first-level cache), and jOOQ doesn't perform any kind of heavy lifting on these objects such as state transitions and auto-flushes. Most of the time, you can use records through the jOOQ API directly since you'll not even need a POJO.

> **Important Note**
>
> In jOOQ, by default, the JDBC result set is fetched into memory eagerly (all data projected by the current query will be stored in memory), but as you'll see in *Chapter 8, Fetching and Mapping*, we can operate on large result sets "lazily" using fetchLazy() and the Cursor type. Mapping the JDBC result set to Result<Record> comes with multiple benefits of which we highlight the following:
>
> a) Result<Record> represents non-type-safe query results, but it can also represent type-safe query results via Record specializations such as table records, updatable records, and degree records up to degree 22 (number 22 is derived from Scala – https://stackoverflow.com/q/6241441/521799).
>
> b) After fully loading Result<Record> into memory, jOOQ frees the resources as early as possible. It is preferable to operate on an in-memory Result<Record> instead of operating on a JDBC result set holding open a connection to the database.
>
> c) Result<Record> can be easily exported to XML, CSV, JSON, and HTML.
>
> d) jOOQ exposes a friendly and comprehensive API for manipulating Result<Record>, therefore, for manipulating the result set.

jOOQ supports a few types of Record as follows:

- **Table records**: These records are implemented via org.jooq.TableRecord and org.jooq.UpdatableRecord (records that can be stored back in the database again). A TableRecord/UpdatableRecord record originates from a single table (or view) having a primary key. Only UpdatableRecords have a (known to jOOQ) primary key The jOOQ Code Generator can produce this type of record on our behalf – for instance (check out our previous applications), the jooq.generated.tables.records package, which contains CustomerRecord, EmployeeRecord, and OfficeRecord. All these table records have been generated via the jOOQ generator and are strongly typed.

- **Records of well-defined degree**: jOOQ defines 22 interfaces that extend `Record` with the purpose of providing type-safety for queries that project custom record types in SQL. The query can contain records originating from a single table or from multiple tables. jOOQ will choose the proper `Record1` ... `Record22` interface and will pick up the correct types to guarantee the type-safety of query results.

> **Important Note**
>
> This kind of type-safety is applied to records for degrees up to 22. This also applies to row value expressions, subselects that are combined by a set operator (for example, `UNION`), `IN` predicates and comparison predicates taking subselects, and `INSERT` and `MERGE` statements that take type-safe `VALUES()` clauses. Beyond degree 22, there is no type-safety.

- **UDT records**: These records are useful for supporting **User-Defined Types (UDTs)** specific to Oracle and PostgreSQL. They are represented in jOOQ via the `org.jooq.UDTRecord` API.
- **Embeddable records**: These records represent synthetic UDTs and they are implemented via `org.jooq.EmbeddableRecord`. This topic is covered in *Chapter 7, Types, Converters, and Bindings*.

Let's see several examples of fetching jOOQ records.

Fetching Result<Record> via plain SQL

In jOOQ, plain SQL, such as an SQL string, returns an anonymous type-safe `Result<Record>`. Here are two examples:

```
/* non type-safe Result<Record> */
Result<Record> result = ctx.fetch(
  "SELECT customer_name, customer_number, credit_limit
   FROM customer");

/* non type-safe Result<Record> */
Result<Record> result = ctx.resultQuery(
  "SELECT customer_name, customer_number, credit_limit
   FROM customer").fetch();
```

Iterating `Result` is like iterating `java.util.List`. Each `Record` can be accessed via a comprehensive API that, among other methods, exposes more than 10 `get()`/`getValue()` methods for retrieving values from records in a non type-safe manner. Consider the following example:

```
/* non type-safe values */
for (Record record : result) {
  // get value by index
  Object r1 = record.get(0);

  // get value by name
  Object r2 = record.get("customer_number");

  // get value by name and type
  BigDecimal r3 = record.getValue(
    "credit_limit", BigDecimal.class);
}
```

Pay attention to `r3`. Our example works just fine, but if the specified type is not the proper one for the specified column (in other words, the data type cannot be converted, but conversion is possible), then we'll get a jOOQ `DataTypeException` or, even worse, you'll silently use the results of an apparently successful conversion that may have an improper representation. Moreover, typos in column names or columns that don't exist will cause `java.lang.IllegalArgumentException`.

In order to avoid such unpleasant cases, from this point forward, we rely on classes obtained via the jOOQ Code Generator. This gives us a tremendous boost in productivity and a wide range of features. Hmmm, have I told you that you should always count on the jOOQ Code Generator? Anyway, let's continue with examples.

Fetching Result<Record> via select()

The parameter-less `select()` method of `DSLContext` results in a projection that includes all columns. It also produces a non-type-safe `Result<Record>`. This time, we use the Java-based schema produced by the jOOQ Code Generator:

```
/* non type-safe Result<Record> */
Result<Record> result = ctx.select().from(CUSTOMER).fetch();
```

Even if `Result<Record>` is non-type-safe, the values of records can be type-safely extracted via the jOOQ generated classes. More precisely, we use the attributes of the generated `Customer` class as follows (`CUSTOMER` is `static`):

```
/* type-safe values */
for (Record r : result) {

  String r1 = r.get(CUSTOMER.CUSTOMER_NAME);
  Long r2 = r.get(CUSTOMER.CUSTOMER_NUMBER);
  BigDecimal r3 = r.get(CUSTOMER.CREDIT_LIMIT);
  ...
}
```

One step further, we can express this non-type-safe `Result<Record>` as a type-safe one.

Mapping org.jooq.Record into a strongly-typed org.jooq.TableRecord

Since we fetch data from a single table having a primary key (`CUSTOMER`), we can use `TableRecord` associated by jOOQ with the database table.

Transforming the previous non-type-safe `Result<Record>` into a type-safe one can be done by mapping `org.jooq.Record` into the corresponding strongly-typed `org.jooq.TableRecord` via the `Record.into(Table<Z> table)` method. In this case, the corresponding strongly-typed `org.jooq.TableRecord` is `CustomerRecord`. Check out the following code:

```
/* type-safe Result<Record> */
Result<CustomerRecord> result = ctx.select().from(CUSTOMER)
  .fetch().into(CUSTOMER);
```

The same thing can be done via `Record.into(Class<? extends E> type)`:

```
/* type-safe Result<Record> */
List<CustomerRecord> result = ctx.select().from(CUSTOMER)
  .fetch().into(CustomerRecord.class);
```

This time, we can use the `CustomerRecord` getters to access the values of records:

```
/* type-safe values */
for (CustomerRecord r : result) {
  String r1 = r.getCustomerName();
  Long r2 = r.getCustomerNumber();
  BigDecimal r3 = r.getCreditLimit();
```

```
    ...
}
```

Let's see what happens if we enrich this query to fetch data from two (or more) tables.

Fetching Result<Record> via select() and join()

Let's enrich `ctx.select().from(CUSTOMER)` with a `JOIN` clause to fetch records from `CUSTOMERDETAIL` as well (there is a one-to-one relationship between `CUSTOMER` and `CUSTOMERDETAIL`):

```
/* non type-safe Result<Record> */
Result<Record> result = ctx.select()
  .from(CUSTOMER)
  .join(CUSTOMERDETAIL)
    .on(CUSTOMER.CUSTOMER_NUMBER
      .eq(CUSTOMERDETAIL.CUSTOMER_NUMBER))
  .fetch();
```

The values of records can be type-safely extracted from the attributes of the generated `Customer` and `Customerdetail` class:

```
/* type-safe values */
for (Record r : result) {
  String r1 = r.get(CUSTOMER.CUSTOMER_NAME);
  Long r2 = r.get(CUSTOMER.CUSTOMER_NUMBER);
  BigDecimal r3 = r.get(CUSTOMER.CREDIT_LIMIT);
  ...
  String r4 = r.get(CUSTOMERDETAIL.CITY);
  String r5 = r.get(CUSTOMERDETAIL.COUNTRY);
  ...
}
```

Re-writing this non-type-safe `Result<Record>` as a type-safe one is a little bit verbose. Let's see how to do it.

Mapping org.jooq.Record into a strongly-typed org.jooq.TableRecord

Transforming the previous non-type-safe `Result<Record>` into a type-safe one can be done via the proper `select(SelectField<T1>, SelectField<T2>... SelectField<T22>)` or `into(Field<T1>, Field<T2> ... Field<T22>)` method and the proper `Record[N]` interface, `N=1..22`. Our schema reveals that the

CUSTOMER and CUSTOMERDETAIL tables contain a total of 15 fields, therefore, the proper Record[N] is Record15 and we use the select(SelectField<T1>, SelectField<T2>... SelectField<T15>) counterpart:

```
/* type-safe Result<Record> via select() */
Result<Record15<Long, String, String, String,
  String, Long, BigDecimal, Integer, Long, String, String,
  String, String, String, String>> result
  = ctx.select(CUSTOMER.CUSTOMER_NUMBER,
    CUSTOMER.CUSTOMER_NAME,CUSTOMER.CONTACT_FIRST_NAME,
    CUSTOMER.CONTACT_LAST_NAME,CUSTOMER.PHONE,
    CUSTOMER.SALES_REP_EMPLOYEE_NUMBER,
    CUSTOMER.CREDIT_LIMIT,CUSTOMER.FIRST_BUY_DATE,
    CUSTOMERDETAIL.CUSTOMER_NUMBER,
    CUSTOMERDETAIL.ADDRESS_LINE_FIRST,
    CUSTOMERDETAIL.ADDRESS_LINE_SECOND,
    CUSTOMERDETAIL.CITY,CUSTOMERDETAIL.COUNTRY,
    CUSTOMERDETAIL.POSTAL_CODE,CUSTOMERDETAIL.STATE)
      .from(CUSTOMER)
      .join(CUSTOMERDETAIL)
      .on(CUSTOMER.CUSTOMER_NUMBER.eq(
        CUSTOMERDETAIL.CUSTOMER_NUMBER))
      .fetch();
```

Or, we can use the into(Field<T1>, Field<T2> ... Field<T15>) counterpart:

```
/* type-safe Result<Record>via into() */
Result<Record15<Long, String, String, String,
         String, Long, BigDecimal, Integer, Long, String,
         String, String, String, String, String>> result =
   ctx.select()
      .from(CUSTOMER)
      .join(CUSTOMERDETAIL)
         .on(CUSTOMER.CUSTOMER_NUMBER
            .eq(CUSTOMERDETAIL.CUSTOMER_NUMBER))
      .fetch()
      .into(CUSTOMER.CUSTOMER_NUMBER,
    CUSTOMER.CUSTOMER_NAME, CUSTOMER.CONTACT_FIRST_NAME,
    CUSTOMER.CONTACT_LAST_NAME,CUSTOMER.PHONE,
    CUSTOMER.SALES_REP_EMPLOYEE_NUMBER, CUSTOMER.CREDIT_LIMIT,
```

```
CUSTOMER.FIRST_BUY_DATE, CUSTOMERDETAIL.CUSTOMER_NUMBER,
CUSTOMERDETAIL.ADDRESS_LINE_FIRST,
CUSTOMERDETAIL.ADDRESS_LINE_SECOND, CUSTOMERDETAIL.CITY,
CUSTOMERDETAIL.COUNTRY, CUSTOMERDETAIL.POSTAL_CODE,
CUSTOMERDETAIL.STATE);
```

Obviously, we have 22 such `select()` and `into()` methods, but we need the one that corresponds to our records' degree.

> **Important Note**
>
> Have you noticed the `Record15<…>` construction? Of course you have! It's hard to miss! Besides the obvious verbosity, it is not that easy to fill up the data types as well. You have to identify and write down each data type of the fetched fields in the correct order. Fortunately, we can avoid this torturous step by using the Java 9 `var` keyword. Once you have practiced the examples from this chapter and you've got familiar with `Record[N]`, consider using `var` whenever you don't have a good reason to manually write down `Record[N]`. On the other hand, if you are using Kotlin/Scala, then you can take advantage of better support for tuple-style data structures and rely on automatic destructuration of `Record[N]` as `val(a, b, c) = select(A, B, C)`. For more details, consider this example: `https://github.com/jOOQ/jOOQ/tree/main/jOOQ-examples/jOOQ-kotlin-example`. So far, in Java, the previous two examples can be expressed using `var` as follows:
>
> ```
> var result = ctx.select(...);
> var result = ctx.select()...into(...);
> ```

The records values can be accessed in the same way via the attributes of the generated `Customer` and `Customerdetail` classes. But, can we access it via the corresponding table records?

Extracting the two TableRecords from Record

Extracting the two individual strongly-typed `TableRecord` types (`CustomerRecord` and `CustomerdetailRecord`) from the denormalized `Record` can be done via the `Record.into(Table<Z> table)` method. I bet you didn't think that this was possible:

```
Result<CustomerRecord> rcr=result.into(CUSTOMER);
Result<CustomerdetailRecord> rcd=result.into(CUSTOMERDETAIL);
```

Further, we can rely on the built-in getters of `CustomerRecord` and `CustomerdetailRecord` to access the corresponding values.

Fetching Result<Record> via selectFrom()

The best approach for selecting, in a type-safe manner, all the columns from a single table into Result<Record> relies on the selectFrom(table) method. In this context, jOOQ returns the record type supplied with the argument table, therefore, it returns TableRecord. Check out the code:

```
/* type-safe Result<Record> */
Result<CustomerRecord> result
  = ctx.selectFrom(CUSTOMER).fetch();
```

Further, the CustomerRecord getters return the values:

```
/* type-safe values */
for (CustomerRecord r : result) {
  String r1 = r.getCustomerName();
  Long r2 = r.getCustomerNumber();
  BigDecimal r3 = r.getCreditLimit();
  ...
}
```

While this is really cool, please consider the following important note as well.

> **Important Note**
>
> Don't consider that select().from(table) and selectFrom(table) are the same thing. The former, select().from(table), returns a non-type-safe Result<Record> and we can use any clause that modifies the type of the table expression (for instance, JOIN). On the other hand, selectFrom(table) returns a type-safe Result<TableRecord> and doesn't permit the usage of any clause that modifies the type of the table expression.

Next, let's tackle ad hoc selects.

Fetching Result<Record> via ad hoc selects

In ad hoc selects, we enlist the needed columns that can originate in one or more tables. As long as we enlist the columns explicitly and rely on Java-based schema, jOOQ will determine the correct types and will prepare a record of a certain degree. Here is an example that selects some columns from a single table:

```
/* type-safe Result<Record> */
Result<Record3<Long, String, BigDecimal>> result = ctx.select(
```

```
    CUSTOMER.CUSTOMER_NUMBER, CUSTOMER.CUSTOMER_NAME,
    CUSTOMER.CREDIT_LIMIT)
  .from(CUSTOMER)
  .fetch();
```

Since we have three columns, jOOQ has picked up the record of degree 3, `Record3`, and automatically inferred the correct Java types, `Long`, `String`, and `BigDecimal`.

Next, let's see an example that fetches five columns originating from two tables:

```
/* type-safe Result<Record> */
Result<Record5<Long, BigDecimal, String, String, String>>
  result = ctx.select(CUSTOMER.CUSTOMER_NUMBER,
      CUSTOMER.CREDIT_LIMIT, CUSTOMERDETAIL.CITY,
      CUSTOMERDETAIL.COUNTRY, CUSTOMERDETAIL.POSTAL_CODE)
    .from(CUSTOMER)
    .join(CUSTOMERDETAIL)
    .on(CUSTOMER.CUSTOMER_NUMBER
      .eq(CUSTOMERDETAIL.CUSTOMER_NUMBER))
    .fetch();
```

This time, jOOQ picked up `Record5<Long, BigDecimal, String, String, String>`. I think you've got the idea!

Accessing the values of records in a type-safe manner can be done via the attributes of the generated classes or you can use `Record.into(Table<Z> table)` to extract the strongly-typed `TableRecords` and rely on the corresponding getters. But, pay attention that only the fields listed/projected in the query have been populated with values from the result set.

Fetching Result<Record> via UDTs

UDTs are ORDBMS features formally supported by Oracle and PostgreSQL and are modeled by jOOQ as `UDTRecord`. Let's consider the following UDT defined in PostgreSQL:

```
/* Define a type using CREATE TYPE */
CREATE TYPE "evaluation_criteria" AS ("communication_ability"
  INT, "ethics" INT, "performance" INT, "employee_input" INT);
```

Next, the `MANAGER` table schema uses this type as follows:

```
CREATE TABLE "manager" (
  ...
  "manager_evaluation" evaluation_criteria DEFAULT NULL
```

```
  ...
);
```

Running the jOOQ Code Generator produces an `org.jooq.UDT` implementation named `EvaluationCriteria.java` (in the `jooq.generated.udt` package). Besides
the `org.jooq.UDT` implementation, an `org.jooq.UDTRecord` implementation is also generated under the name `EvaluationCriteriaRecord.java` (in the `jooq.generated.udt.records` package).

Having these artifacts generated, we can write the following example that returns a type-safe `Result<Record>`:

```
/* type-safe Result<Record> */
Result<Record2<String, EvaluationCriteriaRecord>> result =
  ctx.select(MANAGER.MANAGER_NAME, MANAGER.MANAGER_EVALUATION)
     .from(MANAGER)
     .fetch();
```

Accessing the values of records can be done as follows. Of course, the climax is represented by accessing the UDT record's values:

```
/* type-safe values */
for(Record2 r : result) {
  String r1 = r.get(MANAGER.MANAGER_NAME);
  Integer r2 = r.get(MANAGER.MANAGER_EVALUATION)
    .getCommunicationAbility();
  Integer r3 = r.get(MANAGER.MANAGER_EVALUATION)
    .getEthics();
  Integer r4 = r.get(MANAGER.MANAGER_EVALUATION)
    .getPerformance();
  Integer r5 = r.get(MANAGER.MANAGER_EVALUATION)
    .getEmployeeInput();
}
```

Alternatively, relying on `Record.into(Table<Z> table)` can be done as follows:

```
/* type-safe Result<Record> */
Result<ManagerRecord> result =
  ctx.select(MANAGER.MANAGER_NAME, MANAGER.MANAGER_EVALUATION)
     .from(MANAGER)
     .fetch()
     .into(MANAGER); // or, into(ManagerRecord.class)
```

This time, accessing the values of records can be done via `getManagerEvaluation()`:

```
/* type-safe values */
for(ManagerRecord r : result) {
  String r1 =r.getManagerName();
  Integer r2 =r.getManagerEvaluation()
    .getCommunicationAbility();
  Integer r3 = r.getManagerEvaluation().getEthics();
  Integer r4 = r.getManagerEvaluation().getPerformance();
  Integer r5 = r.getManagerEvaluation().getEmployeeInput();
}
```

Well, this was a brief overview of jOOQ records. I've intentionally skipped `UpdatableRecord` for now since this topic is covered later in *Chapter 9, CRUD, Transactions, and Locking.*

> **Important Note**
>
> When this book was written, attempting to serialize a jOOQ record to JSON/XML via Spring Boot default Jackson features (for instance, by returning `Record` from a REST controller) will result in an exception! Setting `FAIL_ON_EMPTY_BEANS=false` will eliminate the exception but will lead to a weird and useless result. Alternatively, you can return POJOs or rely on jOOQ formatting capabilities – as you'll see later, jOOQ can format a record as JSON, XML, and HTML. And, let's not forget the alternative of using SQL/XML or SQL/JSON features and generating the JSON directly in the database (see *Chapter 8, Fetching and Mapping*). However, if you really want to serialize the jOOQ record, then you can rely on `intoMap()` and `intoMaps()`, as you can see in the bundled code. Meanwhile, you can monitor the progress on this topic here: `https://github.com/jOOQ/jOOQ/issues/11889`.

The examples covered in this section are available for Maven and Gradle in the code bundled with the book under the name *RecordResult*.

Exploring jOOQ query types

jOOQ distinguishes between two main types of queries:

- DML (`INSERT`, `UPDATE`, `DELETE`, and `MERGE`, among others) and DDL (`CREATE`, `ALTER`, `DROP`, `RENAME`, and similar) queries that produce a modification in the database
- DQL (`SELECT`) queries that produce results

DML and DDL queries are represented in jOOQ by the `org.jooq.Query` interface, while DQL queries are represented by the `org.jooq.ResultQuery` interface. The `ResultQuery` interface extends (among others) the `Query` interface.

For instance, the following snippet of code contains two jOOQ queries:

```
Query query = ctx.query("DELETE FROM payment
  WHERE customer_number = 103");

Query query = ctx.deleteFrom(PAYMENT)
  .where(PAYMENT.CUSTOMER_NUMBER.eq(103L));
```

These queries can be executed via jOOQ and they return the number of affected rows:

```
int affectedRows = query.execute();
```

And, here are two result queries: first, a plain SQL query – here, jOOQ cannot infer the `Record` types:

```
ResultQuery<Record> resultQuery = ctx.resultQuery(
  "SELECT job_title FROM employee WHERE office_code = '4'");

Result<Record> fetched = resultQuery.fetch();
List<String> result = fetched.into(String.class);
```

Second, a jOOQ `ResultQuery` expressed via jOOQ generated classes (notice that this time, jOOQ infers the number of `ResultQuery` parameters and types – since we fetch only `JOB_TITLE`, there is `Record1<String>`):

```
ResultQuery<Record1<String>> resultQuery
  = ctx.select(EMPLOYEE.JOB_TITLE)
       .from(EMPLOYEE)
       .where(EMPLOYEE.OFFICE_CODE.eq("4"));
Result<Record1<String>> fetched = resultQuery.fetch();
List<String> result = fetched.into(String.class);
```

Since `ResultQuery` extends `Iterable`, you can just *foreach* your queries in PL/SQL style and do something with each record. For instance, the following snippet of code works like a charm:

```
for (Record2<String, String> customer : ctx.select(
             CUSTOMER.CUSTOMER_NAME, CUSTOMER.PHONE)
       .from(CUSTOMER)) {
```

```
  System.out.println("Customer:\n" + customer);
}

for (CustomerRecord customer : ctx.selectFrom(CUSTOMER)
        .where(CUSTOMER.SALES_REP_EMPLOYEE_NUMBER.eq(1504L))) {
  System.out.println("Customer:\n" + customer);
}
```

There is no need to explicitly call `fetch()`, but you can do it. The examples from this section are grouped in an application named *QueryAndResultQuery*. Next, let's talk about the jOOQ fluent API.

Understanding the jOOQ fluent API

Most of the time spent with jOOQ is about writing fluent code via the jOOQ fluent API. This approach is quite convenient for building fluent SQL expressions that avoid disrupting or chunking the code. Moreover, fluent APIs are easy to enrich with more operations.

Relying on a brilliant implementation of the interface-driven design concept, jOOQ hides most implementations from client code and acts as a *good friend* that is ready to listen regarding the SQL that you need to run. Let's see several usages of the jOOQ fluent API.

Writing fluent queries

So far, we have written several SQL in the jOOQ DSL API fluent style. Let's have another one as follows:

```
DSL.select(
      ORDERDETAIL.ORDER_LINE_NUMBER,
      sum(ORDERDETAIL.QUANTITY_ORDERED).as("itemsCount"),
      sum(ORDERDETAIL.PRICE_EACH
         .mul(ORDERDETAIL.QUANTITY_ORDERED)).as("total"))
   .from(ORDERDETAIL)
   .where((val(20).lt(ORDERDETAIL.QUANTITY_ORDERED)))
   .groupBy(ORDERDETAIL.ORDER_LINE_NUMBER)
   .orderBy(ORDERDETAIL.ORDER_LINE_NUMBER)
   .getSQL();
```

The goal of dissecting to the bone the previous jOOQ query is far away from us, but let's try to have some insights about how this query is seen through jOOQ eyes. This will help you to quickly accumulate the information from the chapters that follow and will increase your confidence in jOOQ.

Roughly, a JOOQ fluent query is composed of two basic building blocks: **column expressions** (or fields) and **table expressions**. These are manipulated via conditions, functions, and constraints to obtain a set of valid query steps that are logically chained and/or nested in the final jOOQ query. Of course, a jOOQ query may contain other parts as well, and all the parts that compose that query are referenced as **query parts** and have the `org.jooq.QueryPart` interface as a common base type. Let's briefly cover column expressions, table expressions, and query steps to better understand this paragraph.

Column expressions

Column expressions or **fields** refer to one or more columns, and they are represented by the `org.jooq.Field` interface. There are many kinds of column expressions and all of them can be used in a variety of SQL statements/clauses to produce fluent queries. For example, in the `SELECT` clause, we have `org.jooq.SelectField` (which is a special `org.jooq.Field` interface for `SELECT`); in the `WHERE` clause, we have `org.jooq.Field`; in the `ORDER BY` clause, we have `org.jooq.OrderField`; in the `GROUP BY` clause, we have `org.jooq.GroupField`; and in conditions and functions, we typically have `org.jooq.Field`.

Column expressions can be arbitrary built via the jOOQ fluent API to shape different query parts such as arithmetic expressions (for example, `column_expression_1.mul(column_expression_2)`), conditions/predicates (`org.jooq.Condition`) used in `WHERE` and `HAVING` (for example, here is an equality condition: `WHERE(column_expression_1.eq(column_expression_2))`), and so on.

When column expressions refer to table columns, they are referenced as **table columns**. Table columns implement a more specific interface called `org.jooq.TableField`. These kinds of column expressions are produced internally by the jOOQ Code Generator and you can see them in each Java class specific to a table. The instances of `TableField` cannot be created directly.

Let's identify the column expressions types from our query using the following figure, which highlights them:

```
DSL.select(ORDERDETAIL.ORDER_LINE_NUMBER,
  sum(ORDERDETAIL.QUANTITY_ORDERED).as("itemsCount"),
  sum(ORDERDETAIL.PRICE_EACH.mul(ORDERDETAIL.QUANTITY_ORDERED)).as("total"))
  .from(ORDERDETAIL)
  .where((val(20).lt(ORDERDETAIL.QUANTITY_ORDERED)))
  .groupBy(ORDERDETAIL.ORDER_LINE_NUMBER)
  .orderBy(ORDERDETAIL.ORDER_LINE_NUMBER)
  .getSQL();
```

Figure 3.2 – Identify the column expressions of this query

First of all, we have some table columns that reference the `ORDERDETAIL` table:

```
Field<Integer> tc1 = ORDERDETAIL.ORDER_LINE_NUMBER;
Field<Integer> tc2 = ORDERDETAIL.QUANTITY_ORDERED;
Field<BigDecimal> tc3 = ORDERDETAIL.PRICE_EACH;
```

We have some extracted as `TableField`:

```
TableField<OrderdetailRecord,Integer>
   tfc1 = ORDERDETAIL.ORDER_LINE_NUMBER;
TableField<OrderdetailRecord,Integer>
   tfc2 = ORDERDETAIL.QUANTITY_ORDERED;
TableField<OrderdetailRecord,BigDecimal>
   tfc3 = ORDERDETAIL.PRICE_EACH;
```

We also have an unnamed column expression:

```
Field<Integer> uc1 = val(20);
```

Just as a quick note, here, the `DSL.val()` method simply creates `Field<Integer>` (gets a bind value as `Param<Integer>`, where `Param` extends `Field`) representing a constant value. We will discuss jOOQ parameters a little bit later in this chapter.

Let's rewrite the query so far using the extracted columns expression:

```
DSL.select(tc1, sum(tc2).as("itemsCount"),
           sum(tc3.mul(tc2)).as("total"))
    .from(ORDERDETAIL)
    .where(uc1.lt(tc2))
    .groupBy(tc1)
```

```
    .orderBy(tc1)
    .getSQL();
```

Next, let's extract the usages of the `sum()` aggregate function. The first usage of `sum()` relies on a table column expression (`tc2`) to produce a function expression:

```
Field<BigDecimal> f1 = sum(tc2); // function expression
```

The second usage of `sum()` wraps an arithmetic expression that uses two table column expressions (`tc3` and `tc2`), therefore, it can be extracted as follows:

```
Field<BigDecimal> m1 = tc3.mul(tc2); // arithmetic expression
Field<BigDecimal> f2 = sum(m1);      // function expression
```

One step further, and we notice that our query uses aliases for `f1` and `f2`, therefore, these can be extracted as aliased expressions:

```
Field<BigDecimal> a1 = f1.as("itemsCount"); // alias expression
Field<BigDecimal> a2 = f2.as("total");      // alias expression
```

Let's rewrite the query again:

```
DSL.select(tc1, a1, a2)
    .from(ORDERDETAIL)
    .where(uc1.lt(tc2))
    .groupBy(tc1)
    .orderBy(tc1)
    .getSQL();
```

Done! At this point, we have identified all column expressions of our query. How about table expressions?

Table expressions

Next to fields, tables also represent the basic building blocks of any query. jOOQ represents a table via `org.jooq.Table`. In our query, there is a single table reference:

```
.from(ORDERDETAIL) // table expression ORDERDETAIL
```

It can be extracted as follows:

```
// non type-safe table expression
Table<?> t1 = ORDERDETAIL;
```

```
// type-safe table expression
Table<OrderdetailRecord> t1 = ORDERDETAIL;
```

This time, the query becomes the following:

```
DSL.select(tc1, a1, a2)
   .from(t1)
   .where(uc1.lt(tc2))
   .groupBy(tc1)
   .orderBy(tc1)
   .getSQL();
```

jOOQ supports a wide range of tables not only database tables, including plain SQL tables, aliased tables, derived tables, **Common Table Expressions** (**CTEs**), temporary tables, and table-valued functions. But, we will discuss these in the upcoming chapters.

So far, notice that we haven't touched `uc1.lt(tc2)`. As you can probably intuit, this is a condition that uses two column expressions and is mapped by jOOQ as `org.jooq.Condition`. It can be extracted as follows:

```
Condition c1 = uc1.lt(tc2); // condition
```

After extracting all these parts, we obtain the following query:

```
DSL.select(tc1, a1, a2)
   .from(t1)
   .where(c1)
   .groupBy(tc1)
   .orderBy(tc1)
   .getSQL();
```

Actually, you could even do the following, but there is no more type-safety:

```
Collection<? extends SelectField> sf = List.of(tc1, a1, a2);
DSL.select(sf) …
```

Obviously, these query parts can be used to form other arbitrary queries as well. After all, in jOOQ, we can write queries that are 100% dynamic.

> **Important Note**
> In jOOQ, even when they look like static queries (due to jOOQ's API design), every SQL is dynamic, therefore, it can be broken up into query parts that can be fluently glued back in any valid jOOQ query. We'll talk about more examples later when we'll tackle dynamic filters.

Finally, let's quickly get an overview of the query steps topic.

Query steps (SelectFooStep, InsertFooStep, UpdateFooStep, and DeleteFooStep)

Continuing to identify the remaining query parts, we have `select`, `from`, `where`, `groupBy`, and `orderBy`. These parts are logically chained to form our query and are represented by jOOQ as query steps. There are many types of query steps, but the ones used by our query can be decomposed as follows:

```
SelectSelectStep s1 = DSL.select(tc1, a1, a2);
SelectJoinStep s2 = s1.from(t1);
SelectConditionStep s3 = s2.where(c1);
SelectHavingStep s4 = s3.groupBy(tc1);
SelectSeekStep1 s5 = s4.orderBy(tc1);

return s5.getSQL();
```

Or, the ones used as type-safe steps are as follows (remember, you can use Java 9 `var` instead of `SelectSelectStep<Record3<Short, BigDecimal, BigDecimal>>`):

```
SelectSelectStep<Record3<Integer, BigDecimal, BigDecimal>>
  s1ts = DSL.select(tc1, a1, a2);
SelectJoinStep<Record3<Integer, BigDecimal, BigDecimal>>
  s2ts = s1ts.from(t1);
SelectConditionStep<Record3<Integer, BigDecimal, BigDecimal>>
  s3ts = s2ts.where(c1);
SelectHavingStep<Record3<Integer, BigDecimal, BigDecimal>>
  s4ts = s3ts.groupBy(tc1);
SelectSeekStep1<Record3<Integer, BigDecimal, BigDecimal>,
  Integer> s5ts = s4ts.orderBy(tc1);

return s5ts.getSQL();
```

Check out the last line of this snippet of code. We return the generated valid SQL as a plain string without executing this query. Execution can happen in the presence of a connection to the database, therefore, we need `DSLContext` configured to accomplish this task. If we have injected `DSLContext`, then all we need to do is to use it as follows:

```
return ctx.fetch(s5); // or, s5ts
```

Or, we can use it like this:

```
SelectSelectStep s1 = ctx.select(tc1, a1, a2);
// or
SelectSelectStep<Record3<Integer, BigDecimal, BigDecimal>>
   s1ts = ctx.select(tc1, a1, a2);
```

This `SelectSelectStep` contains an internal reference to the `DSLContext` configuration, therefore, we can replace the last line as follows:

```
return s5.fetch(); // or, s5ts
```

The complete code is available for Maven and Gradle in the code bundled with this book under the name `FluentQueryParts`. While in this section, you saw how to decompose the query steps, keep in mind that it's almost always a better choice to rely on dynamic SQL queries than referencing these step types. So, as a rule of thumb, *always* try to avoid assigning or referencing the query steps directly.

Obviously, decomposing a query into parts is not a day-to-day task. Most of the time, you'll just use the fluent API, but there are cases when it is nice to know how to do it (for instance, it can be helpful for writing dynamic filters, referencing aliases in different places of a query, re-using a query part in multiple places, and writing correlated subqueries).

Another use of the jOOQ fluent API is focused on the `DSLContext` creation.

Creating DSLContext

Most probably, in Spring Boot applications, we'll prefer to inject the default `DSLContext` as you saw in *Chapter 1, Starting jOOQ and Spring Boot*, and *Chapter 2, Customizing the jOOQ Level of Involvement*. But, in certain scenarios (for instance, wrapping and running a specific query with a custom setting, rendering an SQL in a different dialect than the default one, or needing to trigger an occasional query against a database that is not configured in Spring Boot), we'll prefer to use `DSLContext` as a local variable. This can be done in fluent style via the `DSL.using()` methods as in the following non-exhaustive list of examples.

Creating DSLContext from a data source and a dialect

Having `DataSource` (for instance, injected in your repository), we can create `DSLContext` and execute a query in fluent style as here:

```
private final DataSource ds; // injected DataSource
...
```

```
List<Office> result = DSL.using(ds, SQLDialect.MYSQL)
  .selectFrom(OFFICE)
  .where(OFFICE.TERRITORY.eq(territory))
  .fetchInto(Office.class);
```

This example relies on the `DSL.using(DataSource datasource, SQLDialect dialect)` method.

Creating DSLContext from a data source, a dialect, and some settings

Enabling/disabling some settings to the previous example requires us to instantiate `org.jooq.conf.Settings`. This class exposes a comprehensive fluent API (via the `withFoo()` methods) that influences the way jOOQ renders SQL code. For instance, the following snippet of code inhibits the rendering of the schema name (just look at this nice piece of fluent code):

```
private final DataSource ds; // injected DataSource
...
List<Office> result = DSL.using(ds, SQLDialect.MYSQL,
      new Settings().withRenderSchema(Boolean.FALSE))
  .selectFrom(OFFICE)
  .where(OFFICE.TERRITORY.eq(territory))
  .fetchInto(Office.class);
```

This example relies on the `using(DataSource datasource, SQLDialect dialect, Settings settings)` method.

Alter a setting of the injected DSLContext

In the previous example, we created `DSLContext` that doesn't render the schema name. This setting is applied to all usages of the created `DSLContext`, or in other words, to all queries triggered under the configuration of this `DSLContext`. How can we do the same thing for the default `DSLContext` provided by Spring Boot after it was injected into a repository? The following code provides the answer:

```
private final DSLContext ctx; // injected DSLContext
...
List<Office> result = ctx.configuration()
 .set(new Settings().withRenderSchema(Boolean.FALSE)).dsl()
 .selectFrom(OFFICE)
```

```
  .where(OFFICE.TERRITORY.eq(territory))
  .fetchInto(Office.class);
```

Mainly, we access the current configuration of the injected `DSLContext` via `configuration()`, we set our setting, and call the `dsl()` method to get access back to `DSLContext`. Notice that from this point forward, all usages of `ctx` will not render the schema name unless you don't enable it again. If you prefer to use some specific settings for a certain query, then create `DSLContext` derived from the injected one via `derive()` in place of `set()`. This way, the original `DSLContext` remains unaltered and you can operate on the derived one:

```
private final DSLContext ctx; // injected DSLContext
...
List<Office> result = ctx.configuration()
  .derive(new Settings().withRenderSchema(Boolean.FALSE)).dsl()
  .selectFrom(OFFICE)
  .where(OFFICE.TERRITORY.eq(territory))
  .fetchInto(Office.class);
```

So, in the previous example, `ctx` remains unchanged and jOOQ uses a derived `DSLContext`, which will not render the schema name.

Creating DSLContext from a connection

Creating `DSLContext` from a connection and executing the query in fluent style can be done as follows:

```
try ( Connection conn
  = DriverManager.getConnection(
   "jdbc:mysql://localhost:3306/classicmodels",
   "root", "root")) {

    List<Office> result = DSL.using(conn)
      .selectFrom(OFFICE)
      .where(OFFICE.TERRITORY.eq(territory))
      .fetchInto(Office.class);

    return result;
} catch (SQLException ex) { // handle exception }
```

In such cases, we have to close the connection manually; therefore, we have used the `try-with-resources` technique. This example relies on the `DSL.using (Connection c)` method. If you want to specify the SQL dialect as well, then try out `DSL.using(Connection c, SQLDialect d)`.

Creating DSLContext from a URL, user, and password

For standalone-based scripts, where handling resources is not important since the connection lives as long as the script itself, we can rely on `DSL.using(String url)`, `DSL.using(String url, Properties properties)`, and `DSL.using(String url, String user, String password)`.

If you prefer to use the `DSL.using(String url, String user, String password)` method (or any of the other two) prior to jOOQ 3.14, then you have to explicitly close the connection as well. This can be done by explicitly calling `DSLContext.close()` or by using `try-with-resources`. Starting with jOOQ 3.14, these overloads of `DSL.using()` will produce the new `CloseableDSLContext` type that allows us to write this:

```
try (CloseableDSLContext cdctx = DSL.using(
     "jdbc:mysql://localhost:3306/classicmodels",
     "root", "root")) {

   List<Office> result = cdctx.selectFrom(OFFICE)
     .where(OFFICE.TERRITORY.eq(territory))
     .fetchInto(Office.class);

   return result;
}
```

Next, let's see how to use `DSLContext` without a database connection.

Rendering SQL in a certain dialect

Rendering SQL in a certain dialect (here, MySQL) can be done via this fluent code:

```
String sql = DSL.using(SQLDialect.MYSQL)
                .selectFrom(OFFICE)
                .where(OFFICE.TERRITORY.eq(territory))
                .getSQL();
```

Since there is no connection or data source, there is no interaction with the database. The returned string represents the generated SQL specific to the provided dialect. This example relies on the `DSL.using(SQLDialect dialect)` method.

You can find all these examples in the code bundled with this book under the name *CreateDSLContext*.

Using Lambdas and streams

The jOOQ fluent API and Java 8 Lambdas and streams make a perfect team. Let's look at several examples that demonstrate this.

Using Lambdas

For instance, jOOQ comes with a functional interface named `RecordMapper` used for mapping a jOOQ record to a POJO. Let's assume that we have the following POJOs. First, let's assume we have `EmployeeName`:

```
public class EmployeeName implements Serializable {

  private String firstName;
  private String lastName;

  // constructors, getters, setters,... omitted for brevity
}
```

Next, let's assume we have `EmployeeData`:

```
public class EmployeeData implements Serializable {

  private Long employeeNumber;
  private int salary;
  private EmployeeName employeeName;

  // constructors, getters, setters,... omitted for brevity
}
```

Next, let's assume that we have the following plain SQL:

```
SELECT employee_number, salary, first_name, last_name
FROM employee
```

Executing and mapping this plain SQL is achievable via the `fetch(String sql)` flavor and `map(RecordMapper<? super R,E> rm)` as in the following:

```
List<EmployeeData> result
  = ctx.fetch("SELECT employee_number, first_name,
               last_name, salary FROM employee")
    .map(
      rs -> new EmployeeData(
        rs.getValue("employee_number", Long.class),
        rs.getValue("salary", Integer.class),
        new EmployeeName(
          rs.getValue("first_name", String.class),
          rs.getValue("last_name", String.class))
      )
    );
```

The same thing is applicable if the plain SQL is expressed via the Java-based schema:

```
List<EmployeeData> result
  = ctx.select(EMPLOYEE.EMPLOYEE_NUMBER,
          EMPLOYEE.FIRST_NAME,EMPLOYEE.LAST_NAME,
          EMPLOYEE.SALARY)
    .from(EMPLOYEE)
    .fetch()
    .map(
      rs -> new EmployeeData(
        rs.getValue(EMPLOYEE.EMPLOYEE_NUMBER),
        rs.getValue(EMPLOYEE.SALARY),
        new EmployeeName(rs.getValue(EMPLOYEE.FIRST_NAME),
                         rs.getValue(EMPLOYEE.LAST_NAME))
      )
);
```

It is also applicable if it is more concisely expressed via `fetch(RecordMapper<? super R,E> rm)`:

```
List<EmployeeData> result
  = ctx.select(EMPLOYEE.EMPLOYEE_NUMBER, EMPLOYEE.FIRST_NAME,
         EMPLOYEE.LAST_NAME, EMPLOYEE.SALARY)
       .from(EMPLOYEE)
  .fetch(
     rs -> new EmployeeData(
```

```
      rs.getValue(EMPLOYEE.EMPLOYEE_NUMBER),
      rs.getValue(EMPLOYEE.SALARY),
      new EmployeeName(rs.getValue(EMPLOYEE.FIRST_NAME),
                       rs.getValue(EMPLOYEE.LAST_NAME))
    )
  );
```

If you think that these mappings are too simple for using a custom `RecordMapper`, then you are right. You'll see more proper cases for custom record mappers later on when we'll detail mappings. For this case, both of them can be solved via the built-in `into()` and `fetchInto()` methods by simply enriching the SQLs with hints via aliases. First, we can enrich the plain SQL (for MySQL, we use backticks):

```
List<EmployeeData> result = ctx.fetch("""
  SELECT employee_number, salary,
         first_name AS `employeeName.firstName`,
         last_name AS `employeeName.lastName`
  FROM employee""").into(EmployeeData.class);
```

And then, we can enrich the jOOQ SQL:

```
List<EmployeeData> result
  = ctx.select(EMPLOYEE.EMPLOYEE_NUMBER, EMPLOYEE.SALARY,
           EMPLOYEE.FIRST_NAME.as("employeeName.firstName"),
           EMPLOYEE.LAST_NAME.as("employeeName.lastName"))
       .from(EMPLOYEE)
       .fetchInto(EmployeeData.class);
```

Let's see a few more examples of using Lambdas.

The following snippet of code prints all sales. Since `selectFrom()` returns the record type supplied with the argument table, this code prints each `SaleRecord` (notice that calling `fetch()` is optional):

```
ctx.selectFrom(SALE)
   .orderBy(SALE.SALE_)
   // .fetch() - optional
   .forEach(System.out::println);
```

Mapping the result set (`SaleRecord`) to `List<Double>` containing only the `sale` column can be done as follows via `fetch().map(RecordMapper<? super R,E> rm)`:

```
ctx.selectFrom(SALE)
    .orderBy(SALE.SALE_)
    .fetch()
    .map(SaleRecord::getSale)
    .forEach(System.out::println);
```

Or, it can be done via `fetch(RecordMapper<? super R,E> rm)` as follows:

```
ctx.selectFrom(SALE)
    .orderBy(SALE.SALE_)
    .fetch(SaleRecord::getSale)
    .forEach(System.out::println);
```

It can also be done via a Lambda expression as follows:

```
ctx.selectFrom(SALE)
    .orderBy(SALE.SALE_)
    .fetch(s -> s.getSale())
    .forEach(System.out::println);
```

Or, it can even be done via an anonymous record mapper as follows:

```
return ctx.selectFrom(SALE)
          .orderBy(SALE.SALE_)
          .fetch(new RecordMapper<SaleRecord, Double>() {
             @Override
             public Double map(SaleRecord sr) {
                return sr.getSale();
             }
          });
```

Next, let's see how the jOOQ fluent API can be used with the Java Stream fluent API.

Using the Stream API

Using the jOOQ fluent API and the Stream fluent API as an apparently single fluent API is straightforward. Let's assume that we have this POJO:

```
public class SaleStats implements Serializable {
```

```
  private double totalSale;
  private List<Double> sales;

  // constructor, getters, setters, ... omitted for brevity
}
```

A plain SQL can obtain a `SaleStats` instance as follows:

```
SaleStats result = ctx.fetch(
    "SELECT sale FROM sale") // jOOQ fluent API ends here
  .stream() // Stream fluent API starts here
  .collect(Collectors.teeing(
     summingDouble(rs -> rs.getValue("sale", Double.class)),
     mapping(rs -> rs.getValue("sale", Double.class),
     toList()), SaleStats::new));
```

But, if we use the Java-based schema, then this code can be re-written as follows:

```
SaleStats result = ctx.select(SALE.SALE_)
  .from(SALE)
  .fetch()  // jOOQ fluent API ends here
  .stream() // Stream fluent API starts here
  .collect(Collectors.teeing(
     summingDouble(rs -> rs.getValue(SALE.SALE_)),
     mapping(rs -> rs.getValue(SALE.SALE_), toList()),
     SaleStats::new));
```

It looks like the jOOQ fluent API and the Stream fluent API work together like a charm! All we have to do is call the `stream()` method after `fetch()`. While `fetch()` fetches the entire result set into memory, `stream()` opens a stream on this result set. Fetching the entire result set into memory via `fetch()` allows the JDBC resources (for instance, the connection) to be closed before streaming the result set.

Nevertheless, besides `stream()`, jOOQ also exposes a method named `fetchStream()`, which is tackled later in the chapter, dedicated to lazy loading next to other specific topics. As a quick hint, keep in mind that `fetch().stream()` and `fetchStream()` are not the same thing.

The examples from this section are grouped in the *FunctionalJooq* application.

Fluent programmatic configuration

In the previous chapter, you already had a flavor of constructing the Code Generator configuration via the programmatic fluent API. The following snippet of code is just another example of the jOOQ `Settings` fluent API:

```
List<Office> result = ctx.configuration()
  .set(new Settings().withRenderSchema(Boolean.FALSE)
                     .withMaxRows(5)
                     .withInListPadding(Boolean.TRUE)).dsl()
  .selectFrom(...)
  ...
  .fetchInto(Office.class);
```

These are not the only cases when the jOOQ fluent API rocks. For instance, check the jOOQ JavaFX application for creating a bar chart from a jOOQ `result`. This is available in the jOOQ manual.

Next, let's see how jOOQ emphasizes that our fluent code should respect the SQL syntax correctness.

Highlighting that jOOQ emphasizes SQL syntax correctness

One of the coolest features of jOOQ consists of the fact that jOOQ doesn't allow us to write bad SQL syntax. If you aren't an SQL expert or simply have issues with SQL-specific syntax, then all you have to do is to let jOOQ guide you step by step.

Having a fluent API for chaining methods to obtain a SQL is cool, but having a fluent API that emphasizes SQL syntax correctness is the coolest. jOOQ knows exactly how the query parts fit the puzzle and will help you via your IDE.

For instance, let's assume that we *accidentally* wrote the following bad SQLs. Let's start with an SQL that misses the `ON` clause:

```
ctx.select(EMPLOYEE.JOB_TITLE,
           EMPLOYEE.OFFICE_CODE, SALE.SALE_)
   .from(EMPLOYEE)
   .join(SALE)
   // "on" clause is missing here
   .fetch();
```

The IDE signals this issue immediately, as shown in the following figure:

```
ctx.select(EMPLOYEE.JOB_TITLE,
           EMPLOYEE.OFFICE_CODE, SALE.SALE_)
   .from(EMPLOYEE)
   .join(SALE)
   .fetch();
```

cannot find symbol
symbol: method fetch()
location: interface SelectOnStep<Record3<String,String,Double>>

(Alt-Enter shows hints)

expecting "on"

Figure 3.3 – Wrong SQL

Let's continue with another wrong SQL that uses `JOIN` in an improper place:

```
ctx.select(EMPLOYEE.FIRST_NAME, EMPLOYEE.LAST_NAME)
   .from(EMPLOYEE)
   .union(select(CUSTOMER.CONTACT_FIRST_NAME,
                 CUSTOMER.CONTACT_LAST_NAME)
          .from(CUSTOMER))
   .join(CUSTOMER)
   // "join" is not allowed here
   .on(CUSTOMER.SALES_REP_EMPLOYEE_NUMBER
      .eq(EMPLOYEE.EMPLOYEE_NUMBER))
   .fetch();
```

And, for the last example, let's look at a wrong SQL that misses `over()`:

```
ctx.select(CUSTOMER.CUSTOMER_NAME,
           ORDER.ORDER_DATE,lead(ORDER.ORDER_DATE, 1)
   // missing over()
   .orderBy(ORDER.ORDER_DATE).as("NEXT_ORDER_DATE"))
   .from(ORDER)
   .join(CUSTOMER)
      .on(ORDER.CUSTOMER_NUMBER
        .eq(CUSTOMER.CUSTOMER_NUMBER))
   .fetch();
```

Of course, we can continue like this forever, but I think you get the idea! So, count on jOOQ!

Casting, coercing, and collating

jOOQ was designed to handle most of the casting issues under the hood, including for ultra-strong-typed databases such as DB2. Nevertheless, explicit casting and/or coercing still serve some isolated cases. Most probably, we'll need them when we are not satisfied with the jOOQ automatic mapping (for instance, we consider that jOOQ didn't find the most accurate mapping), or we just need a certain type to respond to a special case. Even if they add a little bit of verbosity, casting and coercing can be used fluently; therefore, the DSL expressions are not disrupted.

Casting

Most of the time, jOOQ finds the most accurate data type mapping between the database and Java. If we look into a jOOQ generated class that mirrors a database table, then we see that, for each column that has a database-specific type (for example, `VARCHAR`), jOOQ has found a Java type correspondent (for example, `String`). If we compare the schema of the `PAYMENT` table with the generated `jooq.generated.tables.Payment` class, then we find the following data type correspondence:

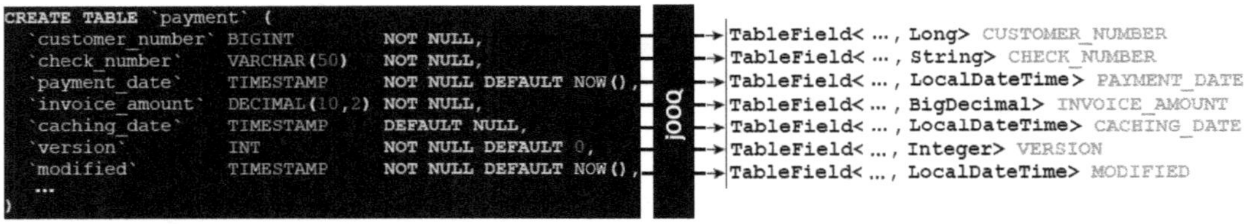

```
CREATE TABLE `payment` (
  `customer_number` BIGINT        NOT NULL,
  `check_number`    VARCHAR(50)   NOT NULL,
  `payment_date`    TIMESTAMP     NOT NULL DEFAULT NOW(),
  `invoice_amount`  DECIMAL(10,2) NOT NULL,
  `caching_date`    TIMESTAMP     DEFAULT NULL,
  `version`         INT           NOT NULL DEFAULT 0,
  `modified`        TIMESTAMP     NOT NULL DEFAULT NOW(),
  ...
)
```

jOOQ

```
TableField< ..., Long> CUSTOMER_NUMBER
TableField< ..., String> CHECK_NUMBER
TableField< ..., LocalDateTime> PAYMENT_DATE
TableField< ..., BigDecimal> INVOICE_AMOUNT
TableField< ..., LocalDateTime> CACHING_DATE
TableField< ..., Integer> VERSION
TableField< ..., LocalDateTime> MODIFIED
```

Figure 3.4 – Type mapping between the database and Java

When the jOOQ mapping is not what we need or jOOQ cannot infer a certain type, then we can rely on the jOOQ casting API, which contains the following methods:

```
// cast this field to the type of another field
<Z> Field<Z> cast(Field<Z> field);

// cast this field to a given DataType
<Z> Field<Z> cast(DataType<Z> type);

// cast this field to the default DataType for a given Class
<Z> Field<Z> cast(Class<? extends Z> type);
```

Besides these methods, the `DSL` class contains these methods:

```
<T> Field<T> cast(Object object, Field<T> field);
<T> Field<T> cast(Object object, DataType<T> type);
<T> Field<T> cast(Object object, Class<? extends T> type);
<T> Field<T> castNull(Field<T> field);
<T> Field<T> castNull(DataType<T> type);
<T> Field<T> castNull(Class<? extends T> type);
```

Let's have some examples against MySQL and let's start with the following query that maps the fetched data to the Java types that jOOQ has automatically chosen:

```
Result<Record2<BigDecimal, LocalDateTime>> result =
ctx.select(PAYMENT.INVOICE_AMOUNT.as("invoice_amount"),
           PAYMENT.CACHING_DATE.as("caching_date"))
   .from(PAYMENT)
   .where(PAYMENT.CUSTOMER_NUMBER.eq(103L))
   .fetch();
```

So, `INVOICE_AMOUNT` is mapped to `BigDecimal`, and `CACHING_DATE` is mapped to `LocalDateTime`. Let's assume that we are in a corner-case scenario that requires us to fetch `INVOICE_AMOUNT` as `String` and `CACHING_DATE` as `LocalDate`. Of course, we can loop the preceding result and perform the conversions of each record in Java, but, at the query level, we can accomplish this via jOOQ `cast()`, as follows:

```
Result<Record2<String, LocalDate>> result =
  ctx.select(
         PAYMENT.INVOICE_AMOUNT.cast(String.class)
           .as("invoice_amount"),
         PAYMENT.CACHING_DATE.cast(LocalDate.class)
           .as("caching_date"))
     .from(PAYMENT)
     .where(PAYMENT.CUSTOMER_NUMBER.eq(103L))
     .fetch();
```

Check out the SQL string generated after using `cast()`:

```
SELECT
  cast(`classicmodels`.`payment`.`invoice_amount` as char)
    as `invoice_amount`,
  cast(`classicmodels`.`payment`.`caching_date` as date)
    as `caching_date`
FROM`classicmodels`.`payment`
WHERE`classicmodels`.`payment`.`customer_number` = 103
```

In the following figure, you can see the result set returned by these two SQLs:

no cast

```
+-------------+-------------------+
|invoice_amount|caching_date      |
+-------------+-------------------+
|      6066.78|2004-10-19T12:30:15|
+-------------+-------------------+
|     14571.44|2003-06-05T12:34:15|
+-------------+-------------------+
|      1676.14|2004-12-18T16:30:15|
+-------------+-------------------+
```

after cast

```
+-------------+-------------------+
|invoice_amount|caching_date      |
+-------------+-------------------+
|      6066.78 | 2004-10-19       |
+-------------+-------------------+
|     14571.44 | 2003-06-05       |
+-------------+-------------------+
|      1676.14 | 2004-12-18       |
+-------------+-------------------+
```

Figure 3.5 – Casting results

Notice that the jOOQ casting operations are rendered in the generated SQL string, therefore, the database is responsible for performing these casts. But, in this scenario, do we really need these clumsy castings or do we actually need data type coercions?

Coercing

Data type coercions act like casting, except that they have no footprint on the actual SQL query being generated. In other words, data type coercions act as an unsafe cast in Java and are not rendered in the SQL string. With data type coercions, we only instruct jOOQ to pretend that a data type is of another data type and to bind it accordingly. Whenever possible, it is preferable to use coercions over casting. This way, we don't risk casting issues and we don't pollute the generated SQL with unnecessary castings. The API consists of several methods:

```
// coerce this field to the type of another field
<Z> Field<Z> coerce(Field<Z> field);

// coerce this field to a given DataType
<Z> Field<Z> coerce(DataType<Z> type);

// coerce this field to the default DataType for a given Class
<Z> Field<Z> coerce(Class<? Extends Z> type);
```

Besides these methods, the `DSL` class contains these methods:

```
<T> Field<T> coerce(Field<?> field,  DataType<T> as)
<T> Field<T> coerce(Field<?> field, Field<T> as)
<T> Field<T> coerce(Field<?> field, Class<T> as)
<T> Field<T> coerce(Object value, Field<T> as)
```

```
<T> Field<T> coerce(Object value, DataType<T> as)
<T> Field<T> coerce(Object value, Field<T> as)
```

In the example from the *Casting* section, we relied on casting from `BigDecimal` to `String` and from `LocalDateTime` to `LocalDate`. This casting was rendered in the SQL string and was performed by the database. But, we can avoid polluting the SQL string with these casts via coercion as follows:

```
Result<Record2<String, LocalDate>> result= ctx.select(
    PAYMENT.INVOICE_AMOUNT.coerce(String.class)
      .as("invoice_amount"),
    PAYMENT.CACHING_DATE.coerce(LocalDate.class)
      .as("caching_date"))
  .from(PAYMENT)
  .where(PAYMENT.CUSTOMER_NUMBER.eq(103L))
  .fetch();
```

The produced result set is the same as in the case of using casting, but the SQL string doesn't reflect coercions and the database didn't perform any casting operations. This is much better and safer:

```
SELECT
  `classicmodels`.`payment`.`invoice_amount`
    as `invoice_amount`,
  `classicmodels`.`payment`.`caching_date`
    as `caching_date`
FROM `classicmodels`.`payment`
WHERE `classicmodels`.`payment`.`customer_number` = 103
```

Starting with version 3.12, jOOQ allows for coercing `ResultQuery<R1>` to a new `ResultQuery<R2>` type as well. For instance, check out this plain SQL:

```
ctx.resultQuery(
  "SELECT first_name, last_name FROM employee").fetch();
```

The result type of this query is `Result<Record>` but we can easily replace `fetch()` with `fetchInto()` to map this result to the generated `Employee` POJO (only the `firstName` and `lastName` fields will be populated) or to a custom POJO containing only the fetched fields. But, how about fetching `Result<Record2<String, String>>`? This can be accomplished via one of the `ResultQuery.coerce()` flavors as follows:

```
Result<Record2<String, String>> result = ctx.resultQuery(
    "SELECT first_name, last_name FROM employee")
```

```
  .coerce(EMPLOYEE.FIRST_NAME, EMPLOYEE.LAST_NAME)
  .fetch();
```

Coercing a result set to a table can be done via `ResultQuery.coerce(Table<X> table)`. You can find an example in the bundled code next to an alternative before jOOQ 3.12. If during coercing, jOOQ finds any `Converter` or `Binding` configurations, then it will apply them (this is covered in *Chapter 7*, *Types, Converters, and Bindings*).

Coercing versus casting

Don't conclude that `coerce()` can replace `cast()` all the time. Check out this example that uses `coerce()`:

```
Result<Record2<BigDecimal, String>> result = ctx.select(
  PRODUCT.BUY_PRICE.coerce(SQLDataType.DECIMAL(10, 5))
    .as("buy_price"),
  PRODUCT.PRODUCT_DESCRIPTION.coerce(SQLDataType.VARCHAR(10))
    .as("prod_desc"))
  .from(PRODUCT)
  .where(PRODUCT.PRODUCT_ID.eq(1L))
  .fetch();
```

So, we pretend that `BUY_PRICE` is `BigDecimal` having a precision of *10* and a scale of `5`, and `PRODUCT_DESCRIPTION` is a string of length *10*. But, coercing cannot do that. In this case, coercing can pretend the `BigDecimal` (`BUY_PRICE` is really treated as a `BigDecimal` value), and `String` (`PRODUCT_DESCRIPTION` is really treated as a `String` value) types, but it cannot pretend the domain constraints.

Let's replace `coerce()` with `cast()`:

```
Result<Record2<BigDecimal, String>> result = ctx.select(
  PRODUCT.BUY_PRICE.cast(SQLDataType.DECIMAL(10, 5))
    .as("buy_price"),
  PRODUCT.PRODUCT_DESCRIPTION.cast(SQLDataType.VARCHAR(10))
    .as("prod_desc"))
  .from(PRODUCT)
  .where(PRODUCT.PRODUCT_ID.eq(1L))
  .fetch();
```

This time, casting is rendered in the generated SQL string. The following figure compares the result of using `coerce()` and `cast()`; this works as expected:

```
coercing
+---------+--------------------------------------------------+
|buy_price|prod_desc                                         |
+---------+--------------------------------------------------+
|    48.81|This replica features working kickstand, front ...|
+---------+--------------------------------------------------+
```

```
casting
+---------+----------+
|buy_price|prod_desc |
+---------+----------+
| 48.81000|This repli|
+---------+----------+
```

Figure 3.6 – Coercing versus casting (1)

Let's have one more example. Check out this example that uses `coerce()`:

```
public void printInvoicesPerDayCoerce(LocalDate day) {

  ctx.select(PAYMENT.INVOICE_AMOUNT)
     .from(PAYMENT)
     .where(PAYMENT.PAYMENT_DATE
       .coerce(LocalDate.class).eq(day))
     .fetch()
     .forEach(System.out::println);
}
```

`PAYMENT.PAYMENT_DATE` is a timestamp, therefore, it is not enough to pretend that it is a date since the time component will fail our predicate. For instance, *2003-04-09 09:21:25* is not equal to *2003-04-09*. In this case, we need an actual cast from timestamp to date as follows:

```
public void printInvoicesPerDayCast(LocalDate day) {

  ctx.select(PAYMENT.INVOICE_AMOUNT)
     .from(PAYMENT)
     .where(PAYMENT.PAYMENT_DATE
       .cast(LocalDate.class).eq(day))
     .fetch()
     .forEach(System.out::println);
}
```

This time, the cast takes place via this SQL (for *2003-04-09*):

```
SELECT `classicmodels`.`payment`.`invoice_amount`
FROM `classicmodels`.`payment`
WHERE cast(`classicmodels`.`payment`.`payment_date` as date)
  = { d '2003-04-09' }
```

The following figure compares the results of using `coerce()` and `cast()`:

```
coercing                casting
+--------------+        +--------------+
|invoice_amount|        |invoice_amount|
+--------------+        +--------------+
                        |      16537.85|
                        |       7674.94|
                        +--------------+
```

Figure 3.7 – Coercing versus casting (2)

Another good example where cast works and coerce does not is when the cast is performed in `GROUP BY`, which isn't uncommon when grouping timestamp columns by `CAST(ts AS DATE)`. Also, when the value being cast is an expression, not a bind variable, the effect is different (although coerce can be used to compare, for instance, `INTEGER` columns with `BIGINT` columns without the database needing to convert anything).

> **Important Note**
> As a rule of thumb, in some cases when both could work (for instance, when you project the expressions), it is best to use `coerce()` rather than `cast()`. This way, you don't risk unsafe or raw-type casting in Java and you don't pollute the generated SQL with unnecessary castings.

Next, let's discuss collations.

Collation

Databases define a character set as a set of symbols and encodings. A collation defines a set of rules for comparing (ordering) characters in a character set. jOOQ allows us to specify a collation via `collation(Collation collation)` for `org.jooq.DateType` and via `collate(String collation)`, `collate(Collation collation)`, and `collate(Name collation)` for `org.jooq.Field`. Here is an example of setting the `latin1_spanish_ci` collation for a field:

```
ctx.select(PRODUCT.PRODUCT_NAME)
   .from(PRODUCT)
   .orderBy(PRODUCT.PRODUCT_NAME.collate("latin1_spanish_ci"))
   .fetch()
   .forEach(System.out::println);
```

All the examples from this section are available in the *CastCoerceCollate* application.

Binding values (parameters)

Binding values is another fundamental topic of jOOQ.

The well-known prepared statements and bind values combination is the preferable approach to express SQL statements in JDBC. Among benefits, this combination provides protection against SQL injections, sustains caching (for instance, most connection pools cache prepared statements across connections or rely on JDBC driver caching capabilities as HikariCP does), and reusability capabilities (re-using execution plans for identical SQL statements, regardless of actual bind values).

Having security and performance packed into this combination makes it preferable against static statements (`java.sql.Statement`) and inlined values, so jOOQ also embraces it as default.

> **Important Note**
>
> By default, jOOQ aligns its support for bind values to JDBC style. In other words, jOOQ relies on `java.sql.PreparedStatement` and indexed bind values or indexed parameters. Moreover, exactly like JDBC, jOOQ uses a `?` (question mark) character for marking the bind value placeholders.
>
> However, in contrast to JDBC, which supports only indexed parameters and the `?` character, jOOQ supports named and inlined parameters as well. Each of them is detailed in this section.

So, in JDBC, the only way to exploit bind values aligns to the following example:

```
Connection conn = ...;
try (PreparedStatement stmt = conn.prepareStatement(
  """SELECT first_name, last_name FROM employee
     WHERE salary > ? AND job_title = ?""")) {

     stmt.setInt(1, 5000);
     stmt.setString(2, "Sales Rep");

     stmt.executeQuery();
}
```

In other words, in JDBC, it is our responsibility to keep track of the number of question marks and their corresponding index. This becomes cumbersome in complex/dynamic queries.

As Lukas Eder highlights, "*The strength of languages such as L/SQL, PL/pgSQL, T-SQL (among other things) is precisely the fact that prepared statements can naturally embed bind values transparently, without the user having to think about the binding logic.*"

Now, let's see how jOOQ tackles bind values via indexed bind values or indexed parameters.

Indexed parameters

Writing the previous query via jOOQ's DSL API can be done as follows:

```
ctx.select(EMPLOYEE.FIRST_NAME, EMPLOYEE.LAST_NAME)
   .from(EMPLOYEE)
   .where(EMPLOYEE.SALARY.gt(5000)
     .and(EMPLOYEE.JOB_TITLE.eq("Sales Rep")))
   .fetch();
```

Even if it looks like we've inlined the values (*5000* and *Sales Rep*), this is not true. jOOQ abstracts away the JDBC frictions and allows us to use indexed parameters exactly where needed (directly in SQL). Since jOOQ takes care of everything, we don't even care about the indexes of the parameters. Moreover, we take advantage of type-safety for these parameters and we don't need to explicitly set their type. The preceding SQL renders the following SQL string (notice the rendered question marks as bind values placeholders):

```
SELECT
  `classicmodels`.`employee`.`first_name`,
  `classicmodels`.`employee`.`last_name`
FROM`classicmodels`.`employee`
WHERE (`classicmodels`.`employee`.`salary` > ?
       and `classicmodels`.`employee`.`job_title` = ?)
```

And, after jOOQ resolves the bind values, we have the following:

```
SELECT
  `classicmodels`.`employee`.`first_name`,
  `classicmodels`.`employee`.`last_name`
FROM `classicmodels`.`employee`
WHERE(`classicmodels`.`employee`.`salary` > 5000
   and `classicmodels`.`employee`.`job_title` = 'Sales Rep')
```

Behind the scene, jOOQ uses a method named `DSL.val(value)` for transforming the given `value` argument (`value` can be `boolean`, `byte`, `String`, `float`, `double`, and so on) into a bind value. This `DSL.val()` method wraps and returns a bind value via the `org.jooq.Param` interface. This interface extends `org.jooq.Field`, therefore, extends a column expression(or field) and can be used accordingly via the jOOQ API. The previous query can also be written by explicitly using `DSL.val()` as follows:

```
ctx.select(EMPLOYEE.FIRST_NAME, EMPLOYEE.LAST_NAME)
   .from(EMPLOYEE)
   .where(EMPLOYEE.SALARY.gt(val(5000))
      .and(EMPLOYEE.JOB_TITLE.eq(val("Sales Rep"))))
   .fetch();
```

But, as you just saw, using `val()` explicitly is not needed in this case. Using `val()` like this is just adding noise to the SQL expression.

In this query, we've used hardcoded values, but, most probably, these values represent user inputs that land in the query via the arguments of the method containing this query. Check out this example, which extracts these hardcoded values as arguments of the method:

```
public void userInputValuesAsIndexedParams(
        int salary, String job) {

  ctx.select(EMPLOYEE.FIRST_NAME, EMPLOYEE.LAST_NAME)
     .from(EMPLOYEE)
     .where(EMPLOYEE.SALARY.gt(salary)
        .and(EMPLOYEE.JOB_TITLE.eq(job)))
     .fetch();
}
```

Of course, mixing hardcoded and user input values in the same query is supported as well. Next, let's tackle a bunch of examples where the explicit usage of `val()` is really needed.

Explicit usage of val()

There are cases when we cannot pass plain values to jOOQ and expect back bind values. There are a few such cases:

- When the bind value is at the left-hand side of an operator
- When `Field` references and `Param` values are mixed

- When the bind value occurs in a clause that doesn't support it (for instance, in `select()`)
- When functions require a `Field<T>` type for one of the parameters

Let's have some examples.

Bind value is at the left-hand side of an operator

Having the plain value at the left-hand side of an operator doesn't allow us to write the needed jOOQ expression since we don't have access to the jOOQ DSL API. For instance, we cannot write `...5000.eq(EMPLOYEE.SALARY)` since the `eq()` method is not available. On the other hand, we should write `...val(5000).eq(EMPLOYEE.SALARY)`. This time, *5000* is wrapped in `Param` (which extends `Field`) via `val(int/Integer value)` and we can continue to exploit the jOOQ DSL API, such as the `eq()` method. Here is another example:

```
ctx.select(PAYMENT.INVOICE_AMOUNT)
   .from(PAYMENT)
   .where(val(LocalDateTime.now())
      .between(PAYMENT.PAYMENT_DATE)
         .and(PAYMENT.CACHING_DATE))
   .fetch();
```

Here is an example where the value is a user input:

```
public void usingValExplicitly(LocalDateTime date) {

ctx.select(PAYMENT.INVOICE_AMOUNT)
     .from(PAYMENT)
     .where(val(date).between(PAYMENT.PAYMENT_DATE)
       .and(PAYMENT.CACHING_DATE))
     .fetch();
}
```

Next, let's see the other case, when `Field` references and `Param` values are mixed.

Field references and Param values are mixed

Let's consider that we want to use the `DSL.concat(Field<?>... fields)` method for concatenating `CUSTOMER.CONTACT_FIRST_NAME`, the whitespace literal (`" "`), and `CUSTOMER.CONTACT_LAST_NAME` (for example, *Joana Nimar*). While `CONTACT_FIRST_NAME` and `CONTACT_LAST_NAME` are fields, the whitespace literal (`" "`) cannot be used in this context as a plain string. But, it can be wrapped in `Param` via the `val()` method, as follows:

```
ctx.select(CUSTOMER.CUSTOMER_NUMBER,
    concat(CUSTOMER.CONTACT_FIRST_NAME, val(" "),
           CUSTOMER.CONTACT_LAST_NAME))
   .from(CUSTOMER)
   .fetch();
```

Here is another example that mixes the jOOQ internal usage of `val()` and our explicit usage of `val()` for wrapping a user input value to add it as a column in the result set:

```
public void usingValExplicitly(float vat) {

    ctx.select(
    EMPLOYEE.SALARY,

         // jOOQ implicit val()
         EMPLOYEE.SALARY.mul(vat).as("vat_salary"),

         // explicit val()
         val(vat).as("vat"))
      .from(EMPLOYEE)
      .fetch();
    }
```

Here is another example of mixing implicit and explicit `val()` usage for writing a simple arithmetic expression, *mod((((10 - 2) * (7 / 3)) / 2), 10)*:

```
ctx.select(val(10).sub(2).mul(val(7).div(3)).div(2).mod(10))
   .fetch();
```

When the same parameter is used multiple times, it is advisable to extract it as in the following example:

```
public void reusingVal(int salary) {

  Param<Integer> salaryParam = val(salary);

  ctx.select(EMPLOYEE.FIRST_NAME, EMPLOYEE.LAST_NAME,
           salaryParam.as("base_salary"))
     .from(EMPLOYEE)
```

```
        .where(salaryParam.eq(EMPLOYEE.SALARY))
          .and(salaryParam.mul(0.15).gt(10000))
        .fetch();
}
```

While we take care of the `salary` value, jOOQ will take care of the *0.15* and *10000* constants. All three will become indexed bind values.

Bind values from string query

If, for some reason, you want to bind values directly from a string query, then you can do it via plain SQL as in the following example:

```
// bind value from string query
ctx.fetch("""
       SELECT first_name, last_name
       FROM employee WHERE salary > ? AND job_title = ?
          """, 5000, "Sales Rep");

// bind value from string query
ctx.resultQuery("""
        SELECT first_name, last_name
        FROM employee WHERE salary > ? AND job_title = ?
                """, 5000, "Sales Rep")
   .fetch();
```

Next, let's talk about the named parameters.

Named parameters

While JDBC support is limited to indexed bind values, jOOQ goes beyond this limit and supports named parameters as well. Creating a jOOQ named parameter is accomplished via the `DSL.param()` methods. Among these methods, we have `param(String name, T value)`, which creates a named parameter with a name and an initial value. Here is an example:

```
ctx.select(EMPLOYEE.FIRST_NAME, EMPLOYEE.LAST_NAME)
   .from(EMPLOYEE)
   .where(EMPLOYEE.SALARY.gt(param("employeeSalary", 5000))
      .and(EMPLOYEE.JOB_TITLE
         .eq(param("employeeJobTitle", "Sales Rep"))))
   .fetch();
```

Here is an example of the values of named parameters being provided as user inputs:

```
public void userInputValuesAsNamedParams(
                      int salary, String job) {

  ctx.select(EMPLOYEE.FIRST_NAME, EMPLOYEE.LAST_NAME)
     .from(EMPLOYEE)
     .where(EMPLOYEE.SALARY
       .gt(param("employeeSalary", salary))
       .and(EMPLOYEE.JOB_TITLE
         .eq(param("employeeJobTitle", job))))
     .fetch();
}
```

While rendering the SQL of the previous queries, you have observed that jOOQ doesn't render the names of these parameters as placeholders. It still renders a question mark as the default placeholder. To instruct jOOQ to render the names of the parameters as placeholders, we call via the `DSL.renderNamedParams()` method that returns a string, as in the following example:

```
String sql = ctx.renderNamedParams(
  ctx.select(EMPLOYEE.FIRST_NAME, EMPLOYEE.LAST_NAME)
     .from(EMPLOYEE)
     .where(EMPLOYEE.SALARY.gt(param("employeeSalary", 5000))
     .and(EMPLOYEE.JOB_TITLE
        .eq(param("employeeJobTitle", "Sales Rep")))))
```

Moreover, we can specify a string to be used as a prefix for each rendered named parameter via `Settings.withRenderNamedParamPrefix()`. You can see an example in the bundled code.

The returned string can be passed to another SQL access abstraction that supports named parameters. For this example, the rendered SQL string is as follows:

```
SELECT
  `classicmodels`.`employee`.`first_name`,
  `classicmodels`.`employee`.`last_name`
FROM `classicmodels`.`employee`
WHERE (`classicmodels`.`employee`.`salary` > : employeeSalary
       and `classicmodels`.`employee`.`job_title`
         = : employeeJobTitle)
```

Next, let's talk about the inline parameters.

Inline parameters

An inline bind value is rendered as the actual plain value via `DSL.inline()`. In other words, while indexed and named parameters render the bind values as placeholders via question marks (or names), inline parameters render their plain values directly. jOOQ automatically replaces the placeholders (`?` or `:name` for named parameters) and will properly escape inline bind values to avoid SQL syntax errors and SQL injection. Nevertheless, be warned that abusing the usage of the inline parameters may lead to poor performance on RDBMSs that have execution plan caches. So, avoid copying and pasting `inline()` everywhere!

Typically, using `inline()` for constants is a good practice. For instance, earlier, we used `val(" ")` to express `concat(CUSTOMER.CONTACT_FIRST_NAME, val(" "), CUSTOMER.CONTACT_LAST_NAME))`. But, since the `" "` string is a constant, it can be inlined:

```
ctx.select(CUSTOMER.CUSTOMER_NUMBER,
      concat(CUSTOMER.CONTACT_FIRST_NAME, inline(" "),
      CUSTOMER.CONTACT_LAST_NAME))
   .from(CUSTOMER)
   .fetch();
```

But, if you know that this is not a constant, then it is better to rely on `val()` to sustain execution plan caches.

At the `Configuration` level, we can use inline parameters by switching from the `PreparedStatement` default to a static `Statement` via jOOQ settings. For example, the following `DSLContext` will use static statements, and all queries triggered in the context of this configuration will use inline parameters:

```
public void inlineParamsViaSettings() {

  DSL.using(ds, SQLDialect.MYSQL,
        new Settings().withStatementType(
              StatementType.STATIC_STATEMENT))
     .select(EMPLOYEE.FIRST_NAME, EMPLOYEE.LAST_NAME)
     .from(EMPLOYEE)
     .where(EMPLOYEE.SALARY.gt(5000)
       .and(EMPLOYEE.JOB_TITLE.eq("Sales Rep")))
```

```
        .fetch();
}
```

Obviously, another option is to rely on `inline()`:

```
ctx.select(EMPLOYEE.FIRST_NAME, EMPLOYEE.LAST_NAME)
   .from(EMPLOYEE)
   .where(EMPLOYEE.SALARY.gt(inline(5000))
   .and(EMPLOYEE.JOB_TITLE.eq(inline("Sales Rep"))))
   .fetch();
```

Of course, the inlined values can be user inputs as well. But, this technique is not recommended since user inputs may vary across executions and this will affect the performance of RDBMSs that rely on execution plan caches.

The previous two examples render the same SQL having the actual plain values inlined:

```
SELECT
  `classicmodels`.`employee`.`first_name`,
  `classicmodels`.`employee`.`last_name`
FORM `classicmodels`.`employee`
WHERE (`classicmodels`.`employee`.`salary` > 5000
   and `classicmodels`.`employee`.`job_title` = 'Sales Rep')
```

Globally, we can choose the type of parameters via `Settings`, as here (indexed parameters (`ParamType.INDEXED`) are used by default):

```
@Bean
public Settings jooqSettings() {
  return new Settings().withParamType(ParamType.NAMED);
}
```

Or, here is the global setting for using static statements and inline parameters:

```
@Bean
public Settings jooqSettings() {
  return new Settings()
    .withStatementType(StatementType.STATIC_STATEMENT)
    .withParamType(ParamType.INLINED);
}
```

Next, let's see a handy approach to rendering a query with different types of parameter placeholders.

Rendering a query with different types of parameter placeholders

Let's assume that we have a query that uses indexed parameters and we need to render it as a certain SQL string having a different type of parameter placeholder (for instance, this may be required by another SQL abstraction):

```
ResultQuery query
  = ctx.select(EMPLOYEE.FIRST_NAME, EMPLOYEE.LAST_NAME)
       .from(EMPLOYEE)
       .where(EMPLOYEE.SALARY.gt(5000)
         .and(EMPLOYEE.JOB_TITLE.eq("Sales Rep")));
```

A handy approach for rendering this query with a different type of parameter placeholder relies on the `Query.getSQL(ParamType)` method as follows:

- `ParamType.INDEXED` (in this example, this is the default behavior):

  ```
  String sql = query.getSQL(ParamType.INDEXED);

  SELECT
    `classicmodels`.`employee`.`first_name`,
    `classicmodels`.`employee`.`last_name`
  FROM `classicmodels`.`employee`
  WHERE (`classicmodels`.`employee`.`salary` > ?
      and `classicmodels`.`employee`.`job_title` = ?)
  ```

- `ParamType.NAMED` (for parameters with names, this produces placeholders of the `:name` type, but for unnamed parameters, it produces `:1`, `:2`, to `:n`, therefore, a combination of colon and index):

  ```
  String sql = query.getSQL(ParamType.NAMED);

  SELECT
    `classicmodels`.`employee`.`first_name`,
    `classicmodels`.`employee`.`last_name`
  FROM `classicmodels`.`employee`
  WHERE (`classicmodels`.`employee`.`salary` > :1
      and `classicmodels`.`employee`.`job_title` = :2)
  ```

- `ParamType.INLINED` and `ParamType.NAMED_OR_INLINED`:

```
String sql = query.getSQL(ParamType.INLINED);
String sql = query.getSQL(ParamType.NAMED_OR_INLINED);

SELECT
  `classicmodels`.`employee`.`first_name`,
  `classicmodels`.`employee`.`last_name`
FROM `classicmodels`.`employee`
WHERE (`classicmodels`.`employee`.`salary` > 5000 and
       `classicmodels`.`employee`.`job_title` = 'Sales Rep')
```

In this case, `ParamType.INLINED` and `ParamType.NAMED_OR_INLINED` produce the same output – inlined plain values. Actually, `ParamType.NAMED_OR_INLINED` generates named parameter placeholders only for parameters that are named explicitly, otherwise, it inlines all unnamed parameters. You can see more examples in the code bundled with the book.

Next, let's see how we can extract jOOQ parameters from the query as `List<Object>`.

Extracting jOOQ parameters from the query

Accessing all types of supported parameters of a query can be accomplished via `Query.getParams()`, while accessing a single parameter can be done by index via `Query.getParam()`, as in the following example, which uses indexed parameters (the same approach can be used for inlined parameters):

```
ResultQuery query
  = ctx.select(EMPLOYEE.FIRST_NAME, EMPLOYEE.LAST_NAME)
       .from(EMPLOYEE)
       .where(EMPLOYEE.SALARY.gt(5000))
       .and(EMPLOYEE.JOB_TITLE.eq("Sales Rep"));

// wrap the value, 5000
Param<?> p1 = query.getParam("1");

// wrap the value, "Sales Rep"
Param<?> p2 = query.getParam("2");
```

If we use named parameters, then those names can be used in place of indexes:

```
ResultQuery query
  = ctx.select(EMPLOYEE.FIRST_NAME, EMPLOYEE.LAST_NAME)
       .from(EMPLOYEE)
       .where(EMPLOYEE.SALARY.gt(
          param("employeeSalary", 5000)))
       .and(EMPLOYEE.JOB_TITLE.eq(
          param("employeeJobTitle", "Sales Rep")));

// wrap the value, 5000
Param<?> p1 = query.getParam("employeeSalary");

// wrap the value, "Sales Rep"
Param<?> p2 = query.getParam("employeeJobTitle");
```

As you'll see soon, parameters can be used to set new binding values. Next, let's see how we can extract indexed and named parameters.

Extracting binding values

Having a parameter, we can extract its underlying bind value via `getValue()`.

But, extracting all the query bind values for indexed and named parameters without interacting with `Param` can be done via `getBindValues()`. This method returns `List<Object>` containing all the bind values of the query represented as a query or any of its subinterfaces such as `ResultQuery`, `Select`, and so on. Here is an example for indexed parameters:

```
public void extractBindValuesIndexedParams() {

  ResultQuery query
    = ctx.select(EMPLOYEE.FIRST_NAME, EMPLOYEE.LAST_NAME)
         .from(EMPLOYEE)
         .where(EMPLOYEE.SALARY.gt(5000))
         .and(EMPLOYEE.JOB_TITLE.eq("Sales Rep"));

  System.out.println("Bind values: "
    + query.getBindValues());
}
```

And, here is an example for named parameters:

```
public void extractBindValuesNamedParams() {

  ResultQuery query = ctx.select(
            EMPLOYEE.FIRST_NAME, EMPLOYEE.LAST_NAME)
    .from(EMPLOYEE)
    .where(EMPLOYEE.SALARY.gt(param("employeeSalary", 5000))
    .and(EMPLOYEE.JOB_TITLE
       .eq(param("employeeJobTitle", "Sales Rep"))));

  System.out.println("Bind values: "
    + query.getBindValues());
}
```

In both examples, the returned list will contain two bind values, [*5000* and *Sales Rep*]. For inline parameters, `getBindValues()` returns an empty list. This is happening because, unlike `getParams()`, which returns all types of supported parameters, `getBindValues()` returns only actual bind values that render an actual placeholder.

We can use the extracted binding values in another SQL abstraction, such as `JdbcTemplate` or JPA. For instance, here is `JdbcTemplate`:

```
Query query = ctx.select(...)
                 .from(PAYMENT)
                 .where(...);

List<DelayedPayment> result = jdbcTemplate.query(query.getSQL(),
   query.getBindValues().toArray(), new BeanPropertyRowMapper
      (DelayedPayment.class));
```

Setting new bind values

We must start this section with the following important note.

> **Important Note**
>
> Conforming to jOOQ documentation, starting with version 4.0, jOOQ plans to make the `Param` class immutable. Modifying `Param` values is strongly discouraged; therefore, use the information from this section carefully.

Nevertheless, modifying bind values via `Param` was still possible when this book was written. For instance, the following example executes an SQL with an initial set of bind values, sets new bind values, and executes the query again. Setting new bind values is done via the deprecated `setConverted()` method:

```
public void modifyingTheBindValueIndexedParam() {

  try ( ResultQuery query
    = ctx.select(EMPLOYEE.FIRST_NAME, EMPLOYEE.LAST_NAME)
        .from(EMPLOYEE)
        .where(EMPLOYEE.SALARY.gt(5000))
        .and(EMPLOYEE.JOB_TITLE.eq("SalesRep"))
        .keepStatement(true)) {

     // lazily create a new PreparedStatement
     Result result1 = query.fetch();
     System.out.println("Result 1: " + result1);

     // set new bind values
     Param<?> p1 = query.getParam("1");
     Param<?> p2 = query.getParam("2");
     p1.setConverted(75000);
     p2.setConverted("VP Marketing");

     // re-use the previous PreparedStatement
     Result result2 = query.fetch();
     System.out.println("Result 2: " + result2);
   }
}
```

The `Query` interface also allows for setting new bind values directly, without explicitly accessing the `Param` type via the `bind()` method as follows (if there are named parameters that refer to them via their names instead of indexes):

```
public void modifyingTheBindValueIndexedParam() {

  try ( ResultQuery query
    = ctx.select(EMPLOYEE.FIRST_NAME, EMPLOYEE.LAST_NAME)
        .from(EMPLOYEE)
        .where(EMPLOYEE.SALARY.gt(5000))
        .and(EMPLOYEE.JOB_TITLE.eq("Sales Rep"))
```

```
        .keepStatement(true)) {

    // lazily create a new PreparedStatement
    Result result1 = query.fetch();
    System.out.println("Result 1: " + result1);

    // set new bind values
    query.bind(1, 75000);
    query.bind(2, "VP Marketing");

    // re-use the previous PreparedStatement
    Result result2 = query.fetch();
    System.out.println("Result 2: " + result2);
  }
}
```

Nevertheless, behind the scene, `bind()` works via `Param.setConverted()`.

For convenience (but, not required), notice that both examples take advantage of the fact that a `PreparedStatement` can be reused with different bind values. First, we ask jOOQ to keep the statement open via `keepStatement(true)`. Second, the `Query` becomes like a resource that must be closed via `Query.close()` or in a `try-with-resources` statement.

In the case of inline parameters, jOOQ will automatically close any underlying `PreparedStatement` in order for new bind values to have an effect; therefore, there is no use in keeping the statements open. The code is straightforward and is available in the code bundled with the book.

Named/unnamed parameters with no initial value

While in the previous examples the parameters have an initial value that was modified later, jOOQ also supports named/unnamed parameters with no initial value.

If you need a named parameter without providing an initial value at its creation, then you may need one of the following `DSL.param()` flavors.

Here is an example of using `DSL.param(String name)` that returns `Param<Object>`:

```
Param<Object> phoneParam = DSL.param("phone");

// set the parameter value
```

```
phoneParam.setValue("(26) 642-7555");

ctx.selectFrom(CUSTOMER)
   .where(phoneParam.eq(CUSTOMER.PHONE))
   .fetch();
```

This is an example of creating a named parameter with a defined class type and no initial value via `param(String name, Class<T> type)`:

```
Param<String> phoneParam = DSL.param("phone", String.class);
phoneParam.setValue("(26) 642-7555");
```

This is how we create a named parameter with a defined data type and no initial value via `param(String name, DataType<T> type)`:

```
Param<String> phoneParam
  = DSL.param("phone", SQLDataType.VARCHAR);
phoneParam.setValue("(26) 642-7555");
```

And, we can create a named parameter with a defined type of another field and no initial value via `param(String name, Field<T> type)`:

```
Param<String> phoneParam = DSL.param("phone", CUSTOMER.PHONE);
phoneParam.setValue("(26) 642-7555");
```

We can also keep a reference to a named parameter having an initial value (for instance, just to not lose the generic type, `<T>`):

```
Param<String> phoneParam
  = DSL.param("phone", "(26) 642-7555");

// changing the value is still possible
phoneParam.setValue("another_value");
```

In addition, jOOQ supports unnamed parameters without initial values but with a defined type. We can create such parameters via `param(Class<T> class)`, `param(DataType<T> dataType)`, and `param(Field<T> field)`.

Moreover, we can create a parameter without a name and initial value with a generic type (`Object`/`SQLDataType.OTHER`) via `param()`. You can find examples in the code bundled with this book.

Rendering unnamed parameters via `Query` with `renderNamedParams()` results in rendering the positions of parameters starting with 1, such as `:1`, `:2`, to `:n`.

> **Important Note**
>
> At the time of writing, jOOQ still supports modifying binding values, but `setValue()` and `setConverted()` are deprecated and probably removed starting with version 4.0 when jOOQ plans to make `Param` immutable.
>
> Also, pay attention to `param()` and `param(String name)`. As a rule of thumb, avoid these methods if you are using any of the following dialects: SQLDialect.DB2, DERBY, H2, HSQLDB, INGRES, and SYBASE. These dialects may have trouble inferring the type of the bind value. In such cases, prefer `param()` flavors that explicitly set a type of the bind value.

All the examples from this section are available in the *BindingParameters* application.

Summary

This was a comprehensive chapter, which covered several fundamental aspects of jOOQ. So far, you have learned how to create `DSLContext`, how the jOOQ fluent API works, how to deal with jOOQ `Result` and `Record`, how to tackle edge cases of casting and coercing, and how to use bind values. As a rule of thumb, having these fundamentals under your tool belt is a major advantage that helps you to make the correct and optimal decisions and will be a great support in the next chapters.

In the next chapter, we will discuss alternatives for building a DAO layer and/or evolving the jOOQ-generated DAO layer.

4
Building a DAO Layer (Evolving the Generated DAO Layer)

At this point, we know how to enable the jOOQ Code Generator and how to express queries via the jOOQ DSL API, and we have a decent level of understanding of how jOOQ works. In other words, we know how to start and configure a Spring Boot application relying on jOOQ for the persistence layer implementation.

In this chapter, we tackle different approaches for organizing our queries in a **Data Access Object** (**DAO**) layer. Being a Spring Boot fan, you are most probably familiar with a DAO layer that is repository-centric, therefore, you'll see how jOOQ fits into this context. By the end of this chapter, you'll be familiar with the following:

- Hooking the DAO layer
- Shaping the DAO design pattern and using jOOQ
- Shaping the generic DAO design pattern and using jOOQ
- Extending the jOOQ built-in DAO

Let's get started!

Technical requirements

The code for this chapter can be found on GitHub at https://github.com/PacktPublishing/jOOQ-Masterclass/tree/master/Chapter04.

Hooking the DAO layer

DAO is a design pattern that stands for **Data Access Object**. Following the separation of logic principle, DAO separates the data persistence logic in a dedicated layer and abstracts away the low-level database operations. Typically, the DAO is sketched around three main components:

- A model representing the data that is transferred between layers (for example, the `Sale` model corresponds to the `SALE` database table)
- An interface containing the API that should be implemented for the model (for example, `SaleDao`, or in Spring terms, `SaleRepository`)
- A concrete implementation of this interface (for example, `SaleDaoImpl`, or in Spring terms, `SaleRepositoryImpl`)

The following diagram represents the relationships between these components using `Sale`, `SaleRepository`, and `SaleRepositoryImpl`:

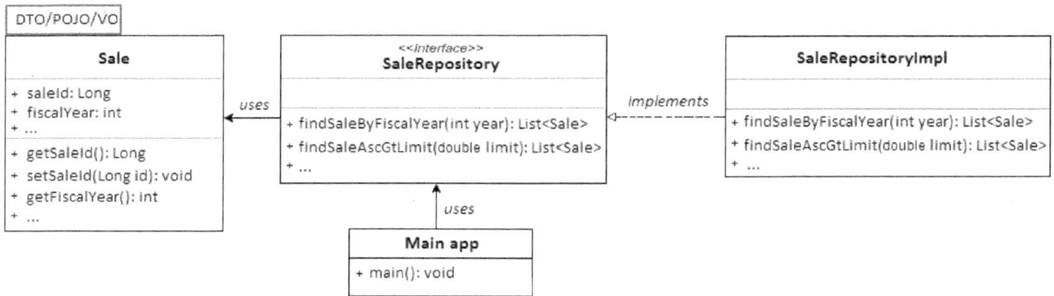

Figure 4.1 – DAO design pattern

If you are a `JdbcTemplate` fan, you most probably recognize this pattern in your own applications. On the other hand, if you are familiar with Spring Data JPA/JDBC, then you can associate `Sale` with a JPA/JDBC entity, `SaleRepository` with an extension of the Spring repository (for instance, `CrudRepository` or `JpaRepository`), and `SaleRepositoryImpl` with the Spring proxy instance automatically created for `SaleRepository`.

A flavor of this design pattern is known as *generic DAO*. In this case, the goal is to isolate the query methods that are common to all repositories (for instance, `fetchAll()`, `fetchById()`, `insert()`, `update()`, and so on) from the query methods that are repository-specific (for instance, `findSaleByFiscalYear()`). This time, we add the common methods in a generic interface (such as `ClassicModelsRepository<>`) and we provide an implementation for it (`ClassicModelsRepositoryImpl<>`).

The following diagrams depict two classical flavors of the generic DAO using the same `Sale`, `SaleRepository` and `SaleRepositoryImpl`:

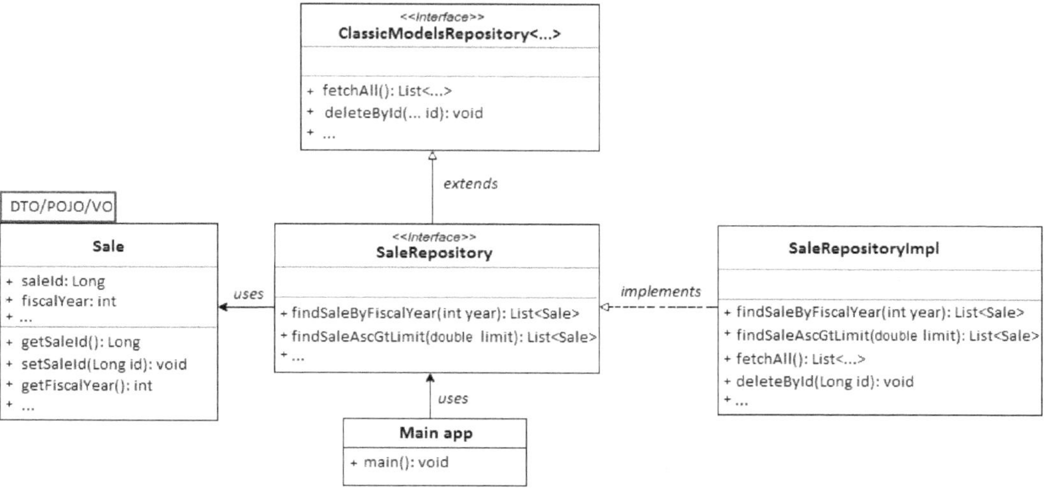

Figure 4.2 – Generic DAO (1)

In *Figure 4.2*, the implementation of `SaleRepository` must provide an implementation of the generic `ClassicModelsRepository` as well. Each repository will follow this technique. To increase the DAO layer flexibility, we add a separate implementation for the generic interface as shown in the following figure:

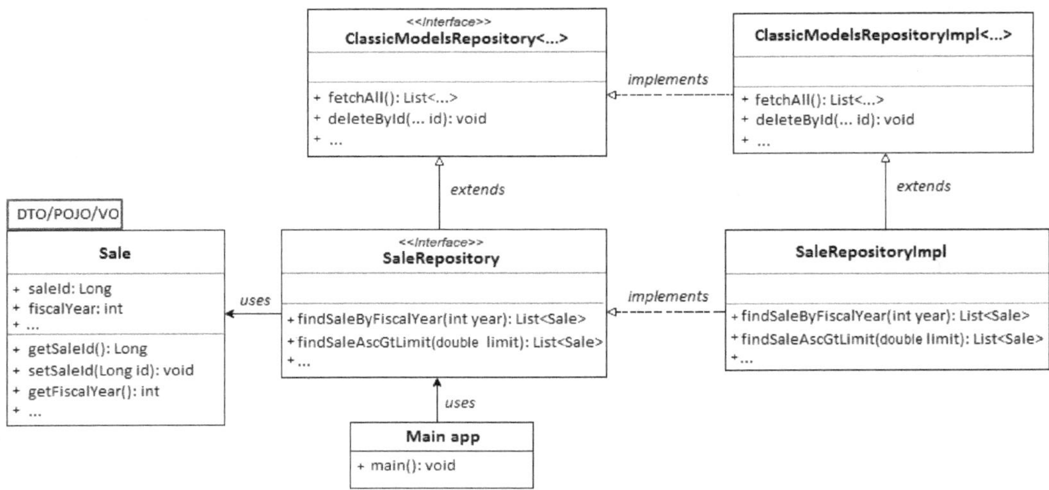

Figure 4.3 – Generic DAO (2)

If you are familiar with Spring Data JPA/JDBC, then you can associate `ClassicModelsRepository` with a Spring built-in repository (for example, `CrudRepository` or `JpaRepository`) and the implementation of this interface, `ClassicModelsRepositoryImpl`, with a Spring built-in implementation such as `SimpleJpaRepository`.

Next, let's see how we can shape these DAO patterns and use jOOQ.

Shaping the DAO design pattern and using jOOQ

Let's assume that we have a bunch of SQLs written in jOOQ for the `SALE` table, and we want to shape a simple DAO implementation around them. This is quite simple because all we have to do is to follow *Figure 4.1* from the previous section.

First of all, the model is provided as POJOs by the jOOQ generator (we can have user-defined POJOs as well), therefore, we already have the `Sale` POJO. Next, we write `SaleRepository`:

```
@Repository
@Transactional(readOnly=true)
```

```
public interface SaleRepository {

  public List<Sale> findSaleByFiscalYear(int year);
  public List<Sale> findSaleAscGtLimit(double limit);
}
```

`SaleRepositoryImpl` provides a jOOQ implementation for these two methods:

```
@Repository
public class SaleRepositoryImpl implements SaleRepository {

  private final DSLContext ctx;

  public SaleRepositoryImpl(DSLContext ctx) {
    this.ctx = ctx;
  }

  @Override
  public List<Sale> findSaleByFiscalYear(int year) {

    return ctx...
  }

  @Override
  public List<Sale> findSaleAscGtLimit(double limit) {

    return ctx...;
  }
}
```

Done! Further, we can simply inject `SaleRepository` and call the query methods:

```
@Service
public class SalesManagementService {

  private final SaleRepository saleRepository;

  public SalesManagementService(
    SaleRepository saleRepository) {
      this.saleRepository = saleRepository;
```

```
  }

  public List<Sale> fetchSaleByFiscalYear(int year) {

    return saleRepository.findSaleByFiscalYear(year);
  }

  public List<Sale> fetchSaleAscGtLimit(double limit) {

    return saleRepository.findSaleAscGtLimit(limit);
  }
}
```

In the same way, we can evolve this DAO layer by adding more repositories and implementations for other models. This application is available for Maven and Gradle as *SimpleDao*.

Moreover, if you have to combine Spring Data JPA DAO with the user-defined jOOQ DAO in a single interface, then simply extend the needed interfaces as in the following example:

```
@Repository
@Transactional(readOnly = true)
public interface SaleRepository
  extends JpaRepository<Sale, Long>, // JPA
  com.classicmodels.jooq.repository.SaleRepository { // jOOQ

  List<Sale> findTop10By(); // Sale is a JPA entity
}
```

Once you inject `SaleRepository`, you'll have access to a Spring Data JPA DAO and the user-defined jOOQ DAO in the same service. This example is named `JpaSimpleDao`.

Shaping the generic DAO design pattern and using jOOQ

Trying to implement the generic DAO from *Figure 4.2* starts with the generic interface, `ClassicModelsRepository`:

```
@Repository
@Transactional(readOnly = true)
```

```
public interface ClassicModelsRepository<T, ID> {

  List<T> fetchAll();

  @Transactional
  void deleteById(ID id);
}
```

While `ClassicModelsRepository` contains the common query methods, `SaleRepository` extends it to add specific query methods, as follows:

```
@Repository
@Transactional(readOnly = true)
public interface SaleRepository
        extends ClassicModelsRepository<Sale, Long> {

  public List<Sale> findSaleByFiscalYear(int year);
  public List<Sale> findSaleAscGtLimit(double limit);
}
```

The implementation of `SaleRepository` provides implementations for methods from both interfaces:

```
@Repository
public class SaleRepositoryImpl implements SaleRepository {

  private final DSLContext ctx;

  public SaleRepositoryImpl(DSLContext ctx) {
    this.ctx = ctx;
  }

  @Override
  public List<Sale> findSaleByFiscalYear(int year) { ... }

  @Override
  public List<Sale> findSaleAscGtLimit(double limit) { ... }

  @Override
  public List<Sale> fetchAll() { ... }
```

```
  @Override
  public void deleteById(Long id) { ... }
}
```

The complete example is named *SimpleGenericDao*. Moreover, if you have to combine Spring Data JPA DAO with the user-defined jOOQ generic DAO in a single interface, then extend the needed interfaces as in *JPASimpleGenericDao*. Once you inject `SaleRepository`, you'll have access to Spring Data JPA DAO and the user-defined jOOQ generic DAO in the same service.

How about implementing the generic DAO from *Figure 4.3*? This is more flexible but not that easy to do it. Because of *genericity* aspects, we can't reference tables and fields directly as we did in the previous case. The query methods from `ClassicModelsRepository` are written in a generic fashion, therefore, the jOOQ queries written via the DSL support in `ClassicModelsRepositoryImpl` must be written in a generic fashion as well.

It is not trivial to intuit how to express jOOQ SQLs in a generic fashion, but you can do it after studying the source code of the jOOQ built-in `DAO` interface and the `DAOImpl` class. For those who want to deep dive into this approach, consider the example named *GenericDao*. If you want to involve Spring Data JPA as well then check out *JpaGenericDao*.

But, as you saw in *Chapter 2, Customizing the jOOQ Level of Involvement*, jOOQ can generate a DAO layer on our behalf. Let's extend it and enrich/customize it as we like.

Extending the jOOQ built-in DAO

Let's assume that you have configured the jOOQ generator to output the generated DAO layer in the `jooq.generated.tables.daos` package. While the generated DAO exposes common query methods such as `insert()`, `update()`, `delete()`, and a few specific queries of the `fetchBy...()` or `fetchRange...()` types, we want to extend it with our own query methods.

> **Important Note**
>
> This is one of my favorite ways of writing a DAO layer in a Spring Boot and jOOQ application.

The jOOQ DAO layer contains a set of generated classes that mirrors the database tables and extends the built-in `org.jooq.impl.DAOImpl` class. For example, the `jooq.generated.tables.daos.SaleRepository` class (or, `jooq.generated.tables.daos.SaleDao` if you keep the default naming strategy used by jOOQ) corresponds to the `SALE` table. In order to extend `SaleRepository`, we have to take a quick look at its source code and highlight a part of it as follows:

```
@Repository
public class SaleRepository extends DAOImpl<SaleRecord,
      jooq.generated.tables.pojos.Sale, Long> {
   ...
  @Autowired
  public SaleRepository(Configuration configuration) {
     super(Sale.SALE, jooq.generated.tables.pojos.Sale.class,
             configuration);
 }
 ...
}
```

The highlighted code represents the climax of extending `SaleRepository`. When we extend `SaleRepository` (or any other jOOQ DAO class), it is our responsibility to pass a jOOQ valid configuration, otherwise, the code will produce `NullPointerException`. This is an easy task that can be accomplished as shown in the following snippet of code (basically, we pass into `SaleRepository` the configuration of `DSLContext` prepared by Spring Boot):

```
@Repository
@Transactional(readOnly = true)
public class SaleRepositoryImpl extends SaleRepository {

  private final DSLContext ctx;

  public SaleRepositoryImpl(DSLContext ctx) {
    super(ctx.configuration());
    this.ctx = ctx;
}
...
}
```

That's all! Now, you can exploit the query methods defined in `SaleRepositoryImpl` and `SaleRepository` as well. In other words, you can use the jOOQ built in and your own DAO as a "single" DAO. Here is an example:

```
@Service
public class SalesManagementService {

  private final SaleRepositoryImpl saleRepository;

  public SalesManagementService(
       SaleRepositoryImpl saleRepository) {
    this.saleRepository = saleRepository;
   }

  // call jOOQ DAO
  @Transactional(readOnly = true)
  public List<Sale> fetchSaleByFiscalYear(int year) {

      return saleRepository.fetchByFiscalYear(year);
  }

   // call your DAO
   public List<Sale> fetchSaleAscGtLimit(double limit) {

      return saleRepository.findSaleAscGtLimit(limit);
   }
}
```

Please consider the following note as well.

> **Important Note**
>
> At the time of writing, jOOQ DAOs work under the following statements:
>
> • jOOQ DAOs can be instantiated as much as you like since they don't have their own state.
>
> • jOOQ DAOs cannot generate methods on DAOs that use the interfaces of POJOs instead of classes. Actually, at the time of writing, the `<interfaces/>` and `<immutableInterfaces/>` features have been proposed to be removed. You can track this here: `https://github.com/jOOQ/jOOQ/issues/10509`.
>
> • jOOQ cannot generate interfaces for DAOs.
>
> • jOOQ DAOs can be annotated with `@Repository` but they are not running by default in a transactional context (they cannot be annotated with `@Transactional` by the jOOQ generator). You can track this here: `https://github.com/jOOQ/jOOQ/issues/10756`.
>
> • The DAO's generated `insert()` method cannot return the newly generated ID from the database or the POJO. It simply returns `void`. You can track this here: `https://github.com/jOOQ/jOOQ/issues/2536` and `https://github.com/jOOQ/jOOQ/issues/3021`.
>
> You don't have to consider these shortcomings as the end of the road. The jOOQ team filters dozens of features in order to choose the most popular ones that fit a significant number of scenarios and deserve an implementation directly in the jOOQ releases. Nevertheless, any corner-case or edge-case feature can be supplied by you via a custom generator, custom strategy, or customer configuration.

The complete example from this section is named *jOOQ DAO.*

Summary

In this chapter, we have covered several approaches to developing, from scratch, a DAO layer or evolving the jOOQ-generated DAO layer in a Spring Boot and jOOQ application. Each of the presented applications can serve as a stub application for your own applications. Just choose the one that is suitable for you, replace the schema, and start developing.

In the next chapter, we'll use jOOQ to express a wide range of queries involving `SELECT`, `INSERT`, `UPDATE`, and `DELETE`.

5
Tackling Different Kinds of SELECT, INSERT, UPDATE, DELETE, and MERGE

A common scenario for jOOQ beginners originates from having a plain valid SQL that should be expressed via the jOOQ DSL API. While the jOOQ DSL API is extremely intuitive and easy to learn, the lack of practice may still lead to scenarios where we simply cannot find or intuit the proper DSL methods that should be chained to express a certain SQL.

This chapter addresses this kind of issue via a comprehensive collection of popular queries, which gives you the chance to practice jOOQ DSL syntax based on the Java-based schema. More precisely, our aim is to express, in jOOQ DSL syntax, a carefully harvested list of `SELECT`, `INSERT`, `UPDATE`, `DELETE`, and `MERGE` statements that are used in our day-to-day job.

This way, by the end of this chapter, you should have funneled a significant number of SQLs through the jOOQ DSL syntax and tried them out against MySQL, PostgreSQL, SQL Server, and Oracle databases in Java applications based on Maven and Gradle. Being dialect-agnostic, jOOQ DSL excels at handling tons of dialect-specific issues by emulating a valid syntax, therefore, this is also a good chance to taste this aspect for these four popular databases.

Notice that, even if you see some performance tips, our focus is not on finding the best SQL or the most optimal SQL for a certain use case. This is not our goal! Our goal is to learn the jOOQ DSL syntax at a decent level that allows writing almost any `SELECT`, `INSERT`, `UPDATE`, `DELETE`, and `MERGE` statement in a productive manner.

In this context, our agenda contains the following:

- Expressing `SELECT` statements
- Expressing `INSERT` statements
- Expressing `UPDATE` statements
- Expressing `DELETE` statements
- Expressing `MERGE` statements

Let's get started!

Technical requirements

The code for this chapter can be found on GitHub at `https://github.com/PacktPublishing/jOOQ-Masterclass/tree/master/Chapter05`.

Expressing SELECT statements

In this section, we will express/write via jOOQ DSL syntax a wide range of `SELECT` statements, including common projections, popular subqueries, scalar and correlated subqueries, unions, and row value expressions. We'll start with the commonly used projections.

Expressing commonly used projections

By *commonly used projections*, we understand the projections written against the well-known dummy table, `DUAL`. As you most probably know, the `DUAL` table is specific to Oracle; it's mostly unnecessary in MySQL (although jOOQ still renders it for MySQL 5.7 compatibility) and doesn't exist in PostgreSQL and SQL Server.

In this context, even if the SQL standard requires a `FROM` clause, jOOQ never requires such a clause and it renders the `DUAL` table whenever it is needed/supported. For example, selecting *0* and *1* can be done via the `selectZero()` and `selectOne()` methods (these statics are available in `org.jooq.impl.DSL`). The latter (`selectOne()`), including some alternatives, is exemplified next:

```
MySQL 8.x                          : select 1 as `one`
PostgreSQL (no dummy-table concept): select 1 as "one"
```

```
ctx.selectOne().fetch();
ctx.select(val(1).as("one")).fetch();
ctx.fetchValue((val(1).as("one")));
```

As a parenthesis, the DSL class also expose three helpers for expressing the commonly used `0 literal (DSL.zero())`, `1 literal (DSL.one())`, and `2 literal (DSL.two())`. So, while selectZero() results in a new DSL subselect for a constant 0 literal, the zero() represents the 0 literal itself. Selecting ad hoc values can be done as follows (since we cannot use plain values in `select()`, we rely on the `val()` method introduced in *Chapter 3, jOOQ Core Concepts*, to obtain the proper parameters):

```
Oracle: select 1 "A", 'John' "B", 4333 "C", 0 "D" from dual
```

```
ctx.select(val(1).as("A"), val("John").as("B"),
           val(4333).as("C"), val(false).as("D")).fetch();
```

Or, it can be done via the `values()` table constructor, which allows us to express in-memory temporary tables. With jOOQ, the `values()` table constructor can be used to create tables that can be used in a `SELECT` statement's `FROM` clause. Notice how we specified the column aliases ("derived column lists") along with the table alias (`"t"`) for the `values()` constructor:

```
MySQL:
select `t`.`A`, ..., `t`.`D`
  from (select null as `A`, ..., null as `D`
          where false
          union all
            select * from
              (values row ('A', 'John', 4333, false)) as `t`
       ) as `t`
```

```
ctx.select().from(values(row("A", "John", 4333, false))
   .as("t", "A", "B", "C", "D")).fetch();
```

Here is another alternative to `selectOne()`:

```
PostgreSQL (no dummy-table concept):
select "t"."one" from (values (1)) as "t" ("one")

ctx.select().from(values(row(1)).as("t", "one")).fetch();
```

We can also specify an explicit `FROM` clause to point out some specific tables. Here is an example:

```
SQL Server:
select 1 [one] from [classicmodels].[dbo].[customer],
                    [classicmodels].[dbo].[customerdetail]

ctx.selectOne().from(CUSTOMER, CUSTOMERDETAIL).fetch();
```

Of course, the purpose of offering `selectOne()` and the like is not really to allow for querying `ctx.selectOne().fetch()`, but to be used in queries where the projection doesn't matter, as in the following example:

```
ctx.deleteFrom(SALE)
   .where(exists(selectOne().from(EMPLOYEE)
// .whereExists(selectOne().from(EMPLOYEE)
      .where(SALE.EMPLOYEE_NUMBER.eq(EMPLOYEE.EMPLOYEE_NUMBER)
        .and(EMPLOYEE.JOB_TITLE.ne("Sales Rep")))))
   .execute();
```

In the code bundled with this book, you can find more examples that are not listed here. Take your time to explore the *CommonlyUsedProjections* application. Next, let's tackle `SELECT` subqueries or subselects.

Expressing SELECT to fetch only the needed data

Starting with jOOQ DSL may target a simple `SELECT` query of the `SELECT all_columns FROM table` or `SELECT * FROM table` type. This kind of query can be written in jOOQ as follows:

```
ctx.select().from(ORDER)
   .where(ORDER.ORDER_ID.eq(10101L)).fetch();
```

```
ctx.selectFrom(ORDER)
   .where(ORDER.ORDER_ID.eq(10101L)).fetch();

ctx.select(ORDER.fields()).from(ORDER)
   .where(ORDER.ORDER_ID.eq(10101L)).fetch();
```

Since we rely on the generated Java-based schema (obtained via the jOOQ generator as you saw in *Chapter 2*, *Customizing the jOOQ Level of Involvement*), jOOQ can infer the fields (columns) of the `ORDER` table and explicitly render them in the generated query. But, if you need to render the `*` itself instead of the list of fields, then you can use the handy `asterisk()` method, as in the following query:

```
ctx.select(asterisk()).from(ORDER)
   .where(ORDER.ORDER_ID.eq(10101L))
   .fetch();
```

As Lukas Eder mentioned: "*Perhaps worth stressing more heavily here that the asterisk (`*`) is not the same thing as the other three ways of querying all columns. The asterisk (`*`) projects all the columns from the live database schema including the ones that jOOQ doesn't know. The other three approaches project all the columns that jOOQ knows but those columns may no longer exist in the live database schema. There may be a mismatch, which is especially important when mapping to records (for instance, using `selectFrom()`, or `into(recordtype)`). Even so, when using *, and when all the tables in `from()` are known to jOOQ, jOOQ will try to expand the asterisk in order to access all converters and data type bindings, and embeddable records, and other things.*"

Moreover, notice that such queries may fetch more data than needed, and relying on `*` instead of a list of columns may come with performance penalties, which are discussed in this article: `https://tanelpoder.com/posts/reasons-why-select-star-is-bad-for-sql-performance/`. When I say that *it may fetch more data than needed*, I refer to the scenarios that process only a subset of the fetched result set, while the rest of it is simply discarded. Fetching data can be an expensive (especially, time-consuming) operation, therefore, fetching data just to discard it is a waste of resources and it can lead to long-running transactions that affect the application's scalability. This is a common scenario in JPA-based applications (for instance, in Spring Boot, `spring.jpa.open-in-view=true` may lead to loading more data than is needed).

Among others, Tanel Poder's article mentions one thing that a lot of beginners overlook. By forcing the database to do "*useless, mandatory work*" (you'll love this article for sure: `https://blog.jooq.org/2017/03/08/many-sql-performance-problems-stem-from-unnecessary-mandatory-work/`) via a `*` projection, it can no longer apply some optimizations to the query, for example, join elimination, which is essential for complex queries (https://tanelpoder.com/posts/reasons-why-select-star-is-bad-for-sql-performance/#some-query-plan-optimizations-not-possible).

Fetching a subset of columns

As a rule of thumb, fetching more data than needed is a common cause of persistence layer performance penalties. Therefore, if all you need is a subset of columns from ORDER, then simply enlist them explicitly in SELECT as select(ORDER.ORDER_ID, ORDER.ORDER_DATE, ORDER.REQUIRED_DATE, ORDER.SHIPPED_DATE, ORDER.CUSTOMER_NUMBER). Sometimes, the needed subset of columns is almost equal to (but not equal to) the total number of fields/columns. In such cases, instead of enlisting the subset of columns as previously, it is more practical to point out the fields/columns that should be excluded via the except() method. Here is an example of fetching all fields/columns from ORDER except ORDER.COMMENTS and ORDER.STATUS:

```
ctx.select(asterisk().except(ORDER.COMMENTS, ORDER.STATUS))
   .from(ORDER)
   .where(ORDER.ORDER_ID.eq(10101L)).fetch();
```

Here is another example that applies the SQL nvl() function to the OFFICE.CITY field. Whenever OFFICE.CITY is null, we fetch the N/A string:

```
ctx.select(nvl(OFFICE.CITY, "N/A"),
               OFFICE.asterisk().except(OFFICE.CITY))
   .from(OFFICE).fetch();
```

If you need to attach an alias to a condition, then we first need to wrap this condition in a field via the field() method. Here is an example:

```
ctx.select(field(SALE.SALE_.gt(5000.0)).as("saleGt5000"),
                SALE.asterisk().except(SALE.SALE_))
   .from(SALE).fetch();
```

And, the result set table-like head of this query looks as here:

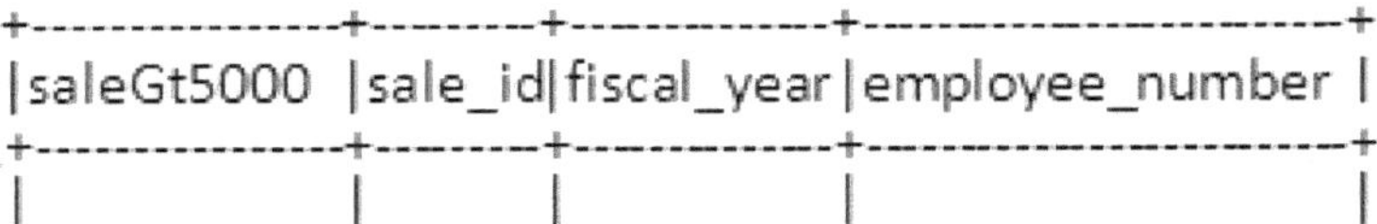

saleGt5000	sale_id	fiscal_year	employee_number

Notice that the `* EXCEPT (...)` syntax is inspired by BigQuery, which also has a `* REPLACE (...)` syntax that jOOQ plans to implement. You can track its progress here: `https://github.com/jOOQ/jOOQ/issues/11198`.

In the code bundled with this book (`SelectOnlyNeededData`), you can see more examples of juggling with the `asterisk()` and `except()` methods to materialize different scenarios that involve one or more tables.

Fetching a subset of rows

Besides using predicates, fetching only a subset of rows is commonly accomplished via the `LIMIT ... OFFSET` clause. Unfortunately, this clause is not part of the SQL standard and it is understood only by a limited number of database vendors such as MySQL and PostgreSQL. Nevertheless, jOOQ allows us to use `LIMIT ... OFFSET` via the `limit()` and `offset()` methods and will handle all aspects of emulating a compatible syntax for the used dialect. Here is an example of rendering `LIMIT 10 OFFSET 5`:

```
ctx.select(EMPLOYEE.FIRST_NAME,
           EMPLOYEE.LAST_NAME, EMPLOYEE.SALARY)
   .from(EMPLOYEE)
   .orderBy(EMPLOYEE.SALARY)
   .limit(10)
   .offset(5)
   .fetch();
```

Here is the same thing (the same query and result) but expressed via the `limit(Number offset, Number numberOfRows)` flavor (pay attention that the offset is the first argument – the order of arguments inherited from MySQL):

```
ctx.select(EMPLOYEE.FIRST_NAME,
           EMPLOYEE.LAST_NAME,EMPLOYEE.SALARY)
   .from(EMPLOYEE)
   .orderBy(EMPLOYEE.SALARY)
   .limit(5, 10)
   .fetch();
```

And, jOOQ will render the following SQL depending on the database vendor:

```
MySQL and PostgreSQL (jOOQ 3.14):
select ... from ...limit 10 offset 5
```

```
PostgreSQL (jOOQ 3.15+):
select ... from ...offset 5 rows fetch first 10 rows only

SQL Server:
select ... from ...offset 5 rows fetch next 10 rows only

Oracle:
select ... from ...offset 5 rows fetch next 10 rows only
```

Commonly, the arguments of `LIMIT` and `OFFSET` are some hard-coded integers. But, jOOQ allows us to use `Field` as well. For instance, here we use a scalar subquery in `LIMIT` (the same thing can be done in `OFFSET`):

```
ctx.select(ORDERDETAIL.ORDER_ID, ORDERDETAIL.PRODUCT_ID,
           ORDERDETAIL.QUANTITY_ORDERED)
   .from(ORDERDETAIL)
   .orderBy(ORDERDETAIL.QUANTITY_ORDERED)
   .limit(field(select(min(ORDERDETAIL.QUANTITY_ORDERED)).
      from(ORDERDETAIL)))
   .fetch();
```

As a note from Lukas Eder, "*Starting with version 3.15, jOOQ generates standard SQL for* `OFFSET` ... `FETCH` *in PostgreSQL, not the vendor-specific* `LIMIT` ... `OFFSET` *anymore. This is to provide native support for* `FETCH NEXT` ... `ROWS WITH TIES`. *Maybe, a future jOOQ will also offer the SQL Standard 2011 Oracle's/SQL Server's syntax:* `OFFSET n ROW[S] FETCH { FIRST | NEXT } m [ PERCENT ] ROW[S] { ONLY | WITH TIES }`." Track it here: `https://github.com/jOOQ/jOOQ/issues/2803`.

Notice that the previous queries use an explicit `ORDER BY` to avoid unpredictable results. If we omit `ORDER BY`, then jOOQ will emulate it on our behalf whenever it is needed. For instance, `OFFSET` (unlike `TOP`) requires `ORDER BY` in SQL Server, and if we omit `ORDER BY`, then jOOQ will render it on our behalf as in the following SQL:

```
SQL Server:
select ... from ...
   order by (select 0)
   offset n rows fetch next m rows only
```

Since we touched on this emulation topic, let's have a note that you should be aware of.

> **Important note**
>
> One of the coolest things about jOOQ is the capability to emulate the valid and optimal SQL syntax when the user database is lacking a specific feature (consider reading this article: `https://blog.jooq.org/2018/03/13/top-10-sql-dialect-emulations-implemented-in-jooq/`). The jOOQ documentation mentions that *"jOOQ API methods which are not annotated with the* `org.jooq.Support` *annotation (*`@Support`*), or which are annotated with the* `@Support` *annotation, but without any SQL dialects can be safely used in all SQL dialects. The aforementioned* `@Support` *annotation does not only designate which databases natively support a feature. It also indicates that a feature is emulated by jOOQ for some databases lacking this feature."* Moreover, whenever jOOQ doesn't support some vendor-specific functionality/syntax, the solution is to use plain SQL templating. This is dissected in a future chapter of this book.

For more examples, please consider the code bundled with this book. You'll find a collection of 15+ examples, including several corner cases as well. For instance, in `EXAMPLE 10.1` and `10.2`, you can see an example of fetching the rows in a certain order via the jOOQ `sortAsc()` method (if you are in such a position, then I suggest you read this article as well: `https://blog.jooq.org/2014/05/07/how-to-implement-sort-indirection-in-sql/`). Or, in `EXAMPLE 11`, you can see how to choose at runtime between `WHERE ... IN` and `WHERE ... NOT IN` statements via the jOOQ `org.jooq.Comparator` API and a Boolean variable. Moreover, in `EXAMPLE 15` and `16`, you can see the usage of the `SelectQuery` API for fetching columns from different tables. Take your time and practice each of these examples. I'm pretty sure that you'll learn a lot of tricks from them. The application is named `SelectOnlyNeededData`. For now, let's talk about expressing subselects in jOOQ.

Expressing SELECT subqueries (subselects)

Roughly speaking, a `SELECT` subquery (or subselect) is represented by a `SELECT` statement nested in another `SELECT` statement. Commonly, they appear in the `WHERE` or the `FROM` clauses, but it is no surprise to see them in the `HAVING` clause or combined with database views.

For example, let's take the following plain SQL containing a subselect in the `WHERE` clause as part of the predicate:

```
SELECT first_name, last_name FROM employee
WHERE office_code IN
    (SELECT office_code FROM office WHERE city LIKE "S%")
```

In jOOQ, we can express this query straightforwardly:

```
ctx.select(EMPLOYEE.FIRST_NAME, EMPLOYEE.LAST_NAME)
   .from(EMPLOYEE)
   .where(EMPLOYEE.OFFICE_CODE.in(
       select(OFFICE.OFFICE_CODE).from(OFFICE)
          .where(OFFICE.CITY.like("S%")))).fetch()
```

Notice how we have used the `IN` statement via the jOOQ `in()` method. In the same manner, you can use other statements supported by jOOQ, such as `NOT IN` (`notIn()`), `BETWEEN` (`between()`), `LIKE` (`like()`), and many others. Always pay attention to using `NOT IN` and that it's a peculiar behavior regarding `NULL` originating from the subquery (`https://www.jooq.org/doc/latest/manual/sql-building/conditional-expressions/in-predicate/`).

Almost any SQL statement has a jOOQ equivalent implementation, therefore, take your time and scan the jOOQ API to cover it as much as possible. This is the right direction in becoming a jOOQ power user.

> **Important Note**
>
> In subselects of type `SELECT foo..., (SELECT buzz)` is a common case to use both `DSLContext.select()` and `DSL.select()` as `ctx.select(foo) ... (select(buzz))`. The `DSLContext.select()` method is used for the outer `SELECT` in order to obtain a reference to a database configured connection, while for the inner or nested `SELECT`, we can use `DSL.select()` or `DSLContext.select()`. However, using `DSL.select()` for the inner `SELECT` is more convenient because it can be statically imported and referenced simply as `select()`. Notice that using `DSL.select()` for both types of `SELECT` or only for the inner `SELECT` leads to an exception of type *Cannot execute query. No Connection configured.* But, of course, you can still execute `DSL.select()` via `DSLContext.fetch(ResultQuery)` and the like.

Next, let's have another plain SQL that has a subselect in the `FROM` clause, as follows:

```
SELECT sale_id, sale
FROM sale,
  (SELECT avg(sale) AS avgs,employee_number AS sen
     FROM sale
     GROUP BY employee_number) AS saleTable
WHERE (employee_number = saleTable.sen
  AND sale < saleTable.avgs);
```

This time, let's express the subselect via a derived table as follows:

```
// Table<Record2<BigDecimal, Long>>
var saleTable = select(avg(SALE.SALE_).as("avgs"),
                       SALE.EMPLOYEE_NUMBER.as("sen"))
   .from(SALE)
   .groupBy(SALE.EMPLOYEE_NUMBER)
   .asTable("saleTable"); // derived table
```

Another approach relies on `table(select(…))`. Practically, `table(Select<R>)` is a synonym for `asTable()`. Choose the one that you find more fluent:

```
var saleTable = table(select(...)).as("saleTable");
```

Next, we can use this derived table to express the outer `SELECT`:

```
ctx.select(SALE.SALE_ID, SALE.SALE_)
   .from(SALE, saleTable)
   .where(SALE.EMPLOYEE_NUMBER
     .eq(saleTable.field("sen", Long.class))
       .and(SALE.SALE_
         .lt(saleTable.field("avgs", Double.class))))
   .fetch();
```

Since jOOQ cannot infer the field types of a user-defined derived table, we can rely on coercing to fill up the expected types of the `sen` and `avgs` fields (for a quick reminder of the *coercing* goal, please revisit *Chapter 3, jOOQ Core Concepts*), or `saleTable.field(name, type)`, as here.

The jOOQ API is so flexible and rich that it allows us to express the same SQL in multiple ways. It depends on us to choose the most convenient approach in a certain scenario. For example, if we consider that the SQL part `WHERE (employee_number = saleTable.sen AND sale < saleTable.avgs)` can be written as `WHERE (employee_number = sen AND sale < avgs)`, then we can extract the following fields as local variables:

```
Field<Double> avgs = avg(SALE.SALE_)
  .coerce(Double.class).as("avgs");
Field<Long> sen = SALE.EMPLOYEE_NUMBER.as("sen");
```

And, we can use them in the derived table and outer `SELECT`:

```
var saleTable = select(avgs, sen)
   .from(SALE)
   .groupBy(SALE.EMPLOYEE_NUMBER)
   .asTable("saleTable"); // derived table

ctx.select(SALE.SALE_ID, SALE.SALE_)
   .from(SALE, saleTable)
   .where(SALE.EMPLOYEE_NUMBER.eq(sen)
      .and(SALE.SALE_.lt(avgs)))
   .fetch();
```

Or, we can eliminate the explicitly derived table and embed the subselect in a fluent style:

```
ctx.select(SALE.SALE_ID, SALE.SALE_)
   .from(SALE, select(avgs, sen)
               .from(SALE)
               .groupBy(SALE.EMPLOYEE_NUMBER))
  .where(SALE.EMPLOYEE_NUMBER.eq(sen)
    .and(SALE.SALE_.lt(avgs)))
  .fetch();
```

Notice that we removed the explicit `saleTable` alias as well. As MySQL complains (and not only), every derived table requires an alias, but we don't have to worry about it. jOOQ knows this and will generate an alias on our behalf (something similar to `alias_25088691`). But, if your benchmarks reveal that alias generation is not negligible, then it is better to supply an explicit alias. As Lukas Eder says, "*However, the generated alias is based deterministically on the SQL string, in order to be stable, which is important for execution plan caches (for instance, Oracle, SQL Server, irrelevant in MySQL, PostgreSQL).*"

Consider more examples in the bundled code. For instance, if you are interested in using `SELECT` nested in`INSERT`, `UPDATE`, and `DELETE`, then you can find examples in the bundled code. The application is named *SampleSubqueries*. Next, let's talk about scalar subqueries.

Expressing scalar subqueries

A scalar subquery selects only one column/expression and returns only one row. It can be used in an SQL query anywhere that a column/expression can be used. For example, let's assume plain SQL that selects the employees with a salary greater than or equal to the average salary plus *25,000*:

```
SELECT first_name, last_name FROM employee
WHERE salary >= (SELECT (AVG(salary) + 25000) FROM employee);
```

In jOOQ, this query is generated by the following code:

```
ctx.select(EMPLOYEE.FIRST_NAME, EMPLOYEE.LAST_NAME)
   .from(EMPLOYEE)
   .where(EMPLOYEE.SALARY.coerce(BigDecimal.class)
      .ge(select(avg(EMPLOYEE.SALARY).plus(25000))
   .from(EMPLOYEE))).fetch();
```

Have you noticed the `...ge(select(...))` construct? In the left-hand of `ge()` we have a `Field`, but in the right-hand we have a `Select`. This is possible thanks to `ge(Select<? extends Record1<T>> select)`, which is a very handy shortcut that saves us from the explicit usage of `field()` as `ge(field(select(...)))`, or from `asField()` as `ge(select(...).asField())`. We can also write conditions such as `select(...).ge(select(...))`. But, what is the difference between `field()` and `asField()`?

Let's have another example that inserts a new `PRODUCT` (I've listed only the relevant part):

```
ctx.insertInto(PRODUCT,
               PRODUCT.PRODUCT_ID, ..., PRODUCT.MSRP)
   .values(...,
     field(select(avg(PRODUCT.MSRP)).from(PRODUCT)))
   .execute();
```

To insert the value of `PRODUCT.MSRP` as the average **MSRPs**, we rely on a scalar subquery that should be transformed in `Field`. This can be done as we did via `field(Select<? extends Record1<T>> s)` or via `asField()`. If we go with `asField()`, then we write something like this:

```
...select(avg(PRODUCT.MSRP)).from(PRODUCT).asField())
```

But, `asField()` returns this result provider as a `Field<?/Object>` object. In other words, `asField()` loses type information and so isn't type-safe. By losing type information, `asField()` allows us to accidentally introduce type-safety-related errors that cannot be detected until runtime, or even worse, that will produce unexpected results.

Here, we have typed `array(PRODUCT.MSRP)` instead of `avg(PRODUCT.MSRP)` but we have no complaints until runtime:

```
...select(array(PRODUCT.MSRP)).from(PRODUCT).asField())
```

Of course, you won't be writing such blabbering, but the idea is that using `asField()` in such contexts is prone to other data type incompatibilities that might be hard to spot and might produce unexpected results. So, let's keep `asField()` for queries as `SELECT b.*, (SELECT foo FROM a) FROM b`, and let's focus on `field()`:

```
... field(select(array(PRODUCT.MSRP)).from(PRODUCT)))
```

Do you think this code will compile? The correct answer is no! Your IDE will immediately signal a data type incompatibility. While `PRODUCT.MSRP` is `BigDecimal`, `(array(PRODUCT.MSRP))` is `Field<BigDecimal[]>`, so `INSERT` is wrong. Replace `array()` with `avg()`. Problem solved!

In the bundled code (`ScalarSubqueries`), you have more examples, including using scalar queries nested in `INSERT`, `UPDATE`, and `DELETE`. Next, let's talk about correlated subqueries.

Expressing correlated subqueries

A correlated subquery (or repeating subquery) uses the values of the outer query for computing its values. Since it depends on the outer query, a correlated subquery can't be executed independently as a standalone subquery. As Lukas Eder mentioned, "*In any case, no RDBMS is forced to naively execute a correlated subquery once for each row evaluated by the outer query (obviously, if such a thing would happen, then this may come with an overhead in performance if the correlated subquery must be executed for a significant number of times). Many RDBMS would optimize a correlated subquery by applying a transformation to a join or semi-join. Other, such as Oracle 11g and later, optimizes correlated subqueries thanks to scalar subquery caching.*" (`https://blogs.oracle.com/oraclemagazine/on-caching-and-evangelizing-sql`)

Let's consider the following plain SQL representing a correlated scalar subquery:

```
SELECT s1.sale, s1.fiscal_year, s1.employee_number
FROM sale AS s1
WHERE s1.sale =
    (SELECT max(s2.sale)
     FROM sale AS s2
     WHERE (s2.employee_number = s1.employee_number
```

```
            AND s2.fiscal_year = s1.fiscal_year))
ORDER BY s1.fiscal_year
```

Expressing this query in jOOQ can be done as follows (notice that we want to preserve the `s1` and `s2` aliases in the rendered SQL):

```
Sale s1 = SALE.as("s1");
Sale s2 = SALE.as("s2");

ctx.select(s1.SALE_, s1.FISCAL_YEAR, s1.EMPLOYEE_NUMBER)
   .from(s1)
   .where(s1.SALE_.eq(select(max(s2.SALE_))
      .from(s2)
      .where(s2.EMPLOYEE_NUMBER.eq(s1.EMPLOYEE_NUMBER)
         .and(s2.FISCAL_YEAR.eq(s1.FISCAL_YEAR)))))
   .orderBy(s1.FISCAL_YEAR)
   .fetch();
```

Did you spot this: `...where(s1.SALE_.eq(select(max(s2.SALE_))`? No `asField()/field()` required! The query didn't need to call `asField()/field()`, because there's a convenience overload provided by jOOQ as `Field<T>.eq(Select<? extends Record1<T>>)`. Yes, I know I already told you about it earlier, but I just wanted to enforce it again.

However, as you probably intuited, this correlated subquery, which relies on repeatedly self-joined tables, can be expressed more efficiently via `GROUP BY` as follows (this time, we don't preserve the aliases):

```
ctx.select(SALE.FISCAL_YEAR,
           SALE.EMPLOYEE_NUMBER, max(SALE.SALE_))
   .from(SALE)
   .groupBy(SALE.FISCAL_YEAR, SALE.EMPLOYEE_NUMBER)
   .orderBy(SALE.FISCAL_YEAR)
   .fetch();
```

In this case, `GROUP BY` is *much* better, because it eliminates that self-join, turning $O(n^2)$ into $O(n)$. As Lukas Eder shared, "*With modern SQL, self-joins are almost never really needed anymore. Beginners might think that these self-joins are the way to go, when they can be quite detrimental,* `https://twitter.com/MarkusWinand/status/1118147557828583424`, *$O(n^2)$ in the worst case.*" So, before jumping in to write such correlated subqueries, try to evaluate some alternatives and compare the execution plans.

Let's look at another example based on the following plain SQL:

```
SELECT product_id, product_name, buy_price
FROM product
WHERE
   (SELECT avg(buy_price)
    FROM product) > ANY
      (SELECT price_each
       FROM orderdetail
       WHERE product.product_id = orderdetail.product_id);
```

So, this query compares the average buy price (or list price) of all products with all the sale prices of each product. If this average is greater than any of the sale prices of a product, then the product is fetched in the result set. In jOOQ, this can be expressed as follows:

```
ctx.select(PRODUCT.PRODUCT_ID,
           PRODUCT.PRODUCT_NAME, PRODUCT.BUY_PRICE)
    .from(PRODUCT)
    .where(select(avg(PRODUCT.BUY_PRICE))
    .from(PRODUCT).gt(any(
         select(ORDERDETAIL.PRICE_EACH)
            .from(ORDERDETAIL)
               .where(PRODUCT.PRODUCT_ID
                  .eq(ORDERDETAIL.PRODUCT_ID)))))
    .fetch();
```

In the bundled code, you have a comprehensive list of examples, including examples containing `WHERE (NOT) EXISTS`, `ALL`, and `ANY`. As Lukas Eder says, "*It is worth mentioning that the name of ALL and ANY is "quantifier" and the comparison is called quantified comparison predicate. These are much more elegant than comparing with* `MIN()` *or* `MAX()`*, or using* `ORDER BY .. LIMIT 1`*, especially when row value expressions are used.*" Moreover, you can check out examples of using correlated subqueries nested in `INSERT`, `UPDATE`, and `DELETE`. The application is named *CorrelatedSubqueries*. Next, let's talk about writing row expressions in jOOQ.

Expressing row expressions

Row value expressions are quite handy for writing elegant multi-row predicates. jOOQ represents row value expressions via the `org.jooq.Row` interface. Its usage is straightforward, as the following plain SQL shows:

```
SELECT customer_number, address_line_first,
       address_line_second, city, state, postal_code, country
```

```
FROM customerdetail
WHERE (city, country) IN(SELECT city, country FROM office)
```

In jOOQ, this can be expressed via `row()`, as follows:

```
ctx.selectFrom(CUSTOMERDETAIL)
   .where(row(CUSTOMERDETAIL.CITY, CUSTOMERDETAIL.COUNTRY)
         .in(select(OFFICE.CITY, OFFICE.COUNTRY).from(OFFICE)))
.fetch();
```

In the bundled code (`RawValueExpression`), you can practice examples of using row value expressions with comparison predicates, `BETWEEN` and `OVERLAPS` predicates (jOOQ supports overlapping dates and arbitrary row value expressions of degree 2 – how cool is that?!), and `NULL`. Next, let's tackle the `UNION` and `UNION ALL` operators.

Expressing the UNION and UNION ALL operators

The `UNION` and `UNION ALL` operators are useful for combining two or more result sets from different `SELECT` statements or `SELECT`s into one result set. `UNION` eliminates duplicate rows from the results of the `SELECT` statements, while `UNION ALL` doesn't do this. To work, the number and order of columns must correspond in both queries and the data types must be the same or at least compatible. Let's consider the following SQL:

```
SELECT concat(first_name, ' ', last_name) AS full_name,
  'Employee' AS contactType
FROM employee
UNION
SELECT concat(contact_first_name, ' ', contact_last_name),
  'Customer' AS contactType
FROM customer;
```

jOOQ renders `UNION` via the `union()` method and `UNION ALL` via the `unionAll()` method. The previous SQL is rendered via `union()` as follows:

```
ctx.select(
      concat(EMPLOYEE.FIRST_NAME, inline(" "),
      EMPLOYEE.LAST_NAME).as("full_name"),
      inline("Employee").as("contactType"))
   .from(EMPLOYEE)
   .union(select(
          concat(CUSTOMER.CONTACT_FIRST_NAME, inline(" "),
          CUSTOMER.CONTACT_LAST_NAME),
```

```
        inline("Customer").as("contactType"))
    .from(CUSTOMER))
    .fetch();
```

Lukas Eder noted that *"*`UNION` *(*`ALL`*) acts differently with respect to* `NULL` *than other operators, meaning that two* `NULL` *values are 'not distinct.' So* `SELECT NULL UNION SELECT NULL` *produces only one row, just like* `SELECT NULL INTERSECT SELECT NULL`*."*

In the bundled code, you can practice more examples, including `UNION` and `ORDER BY`, `UNION` and `LIMIT`, `UNION` and `HAVING`, `UNION` and `SELECT INTO` (MySQL and PostgreSQL), `UNION ALL`, and so on. Unfortunately, there is no space here to list and dissect these examples, therefore, consider the application named *SelectUnions*. Next, let's cover the `INTERSECT` and `EXCEPT` operators.

Expressing the INTERSECT (ALL) and EXCEPT (ALL) operators

The `INTERSECT` operator produces only the values (rows) that are returned by (or common to) both subselects. The `EXCEPT` operator (or `MINUS` in Oracle) produces only the values that occur in the first (or left) subselect and don't occur in the second (or right) subselect. While `INTERSECT` and `EXCEPT` remove duplicates from their results, `INTERSECT ALL` and `EXCEPT ALL` don't do this. Exactly as in the case of `UNION`, to work, the number and order of columns must correspond in both queries and the data types must be the same or at least compatible.

Let's consider the following plain SQL:

```
SELECT buy_price FROM product
INTERSECT
SELECT price_each FROM orderdetail
```

In jOOQ, this can be expressed via `intersect()` as follows (for rendering `INTERSECT ALL`, use the `intersectAll()` method):

```
ctx.select(PRODUCT.BUY_PRICE)
    .from(PRODUCT)
    .intersect(select(ORDERDETAIL.PRICE_EACH)
        .from(ORDERDETAIL))
    .fetch();
```

By replacing the SQL `INTERSECT` with `EXCEPT` and, in jOOQ, the `intersect()` method with `except()` we can obtain an `EXCEPT` use case (for `EXCEPT ALL`, use the `exceptAll()` method). Here is the plain SQL (this time, let's add an `ORDER BY` clause as well):

```
SELECT buy_price FROM product
EXCEPT
SELECT price_each FROM orderdetail
ORDER BY buy_price
```

And the jOOQ code is as follows:

```
ctx.select(PRODUCT.BUY_PRICE)
   .from(PRODUCT)
   .except(select(ORDERDETAIL.PRICE_EACH).from(ORDERDETAIL))
   .orderBy(PRODUCT.BUY_PRICE)
   .fetch();
```

Nevertheless, if your database doesn't support these operators (for example, MySQL), then you have to emulate them. There are several ways to accomplish this and in the application named *IntersectAndExcept* (for MySQL), you can see a non-exhaustive list of solutions that emulate `INTERSECT (ALL)` based on `IN` (useful when no duplicates or `NULL` are present) and `WHERE EXISTS`, and emulate `EXCEPT (ALL)` based on `LEFT OUTER JOIN` and `WHERE NOT EXISTS`. Of course, feel free to check out the examples from *IntersectAndExcept* for PostgreSQL, SQL Server, and Oracle as well. Notice that Oracle 18c (used in our applications) supports only `INTERSECT` and `EXCEPT`, while Oracle20c supports all these four operators. Next, let's tackle the well-known `SELECT DISTINCT` and more.

Expressing distinctness

jOOQ comes with a suite of methods for expressing distinctness in our queries. We have the following:

- `SELECT DISTINCT` via `selectDistinct()`
- `IS (NOT) DISTINCT FROM` via `isDistinctFrom()` and `isNotDistinctFrom()`
- `COUNT (DISTINCT...)` via `countDistinct()`
- `AVG/SUM/MIN/MAX (DISTINCT ...)` via `avg/sum/min/maxDistinct()`
- PostgreSQL `DISTINCT ON` via `selectDistinct().on()` or `distinctOn()`

Let's look at an example of `IS DISTINCT FROM` as follows (in MySQL, `IS DISTINCT FROM` is represented by the <=> operator):

```
SELECT office.office_code, ...
FROM office
JOIN customerdetail
  ON office.postal_code = customerdetail.postal_code
WHERE (not((office.city, office.country) <=>
      (customerdetail.city, customerdetail.country)))
```

And jOOQ renders this query via the following snippet of code:

```
ctx.select()
   .from(OFFICE)
   .innerJoin(CUSTOMERDETAIL)
     .on(OFFICE.POSTAL_CODE.eq(CUSTOMERDETAIL.POSTAL_CODE))
   .where(row(OFFICE.CITY, OFFICE.COUNTRY).isDistinctFrom(
         row(CUSTOMERDETAIL.CITY, CUSTOMERDETAIL.COUNTRY)))
   .fetch();
```

Depending on the used dialect, jOOQ emulates the correct syntax. While for MySQL, jOOQ renders the <=> operator, for Oracle, it relies on `DECODE` and `INTERSECT`, and for SQL Server, it relies on `INTERSECT`. PostgreSQL supports `IS DISTINCT FROM`.

Next, let's see an example of PostgreSQL's `DISTINCT ON`:

```
SELECT DISTINCT ON (product_vendor, product_scale)
    product_id, product_name, ...
FROM product
ORDER BY product_vendor, product_scale
```

And the jOOQ code is as follows:

```
ctx.selectDistinct()
   .on(PRODUCT.PRODUCT_VENDOR, PRODUCT.PRODUCT_SCALE)
   .from(PRODUCT)
   .orderBy(PRODUCT.PRODUCT_VENDOR, PRODUCT.PRODUCT_SCALE)
   .fetch();
```

It is definitely worth mentioning here that jOOQ emulates this PostgreSQL-specific `DISTINCT ON` for MySQL, SQL Server, Oracle, and so on, via the `row_number()` window function.

Of course, we can write jOOQ queries that emulate `DISTINCT ON` as well. For instance, the following example fetches the employee numbers of the maximum sales per fiscal year via jOOQ's `rowNumber()` and `qualify()`:

```
ctx.selectDistinct(SALE.EMPLOYEE_NUMBER,
                   SALE.FISCAL_YEAR, SALE.SALE_)
   .from(SALE)
   .qualify(rowNumber().over(
      partitionBy(SALE.FISCAL_YEAR)
         .orderBy(SALE.FISCAL_YEAR, SALE.SALE_.desc())).eq(1))
   .orderBy(SALE.FISCAL_YEAR)
   .fetch();
```

A classical scenario solved via `DISTINCT ON` relies on selecting some distinct column(s) while ordering by other column(s). For instance, the following query relies on the PostgreSQL `DISTINCT ON` to fetch the distinct employee numbers ordered by minimum sales:

```
ctx.select(field(name("t", "employee_number")))
   .from(select(SALE.EMPLOYEE_NUMBER, SALE.SALE_)
      .distinctOn(SALE.EMPLOYEE_NUMBER)
      .from(SALE)
      .orderBy(SALE.EMPLOYEE_NUMBER, SALE.SALE_).asTable("t"))
   .orderBy(field(name("t", "sale")))
   .fetch();
```

Of course, this can be emulated without `DISTINCT ON` as well. Here is an alternative:

```
ctx.select(field(name("t", "employee_number")))
   .from(select(SALE.EMPLOYEE_NUMBER,
       min(SALE.SALE_).as("sale"))
     .from(SALE)
     .groupBy(SALE.EMPLOYEE_NUMBER).asTable("t"))
   .orderBy(field(name("t", "sale")))
   .fetch();
```

In the bundled code (`SelectDistinctOn`), you can find examples that cover all the bullets of the previous list. Take your time to practice them and get familiar with jOOQ syntax. Moreover, don't stop at these examples; feel free to experiment as much as possible with `SELECT`.

That's all about `SELECT`. Next, let's start talking about inserts.

Expressing INSERT statements

In this section, we will express different kinds of inserts including `INSERT ... VALUES`, `INSERT ... SET`, `INSERT ... RETURNING`, and `INSERT ...DEFAULT VALUES` via the jOOQ DSL syntax. Let's start with the well-known `INSERT ... VALUES` insert, which is supported by most database vendors.

Expressing INSERT ... VALUES

jOOQ supports `INSERT ... VALUES` via the `insertInto()` and `values()` methods. Optionally, we can use the `columns()` method for separating the name of the table in which we insert from the list of fields/columns that we insert. To trigger the actual `INSERT` statement, we have to explicitly call `execute()`; pay attention to this aspect since jOOQ novices tend to forget this call at the end of the insert/update/delete expressions. This method returns the number of rows affected by this `INSERT` statement as an integer value (`0`), which means that nothing happened.

For example, the following jOOQ type-safe expression will render an `INSERT` statement that can be successfully executed against at least MySQL, PostgreSQL, SQL Server, and Oracle (the primary key of the `ORDER` table, `ORDER.ORDER_ID`, is auto-generated, therefore, it can be omitted):

```
ctx.insertInto(ORDER,
      ORDER.COMMENTS, ORDER.ORDER_DATE, ORDER.REQUIRED_DATE,
      ORDER.SHIPPED_DATE, ORDER.STATUS, ORDER.CUSTOMER_NUMBER,
      ORDER.AMOUNT)
   .values("New order inserted...", LocalDate.of(2003, 2, 12),
         LocalDate.of(2003, 3, 1), LocalDate.of(2003, 2, 27),
         "Shipped", 363L, BigDecimal.valueOf(314.44))
   .execute();
```

Or, using `columns()`, it can be expressed like this:

```
ctx.insertInto(ORDER)
   .columns(ORDER.COMMENTS, ORDER.ORDER_DATE,
            ORDER.REQUIRED_DATE, ORDER.SHIPPED_DATE,
            ORDER.STATUS, ORDER.CUSTOMER_NUMBER, ORDER.AMOUNT)
   .values("New order inserted...", LocalDate.of(2003, 2, 12),
         LocalDate.of(2003, 3, 1), LocalDate.of(2003, 2, 27),
         "Shipped", 363L, BigDecimal.valueOf(314.44))
   .execute();
```

If the entire list of columns/fields is omitted (for example, for verbosity reasons), then the jOOQ expression is non-type-safe and you have to explicitly specify a value for each field/column of the table, including for the field/column representing an auto-generated primary key and the fields/columns having default values, otherwise you'll get an exception, as *the number of values must match the number of fields*. Moreover, you have to pay attention to the order of values; jOOQ matches the values to the fields only if you follow the order of arguments defined in the constructor of the `Record` class generated for the table in which we insert (for example, in this case, the order of arguments from the constructor of `OrderRecord`). As Lukas Eder adds, "*The* `Record` *constructor parameter order is also derived, like everything else, from the order of columns as declared in DDL, which is always the source of truth.*" In this context, specifying an explicit dummy value for the auto-generated primary key (or other field) can rely on the almost universal (and standard SQL) way, SQL `DEFAULT`, or `DSL.default_()`/`DSL.defaultValue()` in jOOQ (attempting to use `NULL` instead of SQL `DEFAULT` produces implementation-specific behavior):

```
ctx.insertInto(ORDER)
   .values(default_(), // Oracle, MySQL, PostgreSQL
           LocalDate.of(2003, 2, 12), LocalDate.of(2003, 3, 1),
           LocalDate.of(2003, 2, 27), "Shipped",
           "New order inserted ...", 363L, 314.44)
   .execute();
```

In PostgreSQL, we can use the `ORDER_SEQ.nextval()` call as well; `ORDER_SEQ` is the explicit sequence associated with the `ORDER` table:

```
ctx.insertInto(ORDER)
   .values(ORDER_SEQ.nextval(),
           LocalDate.of(2003, 2, 12), LocalDate.of(2003, 3, 1),
           LocalDate.of(2003, 2, 27), "Shipped",
           "New order inserted ...", 363L, 314.44)
    .execute();
```

Generally speaking, we can use the explicit or auto-assigned sequence (if the primary key is of the (`BIG`)`SERIAL` type) associated with the table and call the `nextval()` method. jOOQ defines a `currval()` method as well, representing the current value of the sequence.

SQL Server is quite challenging because it cannot insert explicit values for the identity column when `IDENTITY_INSERT` is set to `OFF` (`https://github.com/jOOQ/jOOQ/issues/1818`). Until jOOQ comes up with an elegant workaround, you can rely on a batch of three queries: one query sets `IDENTITY_INSERT` to `ON`, one query is `INSERT`, and the last query sets `IDENTITY_INSERT` to `OFF`. But, even so, this is useful only for specifying an explicit valid primary key. The SQL `DEFAULT` or `NULL` values, or any other dummy values, are not allowed as explicit identity values. SQL Server will simply attempt to use the dummy value as the primary key and will end up with an error. As Lukas Eder said, "*In some other RDBMS you'd still get an exception if the auto-generated value is* `GENERATED ALWAYS AS IDENTITY` *(as opposed to* `GENERATED BY DEFAULT AS IDENTITY`*), and you're trying to insert an explicit value.*"

No matter whether the primary key is of an auto-generated type or not, if you specify it explicitly (manually) as a valid value (not a dummy and not a duplicate key), then `INSERT` succeeds in all these four databases (of course, in SQL Server, in the context of `IDENTITY_INSERT` set to `ON`).

> **Important Note**
>
> Omitting the column list was interesting to explain `DEFAULT` and identities/sequences, but it's really not recommended to omit the column list in `INSERT`. So, you'd better strive to use the column list.

For inserting multiple rows, you can simply add a `values()` call per row in fluent style or use a loop to iterate a list of rows (typically, a list of records) and reuse the same `values()` call with different values. In the end, don't forget to call `execute()`. This approach (and not only) is available in the bundled code. But, starting with jOOQ 3.15.0, this can be done via `valuesOfRecords()` or `valuesOfRows()`. For instance, consider a list of records:

```
List<SaleRecord> listOfRecord = List.of( ... );
```

We can insert this list into the database via `valuesOfRecords()` as follows:

```
ctx.insertInto(SALE, SALE.fields())
   .valuesOfRecords(listOfRecord)
   .execute();
```

Here's an example with a list of rows:

```
var listOfRows
  = List.of(row(2003, 3443.22, 1370L,
              SaleRate.SILVER, SaleVat.MAX, 3, 14.55),
              row(...), ...);
```

This time, we can use `valuesOfRows()`:

```
ctx.insertInto(SALE,
               SALE.FISCAL_YEAR, SALE.SALE_,
               SALE.EMPLOYEE_NUMBER, SALE.RATE, SALE.VAT,
               SALE.FISCAL_MONTH, SALE.REVENUE_GROWTH)
   .valuesOfRows(listOfRows)
   .execute();
```

Whenever you need to collect data (for instance, POJOs) into a list or array of `RowN`, you can use the built-in `toRowList()` respectively `toRowArray()` collectors. You can find examples in the bundled code.

In other thoughts, inserting `Record` can be done in several ways; for a quick reminder of jOOQ records, please revisit *Chapter 3, jOOQ Core Concepts*. For now, let's insert the following `SaleRecord` corresponding to the `SALE` table:

```
SaleRecord sr = new SaleRecord();
sr.setFiscalYear(2003); // or, sr.set(SALE.FISCAL_YEAR, 2003);
sr.setSale(3443.22);    // or, sr.set(SALE.SALE_, 3443.22);
sr.setEmployeeNumber(1370L);
sr.setFiscalMonth(3);
sr.setRevenueGrowth(14.55);
```

To insert `sr`, we do the following:

```
ctx.insertInto(SALE)
   .values(sr.getSaleId(), sr.getFiscalYear(), sr.getSale(),
       sr.getEmployeeNumber(), default_(), SaleRate.SILVER,
       SaleVat.MAX, sr.getFiscalMonth(), sr.getFiscalYear(),
       default_())
   .execute()
```

Or, we can do the following:

```
ctx.insertInto(SALE)
   .values(sr.valuesRow().fields())
   .execute();
```

Or, we can even do the following:

```
ctx.executeInsert(sr);
```

Or, we can attach the record to the current configuration via `attach()`:

```
sr.attach(ctx.configuration());
sr.insert();
```

Trying to insert a POJO requires us to wrap it first in the corresponding `Record`. This can be done via the `newRecord()` method, which can load a jOOQ-generated record from your POJO or the jOOQ-generated POJO. Here is an example for the jOOQ-generated `Sale` POJO:

```
// jOOQ Sale POJO
Sale sale = new Sale(null, 2005, 343.22, 1504L,
     null, SaleRate.SILVER, SaleVat.MAX, 4, 15.55, null);
ctx.newRecord(SALE, sale).insert();
```

Another approach relies on the handy `Record.from(POJO)` method, as follows (basically, this time you use an explicit instance of `SaleRecord`):

```
SaleRecord sr = new SaleRecord();
sr.from(sale); // sale is the previous POJO instance

ctx.executeInsert(sr);
```

`Record.from()` comes in several flavors that allow us to populate `Record` from an array or even `Map` of values.

Whenever you need to reset a `Record` primary key (or another field), call the `reset()` method as in the following scenario, which resets the manually assigned primary key and allows the database to generate one on our behalf:

```
Sale sale = new Sale(1L, 2005, 343.22, 1504L, null,
            SaleRate.SILVER, SaleVat.MAX, 6, 23.99, null);
SaleRecord sr = new SaleRecord();
sr.from(sale);

// reset the current ID and allow DB to generate one
sr.reset(SALE.SALE_ID);

ctx.executeInsert(sr);
```

Nevertheless, the `reset()` method resets both the *changed* flag (which tracks record changes) and *value* (in this case, the primary key). If you want to reset only the value (primary key or another field), then you can rely on the `changed(Field<?> field, boolean changed)` method as here:

```
record.changed(SALE.SALE_ID, false);
```

Well, these were just a handful of examples. Many more, including using UDTs (in PostgreSQL and Oracle) and user-defined functions in `INSERT`, are available in the bundled code in the application named *InsertValues*. Next, let's talk about `INSERT ... SET`.

Expressing INSERT ... SET

`INSERT ... SET` is an alternative to `INSERT ... VALUES`, having an `UPDATE`-like syntax, and is commonly used in MySQL (but not only so). Practically, instead of listing columns and values separately, in `INSERT ... SET`, we write field-value pairs via the `set(field, value)` method. This is more readable since we can easily identify the value of each field. Let's look at an example of inserting two rows:

```
ctx.insertInto(SALE)
   .set(SALE.FISCAL_YEAR, 2005)       // first row
   .set(SALE.SALE_, 4523.33)
   .set(SALE.EMPLOYEE_NUMBER, 1504L)
   .set(SALE.FISCAL_MONTH, 3)
   .set(SALE.REVENUE_GROWTH, 12.22)
   .newRecord()
   .set(SALE.FISCAL_YEAR, 2005)       // second row
   .set(SALE.SALE_, 4523.33)
   .set(SALE.EMPLOYEE_NUMBER, 1504L)
   .set(SALE.FISCAL_MONTH, 4)
   .set(SALE.REVENUE_GROWTH, 22.12)
   .execute();
```

This syntax works for `Record` as well:

```
SaleRecord sr = new SaleRecord(...);
ctx.insertInto(SALE).set(sr).execute();
```

Since `INSERT ... SET` and `INSERT ... VALUES` are equivalent, jOOQ emulates `INSERT ... SET` as `INSERT ... VALUES` for all databases supported by jOOQ. The complete application is named *InsertSet*. Next, let's tackle the `INSERT ... RETURNING` syntax.

Expressing INSERT ... RETURNING

The particularity of `INSERT ... RETURNING` relies on the fact that it can return what was inserted (fetch back something that we need further). This may resume returning the primary key of the inserted row(s) or other fields as well (for instance, other sequences, default generated values, and trigger results). PostgreSQL has native support for `INSERT ... RETURNING`. Oracle also supports `INSERT ... RETURNING`, and jOOQ generates a PL/SQL anonymous block for it (not always). SQL Server supports `OUTPUT`, which is almost the same (apart from how trigger-generated values are affected). Other databases have poor support, and jOOQ has to emulate it on our behalf. In such cases, jOOQ relies on the JDBC `getGeneratedKeys()` method to retrieve the inserted primary keys. Moreover, if the generated primary keys (or other columns) cannot be retrieved directly, jOOQ may need to execute an additional `SELECT` to achieve this goal and this may lead to race conditions (for instance, such `SELECT` are needed in MySQL).

The jOOQ API for `INSERT ... RETURNING` contains the `returningResult()` method that comes in different flavors. It comes with different lists of arguments allowing us to specify which fields should be returned. If all fields should be returned, then simply use it without arguments. If only the primary key should be returned (being a popular use case for database auto-generated primary keys such us MySQL's `AUTO_INCREMENT` or PostgreSQL's (`BIG`)`SERIAL`, which automatically produces a sequence), then simply specify it as `returningResult(pk_field)`. If the primary key has multiple fields (a composite primary key), then list all of its fields separated by commas.

Here is an example that returns the primary key of a single insert:

```
// Record1<Long>
var insertedId = ctx.insertInto(SALE,
      SALE.FISCAL_YEAR, SALE.SALE_, SALE.EMPLOYEE_NUMBER,
      SALE.REVENUE_GROWTH, SALE.FISCAL_MONTH)
   .values(2004, 2311.42, 1370L, 10.12, 1)
   .returningResult(SALE.SALE_ID)
   .fetchOne();
```

Since there is a single result, we fetch it via the `fetchOne()` method. Fetching multiple primary keys can be done via `fetch()`, as follows:

```
// Result<Record1<Long>>
var insertedIds = ctx.insertInto(SALE,
      SALE.FISCAL_YEAR,SALE.SALE_, SALE.EMPLOYEE_NUMBER,
      SALE.REVENUE_GROWTH, SALE.FISCAL_MONTH)
   .values(2004, 2311.42, 1370L, 12.50, 1)
   .values(2003, 900.21, 1504L, 23.99, 2)
```

```
    .values(2005, 1232.2, 1166L, 14.65, 3)
    .returningResult(SALE.SALE_ID)
    .fetch();
```

This time, the returned result contains three primary keys. Returning more/other fields can be done as follows (the result looks like an *n-cols x n-rows* table):

```
// Result<Record2<String, LocalDate>>
var inserted = ctx.insertInto(PRODUCTLINE,
      PRODUCTLINE.PRODUCT_LINE, PRODUCTLINE.TEXT_DESCRIPTION,
      PRODUCTLINE.CODE)
     .values(..., "This new line of electric vans ...", 983423L)
     .values(..., "This new line of turbo N cars ...", 193384L)
     .returningResult(PRODUCTLINE.PRODUCT_LINE,
                      PRODUCTLINE.CREATED_ON)
    .fetch();
```

Now, let's look at a more interesting example. In our database schema, the `CUSTOMER` and `CUSTOMERDETAIL` tables are in a one-to-one relationship and share the primary key value. In other words, the `CUSTOMER` primary key is at the same time the primary key and foreign key in `CUSTOMERDETAIL`; this way, there is no need to maintain a separate foreign key. So, we have to use the `CUSTOMER` returned primary key for inserting the corresponding row in `CUSTOMERDETAIL`:

```
ctx.insertInto(CUSTOMERDETAIL)
    .values(ctx.insertInto(CUSTOMER)
       .values(default_(), ..., "Kyle", "Doyle",
               "+ 44 321 321", default_(), default_(), default_())
       .returningResult(CUSTOMER.CUSTOMER_NUMBER).fetchOne()
          .value1(), ..., default_(), "Los Angeles",
           default_(), default_(), "USA")
    .execute();
```

If you are familiar with JPA, then you can recognize an elegant alternative to `@MapsId` here.

The first `INSERT` (inner `INSERT`) will insert a row in `CUSTOMER` and will return the generated primary key via `returningResult()`. Next, the second `INSERT` (outer `INSERT`) will insert a row in `CUSTOMERDETAIL` using this returned primary key as a value for the `CUSTOMERDETAIL.CUSTOMER_NUMBER` primary key.

Moreover, the `returningResult()` method can also return expressions such as `returningResult(A.concat(B).as("C"))`.

As usual, take your time and check out the bundled code, which comes with many more examples. The application is named *InsertReturning*. Next, let's talk about `INSERT ... DEFAULT VALUES`.

Expressing INSERT ... DEFAULT VALUES

The straightforward approach for inserting default values is to omit the fields having default values from `INSERT`. For example, look at the following code:

```
ctx.insertInto(PRODUCT)
   .columns(PRODUCT.PRODUCT_NAME, PRODUCT.PRODUCT_LINE,
            PRODUCT.CODE, PRODUCT.PRODUCT_SCALE,
            PRODUCT.PRODUCT_VENDOR, PRODUCT.BUY_PRICE,
            PRODUCT.MSRP)
   .values("Ultra Jet X1", "Planes", 433823L, "1:18",
           "Motor City Art Classics",
            BigDecimal.valueOf(45.9), BigDecimal.valueOf(67.9))
   .execute();
```

The `PRODUCT` fields that are not listed (`PRODUCT_DESCRIPTION`, `PRODUCT_UID`, `SPECS`, and `QUANTITY_IN_STOCK`) will take advantage of implicit default values.

The jOOQ API comes with `defaultValues()`, `defaultValue()`, and `default_()` methods for explicitly pointing out the fields that should rely on default values. The former is useful for inserting a single row having only default values; if you check the database schema, you can notice that the `MANAGER` table has a default value for each of its columns:

```
ctx.insertInto(MANAGER).defaultValues().execute();
```

On the other hand, the `defaultValue()` method (or `default_()`) allows us to point to the fields that should rely on default values:

```
ctx.insertInto(PRODUCT)
   .columns(PRODUCT.PRODUCT_NAME, PRODUCT.PRODUCT_LINE,
     PRODUCT.CODE, PRODUCT.PRODUCT_SCALE,
     PRODUCT.PRODUCT_VENDOR, PRODUCT.PRODUCT_DESCRIPTION,
     PRODUCT.QUANTITY_IN_STOCK, PRODUCT.BUY_PRICE,
     PRODUCT.MSRP, PRODUCT.SPECS, PRODUCT.PRODUCT_UID)
   .values(val("Ultra Jet X1"), val("Planes"),
           val(433823L),val("1:18"),
```

```
            val("Motor City Art Classics"),
            defaultValue(PRODUCT.PRODUCT_DESCRIPTION),
            defaultValue(PRODUCT.QUANTITY_IN_STOCK),
            val(BigDecimal.valueOf(45.99)),
            val(BigDecimal.valueOf(67.99)),
            defaultValue(PRODUCT.SPECS),
            defaultValue(PRODUCT.PRODUCT_UID))
    .execute();
```

The non-type-safe version of this example is as follows:

```
ctx.insertInto(PRODUCT)
   .values(defaultValue(), "Ultra Jet X1", "Planes", 433823L,
     defaultValue(), "Motor City Art Classics",
     defaultValue(), defaultValue(), 45.99, 67.99,
     defaultValue(), defaultValue())
   .execute();

ctx.insertInto(PRODUCT)
   .values(defaultValue(PRODUCT.PRODUCT_ID),
           "Ultra JetX1", "Planes", 433823L,
           defaultValue(PRODUCT.PRODUCT_SCALE),
           "Motor City Art Classics",
           defaultValue(PRODUCT.PRODUCT_DESCRIPTION),
           defaultValue(PRODUCT.QUANTITY_IN_STOCK),
           45.99, 67.99, defaultValue(PRODUCT.SPECS),
           defaultValue(PRODUCT.PRODUCT_UID))
   .execute()
```

The same result can be obtained by specifying the types of columns. For example, the previous `defaultValue(PRODUCT.QUANTITY_IN_STOCK)` calls can be written as follows:

```
defaultValue(INTEGER) // or, defaultValue(Integer.class)
```

Inserting `Record` with default values can be done quite simply, as in the following examples:

```
ctx.newRecord(MANAGER).insert();

ManagerRecord mr = new ManagerRecord();
ctx.newRecord(MANAGER, mr).insert();
```

Using default values is useful to fill up those fields that will be later updated (for example, via subsequent updates, trigger-generated values, and so on) or if we simply don't have values.

The complete application is named *InsertDefaultValues*. Next, let's talk about jOOQ and `UPDATE` statements.

Expressing UPDATE statements

In this section, we will express different kinds of updates, including `UPDATE ... SET`, `UPDATE ... FROM`, and `UPDATE ... RETURNING`, and update using row value expressions via the jOOQ DSL syntax. At the time of writing, jOOQ supports updates against a single table, while updates against multiple tables represent a work in progress task.

Expressing UPDATE ... SET

The straightforward `UPDATE ... SET` statement can be expressed in jOOQ via the `set(field, value)` method, as in the following example (don't forget to call `execute()` to trigger the update):

```
ctx.update(OFFICE)
   .set(OFFICE.CITY, "Banesti")
   .set(OFFICE.COUNTRY, "Romania")
   .where(OFFICE.OFFICE_CODE.eq("1"))
   .execute();
```

The rendered SQL for MySQL dialect will be as follows:

```
UPDATE `classicmodels`.`office`
SET `classicmodels`.`office'.`city` = ?,
    `classicmodels`.`office`.`country` = ?
WHERE `classicmodels`.`office`.`office_code` = ?
```

Looks like a classic `UPDATE`, right? Notice that jOOQ automatically renders only the updated columns. If you are coming from JPA, then you know that Hibernate JPA renders, by default, all columns and we have to rely on `@DynamicUpdate` to obtain the same thing as jOOQ.

Check out another example for increasing the employee salary by an amount computed based on their sales:

```
ctx.update(EMPLOYEE)
   .set(EMPLOYEE.SALARY, EMPLOYEE.SALARY.plus(
```

```
      field(select(count(SALE.SALE_).multiply(5.75)).from(SALE)
   .where(EMPLOYEE.EMPLOYEE_NUMBER
      .eq(SALE.EMPLOYEE_NUMBER)))))
   .execute();
```

And here is the generated SQL for SQL Server dialect:

```
UPDATE [classicmodels].[dbo].[employee]
SET [classicmodels].[dbo].[employee].[salary] =
  ([classicmodels].[dbo].[employee].[salary] +
   (SELECT (count([classicmodels].[dbo].[sale].[sale]) * ?)
    FROM [classicmodels].[dbo].[sale]
    WHERE [classicmodels].[dbo].[employee].[employee_number]
        = [classicmodels].[dbo].[sale].[employee_number]))
```

Notice that this is an `UPDATE` without a `WHERE` clause and jOOQ will log a message as `A statement is executed without WHERE clause`. This is just friendly information that you can ignore if you have omitted the `WHERE` clause on purpose. But, if you know that this was not done on purpose, then you may want to avoid such situations by relying on the jOOQ `withExecuteUpdateWithoutWhere()` setting. You can choose from several behaviors including throwing an exception, as in this example:

```
ctx.configuration().derive(new Settings()
   .withExecuteUpdateWithoutWhere(ExecuteWithoutWhere.THROW))
   .dsl()
   .update(OFFICE)
   .set(OFFICE.CITY, "Banesti")
   .set(OFFICE.COUNTRY, "Romania")
   .execute();
```

Notice that we use `Configuration.derive()`, not `Configuration.set()`, because if the `DSLContext` is injected, `Configuration` is global and shared. Using `Configuration.set()` will affect the global settings. If this is the desired behavior, then it is better to rely on a separate `@Bean`, as you already saw in this book.

This time, whenever we attempt to execute `UPDATE` without the `WHERE` clause, `UPDATE` doesn't take any action and jOOQ throws an exception of the `org.jooq.exception.DataAccessException` type.

Updating Record is also quite simple. Check this out:

```
OfficeRecord or = new OfficeRecord();
or.setCity("Constanta");
or.setCountry("Romania");

ctx.update(OFFICE)
   .set(or)
   .where(OFFICE.OFFICE_CODE.eq("1")).execute();

// or, like this
ctx.executeUpdate(or, OFFICE.OFFICE_CODE.eq("1"));
```

As you'll see in the bundled code, using DSLContext.newRecord() is also an option.

Expressing UPDATE using row value expressions

Updating using row value expressions is a very handy tool, and jOOQ expresses such updates in a very clean and intuitive way. Check out this example:

```
ctx.update(OFFICE)
    .set(row(OFFICE.ADDRESS_LINE_FIRST,
             OFFICE.ADDRESS_LINE_SECOND, OFFICE.PHONE),
         select(EMPLOYEE.FIRST_NAME, EMPLOYEE.LAST_NAME,
             val("+40 0721 456 322"))
           .from(EMPLOYEE)
           .where(EMPLOYEE.JOB_TITLE.eq("President")))
   .execute();
```

The produced SQL for PostgreSQL is as follows:

```
UPDATE "public"."office"
SET ("address_line_first", "address_line_second", "phone") =
  (SELECT "public"."employee"."first_name",
          "public"."employee"."last_name", ?
   FROM "public"."employee"
   WHERE "public"."employee"."job_title" = ?)
```

Even if row value expressions are particularly useful for writing subselects, as in the previous example, it doesn't mean that you cannot write the following:

```
ctx.update(OFFICE)
   .set(row(OFFICE.CITY, OFFICE.COUNTRY),
        row("Hamburg", "Germany"))
   .where(OFFICE.OFFICE_CODE.eq("1"))
   .execute();
```

This can be useful for reusing fields with minimum verbosity:

```
Row2<String, String> r1 = row(OFFICE.CITY, OFFICE.COUNTRY);
Row2<String, String> r2 = row("Hamburg", "Germany");

ctx.update(OFFICE).set(r1, r2).where(r1.isNull())
   .execute();
```

Next, let's tackle the `UPDATE ... FROM` syntax.

Expressing UPDATE ... FROM

Using the `UPDATE ... FROM` syntax, we can join additional tables to an `UPDATE` statement. Notice that this `FROM` clause is vendor-specific supported in PostgreSQL and SQL Server, but not supported in MySQL and Oracle (however, when you read this book, jOOQ may have already emulated this syntax, so check it out). Here is an example:

```
ctx.update(PRODUCT)
   .set(PRODUCT.BUY_PRICE, ORDERDETAIL.PRICE_EACH)
   .from(ORDERDETAIL)
   .where(PRODUCT.PRODUCT_ID.eq(ORDERDETAIL.PRODUCT_ID))
   .execute();
```

And the SQL rendered for PostgreSQL is as follows:

```
UPDATE "public"."product"
SET "buy_price" = "public"."orderdetail"."price_each"
FROM "public"."orderdetail"
WHERE "public"."product"."product_id"
    = "public"."orderdetail"."product_id"
```

Finally, let's tackle the `UPDATE ... RETURNING` syntax.

Expressing UPDATE ... RETURNING

`UPDATE ... RETURNING` is like `INSERT ... RETURNING` but for `UPDATE`. This is supported natively by PostgreSQL and is emulated by jOOQ for SQL Server and Oracle. In jOOQ DSL, we express `UPDATE ... RETURNING` via `returningResult()` as in the following example:

```
ctx.update(OFFICE)
   .set(OFFICE.CITY, "Paris")
   .set(OFFICE.COUNTRY, "France")
   .where(OFFICE.OFFICE_CODE.eq("1"))
   .returningResult(OFFICE.CITY, OFFICE.COUNTRY)
   .fetchOne();
```

The SQL rendered for PostgreSQL is shown next:

```
UPDATE "public"."office"
SET "city" = ?,
    "country" = ?
WHERE "public"."office"."office_code" = ?
RETURNING "public"."office"."city",
          "public"."office"."country"
```

We can use `UPDATE ... RETURNING` for logically chaining multiple updates. For example, let's assume that we want to increase the salary of an employee with the average of their sales and the credit limit of their customers with the returned salary multiplied by two. We can express these two `UPDATE` statements fluently via `UPDATE ... RETURNING` as follows:

```
ctx.update(CUSTOMER)
   .set(CUSTOMER.CREDIT_LIMIT, CUSTOMER.CREDIT_LIMIT.plus(
      ctx.update(EMPLOYEE)
         .set(EMPLOYEE.SALARY, EMPLOYEE.SALARY.plus(
            field(select(avg(SALE.SALE_)).from(SALE)
                  .where(SALE.EMPLOYEE_NUMBER
                     .eq(EMPLOYEE.EMPLOYEE_NUMBER)))))
         .where(EMPLOYEE.EMPLOYEE_NUMBER.eq(1504L))
         .returningResult(EMPLOYEE.SALARY
            .coerce(BigDecimal.class))
         .fetchOne().value1()
            .multiply(BigDecimal.valueOf(2))))
```

```
    .where(CUSTOMER.SALES_REP_EMPLOYEE_NUMBER.eq(1504L))
    .execute();
```

However, pay attention to potential race conditions, given that there are two round trips hidden in what looks like a single query.

Practicing this example and many others can be done via the application named *UpdateSamples*. Next, let's tackle the DELETE statement.

Expressing DELETE statements

Expressing DELETE statements in jOOQ can be done via the DSLContext.delete() and DSLContext.deleteFrom() API or via DSLContext.deleteQuery() and DSLContext.executeDelete(), respectively. While the first three methods receive an argument of the Table<R> type, the executeDelete() method is useful for deleting a record as TableRecord<?> or UpdatableRecord<?>. As you can see from the following example, delete() and deleteFrom() work exactly the same:

```
ctx.delete(SALE)
    .where(SALE.FISCAL_YEAR.eq(2003))
    .execute();

ctx.deleteFrom(SALE)
    .where(SALE.FISCAL_YEAR.eq(2003))
    .execute();
```

Both of these expressions render this SQL:

```
DELETE FROM `classicmodels`.`sale`
WHERE `classicmodels`.`sale`.`fiscal_year` = ?
```

Combining DELETE and row value expressions is useful for deleting via subselects, as in the following example:

```
ctx.deleteFrom(CUSTOMERDETAIL)
    .where(row(CUSTOMERDETAIL.POSTAL_CODE,
               CUSTOMERDETAIL.STATE).in(
                  select(OFFICE.POSTAL_CODE, OFFICE.STATE)
                    .from(OFFICE)
                    .where(OFFICE.COUNTRY.eq("USA"))))
    .execute();
```

One important aspect of DELETE resumes to cascading deletion from parent to child. Whenever possible, it is a good idea to rely on database support for accomplishing DELETE cascading tasks. For example, you can use ON DELETE CASCADE or a stored procedure that implements the cascading deletion logic. As Lukas Eder highlights, "*The rule of thumb is to* CASCADE *compositions (UML speak) and to* RESTRICT *or* NO ACTION*, or* SET NULL *(if supported) aggregations. In other words, if the child cannot live without the parent (composition), then delete it with the parent. Otherwise, raise an exception (*RESTRICT, NO ACTION*), or set the reference to* NULL*. jOOQ might support* DELETE ... CASCADE *in the future:* https://github.com/jOOQ/jOOQ/issues/7367."

However, if none of these approaches are possible, then you can do it via jOOQ as well. You can write a chain of separate DELETE statements or rely on DELETE ... RETURNING as in the following examples, which delete PRODUCTLINE via cascading (PRODUCTLINE - PRODUCTLINEDETAIL - PRODUCT - ORDERDETAIL). In order to delete PRODUCTLINE, we have to delete all its products from PRODUCT and the corresponding record from PRODUCTLINEDETAIL. To delete all products of PRODUCTLINE from PRODUCT, we have to delete all references for these products from ORDERDETAIL. So, we start deleting from ORDERDETAIL, as follows:

```
ctx.delete(PRODUCTLINE)
  .where(PRODUCTLINE.PRODUCT_LINE.in(
    ctx.delete(PRODUCTLINEDETAIL)
      .where(PRODUCTLINEDETAIL.PRODUCT_LINE.in(
        ctx.delete(PRODUCT)
          .where(PRODUCT.PRODUCT_ID.in(
            ctx.delete(ORDERDETAIL)
              .where(ORDERDETAIL.PRODUCT_ID.in(
                select(PRODUCT.PRODUCT_ID).from(PRODUCT)
                  .where(PRODUCT.PRODUCT_LINE.eq("Motorcycles")
                    .or(PRODUCT.PRODUCT_LINE
                      .eq("Trucks and Buses")))))
            .returningResult(ORDERDETAIL.PRODUCT_ID).fetch()))
          .returningResult(PRODUCT.PRODUCT_LINE).fetch()))
      .returningResult(PRODUCTLINEDETAIL.PRODUCT_LINE).fetch()))
  .execute();
```

This jOOQ fluent expression renders four DELETE statements, which you can check in the bundled code. The challenge here consists of guaranteeing the roll-back functionality if something goes wrong. But, having the jOOQ expression in a Spring Boot @Transactional method, the roll-back functionality is out of the box. This is much better than the JPA cascading via CascadeType.REMOVE or orphanRemoval=true, which are very prone to *N + 1* issues. jOOQ allows us to control both what is deleted, and how this takes place.

In other thoughts, deleting `Record` (`TableRecord` or `UpdatableRecord`) can be done via `executeDelete()`, as in the following examples:

```
PaymentRecord pr = new PaymentRecord();
pr.setCustomerNumber(114L);
pr.setCheckNumber("GG31455");
...

// jOOQ render a WHERE clause based on the record PK
ctx.executeDelete(pr);

// jOOQ render our explicit Condition
ctx.executeDelete(pr,
   PAYMENT.INVOICE_AMOUNT.eq(BigDecimal.ZERO));
```

Exactly as in the case of `UPDATE`, if we attempt to perform `DELETE` without a `WHERE` clause, then jOOQ will inform us in a friendly way via a message. We can take control of what should happen in such cases via the `withExecuteDeleteWithoutWhere()` setting.

In the bundled code, you can see `withExecuteDeleteWithoutWhere()` next to many other examples that have not been listed here. The complete application is named *DeleteSamples*. Next, let's talk about `MERGE` statements.

Expressing MERGE statements

The `MERGE` statement is quite a powerful tool; it allows us to perform `INSERT/UPDATE` and even `DELETE` on a table known as the *target table* from a table known as the *source table*. I strongly suggest you read this article, especially if you need a quick reminder of the `MERGE` statement: `https://blog.jooq.org/2020/04/10/the-many-flavours-of-the-arcane-sql-merge-statement/`.

MySQL and PostgreSQL support a `MERGE` flavor known as `UPSERT` (`INSERT` or `UPDATE`) via `ON DUPLICATE KEY UPDATE`, respectively via `ON CONFLICT DO UPDATE` clauses. You can find examples of these statements next to the well-known `INSERT IGNORE INTO` (MySQL) and `ON CONFLICT DO NOTHING` (PostgreSQL) clauses in the code bundled with this book. By the way, we can use all these statements interchangeably (for example, we can use `onConflictDoNothing()` with MySQL and `onDuplicateKeyIgnore()` with PostgreSQL), since jOOQ will always emulate the correct syntax. We can even use them with SQL Server and Oracle, as jOOQ will emulate them via the `MERGE INTO` syntax.

SQL Server and Oracle have support for `MERGE INTO` with different additional clauses. Here is an example of exploiting the `WHEN MATCHED THEN UPDATE` (jOOQ `whenMatchedThenUpdate()`) and `WHEN NOT MATCHED THEN INSERT` (jOOQ `whenNotMatchedThenInsert()`) clauses:

```
ctx.mergeInto(PRODUCT)
   .usingDual() // or, (ctx.selectOne())
   .on(PRODUCT.PRODUCT_NAME.eq("1952 Alpine Renault 1300"))
   .whenMatchedThenUpdate()
   .set(PRODUCT.PRODUCT_NAME, "1952 Alpine Renault 1600")
   .whenNotMatchedThenInsert(
     PRODUCT.PRODUCT_NAME, PRODUCT.CODE)
    .values("1952 Alpine Renault 1600", 599302L)
   .execute();
```

The rendered SQL for the SQL Server dialect is as follows:

```
MERGE INTO [classicmodels].[dbo].[product] USING
  (SELECT 1 [one]) t ON
    [classicmodels].[dbo].[product].[product_name] = ?
WHEN MATCHED THEN
  UPDATE
  SET [classicmodels].[dbo].[product].[product_name] = ?
WHEN NOT MATCHED THEN
  INSERT ([product_name], [code])
  VALUES (?, ?);
```

Now, let's look at another example using the `WHEN MATCHED THEN DELETE` (jOOQ `whenMatchedThenDelete()`) and `WHEN NOT MATCHED THEN INSERT` (jOOQ `whenNotMatchedThenInsert()`) clauses:

```
ctx.mergeInto(SALE)
   .using(EMPLOYEE)
   .on(EMPLOYEE.EMPLOYEE_NUMBER.eq(SALE.EMPLOYEE_NUMBER))
   .whenMatchedThenDelete()
   .whenNotMatchedThenInsert(SALE.EMPLOYEE_NUMBER,
      SALE.FISCAL_YEAR, SALE.SALE_,
      SALE.FISCAL_MONTH, SALE.REVENUE_GROWTH)
   .values(EMPLOYEE.EMPLOYEE_NUMBER, val(2015),
      coalesce(val(-1.0).mul(EMPLOYEE.COMMISSION), val(0.0)),
      val(1), val(0.0))
   .execute();
```

This works flawlessly in SQL Server, but it doesn't work in Oracle because Oracle doesn't support the `WHEN MATCHED THEN DELETE` clause. But, we can easily obtain the same result by combining `WHEN MATCHED THEN UPDATE` with `DELETE WHERE` (obtained via the jOOQ `thenDelete()`) clause. This works because, in Oracle, you can add a `DELETE WHERE` clause, but only together with an `UPDATE`:

```
ctx.mergeInto(SALE)
   .using(EMPLOYEE)
   .on(EMPLOYEE.EMPLOYEE_NUMBER.eq(SALE.EMPLOYEE_NUMBER))
// .whenMatchedThenDelete() - not supported by Oracle
   .whenMatchedAnd(selectOne().asField().eq(1))
   .thenDelete()
   .whenNotMatchedThenInsert(SALE.EMPLOYEE_NUMBER,
      SALE.FISCAL_YEAR, SALE.SALE_,
      SALE.FISCAL_MONTH, SALE.REVENUE_GROWTH)
   .values(EMPLOYEE.EMPLOYEE_NUMBER, val(2015),
      coalesce(val(-1.0).mul(EMPLOYEE.COMMISSION), val(0.0)),
      val(1), val(0.0))
   .execute();
```

`WHEN MATCHED THEN UPDATE` is obtained via jOOQ's `whenMatchedAnd()`; this is the jOOQ implementation for the `WHEN MATCHED AND <some predicate> THEN` clause, but in this case, it is rendered as `WHEN MATCHED THEN UPDATE`.

Using the `DELETE WHERE` clause in SQL Server and in Oracle works the same. An important aspect of using the `DELETE WHERE` clause consists of which table the `DELETE WHERE` clause references. This clause can target the rows before or after an update. The following `MERGE` example updates all the rows in the *target table* that have a matching row in the *source table*. The `DELETE WHERE` clause deletes only those rows that were matched by `UPDATE` (this is `DELETE` after `UPDATE`):

```
ctx.mergeInto(SALE)
   .using(EMPLOYEE)
   .on(SALE.EMPLOYEE_NUMBER.eq(EMPLOYEE.EMPLOYEE_NUMBER))
   .whenMatchedThenUpdate()
   .set(SALE.SALE_, coalesce(SALE.SALE_
        .minus(EMPLOYEE.COMMISSION), SALE.SALE_))
   .deleteWhere(SALE.SALE_.lt(1000.0))
   .execute();
```

The following example shows that `DELETE WHERE` can match against values of the rows before `UPDATE` as well. This time, `DELETE WHERE` references the *source table*, so the status is checked against the source not against the result of `UPDATE` (this is `DELETE` before `UPDATE`):

```
ctx.mergeInto(SALE)
   .using(EMPLOYEE)
   .on(SALE.EMPLOYEE_NUMBER.eq(EMPLOYEE.EMPLOYEE_NUMBER))
   .whenMatchedThenUpdate()
   .set(SALE.SALE_, coalesce(SALE.SALE_
       .minus(EMPLOYEE.COMMISSION), SALE.SALE_))
   .deleteWhere(EMPLOYEE.COMMISSION.lt(1000))
   .execute();
```

In the bundled code, you can practice more examples. The application is named *MergeSamples*.

Summary

This chapter is a comprehensive resource for examples of expressing popular `SELECT`, `INSERT`, `UPDATE`, `DELETE`, and `MERGE` statements in the jOOQ DSL syntax relying on the Java-based schema.

For brevity, we couldn't list all the examples here, but I strongly recommend you take each application and practice the examples against your favorite database. The main goal is to get you familiar with the jOOQ syntax and to become capable of expressing any plain SQL via the jOOQ API in a productive amount of time.

In the next chapter, we continue this adventure with a very exciting topic: expressing `JOIN` in jOOQ.

6
Tackling Different Kinds of JOINs

The SQL JOIN clause represents one of the most used SQL features. From the well-known INNER and OUTER JOIN clauses, the fictional Semi and Anti Join, to the fancy LATERAL join, this chapter is a comprehensive set of examples meant to help you practice a wide range of JOIN clauses via the jOOQ DSL API.

The topics of this chapter include the following:

- Practicing the most popular types of JOINs (CROSS, INNER, and OUTER)
- The SQL USING and jOOQ onKey() shortcuts
- Practicing more types of JOINs (Implicit, Self, NATURAL, STRAIGHT, Semi, Anti, and LATERAL)

Let's get started!

Technical requirements

The code for this chapter can be found on GitHub at https://github.com/PacktPublishing/jOOQ-Masterclass/tree/master/Chapter06.

Practicing the most popular types of JOINs

By most popular types of JOIN statements we are referring to CROSS JOIN, INNER JOIN, LEFT JOIN, RIGHT JOIN, and FULL JOIN. Let's tackle each of them via the jOOQ DSL API, starting with the most basic type of JOIN.

CROSS JOIN

CROSS JOIN is the most basic type of JOIN that gets materialized in a Cartesian product. Having two tables, A and B, the CROSS JOIN operation between them is represented as A x B, and practically, it means the combination of every row from A with every row from B.

In jOOQ, CROSS JOIN can be rendered by enlisting the tables in the FROM clause (non-ANSI JOIN syntax) or via the crossJoin() method that renders the CROSS JOIN keywords (ANSI JOIN syntax). Here is the first case – let's CROSS JOIN the OFFICE and DEPARTMENT tables:

```
ctx.select().from(OFFICE, DEPARTMENT).fetch();
```

Since this query doesn't expose explicitly or clearly, its intention of using CROSS JOIN is not as friendly as the following one, which uses the jOOQ crossJoin() method:

```
ctx.select().from(OFFICE).crossJoin(DEPARTMENT).fetch();
```

Using the crossJoin() method renders the CROSS JOIN keywords (ANSI JOIN syntax), which clearly communicate our intentions and remove any potential confusion:

```
SELECT `classicmodels`.`office`.`office_code`,
       `classicmodels`.`office`.`city`,
       ...
       `classicmodels`.`department`.`department_id`,
       `classicmodels`.`department`.`name`,
       ...
FROM `classicmodels`.`office`
CROSS JOIN `classicmodels`.`department`
```

Since some offices have NULL values for CITY and/or COUNTRY columns, we can easily exclude them from the OFFICE x DEPARTMENT via a predicate. Moreover, just for fun, we may prefer to concatenate the results as *city, country: department* (for example, *San Francisco, USA: Advertising*):

```
ctx.select(concat(OFFICE.CITY, inline(", "), OFFICE.COUNTRY,
           inline(": "), DEPARTMENT.NAME).as("offices"))
    .from(OFFICE).crossJoin(DEPARTMENT)
```

```
    .where(row(OFFICE.CITY, OFFICE.COUNTRY).isNotNull())
    .fetch();
```

Basically, once we've added a predicate, this becomes `INNER JOIN`, as discussed in the following section. More examples are available in the bundled code as *CrossJoin*.

INNER JOIN

`INNER JOIN` (or simply `JOIN`) represents a Cartesian product filtered by some predicate commonly placed in the `ON` clause. So, with the `A` and `B` tables, `INNERJOIN` returns the rows of `A x B` that validate the specified predicate.

In jOOQ, we render `INNER JOIN` via `innerJoin()` (or simply `join()`, if omitting `INNER` is supported by your database vendor) and the `on()` methods. Here is an example that applies `INNER JOIN` between `EMPLOYEE` and `OFFICE` to fetch employee names and the cities of their offices:

```
ctx.select(EMPLOYEE.FIRST_NAME,
           EMPLOYEE.LAST_NAME, OFFICE.CITY)
   .from(EMPLOYEE)
   .innerJoin(OFFICE)
   .on(EMPLOYEE.OFFICE_CODE.eq(OFFICE.OFFICE_CODE))
   .fetch();
```

The rendered SQL for the MySQL dialect is as follows:

```
SELECT `classicmodels`.`employee`.`first_name`,
       `classicmodels`.`employee`.`last_name`,
       `classicmodels`.`office`.`city`
FROM `classicmodels`.`employee`
JOIN `classicmodels`.`office` ON
   `classicmodels`.`employee`.'office_code'
      = `classicmodels`.`office`.`office_code`
```

By default, jOOQ doesn't render the optional `INNER` keyword. But, you can alter this default via the `withRenderOptionalInnerKeyword()` setting and the argument `RenderOptionalKeyword.ON`.

In jOOQ, chaining multiple `JOIN`s is quite easy. For example, fetching the managers and their offices requires two `INNER JOIN` clauses, since between `MANAGER` and `OFFICE`, we have a many-to-many relationship mapped by the `MANAGER_HAS_OFFICE` junction table:

```
ctx.select()
   .from(MANAGER)
```

```
    .innerJoin(OFFICE_HAS_MANAGER)
       .on(MANAGER.MANAGER_ID
          .eq(OFFICE_HAS_MANAGER.MANAGERS_MANAGER_ID))
    .innerJoin(OFFICE)
       .on(OFFICE.OFFICE_CODE
          .eq(OFFICE_HAS_MANAGER.OFFICES_OFFICE_CODE))
    .fetch();
```

In these examples, we called the jOOQ join method on `org.jooq.SelectFromStep` and the rendered SQL for PostgreSQL dialect is:

```
...
FROM
   "public"."manager"
JOIN "public"."office_has_manager"
ON "public"."manager"."manager_id" =
   "public"."office_has_manager"."managers_manager_id"
JOIN "public"."office"
ON "public"."office"."office_code" =
   "public"."office_has_manager"."offices_office_code"
```

But, for convenience, we can call the join method directly after the `FROM` clause on `org.jooq.Table`. In such case, we obtain a nested fluent code as below (feel free to use the approach that you find most convenient):

```
ctx.select()
    .from(MANAGER
       .innerJoin(OFFICE_HAS_MANAGER
       .innerJoin(OFFICE)
          .on(OFFICE.OFFICE_CODE.eq(
              OFFICE_HAS_MANAGER.OFFICES_OFFICE_CODE)))
       .on(MANAGER.MANAGER_ID.eq(
           OFFICE_HAS_MANAGER.MANAGERS_MANAGER_ID)))
    .fetch();
```

The rendered SQL for the PostgreSQL dialect is as follows:

```
...
FROM
   "public"."manager"
JOIN
(
```

```
    "public"."office_has_manager"
     JOIN "public"."office"
     ON "public"."office"."office_code" =
          "public"."office_has_manager"."offices_office_code"
) ON "public"."manager"."manager_id" =
      "public"."office_has_manager"."managers_manager_id"
```

Next, let's talk about OUTER JOIN.

OUTER JOIN

While INNER JOIN returns only the combinations that pass the ON predicate, OUTER JOIN will also fetch rows that have no match on the left-hand side (LEFT [OUTER] JOIN) or right-hand side (RIGHT [OUTER] JOIN) of the join operation. Of course, we have to mention here FULL [OUTER] JOIN as well. This fetches all rows from both sides of the join operation.

The jOOQ API renders OUTER JOIN via leftOuterJoin(), rightOuterJoin(), and fullOuterJoin(). Since the OUTER keyword is optional, we can omit it via the analogs, leftJoin(), rightJoin(), and fullJoin().

For example, let's fetch all employees (on the left-hand side) and their sales (on the right-hand side). By using LEFT [OUTER] JOIN, we retain all employees, even if they have no sales:

```
ctx.select(EMPLOYEE.FIRST_NAME,
           EMPLOYEE.LAST_NAME, SALE.SALE_)
   .from(EMPLOYEE)
   .leftOuterJoin(SALE)
   .on(EMPLOYEE.EMPLOYEE_NUMBER.eq(SALE.EMPLOYEE_NUMBER))
   .fetch();
```

If we want to retain only the employees that have no sales, then we can rely on an exclusive LEFT [OUTER] JOIN by adding a WHERE clause that excludes all matches:

```
ctx.select(EMPLOYEE.FIRST_NAME,
           EMPLOYEE.LAST_NAME, SALE.SALE_)
   .from(EMPLOYEE)
   .leftOuterJoin(SALE)
   .on(EMPLOYEE.EMPLOYEE_NUMBER.eq(SALE.EMPLOYEE_NUMBER))
   .where(SALE.EMPLOYEE_NUMBER.isNull())
   .fetch();
```

The rendered SQL for the SQL Server dialect is as follows:

```
SELECT
  [classicmodels].[dbo].[employee].[first_name],
  [classicmodels].[dbo].[employee].[last_name],
  [classicmodels].[dbo].[sale].[sale]
FROM
  [classicmodels].[dbo].[employee]
LEFT OUTER JOIN
  [classicmodels].[dbo].[sale]
ON [classicmodels].[dbo].[employee].[employee_number] =
   [classicmodels].[dbo].[sale].[employee_number]
WHERE [classicmodels].[dbo].[sale].[employee_number] IS NULL
```

If you prefer to use the Oracle (+) symbol shorthand for performing `OUTER JOIN` then check this example of an `LEFT [OUTER] JOIN`:

```
ctx.select(EMPLOYEE.FIRST_NAME,
           EMPLOYEE.LAST_NAME, SALE.SALE_)
   .from(EMPLOYEE, SALE)
   .where(SALE.EMPLOYEE_NUMBER.plus()
       .eq(EMPLOYEE.EMPLOYEE_NUMBER))
   .fetch();
```

And, the Oracle SQL is:

```
SELECT
  "CLASSICMODELS"."EMPLOYEE"."FIRST_NAME",
  "CLASSICMODELS"."EMPLOYEE"."LAST_NAME",
  "CLASSICMODELS"."SALE"."SALE"
FROM
  "CLASSICMODELS"."EMPLOYEE",
  "CLASSICMODELS"."SALE"
WHERE
  "CLASSICMODELS"."SALE"."EMPLOYEE_NUMBER"(+) =
     "CLASSICMODELS"."EMPLOYEE"."EMPLOYEE_NUMBER"
```

By default, jOOQ render the optional `OUTER` keyword for both, `leftOuterJoin()` and `leftJoin()`. Alter this default via the `withRenderOptionalOuterKeyword()` setting and the argument `RenderOptionalKeyword.ON`.

In the bundled code, you can practice more examples, including `RIGHT/FULL [OUTER] JOIN`. For MySQL, which doesn't support `FULL [OUTER] JOIN`, we wrote some emulation code based on the `UNION` clause.

> **Important Note**
>
> A special case of `OUTER JOIN` is represented by Oracle's partitioned `OUTER JOIN`.

PARTITIONED OUTER JOIN

A special case of `OUTER JOIN` is represented by the Oracle's partitioned `OUTER JOIN`. Such a join represents an extension of the classical `OUTER JOIN` syntax and is applied to each logical partition defined via an expression in the `PARTITION BY` clause. A partitioned `OUTER JOIN` returns a `UNION` of the outer joins of each of the partitions in the partitioned table (logically partitions) with the table on the other side of the join.

Partitioned outer joins are specific to Oracle and they allow us to do the same "densifying" (fill gaps in sparse data) of data using a quite convenient syntax and an efficient Execution Plan.

A classical scenario where the Oracle's partitioned `OUTER JOIN` can be used sounds like this: write a query returning the sales of every employee (*Sales Representative*) in every fiscal year while taking into account that some employees had no sales in some years - fill gaps in sparse data with 0. For instance, if we try to see the sales of all employees (*Sales Representative*) grouped by fiscal year via a trivial `JOIN` then we obtain some gaps in data as in the following figure:

```
+-----------+----------+---------+----------+
|FISCAL_YEAR|FIRST_NAME|LAST_NAME|  SALES   |
+-----------+----------+---------+----------+
|       2003|Andy      |Fixter   |  8154.18 |
|       2003|Barry     |Jones    |  7063.18 |
|       2003|Gerard    |Hernandez|  5282.64 |
|       2003|Leslie    |Thompson |  3213.0  |
...
```

a

```
+-----------+----------+---------+----------+
|FISCAL_YEAR|FIRST_NAME|LAST_NAME|  SALES   |
+-----------+----------+---------+----------+
|       2003|Andy      |Fixter   | 8154.18  |
|       2003|Barry     |Jones    | 7063.18  |
|       2003|Foon Yue  |Tseng    |     0.0  |
|       2003|George    |Vanauf   |     0.0  |
|       2003|Gerard    |Hernandez| 5282.64  |
|       2003|Julie     |Firrelli |     0.0  |
|       2003|Leslie    |Thompson |  3213.0  |
...
```

b

Figure 6.1 – Fill gaps in sparse data

In figure (a) is what we can easily get from a trivial `JOIN`, while in figure (b) is what we plan to get. So, we want to see all the *Sales Representative* even if they don't have sales in certain years. This is a job for Oracle partitioned `OUTER JOIN` where the logical partition is `FISCAL_YEAR`:

```
ctx.select(SALE.FISCAL_YEAR,
           EMPLOYEE.FIRST_NAME, EMPLOYEE.LAST_NAME,
           sum(nvl(SALE.SALE_, 0.0d)).as("SALES"))
   .from(EMPLOYEE)
   .leftOuterJoin(SALE).partitionBy(SALE.FISCAL_YEAR)
   .on(EMPLOYEE.EMPLOYEE_NUMBER.eq(SALE.EMPLOYEE_NUMBER))
   .where(EMPLOYEE.JOB_TITLE.eq("Sales Rep"))
   .groupBy(SALE.FISCAL_YEAR,
            EMPLOYEE.FIRST_NAME, EMPLOYEE.LAST_NAME)
   .orderBy(1, 2)
   .fetch();
```

Of course, you can express/emulate this query without partitioned `OUTER JOIN`, but for this you have to check out the application *PartitionedOuterJoin*.

The SQL USING and jOOQ onKey() shortcuts

So far, we've covered the typical `JOIN`s that are commonly used in daily work. Before we continue with more types of `JOIN`s, let's introduce two convenient shortcuts that are useful for expressing more concise `JOIN`s.

SQL JOIN ... USING

In certain cases, the SQL `JOIN ... USING` clause can be a convenient alternative to the classical `JOIN ... ON` clause. Instead of specifying a condition in the `JOIN ... ON` clause, we enlist the `JOIN ... USING` clause in the set of fields (columns) whose names are common to both tables – the left-hand side table and right-hand side table of a `JOIN` operation. In jOOQ, the `USING` clause is rendered via the `using()` method, as shown in the following example. The `EMPLOYEE_NUMBER` column mentioned in `using()` is the primary key of the `EMPLOYEE` table and the foreign key of the `SALE` table:

```
ctx.select(EMPLOYEE.FIRST_NAME, EMPLOYEE.LAST_NAME,
           SALE.SALE_)
   .from(EMPLOYEE)
   .innerJoin(SALE)
   .using(EMPLOYEE.EMPLOYEE_NUMBER)
   .fetch();
```

So, `using(EMPLOYEE.EMPLOYEE_NUMBER)` is a less verbose representation of `on(EMPLOYEE.EMPLOYEE_NUMBER.eq(SALE.EMPLOYEE_NUMBER)`, and the rendered SQL for the MySQL dialect is as follows:

```
SELECT `classicmodels`.`employee`.`first_name`,
       `classicmodels`.`employee`.`last_name`,
       `classicmodels`.`sale`.`sale`
FROM `classicmodels`.`employee`
JOIN `classicmodels`.`sale` USING (`employee_number`)
```

But we can use any other field(s). Here is the `USING` clause for a composite primary key:

```
...using(PRODUCTLINE.PRODUCT_LINE, PRODUCTLINE.CODE)
```

Alternatively, this is a `USING` clause for two fields that are not primary/foreign keys:

```
.using(OFFICE.CITY, OFFICE.COUNTRY)
```

Note that `using()` without arguments will render `ON TRUE`, so no filter is applied to the join operation. Practice the complete examples via the *JoinUsing* bundled application. Next, let's introduce a very handy tool from jOOQ named `onKey()`.

However, as I said, `USING` fits only for certain cases. Lukas Eder enforces this statement: "*The* `USING` *clause leads to a bit more difficult to maintain queries when queries get complex, so it's generally not recommended. It's less type-safe (in jOOQ). When you rename a column, your jOOQ code might still compile. It wouldn't if you had been using* `ON`*. When you add a column that accidentally matches a column referenced from* `USING`*, you might get unintended consequences in unrelated queries. Example,* `A JOIN B USING (X) JOIN C USING (Y)`*. This assumes* `A(X)`, `B(X, Y)`, `C(Y)`*. So, what happens if you add* `A(Y)`*? A runtime exception, because* `Y` *is now ambiguous. Or, even worse: What happens if you add* `A(Y)` *but remove* `B(Y)`*? No runtime exception, but possibly (and quietly) wrong query. Moreover, in Oracle, columns referenced from* `USING` *can no longer be qualified in the query. In conclusion,* `USING` *can be useful for quick and dirty ad-hoc querying, just like* `NATURAL`*. But I wouldn't use it in production queries. Especially, because implicit joins work much better in jOOQ.*

*The essence here is always the fact (and this is frequently misunderstood) that joins are *binary* operators between two tables. For instance,* `A JOIN B USING (X) JOIN C USING (Y)` *is just short for* `(A JOIN B USING (X)) JOIN C USING (Y)`*, so* `C` *is joined to* `(A JOIN B USING (X))` *not to B alone. This is also the case for* `onKey()`."

jOOQ onKey()

Whenever we join a well-known foreign key relationship, we can rely on the jOOQ `onKey()` method. Since this is quite easy to understand for a simple foreign key, let's pick up a composite foreign key containing two fields. Check out the following `ON` clause:

```
ctx.select(...)
   .from(PAYMENT)
   .innerJoin(BANK_TRANSACTION)
   .on(PAYMENT.CUSTOMER_NUMBER.eq(
      BANK_TRANSACTION.CUSTOMER_NUMBER)
   .and(PAYMENT.CHECK_NUMBER.eq(
      BANK_TRANSACTION.CHECK_NUMBER)))
```

The (`CUSTOMER_NUMBER`, `CHECK_NUMBER`) represents a composite foreign key in the `BANK_TRANSACTION` table. jOOQ allows us to replace this verbose `ON` clause with the `onKey()` method without arguments, as follows:

```
ctx.select(...)
   .from(PAYMENT)
   .innerJoin(BANK_TRANSACTION)
   .onKey()
   .fetch();
```

Really cool, isn't it? jOOQ infers the `ON` condition on our behalf, and the rendered SQL for MySQL is as follows:

```
SELECT ...
FROM `classicmodels`.`payment`
JOIN `classicmodels`.`bank_transaction`
ON (`classicmodels`.`bank_transaction`.`customer_number`
  = `classicmodels`.`payment`.`customer_number`
AND `classicmodels`.`bank_transaction`.`check_number`
  = `classicmodels`.`payment`.`check_number`)
```

In case of ambiguity caused by multiple keys' potential matches, we can also rely on foreign keys' field references via `onKey(TableField<?,?>... tfs)`, or the generated foreign keys' references via `onKey(ForeignKey<?,?> fk)`. For instance, in order to avoid the `DataAccessException`: *Key ambiguous between tables X and Y* exception, while joining table X with table Y via `onKey()`, we can explicitly indicate the foreign key that should be used as follows (here, via the SQL Server generated foreign key reference, `jooq.generated.Keys.PRODUCTLINEDETAIL_PRODUCTLINE_FK`):

```
ctx.select(...)
   .from(PRODUCTLINE)
   .innerJoin(PRODUCTLINEDETAIL)
   .onKey(PRODUCTLINEDETAIL_PRODUCTLINE_FK)
   .fetch();
```

This time, the rendered SQL is as follows:

```
SELECT ...
FROM [classicmodels].[dbo].[productline]
JOIN
     [classicmodels].[dbo].[productlinedetail]
  ON
   ([classicmodels].[dbo].[productlinedetail].[product_line] =
    [classicmodels].[dbo].[productline].[product_line]
  AND
    [classicmodels].[dbo].[productlinedetail].[code] =
    [classicmodels].[dbo].[productline].[code])
```

But despite its appeal, this method can lead into issues. As Lukas Eder shared here: "*The* `onKey()` *method is not type-safe, and can break in subtle ways, when tables are modified.*"

More examples are available in the application named *JoinOnKey*. For now, let's continue with more types of `JOIN`s.

Practicing more types of JOINs

Next, let's cover more `JOIN`s, such as Implicit/Self Joins, `NATURAL JOIN`, `STRAIGHT JOIN`, Semi/Anti Joins, and `LATERAL` Joins. Let's continue with Implicit/Self Joins.

Implicit and Self Join

Implicit and Self Joins can be easily expressed in jOOQ via type-safe navigation methods produced by the jOOQ generator in classes that mirror the database tables. Let's dissect this aspect of Implicit Joins.

Implicit Join

As an example, an explicit join that fetches a parent table's column from a given child table can be expressed as an Implicit Join. Here is the explicit join:

```
SELECT o.office_code, e.first_name, e.last_name
FROM employee AS e
JOIN office AS o ON e.office_code = o.office_code
```

Here is the less verbose Implicit Join version:

```
SELECT e.office.office_code, e.first_name, e.last_name
FROM employee AS e
```

If we check the generated Java-based schema, then we notice that the `jooq.generated.tables.Employee` class mirroring the `EMPLOYEE` table contains a method named `office()` especially for expressing this syntax. Here is the previous Implicit Join, written via the jOOQ DSL API:

```
ctx.select(EMPLOYEE.office().OFFICE_CODE,
           EMPLOYEE.FIRST_NAME, EMPLOYEE.LAST_NAME)
   .from(EMPLOYEE)
   .fetch();
```

Here is another example that chains several navigation methods to express an Implicit Join, starting from the `ORDERDETAIL` table:

```
ctx.select(
      ORDERDETAIL.order().customer().employee().OFFICE_CODE,
      ORDERDETAIL.order().customer().CUSTOMER_NAME,
      ORDERDETAIL.order().SHIPPED_DATE,
      ORDERDETAIL.order().STATUS,
      ORDERDETAIL.QUANTITY_ORDERED, ORDERDETAIL.PRICE_EACH)
   .from(ORDERDETAIL)
   .orderBy(ORDERDETAIL.order().customer().CUSTOMER_NAME)
   .fetch();
```

The names of these navigation methods correspond to the parent table name. Here is another example of writing an Implicit Join in a m:n relationship. If we think to an m:n relationship from the relationship table then we see two to-one relationships that we exploit as follows (between MANAGER and OFFICE there is a many-to-many relationship):

```
ctx.select(OFFICE_HAS_MANAGER.manager().fields())
   .from(OFFICE_HAS_MANAGER)
   .where(OFFICE_HAS_MANAGER.office().OFFICE_CODE.eq("6"))
   .fetch();
```

Notice that the Implicit Joins covered in this section are *foreign key path-based*. Most probably, you are also familiar with Implicit Joins where you enlist all the tables you want to fetch data from in the FROM clause followed by the WHERE clause having conditions based on primary/foreign keys values for filtering the result. Here is an example of jOOQ code for such an Implicit Join:

```
ctx.select(OFFICE.OFFICE_CODE,
           EMPLOYEE.FIRST_NAME, EMPLOYEE.LAST_NAME)
   .from(OFFICE, EMPLOYEE)
   .where(OFFICE.OFFICE_CODE.eq(EMPLOYEE.OFFICE_CODE))
   .orderBy(OFFICE.OFFICE_CODE)
   .fetch();
```

> **Note**
>
> Nevertheless, note that these kind of Implicit Joins are quite prone to human mistakes, and it is better to rely on the ANSI JOIN syntax by explicitly using the JOIN keyword. Let me take advantage of this context to say that whenever you have old code that should be updated to an ANSI JOIN, you can rely on jOOQ. Besides the jOOQ DSL API, you can check out https://www.jooq.org/translate, and for a quick and neat guide, read this article: https://blog.jooq.org/2020/11/17/automatically-transform-oracle-style-implicit-joins-to-ansi-join-using-jooq/.

In the absence of explicit foreign keys in the schema for whatever reasons (including the tables are actually views), users of the commercial editions can specify synthetic foreign keys to the Code Generator as you can see in *Chapter 11, jOOQ keys*.

Please, consider the jOOQ manual and https://github.com/jOOQ/jOOQ/issues/12037 for covering the limitations of Implicit Joins support. Leaving the context of Implicit Joins, the jOOQ navigation methods are useful for expressing Self Joins as well.

Self Join

Whenever a table is joined with itself, we can rely on Self Joins. Writing a Self Join is done via a navigation method that has the same name as the table itself. For example, here is a Self Join that fetches a result set containing the name of each employee and the name of their boss (`EMPLOYEE.REPORTS_TO`):

```
ctx.select(concat(EMPLOYEE.FIRST_NAME, inline(" "),
           EMPLOYEE.LAST_NAME).as("employee"),
           concat(EMPLOYEE.employee().FIRST_NAME, inline(" "),
           EMPLOYEE.employee().LAST_NAME).as("reports_to"))
   .from(EMPLOYEE)
   .fetch();
```

In the bundled code, *ImplicitAndSelfJoin*, you can practice more examples with implicit and Self Joins.

NATURAL JOIN

Earlier, we used the `JOIN ... USING` syntax by enlisting the fields whose names are common to both tables (the left and right tables of a join operation) and should be rendered in the condition of the `ON` clause. Alternatively, we can rely on `NATURAL JOIN`, which doesn't require any `JOIN` criteria. This leads to a minimalist syntax but also makes our query a sword with two edges.

Basically, `NATURAL JOIN` automatically identifies all the columns that share the same name from both joined tables and use them to define the `JOIN` criteria. This can be quite useful when the primary/foreign keys columns share the same names, as in the following example:

```
ctx.select().from(EMPLOYEE)
   .naturalJoin(SALE)
   .fetch();
```

The jOOQ API for `NATURAL JOIN` relies on the `naturalJoin()` method. Next to this method, we have the methods corresponding to `LEFT/RIGHT/FULL NATURAL OUTER JOIN` as `naturalLeftOuterJoin()`, `naturalRightOuterJoin()`, and `naturalFullOuterJoin()`. Also, you may like to read the article at `https://blog.jooq.org/2020/08/05/use-natural-full-join-to-compare-two-tables-in-sql/` about using `NATURAL FULL JOIN` to compare two tables. You can see all these at work in the bundled code.

For our example, the rendered SQL for the PostgreSQL dialect is as follows:

```
SELECT "public"."employee"."employee_number", ...
       "public"."sale"."sale_id", ...
FROM "public"."employee"
NATURAL JOIN "public"."sale"
```

The `EMPLOYEE` and `SALE` tables share a single column name, `EMPLOYEE_NUMBER` – the primary key in `EMPLOYEE` and the foreign key in `SALE`. This column is used behind the scenes by `NATURAL JOIN` for filtering the result, which is the expected behavior.

But, remember that `NATURAL JOIN` picks up all columns that share the same name, not only the primary/foreign key columns, therefore this `JOIN` may produce undesirable results. For instance, if we join the `PAYMENT` and `BANK_TRANSACTION` tables, then `NATURAL JOIN` will use the common composite key (`CUSTOMER_NUMBER`, `CHECK_NUMBER`) but will also use the `CACHING_DATE` column. If this is not our intention, then `NATURAL JOIN` is not the proper choice. Expecting that only the (`CUSTOMER_NUMBER`, `CHECK_NUMBER`) is used is a wrong assumption, and it is recommended to rely on the `ON` clause or the jOOQ `onKey()` method:

```
ctx.select()
   .from(PAYMENT.innerJoin(BANK_TRANSACTION).onKey())
   .fetch();
```

On the other hand, if we expect that only the `CACHING_DATE` column will be used (which is hard to believe), then the `USING` clause can be a good alternative:

```
ctx.select()
   .from(PAYMENT.innerJoin(BANK_TRANSACTION)
   .using(PAYMENT.CACHING_DATE))
   .fetch();
```

The `USING` clause is useful if we need any custom combination of columns that share the same name. On the other hand, `NATURAL JOIN` is considerably more prone to issues, since any schema changes that lead to a new matching column name will cause `NATURAL JOIN` to combine that new column as well.

It's also worth keeping in mind that Oracle doesn't accept that the columns used by `NATURAL JOIN` for filtering the result have qualifiers (*ORA-25155 – column used in NATURAL join cannot have qualifiers*). In this context, using the jOOQ Java-based schema with default settings comes with some issues. For instance, the expression `ctx.select().from(EMPLOYEE).naturalJoin(SALE)`... results in ORA-25155, since, by default, jOOQ qualifies the columns rendered in `SELECT`, including the common

EMPLOYEE_NUMBER column, which is used by NATURAL JOIN. A quick workaround consists of explicitly rendering * via asterisk() instead of the columns list:

```
ctx.select(asterisk())
   .from(PRODUCT)
   .naturalJoin(TOP3PRODUCT)
   .fetch();
```

Or, we can avoid using Java-based schema and write this:

```
ctx.select()
   .from(table("EMPLOYEE"))
   .naturalJoin(table("SALE"))
   .fetch()
```

Unqualified references to a common column are considered to belong to the left-hand side table if the join is INNER/LEFT OUTER JOIN, or to the right-hand side table if it is RIGHT OUTER JOIN.

Alternatively, the Oracle NATURAL JOIN is the same as the Oracle proprietary Equi Join with a join condition (an Equi Join relies on a join condition containing an equality operator).

As usual, you can practice all these examples and more in the bundled code. The application is named *NaturalJoin*. Next, let's tackle STRAIGHT JOIN.

STRAIGHT JOIN

Right from the start, we have to mention that STRAIGHT JOIN is specific to MySQL.

Basically, STRAIGHT JOIN instructs MySQL to always read the left-hand side table before the right-hand side table of JOIN. In this context, STRAIGHT JOIN may be useful to affect the execution plan chosen by MySQL for a certain JOIN. Whenever we consider that the query optimizer has put the JOIN tables in the wrong order, we can affect this order via STRAIGHT JOIN.

For instance, let's assume that the PRODUCT table has 5,000 rows, the ORDERDETAIL table has 200,000,000 rows, the ORDER table has 3,000 rows, and we have a join, as follows:

```
ctx.select(PRODUCT.PRODUCT_ID, ORDER.ORDER_ID)
   .from(PRODUCT)
   .innerJoin(ORDERDETAIL).on(
       ORDERDETAIL.PRODUCT_ID.eq(PRODUCT.PRODUCT_ID))
```

```
    .innerJoin(ORDER).on(
        ORDER.ORDER_ID.eq(ORDERDETAIL.ORDER_ID))
    .fetch();
```

Now, MySQL may or may not take into account the size of the intersection between `ORDER.ORDER_ID` and `ORDERDETAIL.ORDER_ID` versus `PRODUCT.PRODUCT_ID` and `ORDERDETAIL.PRODUCT_ID`. If the join between `ORDERDETAIL` and `ORDER` returns just as many rows as `ORDERDETAIL`, then this is not an optimal choice. And if starting the join with `PRODUCT` will filter down `ORDERDETAIL` to as many rows as `PRODUCT`, then this will be an optimal choice. This behavior can be enforced via the jOOQ `straightJoin()` method, which renders a `STRAIGHT JOIN` statement, as follows:

```
ctx.select(PRODUCT.PRODUCT_ID, ORDER.ORDER_ID)
    .from(PRODUCT)
    .straightJoin(ORDERDETAIL).on(
        ORDERDETAIL.PRODUCT_ID.eq(PRODUCT.PRODUCT_ID))
    .innerJoin(ORDER).on(
        ORDER.ORDER_ID.eq(ORDERDETAIL.ORDER_ID))
    .fetch();
```

In Oracle, the order of `JOINs` can be altered via `/*+LEADING(a, b)*/` hint. In jOOQ this kind of hints can be passed via `hint()`:

```
ctx.select(PRODUCT.PRODUCT_ID, ORDER.ORDER_ID)
    .hint("/*+LEADING(CLASSICMODELS.ORDERDETAIL
                      CLASSICMODELS.PRODUCT)*/")
    ... // joins come here
```

In SQL Server this can be accomplished via `OPTION (FORCE ORDER)`:

```
ctx.select(PRODUCT.PRODUCT_ID, ORDER.ORDER_ID)
    ... // joins come here
    .option("OPTION (FORCE ORDER)")
    .fetch();
```

Nevertheless, as Lukas Eder shared here: "*MySQL's problems should have been made significantly less severe since they added hash join support. In any case, I think a disclaimer about premature optimization using hints could be added. With reasonable optimizers, hints should almost never be necessary anymore.*"

You can see the rendered SQL by running the *StraightJoin* application available for MySQL. Next, let's cover Semi and Anti Joins.

Semi and Anti Joins

Semi and Anti Joins are two of the relational algebra operators that don't have a direct correspondent in SQL syntax. Apart from the case of using Cloudera Impala, which provides a native syntax for Semi/Anti Joins, we have to rely on workarounds. In this context, Semi Join can be emulated via `EXISTS/IN` and Anti Join via `NOT EXISTS/NOT IN` predicates.

Since Semi/Anti Joins can be emulated via `(NOT) EXISTS/(NOT) IN` predicates, it means that we don't really join the right-hand side. In the case of a Semi Join, we just fetch the rows from the first table (left-hand side table) where there are matches found in the second table (right-hand side table), while in the case of Anti Join, we do exactly the opposite of the Semi Join; we just fetch the rows from the first table (the left-hand side table) where there are no matches found in the second table (the right-hand side table).

For instance, let's fetch the names of all `EMPLOYEE` that have `CUSTOMER`. Accomplishing this via a Semi Join emulated via the `EXISTS` predicate can be done in SQL as follows:

```
SELECT employee.first_name, employee.last_name FROM employee
WHERE EXISTS
    (SELECT 1 FROM customer
     WHERE employee.employee_number
             = customer.sales_rep_employee_number);
```

In the bundled code, you can see how to express this SQL via the jOOQ DSL API. In addition, you can practice this use case emulated via the `IN` predicate. For now, let's use the jOOQ approach, which fills up the gap in expressiveness and enforces the clear intention of using a Semi Join via the `leftSemiJoin()` method. This jOOQ method saves us a lot of headaches – having neat code that is always emulated correctly in different SQL dialects and no brain-teasing in handling complex cases such as nesting `EXISTS/IN` predicates will make you fall in love with this method:

```
ctx.select(EMPLOYEE.FIRST_NAME, EMPLOYEE.LAST_NAME)
   .from(EMPLOYEE)
   .leftSemiJoin(CUSTOMER)
   .on(EMPLOYEE.EMPLOYEE_NUMBER.eq(
      CUSTOMER.SALES_REP_EMPLOYEE_NUMBER))
   .fetch();
```

This is just awesome! Check out the bundled code, *SemiAndAntiJoin*, to see more examples about chaining and/or nesting Semi Joins via the jOOQ DSL API. Every time, check out the rendered SQL and give a big thanks to jOOQ for it!

Next, let's focus on Anti Join. The Anti Join is the opposite of the Semi Join and is emulated via the `NOT EXISTS/NOT IN` predicates. For example, let's write an SQL representing an Anti Join to fetch the names of all `EMPLOYEE` that don't have `CUSTOMER` via `NOT EXISTS`:

```
SELECT employee.first_name, employee.last_name FROM employee
WHERE NOT (EXISTS
       (SELECT 1
       FROM customer
       WHERE employee.employee_number
               = customer.sales_rep_employee_number))
```

In the bundled code, you can see how to express this SQL via the jOOQ DSL API and the same example based on the `NOT IN` predicate. Nevertheless, I strongly encourage you to avoid `NOT IN` and opt for `NOT EXISTS`.

> **Important note**
> Most probably, you already know this, but just as a quick reminder, let's mention that the `EXISTS` and `IN` predicates are equivalent, but the `NOT EXISTS` and `NOT IN` predicates are not because the `NULL` values (if any) lead to undesirable results. For more details, please read this short but essential article: `https://blog.jooq.org/2012/01/27/sql-incompatibilities-not-in-and-null-values/`.

Alternatively, and even better, use the jOOQ Anti Join represented by the `leftAntiJoin()` method:

```
ctx.select(EMPLOYEE.FIRST_NAME, EMPLOYEE.LAST_NAME)
   .from(EMPLOYEE)
   .leftAntiJoin(CUSTOMER)
   .on(EMPLOYEE.EMPLOYEE_NUMBER.eq(
       CUSTOMER.SALES_REP_EMPLOYEE_NUMBER))
   .fetch();
```

Check out the rendered SQL and more examples in the application named *SemiAndAntiJoin*.

A typical problem solved by Anti Joins refers to relational division or simply division. This is another operator of relational algebra without a direct correspondent in SQL syntax. In short, division is the inverse of the `CROSS JOIN` operation.

For instance, let's consider the `ORDERDETAIL` and `TOP3PRODUCT` tables. While `CROSS JOIN` gives us the Cartesian product as `ORDERDETAIL x TOP3PRODUCT`, the division gives us `ORDERDETAIL ÷ TOP3PRODUCT` or `TOP3PRODUCT ÷ ORDERDETAIL`. Let's assume that we want the IDs of all orders that contain at least three products contained in TOP3PRODUCT. This kind of task is a division and is commonly solved via two nested Anti Joins. The jOOQ code that solves this problem is as follows:

```
ctx.select()
   .from(ctx.selectDistinct(ORDERDETAIL.ORDER_ID.as("OID"))
   .from(ORDERDETAIL).asTable("T1")
   .leftAntiJoin(TOP3PRODUCT
     .leftAntiJoin(ORDERDETAIL)
     .on(field("T", "OID")).eq(ORDERDETAIL.ORDER_ID)
       .and(TOP3PRODUCT.PRODUCT_ID
         .eq(ORDERDETAIL.PRODUCT_ID))))
   .on(trueCondition()))
  .fetch();
```

This is cool and much less verbose than writing the same thing via NOT EXISTS. But that's not all! jOOQ comes with an even more elegant solution that can be used to express divisions. This solution uses the divideBy() and returning() methods to express a division in a concise, expressive, and very intuitive way. Check out the following code that can replace the previous code:

```
ctx.select().from(ORDERDETAIL
   .divideBy(TOP3PRODUCT)
   .on(field(TOP3PRODUCT.PRODUCT_ID).eq(
            ORDERDETAIL.PRODUCT_ID))
   .returning(ORDERDETAIL.ORDER_ID))
   .fetch();
```

Check out this example and another one about finding the orders that contain at least the products of a given order in the *BootAntiJoinDivision* application.

As Lukas Eder pointed out here: "*If you want to see how x is the inverse of ÷, you can choose two different tables, for instance A x B = C and C ÷ B = A*".

Next, let's cover the LATERAL/APPLY Join.

LATERAL/APPLY Join

The last topic covered in this chapter refers to the LATERAL/APPLY Join. This is part of standard SQL and is quite similar to a correlated subquery that allows us to return more than one row and/or column or to the Java Stream.flatMap(). Mainly, a lateral inner subquery sits on the right-hand side of JOIN (INNER, OUTER, and so on), and it can be materialized as a classical subquery, a derived table, a function call, an array unnesting, and so on. Its power consists of the fact that it can refer to (or *laterally* access) tables/columns from the left-hand side to determine which rows to retain. A LATERAL Join iterates through each row on the left-hand side, evaluating the inner subquery (the right-hand side) for each row, like a typical for-each loop. The rows returned by the inner subquery are retained to the result of the join with the outer query. The LATERAL keyword is essential because, without it, each subquery is evaluated separately (independently) and can't access columns from the left-hand side (from the FROM clause).

For example, selecting all OFFICE that has DEPARTMENT can be done via the LATERAL Join:

```
ctx.select()
   .from(OFFICE, lateral(select().from(DEPARTMENT)
   .where(OFFICE.OFFICE_CODE.eq(
          DEPARTMENT.OFFICE_CODE))).as("t"))
   .fetch()
```

As you can see, the jOOQ DSL API provides the lateral() method for shaping LATERAL Joins. The SQL rendered for the MySQL dialect is as follows:

```
SELECT `classicmodels`.`office`.`office_code`,...
       `t`.`department_id`,
       ...
FROM `classicmodels`.`office`,
  LATERAL
   (SELECT `classicmodels`.`department`.`department_id`,...
    FROM `classicmodels`.`department`
    WHERE `classicmodels`.`office`.`office_code`
        = `classicmodels`.`department`.`office_code`) AS `t`
```

Without an explicit JOIN, you would expect that CROSS JOIN (INNER JOIN ON true / INNER JOIN IN 1=1) is automatically inferred. Writing the previous query via LEFT OUTER JOIN LATERAL requires a dummy ON true / ON 1=1 clause, as follows:

```
ctx.select()
   .from(OFFICE)
   .leftOuterJoin(lateral(select().from(DEPARTMENT)
      .where(OFFICE.OFFICE_CODE
          .eq(DEPARTMENT.OFFICE_CODE))).as("t"))
      .on(trueCondition())
   .fetch();
```

A LATERAL Join has several use cases where it fits like a glove. For instance, it can be used for lateral unnesting of the array columns, for finding TOP-N per *Foo* (joining TOP-N query to a normal table), and it works nicely in combination with the so-called table-valued functions.

Unnesting the array columns

If you are an Oracle or PostgreSQL fan, then you know about their support for nested arrays (or nested collections). In PostgreSQL, we can declare a column of type array exactly as any other type but suffixed with square brackets – [] (for example, text[]). Since Oracle recognizes only nominal array types, we have to create them first via CREATE TYPE. I will not insist on this pure SQL aspect, since our goal is to jump into jOOQ DSL API usage.

So, let's consider the DEPARTMENT table, which has an array column named TOPIC. For each department, we have a list of topics (area of activities), and more departments may have interleaved topics. For instance, for the Sale department, we have four topics – 'commerce', 'trade', 'sellout', and 'transaction'.

Now, let's assume that we want to fetch the departments that have in common the 'commerce' and 'business' topics. For this, we can write a LATERAL Join via the jOOQ DSL API using the lateral() method, and we unnest the array (transform the array into a useable/queryable table) via the unnest() method, as follows:

```
ctx.select()
   .from(DEPARTMENT, lateral(select(field(name("t", "topic")))
      .from(unnest(DEPARTMENT.TOPIC).as("t", "topic"))
      .where(field(name("t", "topic"))
         .in("commerce", "business"))).as("r"))
   .fetch();
```

For the PostgreSQL dialect, the rendered SQL is as follows:

```
SELECT
  "public"."department"."department_id",
  ...
  "public"."department"."accrued_liabilities",
  "r"."topic"
FROM
  "public"."department",
 LATERAL (SELECT
           "t"."topic"
```

```
          FROM
           unnest("public"."department"."topic")
              AS "t" ("topic")
          WHERE
           "t"."topic" IN (?, ?)) AS "r"
```

Note that MySQL and SQL Server don't have support for array (collection) columns, but we can still declare anonymously typed arrays that can be unnested via the same jOOQ `unnest()` method. Next, let's talk about solving TOP-N per *Foo* tasks.

Solving TOP-N per Foo

While solving TOP-N problems over the entire dataset can be quite challenging, solving TOP-N per *Foo* problems can be really hard to digest. Fortunately, the `LATERAL` Join fits perfectly for these kinds of problems. For instance, fetching TOP-3 sales per employee can be expressed in jOOQ as follows:

```
ctx.select(EMPLOYEE.EMPLOYEE_NUMBER, EMPLOYEE.FIRST_NAME,
           EMPLOYEE.LAST_NAME, field(name("t", "sales")))
   .from(EMPLOYEE,
         lateral(select(SALE.SALE_.as("sales"))
                    .from(SALE)
                    .where(EMPLOYEE.EMPLOYEE_NUMBER
                       .eq(SALE.EMPLOYEE_NUMBER))
                    .orderBy(SALE.SALE_.desc())
                    .limit(3).asTable("t")))
   .orderBy(EMPLOYEE.EMPLOYEE_NUMBER)
   .fetch();
```

The fact that the `LATERAL` Join allows us to access the `EMPLOYEE.EMPLOYEE_NUMBER` field/column does all the magic! The rendered SQL for MySQL dialect is as follows:

```
SELECT
  `classicmodels`.`employee`.`employee_number`,
  `classicmodels`.`employee`.`first_name`,
  `classicmodels`.`employee`.`last_name`,
  `t`.`sales`
FROM `classicmodels`.`employee`,
LATERAL (SELECT `classicmodels`.`sale`.`sale` as `sales`
         FROM `classicmodels`.`sale`
         WHERE `classicmodels`.`employee`.`employee_number`
```

```
            = `classicmodels`.`sale`.`employee_number`
        ORDER BY `classicmodels`.`sale`.`sale` desc limit ?)
  as `t`
ORDER BY `classicmodels`.`employee`.`employee_number`
```

If we think of the derived table obtained via the inner `SELECT` as a table-valued function that has the employee number as an argument, then, in Oracle, we can write this:

```
CREATE TYPE "TABLE_RES_OBJ" AS OBJECT (SALES FLOAT);
CREATE TYPE "TABLE_RES" AS TABLE OF TABLE_RES_OBJ;

CREATE OR REPLACE NONEDITIONABLE FUNCTION
 "TOP_THREE_SALES_PER_EMPLOYEE" ("employee_nr" IN NUMBER)
  RETURN TABLE_RES IS
    "table_result" TABLE_RES;
  BEGIN
    SELECT
        TABLE_RES_OBJ("SALE"."SALE") "sales"
    BULK COLLECT
    INTO "table_result"
    FROM
        "SALE"
    WHERE
        "employee_nr" = "SALE"."EMPLOYEE_NUMBER"
    ORDER BY
        "SALE"."SALE" DESC
    FETCH NEXT 3 ROWS ONLY;

    RETURN "table_result";
  END;
```

Next, we can use a `LATERAL` Join to call this function. The jOOQ code is as follows:

```
ctx.select(EMPLOYEE.EMPLOYEE_NUMBER, EMPLOYEE.FIRST_NAME,
           EMPLOYEE.LAST_NAME, field(name("T", "SALES")))
   .from(EMPLOYEE, lateral(select().from(
          TOP_THREE_SALES_PER_EMPLOYEE
             .call(EMPLOYEE.EMPLOYEE_NUMBER)).asTable("T")))
   .orderBy(EMPLOYEE.EMPLOYEE_NUMBER)
   .fetch();
```

The rendered SQL for Oracle is as follows:

```
SELECT
  "CLASSICMODELS"."EMPLOYEE"."EMPLOYEE_NUMBER",
  "CLASSICMODELS"."EMPLOYEE"."FIRST_NAME",
  "CLASSICMODELS"."EMPLOYEE"."LAST_NAME",
  "T"."SALES"
FROM "CLASSICMODELS"."EMPLOYEE",
     LATERAL (SELECT
       "TOP_THREE_SALES_PER_EMPLOYEE"."SALES"
     FROM
       table("CLASSICMODELS"."TOP_THREE_SALES_PER_EMPLOYEE"
            ("CLASSICMODELS"."EMPLOYEE"."EMPLOYEE_NUMBER"))
             "TOP_THREE_SALES_PER_EMPLOYEE") "T"
ORDER BY "CLASSICMODELS"."EMPLOYEE"."EMPLOYEE_NUMBER"
```

The `TOP_THREE_SALES_PER_EMPLOYEE` static field was generated by the jOOQ generator and is basically an ordinary table placed in the `jooq.generated.tables` package under the name `TopThreeSalesPerEmployee`. It can be used in the `FROM` clause of `SELECT` like any other table. Nevertheless, note that we have access to a method named `call()`, which is used for calling (with arguments) this table-valued function. However, while most databases treat table-valued functions as ordinary tables, in Oracle, it is quite common to treat them as standalone routines. In this context, jOOQ has a flag setting that allows us to indicate whether table-valued functions should be treated as ordinary tables (`true`) or as plain routines (`false`). Depending on this setting, jOOQ places the generated code in tables-section or the routines-section. This setting is set to `true` in all supported databases except Oracle. To enable this, we have to set the following:

```
Maven: <tableValuedFunctions>true</tableValuedFunctions>
Gradle: database { tableValuedFunctions = true }
```

Or, programmatic:

```
...withDatabase(new Database()
     .withTableValuedFunctions(true)
```

While the `LATERAL` keyword (which, by the way, is a pretty confusing word) can be used in MySQL, PostgreSQL, and Oracle, it cannot be used in SQL Server. Actually, SQL Server and Oracle have support for `CROSS APPLY` and `OUTER APPLY` via the `APPLY` keyword.

CROSS APPLY and OUTER APPLY

Specific to T-SQL, `CROSS APPLY`, and `OUTER APPLY` use the more suggestive `APPLY` keyword, which suggests that we apply a function to each table row. Mainly, `CROSS APPLY` is the same thing as `CROSS JOIN LATERAL`, and `OUTER APPLY` is the same thing as `LEFT OUTER JOIN LATERAL`. This is exactly how jOOQ will emulate `CROSS/OUTER APPLY` when they are not supported (for example, in PostgreSQL).

I guess everyone agrees with Lukas Eder's statement: "*I find APPLY much more intuitive, especially when cross applying a table valued function. T CROSS APPLY F (T.X) means we're applying F to each row in T and create the cross product between T and the result of F. On the other hand, LATERAL is so weird, syntactically, especially this stupid requirement of writing ON TRUE all the time.*"

At the beginning of this section, we wrote a `LATERAL` Join to select all `OFFICE` that has `DEPARTMENT`. Writing the same thing but using `CROSS APPLY` can be done via the jOOQ `crossApply()` method, like this:

```
ctx.select()
   .from(OFFICE).crossApply(select()
                 .from(DEPARTMENT)
                 .where(OFFICE.OFFICE_CODE
                   .eq(DEPARTMENT.OFFICE_CODE)).asTable("t"))
   .fetch();
```

The render SQL for SQL Server is as follows:

```
SELECT [classicmodels].[dbo].[office].[office_code], ...
       [t].[department_id], ...
FROM [classicmodels].[dbo].[office] CROSS APPLY
  (SELECT [classicmodels].[dbo].[department].[department_id],
          ...
   FROM [classicmodels].[dbo].[department]
   WHERE [classicmodels].[dbo].[office].[office_code]
       = [classicmodels].[dbo].[department].[office_code] ) [t]
```

Writing the previous query via `LEFT OUTER JOIN LATERAL` requires a dummy `ON true/1=1` clause, but using `OUTER APPLY` via the jOOQ `outerApply()` method eliminates this little inconvenience:

```
ctx.select()
   .from(OFFICE)
   .outerApply(select()
```

```
                  .from(DEPARTMENT)
                  .where(OFFICE.OFFICE_CODE
                     .eq(DEPARTMENT.OFFICE_CODE)).asTable("t"))
   .fetch();
```

And the SQL rendered for the SQL Server dialect is as follows:

```
SELECT [classicmodels].[dbo].[office].[office_code], ...
       [t].[department_id], ...
FROM [classicmodels].[dbo].[office] OUTER APPLY
  (SELECT [classicmodels].[dbo].[department].[department_id],
   ...
   FROM [classicmodels].[dbo].[department]
   WHERE [classicmodels].[dbo].[office].[office_code]
     = [classicmodels].[dbo].[department].[office_code] ) [t]
```

Done! In the bundled code, you can practice examples of `cross/outerApply()` and table-valued functions as well.

> **Important note**
>
> In the examples of this chapter, we have used `fooJoin(TableLike<?> table)` and `cross/outerApply(TableLike<?> table)`, but the jOOQ API also contains other flavors, such as `fooJoin(String sql)`, `cross/outerApply(String sql)`, `fooJoin(SQL sql)`, `cross/outerApply(SQL sql)`, `fooJoin(String sql, Object... bindings)`, `cross/outerApply(String sql, Object... bindings)`, `fooJoin(String sql, QueryPart... parts)`, and `cross/outerApply(String sql, QueryPart... parts)`. All of them are available in the jOOQ documentation and are marked with `@PlainSQL`. This annotation points out methods/types that allow us to produce a `QueryPart` that renders "plain SQL" inside of an AST, which are covered in *Chapter 16*, *Tackling Aliases and SQL Templating*.

All the examples (and more) from this chapter can be found in the *LateralJoin* application. Take your time to practice each example.

Summary

In this chapter, we have covered a comprehensive list of SQL `JOIN`s and how they can be expressed via the jOOQ DSL API. We started with the well-known `INNER/OUTER/CROSS JOIN`, continued with Implicit Joins, Self Joins, `NATURAL` and `STRAIGHT JOIN`s, and ended with Semi/Anti Joins, `CROSS APPLY`, `OUTER APPLY`, and `LATERAL` Joins. Also, among others, we covered the `USING` clause and the amazing jOOQ `onKey()` method.

In the next chapter, we tackle the jOOQ types, converters, and bindings.

7
Types, Converters, and Bindings

Data types, converters, and bindings represent major aspects of working with a database via a Java-based **Domain-Specific Language** (**DSL**) **Application Programming Interface** (**API**). Sooner or later, standard **Structured Query Language** (**SQL**)/**Java Database Connectivity** (**JDBC**) data types will not be enough, or the default mappings between Java types and JDBC types will raise some shortcomings in your specific scenarios. At that moment, you'll be interested in creating new data types, working with custom data types, type conversion, and type-binding capabilities of your DSL API. Fortunately, the **jOOQ Object Oriented Querying** (**jOOQ**) DSL provides versatile and easy-to-use APIs dedicated to the following agenda that represents the subject of this chapter:

- Default data type conversion
- Custom data types and type conversion
- Custom data types and type binding
- Manipulating enums
- Data type rewrites
- Handling embeddable types

Let's get started!

Technical requirements

The code for this chapter can be found on GitHub at `https://github.com/PacktPublishing/jOOQ-Masterclass/tree/master/Chapter07`.

Default data type conversion

One of the aspects of jOOQ that allows us to use it in a smooth manner is its *default data type conversion*. Most of the time, jOOQ hides from us the ceremony of converting between JDBC and Java types. For instance, have you wondered how the following explicit conversions work? Take a look:

```
Record1<Integer> fiscalYear = ctx.select(field("fiscal_year",
  Integer.class)).from(table("sale")).fetchAny();

// Offtake is a POJO
Offtake offtake = ctx.select(field("fiscal_year"),
  field("sale"), field("employee_number")).from(table("sale"))
  .fetchAnyInto(Offtake.class);
```

Both conversions are resolved via *default data type conversion* or *auto-conversions*. Behind the scenes, jOOQ relies on its own API that is capable of performing soft type-safe conversions for `Object` types, arrays, and collections.

You can check out this example in the *ConvertUtil* application.

Custom data types and type conversion

In jOOQ, the common interface for all dialect-specific data types is named `org.jooq.DataType<T>`, where `T` represents the Java type associated with an SQL data type. Each association of a `T` Java data type with an SQL data type (generic SQL types, called standard JDBC types) represented by `java.sql.Types` is present in jOOQ's `org.jooq.impl.SQLDataType` API. The jOOQ Code Generator automatically maps Java types to this `SQLDataType` API, which has an almost 1:1 matching to databases' data types for most dialects. Of course, we are not including here some of the vendor-specific data types, such as spatial data types, PostgreSQL's `INET`/`HSTORE`, nor other non-standard JDBC types (data types not explicitly supported by JDBC).

Roughly, any data type that is not associated in the jOOQ API with a standard JDBC type is considered and treated as a *custom data type*. However, as Lukas Eder mentions: "*There are some data types that I think *should* be standard JDBC types, but are not. They're also listed in* `SQLDataType`, *including:* `JSON`, `JSONB`, `UUID`, `BigInteger` *(!), unsigned numbers, intervals. These don't require custom data types.*"

Whenever your custom data type needs to be mapped onto a standard JDBC type—that is, an `org.jooq.impl.SQLDataType` type—you need to provide and explicitly specify an `org.jooq.Converter` implementation. This converter does the hard work of performing a conversion between the involved types.

> **Important Note**
> When we want to map a type onto a non-standard JDBC type (a type that is not in `org.jooq.impl.SQLDataType`), we need to focus on the `org.jooq.Binding` API, which is covered later. So, if this is your case, don't try to shoehorn your conversion logic onto a `Converter`. Just use a `Binding` (we'll see this later in this chapter).

Pay attention that attempting to insert values/data of a custom data type without passing through a converter may result in inserting `null` values in the database (as Lukas Eder shared: "*This null behavior is an old design flaw. A long time ago, I've not followed a fail-early strategy throwing exceptions*"), while trying to fetch data of custom data type without a converter may lead to `org.jooq.exception.DataTypeException`, *no converter found for types Foo and Buzz.*

Writing an org.jooq.Converter interface

`org.jooq.Converter` is an interface that represents a conversion between two types that are generically denoted as `<T>` and `<U>`. By `<T>`, we represent the database type, and by `<U>`, we represent the **User-Defined Type** (**UDT**) or the type used in the application. Converting from `<T>` to `<U>` is accomplished in a method named `U from(T)`, and converting from `<U>` to `<T>` is accomplished in a method named `T to(U)`.

If you find it hard to remember which direction is "`from()`" and which direction is "`to()`", then think that the former can be read as "*FROM the database to the client*" and the latter as "*from the client TO the database*". Also, pay attention to not confuse `T` and `U` because you risk spending hours staring at compilation errors in generated code.

In other words, via `U from(T)`, we convert from a database type to a UDT (for example, this is useful in `SELECT` statements), and via `T to(U)`, we convert from a UDT to a database type (for example, this is useful in `INSERT`, `UPDATE`, and `DELETE` statements). Moreover, a `T to(U)` direction is used wherever bind variables are used, so also in `SELECT` when writing predicates—for instance, `T.CONVERTED.eq(u)`. The stub of `org.jooq.Converter` is listed here:

```
public interface Converter<T, U> {

   U from(T databaseObject);   // convert to user-defined type
   T to(U userDefinedObject); // convert to database type

   // Class instances for each type
   Class<T> fromType();
   Class<U> toType();
}
```

jOOQ comes with an abstract implementation of this interface (`AbstractConverter`) and a few concrete extensions (converters) of this abstraction that you can explore here: `https://www.jooq.org/javadoc/latest/org.jooq/org/jooq/impl/AbstractConverter.html`. But as you'll see next, we can write our own converters.

If, for instance, you want to use Java 8's `java.time.YearMonth` type in the application but store it as an SQL `INTEGER` type in the database, you write a converter like this:

```
public class YearMonthConverter
             implements Converter<Integer, YearMonth> {

  @Override
  public YearMonth from(Integer t) {

    if (t != null) {
      return YearMonth.of(1970, 1)
               .with(ChronoField.PROLEPTIC_MONTH, t);
    }

    return null;
  }

  @Override
  public Integer to(YearMonth u) {
```

```
    if (u != null) {

      return (int) u.getLong(ChronoField.PROLEPTIC_MONTH);
    }

    return null;
  }

  @Override
  public Class<Integer> fromType() {
    return Integer.class;
  }

  @Override
  public Class<YearMonth> toType() {
    return YearMonth.class;
  }
}
```

Use this converter via `new YearMonthConverter()` or define a handy `static` type, like so:

```
public static final Converter<Integer, YearMonth>
  INTEGER_YEARMONTH_CONVERTER = new YearMonthConverter();
```

Moreover, using this converter for arrays can be done via the following `static` type, like so:

```
public static final Converter<Integer[], YearMonth[]>
  INTEGER_YEARMONTH_ARR_CONVERTER
    = INTEGER_YEARMONTH_CONVERTER.forArrays();
```

Once we have a converter, we can define a new data type. More precisely, we define our own `DataType` type programmatically by calling `asConvertedDataType(Converter)` or `asConvertedDataType(Binding)`. For example, here, we define a `YEARMONTH` data type that can be used as any other data type defined in `SQLDataType`:

```
public static final DataType<YearMonth> YEARMONTH
  = INTEGER.asConvertedDataType(INTEGER_YEARMONTH_CONVERTER);
```

Here, `INTEGER` is the `org.jooq.impl.SQLDataType.INTEGER` data type.

In the `CUSTOMER` table, we have a field named `FIRST_BUY_DATE` of type `INT`. When a customer makes their first purchase, we store the date (year-month) as an integer. For example, the date *2020-10* is stored as *24249* (we manually applied the `Integer to(YearMonth u)` method). Without a converter, we have to insert *24249* explicitly; otherwise, the code will not compile (for example, a type-safe `INSERT` statement will not compile) or we'll get an invalid insert (for example, a non-type-safe `INSERT` statement may store `null`). Relying on our converter, we can write the following type-safe `INSERT` statement:

```
ctx.insertInto(CUSTOMER, CUSTOMER.CUSTOMER_NAME, ... ,
               CUSTOMER.FIRST_BUY_DATE)
   .values("Atelier One", ...,
      INTEGER_YEARMONTH_CONVERTER.to(YearMonth.of(2020, 10)))
   .execute();
```

Next, fetching all the `FIRST_BUY_DATE` values of *Atelier One* without using the converter will result in an array or list of integers. To fetch an array/list of `YearMonth`, we can use the converter, as follows:

```
List<YearMonth> resultListYM
  = ctx.select(CUSTOMER.FIRST_BUY_DATE).from(CUSTOMER)
       .where(CUSTOMER.CUSTOMER_NAME.eq("Atelier One"))
       .fetch(CUSTOMER.FIRST_BUY_DATE,
              INTEGER_YEARMONTH_CONVERTER);
```

In the bundled code (*YearMonthConverter*, available for MySQL and PostgreSQL), you can see more examples, including the usage of the `YEARMONTH` data type for coercing and casting operations.

Writing a converter having its own class is useful when the converter is used sporadically across different places/classes. If you know that the converter is used only in a single class, then you can define it locally in that class via `Converter.of()`/`ofNullable()`, as follows (the difference between them consists of the fact that `Converter.ofNullable()` always returns `null` for `null` inputs):

```
Converter<Integer, YearMonth> converter =
  Converter.ofNullable(Integer.class, YearMonth.class,
    (Integer t) -> {
      return YearMonth.of(1970, 1)
             .with(ChronoField.PROLEPTIC_MONTH, t);
```

```
    },
    (YearMonth u) -> {
      return (int) u.getLong(ChronoField.PROLEPTIC_MONTH);
    }
);
```

Moreover, starting with jOOQ 3.15+, we can use a so-called *ad hoc converter*. This type of converter is very handy for attaching a converter to a certain column just for one query or a few local queries. For instance, having a converter (`INTEGER_YEARMONTH_CONVERTER`), we can use it for a single column, as follows:

```
ctx.insertInto(CUSTOMER, CUSTOMER.CUSTOMER_NAME, ...,
 CUSTOMER.FIRST_BUY_DATE.convert(INTEGER_YEARMONTH_CONVERTER))
   .values("Atelier One", ..., YearMonth.of(2020, 10))
   .execute();
```

For convenience, jOOQ provides—next to the ad hoc `convert()` function (allows you to turn a `Field<T>` type into a `Field<U>` type and vice versa)—`convertTo()` (allows you to turn a `Field<U>` type into a `Field<T>` type) and `convertFrom()` (allows you to turn a `Field<T>` type into a `Field<U>` type) ad hoc flavors. Since our `INSERT` statement cannot take advantage of both directions of the converter, we can revert to `convertTo()`, as follows:

```
ctx.insertInto(CUSTOMER, CUSTOMER.CUSTOMER_NAME, ...,
 CUSTOMER.FIRST_BUY_DATE.convertTo(YearMonth.class,
            u -> INTEGER_YEARMONTH_CONVERTER.to(u)))
   .values("Atelier One", ..., YearMonth.of(2020, 10))
   .execute();
```

Or, in the case of a `SELECT` statement, you may wish to use `converterFrom()`, as follows:

```
List<YearMonth> result = ctx.select(
  CUSTOMER.FIRST_BUY_DATE.convertFrom(
       t -> INTEGER_YEARMONTH_CONVERTER.from(t)))
  .from(CUSTOMER)
  .where(CUSTOMER.CUSTOMER_NAME.eq("Atelier One"))
  .fetchInto(YearMonth.class);
```

Of course, you don't even need to define the converter's workload in a separate class. You can simply inline it, as we've done here:

```
ctx.insertInto(CUSTOMER, ...,
    CUSTOMER.FIRST_BUY_DATE.convertTo(YearMonth.class,
      u -> (int) u.getLong(ChronoField.PROLEPTIC_MONTH)))
   .values(..., YearMonth.of(2020, 10)) ...;

List<YearMonth> result = ctx.select(
    CUSTOMER.FIRST_BUY_DATE.convertFrom(
      t -> YearMonth.of(1970, 1)
        .with(ChronoField.PROLEPTIC_MONTH, t)))
   .from(CUSTOMER) ...;
```

You can check the examples for ad hoc converters in *YearMonthAdHocConverter* for MySQL and PostgreSQL.

Going further, converters can be nested by nesting the calls of the `to()`/`from()` methods and can be chained via the `<X> Converter<T,X> andThen(Converter<? super U, X> converter)` method. Both nesting and chaining are exemplified in the bundled code (*YearMonthConverter*) by using a second converter that converts between `YearMonth` and `Date`, named `YEARMONTH_DATE_CONVERTER`.

Moreover, if you want to inverse a converter from `<T, U>` to `<U, T>`, then rely on the `Converter.inverse()` method. This can be useful when nesting/chaining converters that may require you to inverse `T` with `U` in order to obtain a proper match between data types. This is also exemplified in the bundled code.

The new data type can be defined based on `converter`, as follows:

```
DataType<YearMonth> YEARMONTH
   = INTEGER.asConvertedDataType(converter);
```

The new data type can be defined without an explicit `Converter` as well. Just use the `public default <U> DataType<U> asConvertedDataType(Class<U> toType, Function<? super T,? extends U> from, Function<? super U,? extends T> to)` flavor, as in the bundled code, and jOOQ will use behind the scenes `Converter.of(Class, Class, Function, Function)`.

On the other hand, if a converter is heavily used, then it is better to allow jOOQ to apply it automatically without an explicit call, as in the previous examples. To accomplish this, we need to perform the proper configurations of the jOOQ Code Generator.

Hooking forced types for converters

By using so-called *forced types* (`<forcedTypes/>`), we can instruct the jOOQ Code Generator to override the column data type. One way to accomplish this consists of mapping the column data type to a user-defined data type via `org.jooq.Converter`.

This configuration step relies on using the `<forcedTypes/>` tag, which is a child of the `<database/>` tag. Under the `<forcedTypes/>` tag, we can have one or multiple `<forcedType/>` tags, and each of these tags wraps a specific case of overriding the column's data types. Each such case is defined via several tags. First, we have the `<userType/>` and `<converter/>` tags, used to link the UDT and the proper `Converter`. Second, we have several tags used for identifying a certain column (or multiple columns) by name and/or type. While you can find all these tags described in the jOOQ manual (`https://www.jooq.org/doc/latest/manual/code-generation/codegen-advanced/codegen-config-database/codegen-database-forced-types/`), let's mention here two of the most used: `<includeExpression/>` and `<includeTypes/>`. `<includeExpression/>` contains a Java **regular expression (regex)** matching the fully qualified columns (or attributes/parameters), while `<includeTypes/>` contains a Java regex matching the data types that should be forced to have this type (the `<userType/>` type). In case of multiple regexes, use the pipe operator (`|`) to separate them, and if `<includeExpression/>` and `<includeTypes/>` are present in the same `<forcedType/>` tag, then keep in mind that they must match.

For instance, the `<forcedType/>` type for `YearMonthConverter` looks like this:

```
<forcedTypes>
  <forcedType>
    <!-- The Java type of the custom data type.
         This corresponds to the Converter's <U> type. -->
    <userType>java.time.YearMonth</userType>

    <!-- Associate that custom type with our converter. -->
    <converter>
      com.classicmodels.converter.YearMonthConverter
    </converter>

    <!-- Match the fully-qualified column. -->
    <includeExpression>
      classicmodels\.customer\.first_buy_date
    </includeExpression>
```

```
    <!-- Match the data type to be forced. -->
    <includeTypes>INT</includeTypes>
  </forcedType>
</forcedTypes>
```

Notice how we identified the `first_buy_date` column via an expression containing the schema, table, and column name. In other cases, you may wish to use less restrictive expressions; therefore, here are some popular examples:

```
<!-- All 'first_buy_date' fields in any 'customer' table,
     no matter the schema -->
.*\.customer\.first_buy_date

<!-- All 'first_buy_date' fields,
     no matter the schema and the table -->
.*\.first_buy_date

<!-- All fields containing 'first_buy_' -->
.*\.first_buy_.*

<!-- Case-insensitive expressions -->
(?i:.*\.customer\.first_buy_date)
(?i:classicmodels\.customer\.first_buy_date)
```

> **Important Note**
>
> Notice that all regexes in the jOOQ Code Generator match any of the following:
>
> 1) **Fully qualified object names** (**FQONs**)
>
> 2) Partially qualified object names
>
> 3) Unqualified object names
>
> So, instead of `.*\.customer\.first_buy_date`, you can also just write `customer\.first_buy_date`.
>
> Moreover, keep in mind that, by default, regexes are case-sensitive. This is important when you're using more than one dialect (for instance, Oracle **identifiers** (**IDs**) are *UPPER_CASE*, in PostgreSQL, they are *lower_case*, and in SQL Server, they are *PascalCase*).

Furthermore, matching any type is done via `<includeTypes>.*</includeTypes>`, while matching a certain type such as `NVARCHAR(4000)` is done via `NVARCHAR\(4000\)`, and a type such as `NUMBER(1, 0)` via `NUMBER\(1,\s*0\)`. A more verbose version of this example with detailed comments is available in the bundled code.

This time, the `FIRST_BUY_DATE` field is not mapped to `java.lang.Integer`. If we check the generated table class that mirrors the `CUSTOMER` table (`jooq.generated.tables.Customer`), then we see the following declaration:

```
public final TableField<CustomerRecord, YearMonth>
 FIRST_BUY_DATE = createField(DSL.name("first_buy_date"),
  SQLDataType.INTEGER, this, "", new YearMonthConverter());
```

So, `FIRST_BUY_DATE` is mapped to `YearMonth`, therefore our previous `INSERT` and `SELECT` statements will now look like this:

```
ctx.insertInto(CUSTOMER, CUSTOMER.CUSTOMER_NAME, ...,
               CUSTOMER.FIRST_BUY_DATE)
   .values("Atelier One", ..., YearMonth.of(2020, 10))
   .execute();
```

And the `SELECT` statement will then look like this:

```
List<YearMonth> ymList = ctx.select(CUSTOMER.FIRST_BUY_DATE)
   .from(CUSTOMER)
   .where(CUSTOMER.CUSTOMER_NAME.eq("Atelier One"))
   .fetch(CUSTOMER.FIRST_BUY_DATE);
```

jOOQ applies our converter automatically, so there's no need to call it explicitly. It even works when we perform a coercing operation of `ResultQuery<R1>` to `ResultQuery<R2>`, like so:

```
Result<Record2<String, YearMonth>> result = ctx.resultQuery(
  "SELECT customer_name, first_buy_date FROM customer")
  .coerce(CUSTOMER.CUSTOMER_NAME, CUSTOMER.FIRST_BUY_DATE)
.fetch();
```

In other words, jOOQ uses our converter automatically for binding variables and for fetching data from `java.util.ResultSet`. In queries, we just treat `FIRST_BUY_DATE` as of type `YEARMONTH`. The code is named *YearMonthConverterForcedTypes* and is available for MySQL.

Defining an inline converter via Converter.of() or Converter.ofNullable()

In the previous section, our converter was written as a Java class, and we referenced that class in the configuration of the jOOQ Code Generator. But instead of writing this class, we can associate the custom data type with an *inline converter*, which is a converter written directly into the configuration. For this, we use the `<converter/>` tag, as follows:

```
<forcedTypes>
 <forcedType>
  <userType>java.time.YearMonth</userType>
  ...
  <converter>
   <![CDATA[
    org.jooq.Converter.ofNullable(
       Integer.class, YearMonth.class,
     (Integer t) -> { return YearMonth.of(1970, 1).with(
      java.time.temporal.ChronoField.PROLEPTIC_MONTH, t); },
     (YearMonth u) -> { return (int) u.getLong(
      java.time.temporal.ChronoField.PROLEPTIC_MONTH); }
    )
   ]]>
  </converter>
  ...
 </forcedType>
</forcedTypes>
```

The usage part of this converter remains unchanged. The complete code is named *InlineYearMonthConverter*, and the programmatic version is named *ProgrammaticInlineYearMonthConverter*. Both applications are available for MySQL.

Defining an inline converter via lambda expressions

A more concise inline converter can be written via `<lambdaExpression/>`. This tag saves us from the explicit usage of `Converter.of()`/`Converter.ofNullable()` and allows us to simply specify a lambda expression that converts from the database type via the `<from/>` tag, and a lambda expression that converts to the database type via the `<to/>` tag. Let's exemplify this in our converter, as follows:

```
<forcedTypes>
 <forcedType>
```

```
  <userType>java.time.YearMonth</userType>
  ...
  <lambdaConverter>
   <from>
    <![CDATA[(Integer t) -> { return YearMonth.of(1970, 1)
    .with(java.time.temporal.ChronoField.PROLEPTIC_MONTH, t);
    }]]>
   </from>
   <to>
    <![CDATA[(YearMonth u) -> { return (int)
    u.getLong(java.time.temporal.ChronoField.PROLEPTIC_MONTH);
    }]]>
   </to>
  </lambdaConverter>
  ...
  </forcedType>
</forcedTypes>
```

Again, the usage part of this converter remains unchanged. The complete code is named *LambdaYearMonthConverter*, and the programmatic version is named *ProgrammaticLambdaYearMonthConverter*. Both applications are available for MySQL.

Matching forced types via SQL

In the previous sections, we matched the column names by using regexes in `<includeExpression/>` and `<includeTypes/>` tags. Whenever we need more complex criteria for matching column names, we can rely on the `<sql/>` tag. The body of this tag is an SQL query that executes against the dictionary views of our database. For instance, matching all columns of type `TIMESTAMP` from our MySQL `classicmodels` database can be achieved like so:

```
<sql>
 SELECT concat('classicmodels.', TABLE_NAME, '.', COLUMN_NAME)
 FROM INFORMATION_SCHEMA.COLUMNS
 WHERE TABLE_SCHEMA = 'classicmodels'
  AND TABLE_NAME != 'flyway_schema_history'
  AND DATA_TYPE = 'timestamp'
</sql>
```

This should return several columns, among them being two from the `PAYMENT` table and one from the `BANK_TRANSACTION` table: `PAYMENT.PAYMENT_DATE`, `PAYMENT.CACHING_DATE`, and `BANK_TRANSACTION.CACHING_DATE`. For these columns, jOOQ will apply `Converter<LocalDateTime, JsonNode>` developed in the bundled code. But these are not the only columns returned by our query, and jOOQ will apply this converter to `PAYMENT.MODIFIED` and `TOKEN.UPDATED_ON`, which are also of type `TIMESTAMP`. Now, we have two options to avoid this—we can tune our query predicate accordingly or we can quickly add `<excludeExpression/>`, as follows:

```
<excludeExpression>
    classicmodels\.payment\.modified
  | classicmodels\.token\.updated_on
</excludeExpression>
```

You can find the example for MySQL under the name *SqlMatchForcedTypes*.

I'm pretty sure that you got the idea, and you know how to write such queries for your favorite database.

JSON converters

Whenever jOOQ detects that the database uses **JavaScript Object Notation** (**JSON**) data (for instance, a MySQL/PostgreSQL `JSON` type), it maps the database type to the `org.jooq.JSON` class. This is a very handy class that represents a neat JSON wrapper type for JSON data fetched from the database. Its API consists of the `JSON.data()` method that returns a `String` representation of `org.jooq.JSON` and a `JSON.valueOf(String data)` method that returns `org.jooq.JSON` from the `String` representation. Typically, `org.jooq.JSON` is all you need, but if you want to manipulate the fetched JSON via dedicated APIs (Jackson, Gson, **JSON Binary** (**JSONB**), and so on), then you need a converter.

So, in order to practice more examples, the bundled code with this book comes with a JSON converter (`JsonConverter`), as explained in more detail next.

For MySQL and PostgreSQL, which have the `JSON` data type. The converter converts between `org.jooq.JSON` and `com.fasterxml.jackson.databind.JsonNode`, therefore it implements `Converter<JSON, JsonNode>`. Of course, you can use this as an example, and replace Jackson's `JsonNode` with `com.google.gson.Gson`, `javax/jakarta.json.bind.Jsonb`, and so on. The code available for MySQL and PostgreSQL is named *JsonConverterForcedTypes*. A programmatic version of this application is available only for MySQL (but you can easily adapt it for any other dialect) and is named *ProgrammaticJsonConverter*.

For Oracle 18c, which doesn't have a dedicated JSON type (however, this type is available starting with Oracle 21c; see `https://oracle-base.com/articles/21c/json-data-type-21c`), it's common to use `VARCHAR2(4000)` for relatively small JSON data and `BLOB` for large JSON data. In both cases, we can add a `CHECK ISJSON()` constraint to ensure the JSON data validity. Astonishingly, jOOQ detects that JSON data is present, and it maps such columns to the `org.jooq.JSON` type. Our converter converts between `org.jooq.JSON` and `com.fasterxml.jackson.databind.JsonNode`. Consider the applications named *ConverterJSONToJsonNodeForcedTypes*.

For SQL Server, which doesn't have a dedicated JSON type, it's common to use `NVARCHAR` with a `CHECK ISJSON()` constraint. jOOQ doesn't have support to detect the usage of JSON data (as in the case of Oracle) and maps this type to `String`. In this context, we have a converter in *JsonConverterVarcharToJSONForcedTypes* that converts between `NVARCHAR` and `org.jooq.JSON`, and one between `NVARCHAR` and `JsonNode` in *JsonConverterVarcharToJsonNodeForcedTypes*.

Take your time and practice these examples in order to get familiar with jOOQ converters. Next, let's tackle UDT converters.

UDT converters

As you know from *Chapter 3, jOOQ Core Concepts*, Oracle and PostgreSQL support UDTs, and we have a UDT in our schema named `EVALUATION_CRITERIA`. This UDT is the data type of the `MANAGER.MANAGER_EVALUATION` field, and in Oracle, it looks like this:

```
CREATE OR REPLACE TYPE "EVALUATION_CRITERIA" AS OBJECT (
  "communication_ability" NUMBER(7),
  "ethics" NUMBER(7),
  "performance" NUMBER(7),
  "employee_input" NUMBER(7),

  // the irrelevant part was skipped
);
```

We already know that the jOOQ Code Generator automatically maps the fields of the `evaluation_criteria` UDT via `jooq.generated.udt.EvaluationCriteria`, and the `jooq...pojos.EvaluationCriteria` **Plain Old Java Object** (**POJO**), respectively, maps the `jooq...udt.EvaluationCriteria.EvaluationCriteriaRecord` record.

But if we assume that our application needs to manipulate this type as JSON, then we need a converter that converts between `EvaluationCriteriaRecord` and JSON types (for instance, Jackson `JsonNode`). The `JsonConverter` stub looks like this:

```
public class JsonConverter implements
        Converter<EvaluationCriteriaRecord, JsonNode> {

   @Override
   public JsonNode from(EvaluationCriteriaRecord t) { ... }

   @Override
   public EvaluationCriteriaRecord to(JsonNode u) { ... }
   ...
}
```

Next, we configure this converter, as follows:

```
<forcedTypes>
 <forcedType>
  <userType>com.fasterxml.jackson.databind.JsonNode</userType>
  <converter>com...converter.JsonConverter</converter>
  <includeExpression>
   CLASSICMODELS\.MANAGER\.MANAGER_EVALUATION
  </includeExpression>
  <includeTypes>EVALUATION_CRITERIA</includeTypes>
 </forcedType>
</forcedTypes>
```

Having this set, we can express an `INSERT` statement, as follows:

```
JsonNode managerEvaluation = "{...}";

ctx.insertInto(MANAGER,
          MANAGER.MANAGER_NAME, MANAGER.MANAGER_EVALUATION)
   .values("Mark Joy", managerEvaluation)
   .execute();
```

And we can express a `SELECT` statement, as follows:

```
List<JsonNode> managerEvaluation = ctx.select(
      MANAGER.MANAGER_EVALUATION)
```

```
    .from(MANAGER)
    .fetch(MANAGER.MANAGER_EVALUATION);
```

The bundled code is named *ConverterUDTToJsonNodeForcedTypes* and is available for Oracle and PostgreSQL.

Custom data types and type binding

Roughly, when we want to map a type onto a non-standard JDBC type (a type that is not in `org.jooq.impl.SQLDataType`), we need to focus on the `org.jooq.Binding` API, as illustrated in the following code snippet:

```
public interface Binding<T, U> extends Serializable { ... }
```

For instance, binding the non-standard vendor-specific PostgreSQL `HSTORE` data type to some Java data type (for instance, `HSTORE` can be mapped quite conveniently to Java `Map<String, String>`) needs to take advantage of the `Binding` API, which contains the following methods (please read the comments):

```
// A converter that does the conversion between
// the database type T and the user type U
Converter<T, U> converter();

// A callback that generates the SQL string for bind values of
// this binding type. Typically, just ?, but also ?::json, ...
void sql(BindingSQLContext<U> ctx) throws SQLException;

// Register a type for JDBC CallableStatement OUT parameters
ResultSet void register(BindingRegisterContext<U> ctx)
          throws SQLException;

// Convert U to a type and set in on a JDBC PreparedStatement
void set(BindingSetStatementContext<U> ctx)
          throws SQLException;

// Get a type from JDBC ResultSet and convert it to U
void get(BindingGetResultSetContext<U> ctx)
          throws SQLException;
```

```
// Get a type from JDBC CallableStatement and convert it to U
void get(BindingGetStatementContext<U> ctx)
          throws SQLException;

// Get a value from JDBC SQLInput (useful for Oracle OBJECT)
void get(BindingGetSQLInputContext<U> ctx)
          throws SQLException;

// Get a value from JDBC SQLOutput (useful for Oracle OBJECT)
void set(BindingSetSQLOutputContext<U> ctx)
          throws SQLException;
```

For instance, let's consider that we already have an `org.jooq.Converter` implementation between `Map<String, String>` and `HSTORE` named `HstoreConverter`, and we continue by adding an `org.jooq.Binding` implementation named `HstoreBinding` that starts like this:

```
public class HstoreBinding implements
          Binding<Object, Map<String, String>> {

   private final HstoreConverter converter
      = new HstoreConverter();

   @Override
   public final Converter<Object, Map<String, String>>
                                                    converter() {
      return converter;
   }
   ...
}
```

On the other hand, for a MySQL vendor-specific `POINT` type, we may have a converter named `PointConverter`, and we need a `PointBinding` class as follows—the `POINT` type maps well to the Java `Point2D.Double` type:

```
public class PointBinding implements Binding<Object,Point2D> {

   private final PointConverter converter
      = new PointConverter();

   @Override
```

```
    public final Converter<Object, Point2D> converter() {
        return converter;
    }
    ...
}
```

Next, we focus on implementing the `Binding` SPI for PostgreSQL `HSTORE` and MySQL `POINT`. An important aspect of this is rendering a bind variable for the binding context's value and casting it to the `HSTORE` type. This is done in the `sql()` method, as follows:

```
@Override
public void sql(BindingSQLContext<Map<String, String>> ctx)
                                       throws SQLException {

    if (ctx.render().paramType() == ParamType.INLINED) {
        ctx.render().visit(inline(
            ctx.convert(converter()).value())).sql("::hstore");
    } else {
        ctx.render().sql("?::hstore");
    }
}
```

Notice that for the jOOQ inlined parameters (for details, check *Chapter 3, jOOQ Core Concepts*), we don't need to render a placeholder (`?`); therefore, we render only the PostgreSQL specific syntax, `::hstore`. Depending on the database-specific syntax, you have to render the expected SQL. For instance, for the PostgreSQL `INET` data type, you'll render `?::inet` (or, `::inet`), while for the MySQL `POINT` type, you'll render `ST_PointFromText(?)` as follows (`Point2D` is `java.awt.geom.Point2D`):

```
@Override
public void sql(BindingSQLContext<Point2D> ctx)
        throws SQLException {

   if (ctx.render().paramType() == ParamType.INLINED) {
       ctx.render().sql("ST_PointFromText(")
          .visit(inline(ctx.convert(converter()).value()))
          .sql(")");
   } else {
       ctx.render().sql("ST_PointFromText(?)");
   }
}
```

Next, we focus on registering a compatible/proper type for JDBC `CallableStatement` OUT parameters. Usually, `VARCHAR` is a proper choice (for instance, `VARCHAR` is a good choice for `HSTORE`, `INET`, or `JSON` types). The code is illustrated in the following snippet:

```
@Override
public void register(BindingRegisterContext
          <Map<String, String>> ctx) throws SQLException {
   ctx.statement().registerOutParameter(
      ctx.index(), Types.VARCHAR);
}
```

But since by default MySQL returns a `POINT` as binary data (as long as we don't use any MySQL function such as `ST_AsText(g)` or `ST_AsWKT(g)` for converting geometry values from an internal geometry format to a **Well-Known Text** (**WKT**) format), we can use `java.sql.Blob`, as illustrated in the following code snippet:

```
@Override
public void register(BindingRegisterContext<Point2D> ctx)
                                       throws SQLException {
   ctx.statement().registerOutParameter(
      ctx.index(), Types.BLOB);
}
```

Next, we convert `Map<String, String>` to a `String` value and set it on a JDBC `PreparedStatement` (for the MySQL `POINT` type, we convert `Point2D` to `String`), like so:

```
@Override
public void set(BindingSetStatementContext
            <Map<String, String>> ctx) throws SQLException {
   ctx.statement().setString(ctx.index(), Objects.toString(
     ctx.convert(converter()).value(), null));
}
```

Further, for PostgreSQL `HSTORE`, we get a `String` value from JDBC `ResultSet` and convert it to `Map<String, String>`, like so:

```
@Override
public void get(BindingGetResultSetContext
           <Map<String, String>> ctx) throws SQLException {
```

```
    ctx.convert(converter()).value(
      ctx.resultSet().getString(ctx.index()));
}
```

While for MySQL `POINT`, we get a `Blob` (or an `InputStream`) from JDBC `ResultSet` and convert it to `Point2D`, like so:

```
@Override
public void get(BindingGetResultSetContext<Point2D> ctx)
                                      throws SQLException {
   ctx.convert(converter()).value(ctx.resultSet()
      .getBlob(ctx.index())); // or, getBinaryStream()
}
```

Next, we do the same thing for JDBC `CallableStatement`. For the `HSTORE` type, we have the following:

```
@Override
public void get(BindingGetStatementContext
           <Map<String, String>> ctx) throws SQLException {
   ctx.convert(converter()).value(
      ctx.statement().getString(ctx.index()));
}
```

And for the `POINT` type, we have this:

```
@Override
public void get(BindingGetStatementContext<Point2D> ctx)
                                       throws SQLException {
   ctx.convert(converter()).value(
      ctx.statement().getBlob(ctx.index()));
}
```

Finally, we override the `get(BindingGetSQLInputContext<?> bgsqlc)` and `set(BindingSetSQLOutputContext<?> bsqlc)` methods. Since, we don't need them for `HSTORE`/`POINT`, we just throw an `SQLFeatureNotSupportedException` exception. For brevity, we skipped this code.

Once the `Binding` is ready, we have to configure it in the jOOQ Code Generator. This is quite similar to the configuration of a `Converter` only that, instead of using the `<converter/>` tag, we use the `<binding/>` tag as follows—here, we configure `HstoreBinding` (the configuration of `PointBinding` is available in the bundled code):

```
<forcedTypes>
  <forcedType>
    <userType>java.util.Map&lt;String, String&gt;</userType>
    <binding>com.classicmodels.binding.HstoreBinding</binding>
    <includeExpression>
      public\.product\.specs
    </includeExpression>
    <includeTypes>HSTORE</includeTypes>
  </forcedType>
</forcedTypes>
```

Now, we can test `HstoreBinding`. For instance, the `PRODUCT` table has a field named `SPECS` of type `HSTORE`. The following code inserts a new product with some specifications:

```
ctx.insertInto(PRODUCT, PRODUCT.PRODUCT_NAME,
               PRODUCT.PRODUCT_LINE, PRODUCT.SPECS)
   .values("2002 Masserati Levante", "Classic Cars",
      Map.of("Length (in)", "197", "Width (in)", "77.5",
             "Height (in)", "66.1", "Engine", "Twin Turbo
             Premium Unleaded V-6"))
   .execute();
```

Here's what the rendered SQL looks like:

```
INSERT INTO "public"."product" (
            "product_name", "product_line", "specs")
VALUES (?, ?, ?::hstore)
```

After resolving the `?` placeholders, the SQL looks like this:

```
INSERT INTO "public"."product" (
            "product_name", "product_line", "specs")
VALUES ('2002 Masserati Levante', 'Classic Cars',
         '"Width (in)"=>"77.5", "Length (in)"=>"197",
          "Height (in)"=>"66.1",
          "Engine"=>"Twin Turbo Premium Unleaded V-6"'::hstore)
```

At INSERT (UPDATE, DELETE, and so on), HstoreConverter converts from Java Map<String, String> to an HSTORE type. At SELECT, the same converter converts HSTORE to Map<String, String>. So, our SELECT statement could look like this:

```
List<Map<String, String>> specs = ctx.select(PRODUCT.SPECS)
   .from(PRODUCT)
   .where(PRODUCT.PRODUCT_NAME.eq("2002 Masserati Levante"))
   .fetch(PRODUCT.SPECS);
```

Notice that we don't use explicitly any Binding or Converter and we don't *touch* the HSTORE type. For us, in the application, SPECS is of the type Map<String, String>.

Notice that, starting with jOOQ 3.15, we have access to the *jOOQ-postgres-extensions* module (https://github.com/jOOQ/jOOQ/issues/5507), which supports HSTORE as well.

Bindings and converters can be used to write different helper methods. For instance, the following method can be used to convert any Param to its database data type:

```
static <T> Object convertToDatabaseType(Param<T> param) {
   return param.getBinding().converter().to(param.getValue());
}
```

But what's happening without Binding? Is everything lost?

Understanding what's happening without Binding

When jOOQ detects a non-standard JDBC type that doesn't have an associated Binding, it will mark the corresponding field with *@deprecated Unknown data type*, and with the message, *Please define an explicit {@link org.jooq.Binding} to specify how this type should be handled. Deprecation can be turned off using {@literal <deprecationOnUnknownTypes/>} in your Code Generator configuration.*

As a rule of thumb, relying on Bindings is the way to go, but as a workaround, we can also use explicit mapping for SELECT statements and public static <T> Field<T> field(String sql, Class<T> type, Object... bindings), or another field() flavor that fits better, for INSERT, UPDATE, and so on.

However, using a non-standard JDBC type in INSERT statements (UPDATE statements, and so on) just like that leads to jOOQ's SQLDialectNotSupportedException exception, *Type Foo is not supported in dialect Buzz, and in SELECT statements, to jOOQ's DataTypeException, No Converter found for types Foo and Buzz.*

You can check the HSTORE examples from this section in the application named *HstoreBinding*, and the POINT examples in the application named *PointGeometryBinding*.

In addition, the bundled code contains *InetBinding* for the PostgreSQL `INET` type, *JsonBinding* for the PostgreSQL `JSON` type, and *ProgrammaticInetBinding* representing the programmatic configuration of `Binding` for the PostgreSQL `INET` type. Next, let's discuss enums and how to convert these.

Manipulating enums

jOOQ represents an SQL enum type (for example, the MySQL `enum` or PostgreSQL `enum` data type created via `CREATE TYPE`) via an interface named `org.jooq.EnumType`. Whenever the jOOQ Java Code Generator detects the usage of an SQL enum type, it automatically generates a Java enum that implements `EnumType`. For instance, the MySQL schema of the `SALE` table contains the following `enum` data type:

```
'vat' ENUM ('NONE', 'MIN', 'MAX') DEFAULT NULL
```

For `vat`, the jOOQ generator renders the `jooq.generated.enums.VatType` enum, as follows:

```
public enum VatType implements EnumType {

    NONE("NONE"), MIN("MIN"), MAX("MAX");

    private final String literal;

    private VatType(String literal) {
        this.literal = literal;
    }

    @Override
    public Catalog getCatalog() {
        return null;
    }

    @Override
    public Schema getSchema() {
        return null;
    }

    @Override
    public String getName() {
        return "sale_vat";
```

```
    }

    @Override
    public String getLiteral() {
        return literal;
    }

    public static VatType lookupLiteral(String literal) {
        return EnumType.lookupLiteral(VatType.class, literal);
    }
}
```

By default, the name of such a class is composed of the table name and the column name in *PascalCase*, which means that the name of the preceding class should be `SaleVat`. But whenever we want to modify the default name, we can rely on jOOQ *generator strategies* and regexes, as we did in *Chapter 2, Customizing the jOOQ Level of Involvement*. For instance, we've customized the preceding class name as `VatType` via the following strategy:

```
<strategy>
  <matchers>
    <enums>
      <enum>
        <expression>sale_vat</expression>
        <enumClass>
          <expression>VatType</expression>
          <transform>AS_IS</transform>
        </enumClass>
      </enum>
    </enums>
  </matchers>
</strategy>
```

Having these pieces of knowledge is enough to start writing queries based on jOOQ-generated enums—for instance, an `INSERT` statement into the `SALE` table and a `SELECT` statement from it, as illustrated in the following code snippet:

```
import jooq.generated.enums.VatType;
...
ctx.insertInto(SALE, SALE.FISCAL_YEAR, ..., SALE.VAT)
   .values(2005, ..., VatType.MAX)
```

```
    .execute();

List<VatType> vats = ctx.select(SALE.VAT).from(SALE)
    .where(SALE.VAT.isNotNull())
    .fetch(SALE.VAT);
```

Of course, the `Sale`-generated POJO (or user-defined POJOs) and `SaleRecord` take advantage of `VatType`, as with any other type.

Writing enum converters

Whenever jOOQ-generated jOOQ enums are not enough, we focus on enum converters. Here is a non-exhaustive list of scenarios that may require some kind of enum converting to be done:

- Using your own Java enum for a database enum type
- Using your own Java enum for a database non-enum type (or enum-like type)
- Using a Java non-enum type for a database enum
- Always converting to a Java enum and occasionally to another Java enum

To simplify enum conversion tasks, jOOQ provides a built-in default converter named `org.jooq.impl.EnumConverter`. This converter can convert `VARCHAR` values to enum literals (and vice versa), or `NUMBER` values to enum ordinals (and vice versa). You can also instantiate it explicitly, as has been done here:

```
enum Size { S, M, XL, XXL; }
Converter<String, Size> converter
    = new EnumConverter<>(String.class, Size.class);
```

Next, let's tackle the previous list of enum scenarios.

Using your own Java enum for a database enum type

Of our four databases, only MySQL and PostgreSQL have dedicated types for enums. MySQL has the `enum` type and PostgreSQL has the `CREATE TYPE foo AS enum( ...)` syntax. In both cases, jOOQ generates enum classes on our behalf, but let's suppose that we'd prefer to use our own Java enums. For instance, let's focus on the MySQL schema of the `SALE` table, which contains these two enums:

```
`rate` ENUM ('SILVER', 'GOLD', 'PLATINUM') DEFAULT NULL
`vat` ENUM ('NONE', 'MIN', 'MAX') DEFAULT NULL
```

The same enums in PostgreSQL are declared like this:

```
CREATE TYPE rate_type AS enum('SILVER', 'GOLD', 'PLATINUM');
CREATE TYPE vat_type AS enum('NONE', 'MIN', 'MAX');
...
rate rate_type DEFAULT NULL,
vat vat_type DEFAULT NULL,
```

And let's assume that for `vat`, we still rely on a jOOQ-generated Java enum-class (as in the previous section, `VatType`), while for `rate`, we have written the following Java enum:

```
public enum RateType { SILVER, GOLD, PLATINUM }
```

In order to automatically map the `rate` column to the `RateType` enum, we rely on the `<forcedType/>` and `<enumConverter/>` flag tags, as illustrated here:

```
<forcedTypes>
  <forcedType>
    <userType>com.classicmodels.enums.RateType</userType>
    <enumConverter>true</enumConverter>
    <includeExpression>
      classicmodels\.sale\.rate # MySQL
      public\.sale\.rate        # PostgreSQL
    </includeExpression>
    <includeTypes>
      ENUM      # MySQL
      rate_type # PostgreSQL
    </includeTypes>
  </forcedType>
</forcedTypes>
```

By enabling `<enumConverter/>`, we instruct jOOQ to automatically apply the built-in `org.jooq.impl.EnumConverter` converter whenever `SALE.RATE` is used. Done! From this point forward, we can treat the `SALE.RATE` field as of type `RateType`, and jOOQ will handle the conversion aspects of the mapped field (listed here for MySQL), as follows:

```
public final TableField<SaleRecord, RateType> RATE
  = createField(DSL.name("rate"), SQLDataType.VARCHAR(8),
    this, "", new EnumConverter<String, RateType>
                              (String.class, RateType.class));
```

The application named *SimpleBuiltInEnumConverter* contains the complete example for MySQL and PostgreSQL.

This is a very convenient approach and works the same in MySQL and PostgreSQL, but if we don't employ this automatic conversion, we still can use our `RateType` Java enum manually or explicitly. Let's see how!

First, we configure the jOOQ Code Generator to exclude enum generation for the `sale_rate` (MySQL)/`rate_type` (PostgreSQL) types; otherwise, the `SALE.RATE` field will be automatically mapped to the generated Java enum. The code is illustrated in the following snippet:

```
<database>
   <excludes>
     sale_rate (MySQL) / rate_type (PostgreSQL)
   </excludes>
</database>
```

In this context, jOOQ maps `SALE.RATE` to `String` in MySQL, and to `Object` in PostgreSQL. In PostgreSQL, the field is annotated as *@deprecated Unknown data type*, but we turn off this deprecation via the `<deprecationOnUnknownTypes/>` configuration, as follows:

```
<generate>
  <deprecationOnUnknownTypes>false</deprecationOnUnknownTypes>
</generate>
```

Next, in MySQL, we can write an `INSERT` statement, as follows (`SALE.RATE` is of type `String`):

```
ctx.insertInto(SALE, SALE.FISCAL_YEAR, ..., SALE.RATE)
   .values(2005, ..., RateType.PLATINUM.name())
   .execute();
```

And we can write a `SELECT` statement, as follows:

```
List<RateType> rates = ctx.select(SALE.RATE)
   .from(SALE)
   .where(SALE.RATE.isNotNull())
   .fetch(SALE.RATE, RateType.class);
```

While for MySQL this is quite smooth, for PostgreSQL it's a little bit tricky. The PostgreSQL syntax requires us to render at `INSERT` something like `?::"public"."rate_type"`, as illustrated in the following code snippet:

```
ctx.insertInto(SALE, SALE.FISCAL_YEAR, ..., SALE.RATE)
   .values(2005, ..., field("?::\"public\".\"rate_type\"",
           RateType.PLATINUM.name()))
   .execute();
```

And at `SELECT`, we need an explicit coercing of `Object` to `String`, as illustrated in the following code snippet:

```
List<RateType> rates = ctx.select(SALE.RATE)
   .from(SALE)
   .where(SALE.RATE.isNotNull())
   .fetch(SALE.RATE.coerce(String.class), RateType.class);
```

The application named *MyEnumBuiltInEnumConverter* contains complete examples for MySQL and PostgreSQL. If we don't suppress the jOOQ enum generation, then another approach consists of writing an explicit converter (by extending the jOOQ built-in `org.jooq.impl.EnumConverter` converter) between the jOOQ generated enum and our enum. Of course, this converter must be called explicitly in your queries. You can find such an example for the `vat` enum in the application mentioned earlier.

Using your own Java enum for a database non-enum type (or enum-like type)

Let's consider a legacy database containing a column that takes only certain values but was declared as `VARCHAR` (or `NUMBER`)—for instance, the `SALE` table has a `TREND` field of type `VARCHAR` that takes only the values *UP*, *DOWN*, and *CONSTANT*. In this context, it would be more practical to enforce the usage of this field via an enum, as shown here:

```
public enum TrendType { UP, DOWN, CONSTANT }
```

But now, we have to handle the conversion between `TrendType` and `VARCHAR`. This can be done automatically by jOOQ if we add the following `<forcedType/>` tag (here, for Oracle):

```
<forcedType>
  <userType>com.classicmodels.enums.TrendType</userType>
  <enumConverter>true</enumConverter>
```

```
  <includeExpression>
    CLASSICMODELS\.SALE\.TREND
  </includeExpression>
  <includeTypes>VARCHAR2\(10\)</includeTypes>
</forcedType>
```

In the *SimpleBuiltInEnumConverter* application, you can see a complete example next to other examples for all four databases.

Since SQL Server and Oracle don't have an enum type, we have used an alternative. Among others, a common alternative relies on a `CHECK` constraint to obtain an enum-like behavior. These enum-like types can take advantage of `<enumConverter/>` exactly as shown previously. Here, it is the `SALE.VAT` field in Oracle:

```
vat VARCHAR2(10) DEFAULT NULL
  CHECK (vat IN('NONE', 'MIN', 'MAX'))
```

And here, it is the `<forcedType/>` tag:

```
<forcedType>
  <userType>com.classicmodels.enums.VatType</userType>

  <enumConverter>true</enumConverter>
  <includeExpression>
    CLASSICMODELS\.SALE\.VAT
  </includeExpression>
  <includeTypes>VARCHAR2\(10\)</includeTypes>
</forcedType>
```

If we don't want to rely on automatic conversion, then we can use an explicit converter, as follows:

```
public class SaleStrTrendConverter
           extends EnumConverter<String, TrendType> {

  public SaleStrTrendConverter() {
     super(String.class, TrendType.class);
  }
}
```

In the *BuiltInEnumConverter* application, you can find a complete example next to other examples for all four databases.

Using a Java non-enum type for a database enum

Sometimes, we need a non-enum type for a database enum. For instance, let's assume that we want to use some integers in place of the `VatType` enum (*0* for *NONE*, *5* for *MIN*, and *19* for *MAX*) because we might need these integers in different computations. Maybe the best idea is to write a converter that starts like this:

```
public class SaleVatIntConverter
   extends EnumConverter<VatType, Integer> { … }
```

But this doesn't work, because the `EnumConverter` signature is actually of type `EnumConverter<T,U extends Enum<U>>`. Obviously, `Integer` doesn't pass this signature since it doesn't extend `java.lang.Enum`, hence we can rely on a regular converter (as you saw in the previous section), as illustrated here:

```
public class SaleVatIntConverter
   implements Converter<VatType, Integer> { … }
```

The *BuiltInEnumConverter* application contains this example next to other examples. Of course, you can try to write this converter as an inline converter via `Converter.of()`/`ofNullable()` or lambda expressions as well.

Always converting to a Java enum and occasionally to another Java enum

Always converting to a Java enum and occasionally to another Java enum is most probably not such a popular task, but let's use it as a pretext to condense what we've learned so far about enum conversions.

Let's consider the well-known `SALE.RATE` enum field in MySQL. First, we want to always/automatically convert `SALE.RATE` to our `RateType` Java enum, shown here:

```
public enum RateType { SILVER, GOLD, PLATINUM }
```

For this, we write the following `<forcedType/>` tag:

```
<forcedType>
  <userType>com.classicmodels.enums.RateType</userType>
  <enumConverter>true</enumConverter>
  <includeExpression>
    classicmodels\.sale\.rate
  </includeExpression>
  <includeTypes>ENUM</includeTypes>
</forcedType>
```

So far, we can refer in queries to `SALE.RATE` as a `RateType` enum, but let's assume that we also have the following `StarType` enum:

```
public enum StarType { THREE_STARS, FOUR_STARS, FIVE_STARS }
```

Basically, `StarType` is an alternative to `RateType` (*THREE_STARS* corresponds to *SILVER*, *FOUR_STARS* to *GOLD*, and *FIVE_STARS* to *PLATINUM*). Now, we may occasionally want to use `StarType` in queries instead of `RateType`, therefore we need a converter, as follows:

```
public class SaleRateStarConverter extends
                    EnumConverter<RateType, StarType> {

   public final static SaleRateStarConverter
      SALE_RATE_STAR_CONVERTER = new SaleRateStarConverter();

   public SaleRateStarConverter() {
      super(RateType.class, StarType.class);
   }

   @Override
   public RateType to(StarType u) {

      if (u != null) {
         return switch (u) {
                case THREE_STARS -> RateType.SILVER;
                case FOUR_STARS -> RateType.GOLD;
                case FIVE_STARS -> RateType.PLATINUM;
         };
      }

      return null;
   }
}
```

Since `RateType` and `StarType` don't contain the same literals, we have to override the `to()` method and define the expected matches. Done!

Expressing an `INSERT` statement that uses `RateType` looks like this:

```
// rely on <forcedType/>
ctx.insertInto(SALE, SALE.FISCAL_YEAR, ,..., SALE.RATE)
   .values(2005, ..., RateType.PLATINUM)
   .execute();
```

And whenever we want to use `StarType` instead of `RateType`, we rely on the static `SALE_RATE_STAR_CONVERTER` converter, as shown here:

```
// rely on SALE_RATE_STAR_CONVERTER
ctx.insertInto(SALE, SALE.FISCAL_YEAR, ..., SALE.RATE)
   .values(2005, ...,
           SALE_RATE_STAR_CONVERTER.to(StarType.FIVE_STARS))
   .execute();
```

The *BuiltInEnumConverter* application contains this example, along with other examples.

Via `classicmodels\.sale\.rate`, we nominated a certain column (`CLASSICMODELS.SALE.RATE`), but we may want to pick up all columns of this enum type. In such cases, an SQL query is more proper than a regex. Here is such a query for Oracle:

```
SELECT 'CLASSICMODELS.' || tab.table_name || '.'
                        || cols.column_name
FROM sys.all_tables tab
JOIN sys.all_constraints con ON tab.owner = con.owner
   AND tab.table_name = con.table_name
JOIN sys.all_cons_columns cols ON cols.owner = con.owner
   AND cols.constraint_name = con.constraint_name
   AND cols.table_name = con.table_name
WHERE constraint_type = 'C'
   AND tab.owner in ('CLASSICMODELS')
   AND search_condition_vc
      = q'[rate IN('SILVER', 'GOLD', 'PLATINUM')]'
```

You can find this example for MySQL and Oracle as *BuiltInEnumSqlConverter*.

In the bundled code, there are more applications, such as *EnumConverter*, which has examples of plain `org.jooq.Converter` types for enums; *EnumConverterForceTypes*, which has `<forcedType/>` and enum examples; and *InsertEnumPlainSql*, which has `INSERT` and enum examples when the jOOQ Code Generator is not used.

Retrieving the DataType<T> tag for a given enum data type

Retrieving the `DataType<T>` tag for a given enum data type can be done as in the following three examples that speak for themselves:

```
DataType<RateType> RATETYPE = SALE.RATE.getDataType();

DataType<VatType> VATTYPE
  = VARCHAR.asEnumDataType(VatType.class);

DataType<com.classicmodels.enums.VatType> VATTYPE
  = VARCHAR.asEnumDataType(jooq.generated.enums.VatType.class)
    .asConvertedDataType(VAT_CONVERTER);
```

Now, you can use this data type as any other data type. Next, let's tackle the topic of data type rewrites.

Data type rewrites

Another utility of `<forcedTypes/>` is data type rewrites. This allows us to explicitly choose the SQL data type (supported by the database, or unsupported but present in `org.jooq.impl.SQLDataType`) that should be used in Java.

For instance, in Oracle, a common use case is to map the missing `BOOLEAN` type as `NUMBER(1,0)` or `CHAR(1)`, as follows:

```
CREATE TABLE sale (
  ...
  hot NUMBER(1,0) DEFAULT 0
  hot CHAR(1) DEFAULT '1' CHECK (hot IN('1', '0'))
  ...
}
```

But this means that the jOOQ Code Generator will map fields of type `NUMBER(1, 0)` to the `SQLDataType.TINYINT` SQL data type and the `java.lang.Byte` type and, respectively, the fields of type `CHAR(1)` to the `SQLDataType.CHAR` SQL data type and the `String` Java type.

But the Java `String` type is commonly associated with text data manipulation, while the `Byte` type is commonly associated with binary data manipulations (for example, reading/writing a binary file) and the Java `Boolean` type clearly communicates the intention of using flag-type data. Moreover, the Java `Boolean` type has an SQL type (standard JDBC type) homologous to `SQLDataType.BOOLEAN`.

jOOQ allows us to force the type of columns, therefore we can force the type of `SALE.HOT` to `BOOLEAN`, as follows:

```
<forcedType>
  <name>BOOLEAN</name>
  <includeExpression>CLASSICMODELS\.SALE\.HOT</includeExpression>
  <includeTypes>NUMBER\(1,\s*0\)</includeTypes>
  <includeTypes>CHAR\(1\)</includeTypes>
</forcedType>
```

Done! Now, we can treat `SALE.HOT` as a Java `Boolean` type. Here is an `INSERT` example:

```
ctx.insertInto(SALE, ..., SALE.HOT)
   .values(2005,..., Boolean.FALSE)
   .execute();
```

Depending on `NUMBER` precision, jOOQ will map this data type to `BigInteger`, `Short`, or even `Byte` (as you just saw). If you find it cumbersome to use such Java types and you know that your data fits better for `Long` or `Integer` types, then you have two options: adjust the `NUMBER` precision accordingly, or rely on jOOQ type rewriting. Of course, you can apply this technique to any other type and dialect.

A complete example can be found in *DataTypeRewriting*. The programmatic version of this example is called *ProgrammaticDataTypeRewriting*. Next, let's understand how you can handle jOOQ embeddable types.

Handling embeddable types

Embeddable types represent a powerful feature introduced in jOOQ 3.14. Roughly, this feature gets materialized in synthetic UDTs that can be used with all databases supported by jOOQ. While PostgreSQL and Oracle support UDTs (we can use UDTs directly in **Data Definition Language** (**DDL**)), other databases including MySQL and SQL Server don't support UDTs. But via jOOQ embeddable types, we can work at the application level with synthetic UDTs for any database, and jOOQ will take care of the underlying aspects of mapping these types to the database.

An embeddable type mimics a UDT by synthetically wrapping one (usually more) database column in a generated `org.jooq.EmbeddableRecord`. For instance, we can wrap `OFFICE.CITY`, `OFFICE.STATE`, `OFFICE.COUNTRY`, `OFFICE.TERRITORY`, and `OFFICE.ADDRESS_LINE_FIRST` under an embeddable type named `OFFICE_FULL_ADDRESS` via the following configuration in the jOOQ Code Generator (here, for MySQL):

```
<embeddable>
 <!-- The optional catalog of the embeddable type -->
 <catalog/>

 <!-- The optional schema of the embeddable type -->
 <schema>classicmodels</schema>

 <!-- The name of the embeddable type -->
 <name>OFFICE_FULL_ADDRESS</name>

 <!-- An optional, defining comment of an embeddable -->
 <comment>The full address of an office</comment>

 <!-- The name of the reference to the embeddable type -->
 <referencingName/>

 <!-- An optional, referencing comment of an embeddable -->
 <referencingComment/>
```

And we continue with the settings for matching tables and fields, as follows:

```
 <!-- A regular expression matching qualified/unqualified
      table names to which to apply this embeddable. If left
      blank, this will apply to all tables -->
 <tables>.*\.office</tables>

 <!-- A list of fields to match to an embeddable. Each field
      must match exactly one column in each matched table. A
      mandatory regular expression matches field names, and
      an optional name can be provided to define the
      embeddable attribute name. If no name is provided, then
      the first matched field's name will be taken -->
 <fields>
  <field><expression>CITY</expression></field>
  <field><expression>ADDRESS_LINE_FIRST</expression></field>
```

```
    <field><expression>STATE</expression></field>
    <field><expression>COUNTRY</expression></field>
    <field><expression>TERRITORY</expression></field>
  </fields>
</embeddable>
```

Next, jOOQ generates `jooq...records.OfficeFullAddressRecord`, which extends `EmbeddableRecordImpl` and `jooq...pojos.OfficeFullAddress`. Moreover, in the generated `Office` table, we observe a new `OFFICE_FULL_ADDRESS` field that can be used as in the following `INSERT` statement:

```
ctx.insertInto(OFFICE, ... ,
               OFFICE.ADDRESS_LINE_SECOND, ...)
   .values(...,
           new OfficeFullAddressRecord("Naples", "Giuseppe
              Mazzini", "Campania", "Italy", "N/A"),
           ...)
   .execute();
```

Obviously, the `OFFICE_FULL_ADDRESS` column can be used in all types of statements, including `INSERT`, `UPDATE`, `DELETE`, and `SELECT`. Here, it is used in a `SELECT` statement:

```
Result<Record1<OfficeFullAddressRecord>> result
       = ctx.select(OFFICE.OFFICE_FULL_ADDRESS).from(OFFICE)
   .fetch();
```

Or it can be fetched into the `OfficeFullAddress` POJO, like this:

```
List<OfficeFullAddress> result
   = ctx.select(OFFICE.OFFICE_FULL_ADDRESS).from(OFFICE)
        .fetchInto(OfficeFullAddress.class);
```

In the bundled code, for MySQL, we have *EmbeddableType*, which contains the previous example, and for PostgreSQL, we have *ProgrammaticEmbeddableType*, which is the programmatic version of the previous example.

Replacing fields

At this point, we have access to (we can use) the embeddable type, but we still have direct access to fields wrapped in this embeddable type. For instance, these fields can be used in `INSERT` statements, `SELECT` statements, and so on, and they appear in the **Integrated Development Environment's** (**IDE's**) autocompletion list.

The *replacing fields* feature means to signal to jOOQ to disallow direct access to fields that are part of an embeddable type. These fields will not appear in the IDE's autocompletion list anymore, and the result set of `SELECT` statements will not contain these fields. Enabling this feature can be done via the `<replacesFields/>` flag, as follows:

```
<embeddable>
 ...
 <replacesFields>true</replacesFields>
</embeddable>
```

The *EmbeddableTypeReplaceFields* application contains this example for Oracle, while *ProgrammaticEmbeddableTypeReplaceFields* contains a programmatic version of this example for SQL Server.

Converting embeddable types

Converting an embeddable type can be done via `org.jooq.Converter`, as for any other type. For example, converting between `JsonNode` and `OFFICE_FULL_ADDRESS` can be done via a `Converter` that starts like this:

```
public class JsonConverter implements
           Converter<OfficeFullAddressRecord, JsonNode> {

   public static final JsonConverter JSON_CONVERTER
      = new JsonConverter();

   @Override
   public JsonNode from(OfficeFullAddressRecord t) { ... }

   @Override
   public OfficeFullAddressRecord to(JsonNode u) { ... }
   ...
}
```

And here, it is a `SELECT` statement that fetches `OFFICE.OFFICE_FULL_ADDRESS` as `JsonNode` via `JSON_CONVERTER`:

```
List<JsonNode> result = ctx.select(OFFICE.OFFICE_FULL_ADDRESS)
   .from(OFFICE)
   .fetch(OFFICE.OFFICE_FULL_ADDRESS, JSON_CONVERTER);
```

The *ConvertEmbeddableType* application for MySQL contains this example.

Embedded domains

Quite popular in PostgreSQL, *domain* types represent UDTs built on top of other types and containing optional constraints. For instance, in our PostgreSQL schema, we have the following domain:

```
CREATE DOMAIN postal_code AS varchar(15)
CHECK(
   VALUE ~ '^\d{5}$'
OR VALUE ~ '^[A-Z]{2}[0-9]{3}[A-Z]{2}$'
);
```

And it is used in the `office` table, as shown here:

```
CREATE TABLE office (
  ...
  "postal_code" postal_code NOT NULL,
  ...
);
```

jOOQ can generate a Java type for each domain type if we turn on this feature, as has been done here:

```
// Maven and standalone
<database>
  ...
  <embeddableDomains>.*</embeddableDomains>
</database>

// Gradle
database {
  embeddableDomains = '.*'
}

// programmatic
withEmbeddableDomains(".*")
```

While `.*` matches all domain types, you can use more restrictive regexes to match exactly the domains that will be replaced by embeddable types.

The jOOQ Code Generator generates an embeddable type named (by default) `PostalCodeRecord` (in `jooq.generated.embeddables.records`). We can use it for creating semantically type-safe queries, as in these examples:

```
ctx.select(OFFICE.CITY, OFFICE.COUNTRY)
   .from(OFFICE)
   .where(OFFICE.POSTAL_CODE.in(
        new PostalCodeRecord("AZ934VB"),
        new PostalCodeRecord("DT975HH")))
   .fetch();

ctx.insertInto(OFFICE, ..., OFFICE.POSTAL_CODE, ...)
   .values(..., new PostalCodeRecord("OP909DD"), ...)
   .execute();
```

The complete code for PostgreSQL is named *Domain*.

Well, we've reached the end of this section and the end of this chapter. Notice that we intentionally skipped the topic of embeddable types and embeddable keys (including composite keys) since this topic is covered later in *Chapter 11, jOOQ Keys*.

Summary

This chapter is a must-have in your jOOQ arsenal. Mastering the topics covered here—such as custom data types, converters, bindings, database vendor-specific data types, enums, embeddable types, and so on—will help you to shape the interaction between Java and database data types to fit your non-trivial scenarios. In the next chapter, we cover the topics of fetching and mapping.

8
Fetching and Mapping

Fetching result sets and mapping them in the shape and format expected by the client is one of the most important tasks of querying a database. jOOQ excels in this area and provides a comprehensive API for fetching data and mapping it to scalars, arrays, lists, sets, maps, POJO, Java 16 records, JSON, XML, nested collections, and more. As usual, the jOOQ API hides the friction and challenges raised by different database dialects along with the boilerplate code necessary to map the result set to different data structures. In this context, our agenda covers the following topics:

- Simple fetching/mapping
- Fetching one record, a single record, or any record
- Fetching arrays, lists, sets, and maps
- Fetching groups
- Fetching via JDBC `ResultSet`
- Fetching multiple result sets
- Fetching relationships
- Hooking POJOs
- jOOQ record mapper

- The mighty SQL/JSON and SQL/XML support
- Nested collections via the astonishing `MULTISET`
- Lazy fetching
- Asynchronous fetching
- Reactive fetching

Let's get started!

Technical requirements

The code for this chapter can be found on GitHub at `https://github.com/PacktPublishing/jOOQ-Masterclass/tree/master/Chapter08`.

Simple fetching/mapping

By *simple* fetching/mapping, we refer to the jOOQ fetching techniques that you learned earlier in this book (for instance, the ubiquitous `into()` methods) but also to the new jOOQ utility, `org.jooq.Records`. This utility is available from jOOQ 3.15 onward, and it contains two types of utility methods, as we will discuss next.

Collector methods

The collector methods are named `intoFoo()`, and their goal is to create a collector (`java.util.stream.Collector`) for collecting records (`org.jooq.Record[N]`) into arrays, lists, maps, groups, and more. These collectors can be used in `ResultQuery.collect()` as any other collector. `ResultQuery<R>` implements `Iterable<R>` and comes with convenience methods such as `collect()` on top of it. Besides the fact that `collect()` handles resources internally (there is no need to use try-with-resources), you can use it for any collectors such as standard JDK collectors, jOOλ collectors, `Records` collectors, or your own collectors. For instance, here is an example of collecting into `List<String>`:

```
List<String> result = ctx.select(CUSTOMER.CUSTOMER_NAME)
    .from(CUSTOMER)
    .collect(intoList()); // or, Java's Collectors.toList()
```

And, here is an example of collecting into `Map<Long, String>`:

```
Map<Long, String> result = ctx.select(
           CUSTOMER.CUSTOMER_NUMBER, CUSTOMER.PHONE)
```

```
   .from(CUSTOMER)
   .collect(intoMap());
```

Note that, while the ubiquitous `into()` methods use reflection, these utilities are a pure declarative mapping of jOOQ results/records without using reflection.

Mapping methods

The mapping methods are actually multiple flavors of the mapping (`Function[N]`) method. A mapping method creates a `RecordMapper` parameter that can map from `Record[N]` to another type (for instance, POJO and Java 16 records) in a type-safe way. For instance, you can map to a Java record as follows:

```
public record PhoneCreditLimit(
   String phone, BigDecimal creditLimit) {}

List<PhoneCreditLimit> result = ctx.select(
     CUSTOMER.PHONE, CUSTOMER.CREDIT_LIMIT)
   .from(CUSTOMER)
   .fetch(mapping(PhoneCreditLimit::new));
```

When mapping nested rows (for instance, `LEFT JOIN`) you can achieve `null` safety by combining `mapping()` with `Functions.nullOnAllNull(Function1)` or `Functions.nullOnAnyNull(Function1)`. Here is an example:

```
List<SalarySale> result = ctx.select(
       EMPLOYEE.SALARY, SALE.SALE_)
   .from(EMPLOYEE)
   .leftJoin(SALE)
   .on(EMPLOYEE.EMPLOYEE_NUMBER.eq(SALE.EMPLOYEE_NUMBER))
   .fetch(mapping(nullOnAnyNull(SalarySale::new)));
```

So, how does this work? For instance, when an employee has no sale (or you have an orphan sale), you'll obtain a `null` value instead of an instance of `SalarySale` having the sale as `null`, `SalarySale[salary=120000, sale=null]`.

Many more examples are available for MySQL/PostgreSQL in the bundle code, *Records*.

Simple fetching/mapping continues

Next, let's see other techniques of fetching/mapping data that can be used quite intuitively and effortlessly. Since the jOOQ manual is filled to the brim with examples, let's try to niche several things in this section. For instance, a simple fetch can be done via `DSLContext.resultQuery()` and plain SQL, as follows:

```
Result<Record> result = ctx.resultQuery(
   "SELECT customer_name FROM customer").fetch();

List<String> result = ctx.resultQuery(
   "SELECT customer_name FROM customer")
      .fetchInto(String.class);

List<String> result = ctx.resultQuery(
   "SELECT customer_name FROM customer")
      .collect(intoList(r -> r.get(0, String.class)));
```

Another approach might rely on `DSLContext.fetch()` and plain SQL, as follows:

```
Result<Record> result = ctx.fetch(
   "SELECT customer_name FROM customer");

List<String> result = ctx.fetch(
   "SELECT customer_name FROM customer").into(String.class);

List<String> result = ctx.fetch(
   "SELECT customer_name FROM customer")
      .collect(intoList(r -> r.get(0, String.class)));
```

So, the idea is quite simple. Whenever you have to execute a plain SQL that you can't (or don't want to) express via a jOOQ-generated Java-based schema, then simply rely on `ResultQuery.collect(collector)` or the `resultQuery()` ... `fetch()`/`fetchInto()` combination. Alternatively, simply pass it to the `fetch()` method and call the proper `into()` method or the `intoFoo()` method to map the result set to the necessary data structure. There are plenty of such methods that can map a result set to scalars, arrays, lists, sets, maps, POJO, XML, and more.

On the other hand, using the Java-based schema (which is, of course, the recommended way to go) leads to the following less popular but handy query:

```
List<String> result = ctx.fetchValues(CUSTOMER.CUSTOMER_NAME);
```

This is a shortcut for fetching a single field and obtaining the mapped result (values) without explicitly calling an `into()` method or an `intoFoo()` method. Essentially, jOOQ automatically maps the fetched field to the Java type associated with it when the Java-based schema was generated by the jOOQ generator.

Whenever you need to fetch a single value, you can rely on `fetchValue()`:

```
Timestamp ts = ctx.fetchValue(currentTimestamp());
```

The `<T> T fetchValue(Field<T> field)` and `<T> List<T> fetchValues(TableField<?,T> tf)` methods are just two of the many flavors of methods that are available. Check out the jOOQ documentation to see the rest of them.

However, since you have made it this far in this book, I'm sure that you think of this query as a shortcut for the following four, more popular, approaches:

```
List<String> result = ctx.select(CUSTOMER.CUSTOMER_NAME)
   .from(CUSTOMER).fetch(CUSTOMER.CUSTOMER_NAME);

List<String> result = ctx.select(CUSTOMER.CUSTOMER_NAME)
   .from(CUSTOMER).fetchInto(String.class)

List<String> result = ctx.select(CUSTOMER.CUSTOMER_NAME)
   .from(CUSTOMER).collect(intoList());

// or, mapping to Result<Record1<String>>
var result = ctx.select(CUSTOMER.CUSTOMER_NAME)
   .from(CUSTOMER).fetch();
```

And you are right, as long as you don't also think of the following, too:

```
List<String> result = ctx.select().from(CUSTOMER)
   .fetch(CUSTOMER.CUSTOMER_NAME);

List<String> result = ctx.selectFrom(CUSTOMER)
   .fetch(CUSTOMER.CUSTOMER_NAME);
```

All six of these queries project the same result, but they are not the same. As a jOOQ novice, it is understandable that you might a bad choice and go for the last two queries. Therefore, let's clarify this concern by looking at the generated SQLs. The first four queries produce the following SQL:

```
SELECT `classicmodels`.`customer`.`customer_name`
FROM `classicmodels`.`customer`
```

In contrast, the last two queries produce the following SQL:

```
SELECT `classicmodels`.`customer`.`customer_number`,
       `classicmodels`.`customer`.`customer_name`,
       ...
       `classicmodels`.`customer`.`first_buy_date`
FROM `classicmodels`.`customer`
```

Now, it is obvious that the last two queries perform unnecessary work. We only need the `CUSTOMER_NAME` field, but these queries will fetch all fields, and this is pointless work that negatively impacts performance. In such cases, don't blame jOOQ or the database because both of them did exactly what you asked!

> **Important Note**
> As a rule of thumb, when you don't need to fetch all fields, rely on the first four approaches from earlier and enlist the necessary fields in the `SELECT` statement. In this context, allow me to reiterate the *SelectOnlyNeededData* application from *Chapter 5, Tackling Different Kinds of SELECT, INSERT, UPDATE, DELETE, and MERGE Statements.*

When you fetch more than one field, but not all fields, you should write something like this:

```
// Result<Record2<String, BigDecimal>>
var result = ctx.select(
       CUSTOMER.CUSTOMER_NAME, CUSTOMER.CREDIT_LIMIT)
   .from(CUSTOMER).fetch();

ExpectedType result = ctx.select(
       CUSTOMER.CUSTOMER_NAME, CUSTOMER.CREDIT_LIMIT)
   .from(CUSTOMER)
   .fetchInto(…) // or, collect(), fetch(mapping(…)), ...
```

Now, let's consider another simple fetching method based on the following two POJOs:

```
class NamePhone {String customerName; String phone;}
class PhoneCreditLimit {String phone; BigDecimal creditLimit;}
```

Populating these POJOs can be done via two `SELECT` statements, as follows:

```
List<NamePhone> result1 = ctx.select(
      CUSTOMER.CUSTOMER_NAME, CUSTOMER.PHONE)
   .from(CUSTOMER).fetchInto(NamePhone.class);

List<PhoneCreditLimit> result2 = ctx.select(
      CUSTOMER.PHONE, CUSTOMER.CREDIT_LIMIT)
   .from(CUSTOMER).fetchInto(PhoneCreditLimit.class);
```

However, here, jOOQ allows us to map `Result<Record>` into multiple results. In other words, we can obtain the same result and trigger a single `SELECT` statement, as follows:

```
// Result<Record3<String, String, BigDecimal>>
var result = ctx.select(CUSTOMER.CUSTOMER_NAME,
                        CUSTOMER.PHONE, CUSTOMER.CREDIT_LIMIT)
   .from(CUSTOMER).fetch();

List<NamePhone> r1=result.into(NamePhone.class);
List<PhoneCreditLimit> r2=result.into(PhoneCreditLimit.class);
```

Nice! Of course, this doesn't only apply when mapping result sets to POJOs. In the code bundle of this book, *SimpleFetch* (which is available for MySQL), you can see a result set produced by a single `SELECT` statement formatted entirely as JSON, while a part of it is mapped to a Java `Set`. Next, let's dive into the `fetchOne()`, `fetchSingle()`, and `fetchAny()` methods.

Fetching one record, a single record, or any record

jOOQ has come with three handy methods named `fetchOne()`, `fetchSingle()`, and `fetchAny()`. All three are capable of returning a resulting record, but each of them will do this under certain coordinates. So, let's go through each method in detail.

Using fetchOne()

For instance, the `fetchOne()` method returns, at most, one resulting record. In other words, if the fetched result set has more than one record, then `fetchOne()` throws a jOOQ-specific `TooManyRowsException` exception. But if the result set has no records, then `fetchOne()` returns `null`. In this context, `fetchOne()` can be useful for fetching a record by a primary key, other unique keys, or a predicate that guarantees uniqueness, while you prepare to handle potentially `null` results. Here is an example of using `fetchOne()`:

```
EmployeeRecord result = ctx.selectFrom(EMPLOYEE)
   .where(EMPLOYEE.EMPLOYEE_NUMBER.eq(1370L))
   .fetchOne();
```

Alternatively, you can fetch directly into the `Employee` POJO via `fetchOneInto()`:

```
Employee result = ctx.selectFrom(EMPLOYEE)
   .where(EMPLOYEE.EMPLOYEE_NUMBER.eq(1370L))
   .fetchOneInto(Employee.class);
```

However, pay attention. Remember that `fetchOneInto(Employee.class)` is not the same thing as `fetchOne().into(Employee.class)` since the latter is prone to throw `NullPointerException` exceptions. So, it is better to avoid writing something like this:

```
Employee result = ctx.selectFrom(EMPLOYEE)
   .where(EMPLOYEE.EMPLOYEE_NUMBER.eq(1370L))
   .fetchOne().into(Employee.class);
```

If there is no `EMPLOYEE` POJO with a primary key of *1370*, then this code throws an NPE exception.

Also, avoid chaining the `component[N]()` and `value[N]()` methods, as follows (this code is also prone to throw `NullPointerException`):

```
String result = ctx.select(EMPLOYEE.EMAIL).from(EMPLOYEE)
   .where(EMPLOYEE.EMPLOYEE_NUMBER.eq(1370L))
   .fetchOne().value1();
```

Also, prefer fetching into a proper type (here, it is `String`):

```
String result = ctx.select(EMPLOYEE.EMAIL).from(EMPLOYEE)
   .where(EMPLOYEE.EMPLOYEE_NUMBER.eq(1370L))
   .fetchOneInto(String.class);
```

Of course, an NPE check is still needed before using `result`, but you can wrap this check via `Objects.requireNonNullElseGet()`, as follows:

```
String result = Objects.requireNonNullElseGet(
   ctx.select(EMPLOYEE.EMAIL).from(EMPLOYEE)
      .where(EMPLOYEE.EMPLOYEE_NUMBER.eq(1370L))
      .fetchOneInto(String.class), () -> "");
```

Alternatively, simply wrap it into an `Optional` type via the jOOQ's `fetchOptional()` method:

```
Optional<EmployeeRecord> result = ctx.selectFrom(EMPLOYEE)
   .where(EMPLOYEE.EMPLOYEE_NUMBER.eq(1370L))
   .fetchOptional();
```

Alternatively, you may prefer `fetchOptionalInto()`:

```
Optional<Employee> result = ctx.selectFrom(EMPLOYEE)
   .where(EMPLOYEE.EMPLOYEE_NUMBER.eq(1370L))
   .fetchOptionalInto(Employee.class);
```

As usual, `fetchOne()` comes in many flavors, all of which are available in the documentation. For instance, you can use `DSLContext.fetchOne()` as follows:

```
EmployeeRecord result = ctx.fetchOne(EMPLOYEE,
   EMPLOYEE.EMPLOYEE_NUMBER.eq(1370L));
```

Or you can fetch a record and convert it based on a user-defined converter (this converter was introduced in *Chapter 7, Types, Converters, and Bindings*):

```
YearMonth result = ctx.select(CUSTOMER.FIRST_BUY_DATE)
   .from(CUSTOMER)
   .where(CUSTOMER.CUSTOMER_NUMBER.eq(112L))
   .fetchOne(CUSTOMER.FIRST_BUY_DATE,
             INTEGER_YEARMONTH_CONVERTER);
```

Many other examples are available in the bundled code for MySQL, *FetchOneAnySingle*.

Using fetchSingle()

The `fetchSingle()` method returns exactly one resulting record. In other words, if the fetched result set contains more than one record, then `fetchSingle()` throws the jOOQ-specific `TooManyRowsException` error. And if it doesn't contain any records, then it throws the jOOQ-specific `NoDataFoundException` error.

Essentially, `fetchSingle()` is similar to `fetchOne()`, except that it throws an exception instead of returning `null` when the fetched result set doesn't contain any records. This means that `fetchSingle()` is useful for fetching a record by a primary key, other unique keys, or a predicate that guarantees uniqueness when you are not expecting `null` results. For example, see the following code block:

```
Employee result = ctx.selectFrom(EMPLOYEE)
   .where(EMPLOYEE.EMPLOYEE_NUMBER.eq(1370L))
   .fetchSingleInto(Employee.class);
```

Or you might only fetch the email of this employee, as follows:

```
String result = ctx.select(EMPLOYEE.EMAIL).from(EMPLOYEE)
   .where(EMPLOYEE.EMPLOYEE_NUMBER.eq(1370L))
   .fetchSingle().value1(); // fetchSingleInto(String.class)
```

Many other examples are available in the bundled code.

Using fetchAny()

The `fetchAny()` method returns the first resulting record. In other words, if the fetched result set contains more than one record, then `fetchAny()` returns the first one. And, if it doesn't contain any records, then it returns `null`. This is similar to …`limit(1).fetchOne();`. So, pay attention to avoid any usages that are prone to throw a `NullPointerException` exception. Here's an example:

```
SaleRecord result = ctx.selectFrom(SALE)
   .where(SALE.EMPLOYEE_NUMBER.eq(1370L))
   .fetchAny();
```

Let's see another example:

```
String result = ctx.select(SALE.TREND).from(SALE)
   .where(SALE.EMPLOYEE_NUMBER.eq(1370L))
   .fetchAnyInto(String.class);
```

Many other examples are available for MySQL in *FetchOneAnySingle*.

Fetching arrays, lists, sets, and maps

jOOQ reduces the code that is needed for fetching `Result<Record>` as an array, list, set, or map down to a simple call of its amazing API.

Fetching arrays

Fetching arrays can be done via a comprehensive set of jOOQ methods, including `fetchArray()` (along with its flavors), `fetchOneArray()`, `fetchSingleArray()`, `fetchAnyArray()`, `fetchArrays()`, and `intoArray()`. For instance, fetching all the `DEPARTMENT` fields as an array of `Record` can be done as follows:

```
Record[] result = ctx.select().from(DEPARTMENT).fetchArray();
```

In comparison, you can just fetch `DEPARTMENT.NAME` as a `String[]` as follows:

```
String[] result = ctx.select(DEPARTMENT.NAME).from(DEPARTMENT)
   .fetchArray(DEPARTMENT.NAME);

String[] result = ctx.select(DEPARTMENT.NAME).from(DEPARTMENT)
   .collect(intoArray(new String[0]));
```

Alternatively, fetching all `CUSTOMER.FIRST_BUY_DATE` fields as an array of the `YearMonth` type can be done via `fetchArray(Field<T> field, Converter<? super T,? extends U> cnvrtr)`, as follows (note that the `INTEGER_YEARMONTH_CONVERTER` converter was introduced in *Chapter 7, Types, Converters, and Bindings*):

```
YearMonth[] result = ctx.select(CUSTOMER.FIRST_BUY_DATE)
   .from(CUSTOMER)
   .fetchArray(CUSTOMER.FIRST_BUY_DATE,
               INTEGER_YEARMONTH_CONVERTER);
```

What do you think about fetching a database array into a Java array, such as the `DEPARTMENT.TOPIC` field that was defined in our PostgreSQL schema? Well, the result, in this case, is `String[][]`:

```
String[][] result = ctx.select(DEPARTMENT.TOPIC)
   .from(DEPARTMENT).fetchArray(DEPARTMENT.TOPIC);
```

If we return this `String[][]` from a Spring Boot REST controller, the result will be a JSON array:

```
[
  ["publicity", "promotion"],
  ["commerce","trade","sellout","transaction"],
  ...
]
```

What about fetching a UDT type into a Java array? In our PostgreSQL schema, we have the `MANAGER.MANAGER_EVALUATION` UDT type, so let's give it a try and fetch it as an array next to the `MANAGER_NAME` type:

```
// Record2<String, EvaluationCriteriaRecord>[]
var result = ctx.select(MANAGER.MANAGER_NAME,
                        MANAGER.MANAGER_EVALUATION)
   .from(MANAGER).fetchArray();
```

Let's print out the first manager name and their evaluation:

```
System.out.println(result[0].value1()+"\n"
        + result[0].value2().format());
```

Here is the output (the `format()` method formats `EvaluationCriteriaRecord` as a tabular text):

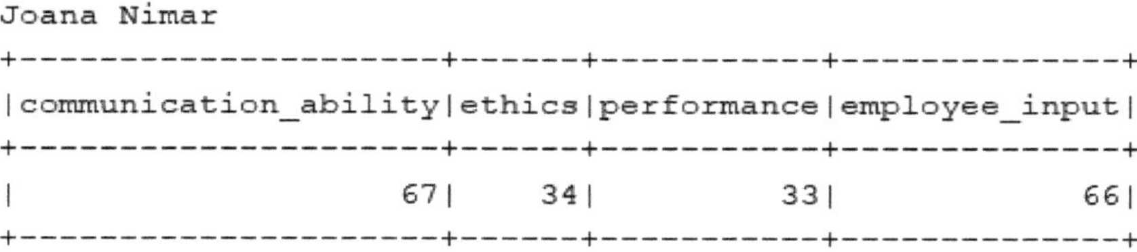

```
Joana Nimar
+---------------------+------+-----------+--------------+
|communication_ability|ethics|performance|employee_input|
+---------------------+------+-----------+--------------+
|                   67|    34|         33|            66|
+---------------------+------+-----------+--------------+
```

Figure 8.1– Printing the first manager and their evaluation

Finally, let's try fetching an embeddable type as an array, too:

```
OfficeFullAddressRecord[] result = ctx.select(
      OFFICE.OFFICE_FULL_ADDRESS).from(OFFICE)
   .fetchArray(OFFICE.OFFICE_FULL_ADDRESS);

OfficeFullAddressRecord[] result = ctx.select(
      OFFICE.OFFICE_FULL_ADDRESS).from(OFFICE)
   .collect(intoArray(new OfficeFullAddressRecord[0]));
```

The last example from this section relies on `fetchArrays()`:

```
Object[][] result = ctx.select(DEPARTMENT.DEPARTMENT_ID,
         DEPARTMENT.OFFICE_CODE, DEPARTMENT.NAME)
   .from(DEPARTMENT).fetchArrays();
```

If we return this `Object[][]` from a Spring Boot REST controller, then the result will be a JSON array of arrays:

```
[
  [1, "1", "Advertising"],
  [2, "1", "Sales"],
  [3, "2", "Accounting"],
  [4, "3", "Finance"]
]
```

In the bundled code, you can find over 15 examples of fetching jOOQ results as arrays.

Fetching lists and sets

So far, most examples fetch the result set in `java.util.List` or `org.jooq.Result` (that is, the jOOQ wrappers of `List`), so there is no mystery regarding how the following examples work:

```
List<String> result = ctx.select(DEPARTMENT.NAME)
   .from(DEPARTMENT).fetch(DEPARTMENT.NAME);

List<String> result = ctx.select(DEPARTMENT.NAME)
   .from(DEPARTMENT).collect(intoList());

List<Department> result = ctx.select(DEPARTMENT.DEPARTMENT_ID,
       DEPARTMENT.OFFICE_CODE, DEPARTMENT.NAME)
   .from(DEPARTMENT).fetchInto(Department.class);
```

So, let's focus on more interesting cases, such as how to fetch the `DEPARTMENT.TOPIC` array field defined in our PostgreSQL schema:

```
List<String[]> result = ctx.select(DEPARTMENT.TOPIC)
   .from(DEPARTMENT)
   .fetch(DEPARTMENT.TOPIC, String[].class);
```

Instead of calling `fetch()`, which will return `Result<Record1<String[]>>`, we prefer to call `fetch(Field<?> field, Class<? extends U> type)`. This allow us to return a `List<String[]>`.

Trying to fetch DEPARTMENT.TOPIC as a Set<String[]> can be done via the jOOQ fetchSet() method (check out the documentation to see all the flavors of this method):

```
Set<String[]> result = ctx.select(DEPARTMENT.TOPIC)
   .from(DEPARTMENT).fetchSet(DEPARTMENT.TOPIC);
```

Consider the bundled code, which contains more examples of fetching lists and sets, including fetching UDT and embeddable types.

Fetching maps

jOOQ comes with a set of fetchMap()/intoMap() methods that allow us to split a result set into key-value pairs of a java.util.Map wrapper. There are more than 20 such methods, but we can primarily distinguish between the fetchMap(key)/ intoMap(Function keyMapper) methods. These methods allow us to specify the field(s) representing the key, while the value is inferred from the SELECT result, and the fetchMap(key, value)/intoMap(Function keyMapper, Function valueMapper) methods in which we specify the field(s) that represents the key and the value, respectively. The Records.intoMap() method without any arguments is only useful if you have a two-column ResultQuery and you want to map the first column as a key and the second column as a value.

For instance, let's fetch a Map that has DEPARTMENT_ID as the key (so, the DEPARTMENT primary key) and DepartmentRecord as the value:

```
Map<Integer, DepartmentRecord>
       result = ctx.selectFrom(DEPARTMENT)
   .fetchMap(DEPARTMENT.DEPARTMENT_ID);

Map<Integer, DepartmentRecord>
       result = ctx.selectFrom(DEPARTMENT)
   .collect(intoMap(r -> r.get(DEPARTMENT.DEPARTMENT_ID)));
```

Alternatively, let's instruct jOOQ that the map value should be a Department POJO (generated by jOOQ) instead of DepartmentRecord:

```
Map<Integer, Department> result = ctx.selectFrom(DEPARTMENT)
   .fetchMap(DEPARTMENT.DEPARTMENT_ID, Department.class);
```

Do you think this is impressive? How about mapping a one-to-one relationship between the CUSTOMER and CUSTOMERDETAIL tables? Here is the magical code:

```
Map<Customer, Customerdetail> result = ctx.select()
   .from(CUSTOMER)
   .join(CUSTOMERDETAIL)
   .on(CUSTOMER.CUSTOMER_NUMBER
       .eq(CUSTOMERDETAIL.CUSTOMER_NUMBER))
   .fetchMap(Customer.class, Customerdetail.class);
```

In order to obtain a correct mapping, you have to provide explicit equals() and hashCode() methods for the involved POJOs.

Simply returning this Map from a REST controller will result in the following JSON code:

```
{
   "Customer (99, Australian Home, Paoule, Sart,
    40.11.2555, 1370, 21000.00, 20210)":
      {
      "customerNumber": 99, "addressLineFirst": "43 Rue 2",
      "addressLineSecond": null, "city": "Paris", "state":
       null, "postalCode": "25017", "country": "France"
      },
   ...
```

Alterantively, you might want to fetch this one-to-one relationship by only using a subset of fields:

```
Map<Record, Record> result = ctx.select(
      CUSTOMER.CONTACT_FIRST_NAME, CUSTOMER.CONTACT_LAST_NAME,
      CUSTOMERDETAIL.CITY, CUSTOMERDETAIL.COUNTRY)
   .from(CUSTOMER)
   .join(CUSTOMERDETAIL)
   .on(CUSTOMER.CUSTOMER_NUMBER
      .eq(CUSTOMERDETAIL.CUSTOMER_NUMBER))
   .fetchMap(new Field[]{CUSTOMER.CONTACT_FIRST_NAME,
                         CUSTOMER.CONTACT_LAST_NAME},
             new Field[]{CUSTOMERDETAIL.CITY,
                         CUSTOMERDETAIL.COUNTRY});
```

In the bundled code, *ArrListMap* (which is available for PostgreSQL), you can see more examples, including mapping a flattened one-to-many relationship, mapping arrays, UDTs and embeddable types, and using `fetchMaps()`, `fetchSingleMap()`, `fetchOneMap()`, and `fetchAnyMap()`. Next, let's talk about fetching groups.

Fetching groups

The jOOQ fetching groups feature is similar to fetching maps, except that it allows us to fetch a list of records as the value of each key-value pair. There are over 40 flavors of the `fetchGroups()`, `intoGroups()`, and `intoResultGroup()` methods; therefore, take your time to practice (or, at the very least, read about) each of them.

We can distinguish between the `fetchGroups(key)` and `intoGroups(Function keyMapper)` methods that allow us to specify the field(s) representing the key, while the value is inferred from the `SELECT` result as the `Result<Record>/List<Record>` and `fetchGroups(key, value)/intoGroups(Function keyMapper, Function valueMapper)` methods in which we specify the field(s) that represents the key and the value, respectively, which could be `Result<Record>`, `List<POJO>`, `List<scalar>`, and more. The `Records.intoGroups()` method without any arguments is only useful if you have a two-column `ResultQuery`, and you want to map the first column as a key and the second column as a value. Additionally, the `intoResultGroup()` method returns a collector that collects a jOOQ `Record`, which results from a `ResultQuery` in a `Map` using the result of the `RecordMapper` parameter as a key to collect the records themselves into a jOOQ `Result`.

For instance, you can fetch all the `OrderRecord` values and group them by customer (`CUSTOMER_NUMBER`) as follows:

```
Map<Long, Result<OrderRecord>> result = ctx.selectFrom(ORDER)
   .fetchGroups(ORDER.CUSTOMER_NUMBER);

Map<Long, List<OrderRecord>> result = ctx.selectFrom(ORDER)
   .collect(intoGroups(r -> r.get(ORDER.CUSTOMER_NUMBER)));
```

Or you can group all bank transfers (`BANK_TRANSACTION.TRANSFER_AMOUNT`) by customer (`BANK_TRANSACTION.CUSTOMER_NUMBER`) into `Map<Long, List<BigDecimal>>`:

```
Map<Long, List<BigDecimal>> result = ctx.select(
      BANK_TRANSACTION.CUSTOMER_NUMBER,
      BANK_TRANSACTION.TRANSFER_AMOUNT)
   .from(BANK_TRANSACTION)
```

```
    .fetchGroups(BANK_TRANSACTION.CUSTOMER_NUMBER,
                 BANK_TRANSACTION.TRANSFER_AMOUNT);

Map<Long, List<BigDecimal>> result = ctx.select(
      BANK_TRANSACTION.CUSTOMER_NUMBER,
      BANK_TRANSACTION.TRANSFER_AMOUNT)
    .from(BANK_TRANSACTION)
    .collect(intoGroups());
    // or, …
    .collect(intoGroups(
        r -> r.get(BANK_TRANSACTION.CUSTOMER_NUMBER),
        r -> r.get(BANK_TRANSACTION.TRANSFER_AMOUNT)));
```

You can group them into `Map<Long, List<Record2<Long, BigDecimal>>>` or `Map<Long, Result<Record2<Long, BigDecimal>>>`, respectively:

```
Map<Long, List<Record2<Long, BigDecimal>>> result
   = ctx.select(BANK_TRANSACTION.CUSTOMER_NUMBER,
                BANK_TRANSACTION.TRANSFER_AMOUNT)
        .from(BANK_TRANSACTION)
        .collect(intoGroups(r ->
            r.get(BANK_TRANSACTION.CUSTOMER_NUMBER)));

Map<Long, Result<Record2<Long, BigDecimal>>> result
   = ctx.select(BANK_TRANSACTION.CUSTOMER_NUMBER,
                BANK_TRANSACTION.TRANSFER_AMOUNT)
        .from(BANK_TRANSACTION)
        .collect(intoResultGroups(r ->
            r.get(BANK_TRANSACTION.CUSTOMER_NUMBER)));
```

As you've probably intuited already, `fetchGroups()` is very handy for fetching and mapping one-to-many relationships. For instance, each product line (`PRODUCTLINE`) has multiple products (`PRODUCT`), and we can fetch this data as follows:

```
Map<Productline, List<Product>> result = ctx.select()
    .from(PRODUCTLINE)
    .innerJoin(PRODUCT)
    .on(PRODUCTLINE.PRODUCT_LINE.eq(PRODUCT.PRODUCT_LINE))
    .fetchGroups(Productline.class, Product.class);
```

Returning this map from a REST controller results in the following JSON:

```
{
  "Productline (Motorcycles, 599302, Our motorcycles ...)": [
    {
      "productId": 1,
      "productName": "1969 Harley Davidson Ultimate Chopper",
      ...
    },
    {
      "productId": 3,
      "productName": "1996 Moto Guzzi 1100i",
      ...
    },
    ...
  ],
  "Productline (Classic Cars, 599302 ... )": [
  ...
  ]
}
```

Of course, relying on user-defined POJOs/Java records is also possible. For instance, let's say you just need the code and name of each product line, along with the product ID and buy price of each product. Having the proper POJOs named `SimpleProductline` and `SimpleProduct`, we can map the following one-to-many relationship:

```
Map<SimpleProductline, List<SimpleProduct>> result =
  ctx.select(PRODUCTLINE.PRODUCT_LINE, PRODUCTLINE.CODE,
             PRODUCT.PRODUCT_ID, PRODUCT.BUY_PRICE)
  .from(PRODUCTLINE)
  .innerJoin(PRODUCT)
  .on(PRODUCTLINE.PRODUCT_LINE.eq(PRODUCT.PRODUCT_LINE))
  .fetchGroups(SimpleProductline.class, SimpleProduct.class);
```

In order to obtain a correct mapping, you have to provide explicit `equals()` and `hashCode()` methods for the involved POJOs. For the jOOQ-generated POJO, this is a configuration step that can be accomplished via `<pojosEqualsAndHashCode/>`, as follows:

```
<generate>
   <pojosEqualsAndHashCode>true</pojosEqualsAndHashCode>
</generate>
```

Notice that using `fetchGroups()` works as expected for `INNER JOIN`, but not for `LEFT JOIN`. If the fetched parent doesn't have children, then instead of an empty list, you'll get a list containing a single `NULL` item. So, if you want to use `LEFT JOIN` (at least until `https://github.com/jOOQ/jOOQ/issues/11888` is resolved), you can rely on the mighty `ResultQuery.collect()` collector, as follows:

```
Map<Productline, List<Product>> result = ctx.select()
   .from(PRODUCTLINE)
   .leftOuterJoin(PRODUCT)
   .on(PRODUCTLINE.PRODUCT_LINE.eq(PRODUCT.PRODUCT_LINE))
   .collect(groupingBy(
       r -> r.into(Productline.class),
          filtering(
             r -> r.get(PRODUCT.PRODUCT_ID) != null,
             mapping(
                 r -> r.into(Product.class),
                 toList()
             )
          )
   ));
```

This time, a parent with no children produces an empty list.

Fetching and mapping a many-to-many relationship is also possible. We can do it elegantly via `CROSS APPLY` (for additional details, check out *Chapter 6, Tackling Different Kinds of JOIN Statements*). For instance, we have a many-to-many relationship between `OFFICE` and `MANAGER` via the `OFFICE_HAS_MANAGER` junction table, and we can map it via `fetchGroups()`, as follows:

```
Map<Manager, List<Office>> result = ctx.select().from(MANAGER)
    .crossApply(select().from(OFFICE).join(OFFICE_HAS_MANAGER)
      .on(OFFICE.OFFICE_CODE
         .eq(OFFICE_HAS_MANAGER.OFFICES_OFFICE_CODE))
   .where(MANAGER.MANAGER_ID
      .eq(OFFICE_HAS_MANAGER.MANAGERS_MANAGER_ID)))
   .fetchGroups(Manager.class, Office.class);
```

Passing this map through a REST controller produces the necessary JSON. Of course, mapping a one-to-many relationship with a junction table is quite obvious based on the previous examples.

However, please consider Lukas Eder's note:

"When talking about fetchGroups(), I think it's always worth pointing out that RDBMS can often do this natively as well, using ARRAY_AGG(), JSON_ARRAYAGG(), or XMLAGG(). Chances are (to be verified), that this may be faster, as less data has to be transferred over the wire."

In the bundled code, you can practice many more examples of how to use `fetchGroups()`. The application is named *FetchGroups* (and is available for PostgreSQL).

Fetching via JDBC ResultSet

jOOQ is an extremely versatile and transparent tool. For instance, jOOQ acts as a wrapper for JDBC `ResultSet` but also allows us to access it directly and even provide support to do this smoothly and *painlessly*. Practically, we can do the following:

- Execute a `ResultQuery` with jOOQ, but return a JDBC `ResultSet` (this relies on the `fetchResultSet()` method).
- Transform the jOOQ `Result` object into a JDBC `ResultSet` (this relies on the `intoResultSet()` method).
- Fetch data from a legacy `ResultSet` using jOOQ.

All three of these bullets are exemplified in the bundled code. However, here, let's consider the second bullet that starts with the following jOOQ query:

```
// Result<Record2<String, BigDecimal>>
var result = ctx.select(CUSTOMER.CUSTOMER_NAME,
   CUSTOMER.CREDIT_LIMIT).from(CUSTOMER).fetch();
```

We understand that the returned result is a jOOQ-specific `Result` that was built automatically from the underlying `ResultSet`. So, can we reverse this operation and obtain the `ResultSet` from the jOOQ `Result`? Yes, we can! We can do this via `intoResultSet()`, as follows:

```
ResultSet rsInMem = result.intoResultSet();
```

The important thing to note is that this magic happens without an active connection to the database. By default, jOOQ closes the database connection after the jOOQ `Result` is fetched. This means that, when we call `intoResultSet()` to obtain this in-memory `ResultSet`, there is no active connection to the database. jOOQ mirrors the `Result` object back into a `ResultSet` without interacting with the database. Next, processing this `ResultSet` is straightforward:

```
while (rsInMem.next()) {
   ...
}
```

This matters because, typically, operating on a JDBC `ResultSet` can be done as long as you hold an open connection to your database. Check out the complete code next to the other two bullets in the bundled application named *ResultSetFetch* (which is available for MySQL).

Fetching multiple result sets

Some RDBMSs (for instance, SQL Server and MySQL after appending the `allowMultiQueries=true` property to the JDBC URL) can return multiple result sets. Here is such a jOOQ query for MySQL:

```
ctx.resultQuery(
  "SELECT * FROM employee LIMIT 10;
   SELECT * FROM sale LIMIT 5");
```

To fetch multiple result sets in jOOQ, call `fetchMany()`. This method returns an object of the `org.jooq.Results` type, as shown in the following snippet (notice the pluralization to avoid any confusion with `org.jooq.Result`):

```
Results results = ctx.resultQuery(
   "SELECT * FROM employee LIMIT 10;
    SELECT * FROM sale LIMIT 5")
  .fetchMany();
```

Next, you can map each result set to its POJO:

```
List<Employee> employees =results.get(0).into(Employee.class);
List<Sale> sales = results.get(1).into(Sale.class);
```

Lukas Eder says:

"Perhaps out of scope, but the Results type also allows for accessing interleaved update counts and exceptions, which is something that is done frequently in T-SQL databases, like SQL Server or Sybase."

Done! In the *FetchMany* application (which is available for MySQL and SQL Server), you can check out this example next to another one that returns two result sets from a query that combines `DELETE` and `SELECT`.

Fetching relationships

I'm pretty sure that you're familiar with the one-to-one, one-to-many, and many-to-many relationships. An emblematic mapping of unidirectional one-to-many roughly looks like this:

```
public class SimpleProductLine implements Serializable {
   ...
   private List<SimpleProduct> products = new ArrayList<>();
}

public class SimpleProduct implements Serializable { ... }
```

Moreover, when `SimpleProduct` contains a reference to `SimpleProductLine`, this is considered a bidirectional one-to-many relationship:

```
public class SimpleProduct implements Serializable {
   ...
   private SimpleProductLine productLine;
}
```

If we have this POJO model, can we map the corresponding result set to it via the jOOQ API? The answer is definitely yes, and this can be done in several ways. From the `fetchInto()`, `fetchMap()`, and `fetchGroups()` methods that you already saw to the record mappers, the mighty SQL JSON/XML mapping, and the astonishing MULTISET value constructor operator, jOOQ provides so many fetching modes that it is almost impossible to not find a solution.

Anyway, let's not deviate too much from the subject. Let's consider the following query:

```
// Map<Record, Result<Record>>
var map = ctx.select(PRODUCTLINE.PRODUCT_LINE,
     PRODUCTLINE.TEXT_DESCRIPTION,PRODUCT.PRODUCT_NAME,
     PRODUCT.PRODUCT_VENDOR, PRODUCT.QUANTITY_IN_STOCK)
   .from(PRODUCTLINE)
   .join(PRODUCT)
   .on(PRODUCTLINE.PRODUCT_LINE
      .eq(PRODUCT.PRODUCT_LINE))
```

```
    .orderBy(PRODUCTLINE.PRODUCT_LINE).limit(3)
    .fetchGroups(new Field[]{PRODUCTLINE.PRODUCT_LINE,
           PRODUCTLINE.TEXT_DESCRIPTION},
                 new Field[]{PRODUCT.PRODUCT_NAME,
           PRODUCT.PRODUCT_VENDOR, PRODUCT.QUANTITY_IN_STOCK});
```

With `Map<Record, Result<Record>>` (which, most of the time, is all you need), we can populate our bidirectional domain model, as follows:

```
List<SimpleProductLine> result = map.entrySet()
    .stream()
    .map((e) -> {
       SimpleProductLine productLine
          = e.getKey().into(SimpleProductLine.class);
       List<SimpleProduct> products
          = e.getValue().into(SimpleProduct.class);

       productLine.setProducts(products);
       products.forEach(p ->
         ((SimpleProduct) p).setProductLine(productLine));

       return productLine;
    }).collect(Collectors.toList());
```

If you want to avoid passing through `fetchGroups()`, then you can rely on `ResultQuery.collect()` and `Collectors.groupingBy()`. This is especially useful if you want to run a `LEFT JOIN` statement since `fetchGroups()` has the following issue: `https://github.com/jOOQ/jOOQ/issues/11888`. Another approach is to map from `ResultSet`. You can see these approaches along with other approaches for unidirectional/bidirectional one-to-one and many-to-many relationships in the bundled code in the *OneToOne*, *OneToMany*, and *ManyToMany* applications (which are available for MySQL).

Hooking POJOs

You already know that jOOQ can generate POJOs on our behalf and it can handle user-defined POJOs, too. Moreover, you saw a significant number of mappings of a jOOQ result into POJOs (typically, via `fetchInto()`); therefore, this is not a brand new topic for you. However, in this section, let's take a step further and really focus on different types of POJOs that are supported by jOOQ.

If all we configure is `<pojos>true</pojos>` (here, Maven), then jOOQ generates POJOs with `private` fields, empty constructors, constructors with arguments, getters and setters, and `toString()`. However, jOOQ can also handle a very simple user-defined POJO such as this one:

```
public class SimplestCustomer {
   public String customerName;
   public String customerPhone;
}
```

Here is a query that populates this POJO:

```
List<SimplestCustomer> result = ctx.select(
   CUSTOMER.CUSTOMER_NAME, CUSTOMER.PHONE.as("customerPhone"))
   .from(CUSTOMER).fetchInto(SimplestCustomer.class);
```

Pay attention to the `as("customerPhone")` alias. This is needed to map `CUSTOMER.PHONE` to POJO's `customerPhone` field; otherwise, this POJO field will be left `null` since jOOQ cannot find the proper match. Another approach is to add a constructor with arguments, as shown in the following POJO:

```
public class SimpleDepartment {

   private String depName;
   private Short depCode;
   private String[] depTopic;

   public SimpleDepartment(String depName,
          Short depCode, String[] depTopic) {
      this.depName = depName;
      this.depCode = depCode;
      this.depTopic = depTopic;
   }
   ...
}
```

Even if none of the POJO's field names match the names of the fetched fields, the POJO is correctly populated by jOOQ based on this constructor with arguments:

```
List<SimpleDepartment> result = ctx.select(
      DEPARTMENT.NAME, DEPARTMENT.CODE, DEPARTMENT.TOPIC)
```

```
    .from(DEPARTMENT).fetchInto(SimpleDepartment.class);

List<SimpleDepartment> result = ctx.select(
      DEPARTMENT.NAME, DEPARTMENT.CODE, DEPARTMENT.TOPIC)
    .from(DEPARTMENT)
    .fetch(mapping(SimpleDepartment::new));
```

User-defined POJOs are useful for mapping jOOQ results that contain fields from multiple tables. For example, a POJO can be used to flatten a one-to-many relationship, as shown here:

```
public class FlatProductline {

    private String productLine;
    private Long code;
    private String productName;
    private String productVendor;
    private Integer quantityInStock;

    // constructors, getters, setters, toString()
}
```

And, here's the jOOQ query:

```
List<FlatProductline> result = ctx.select(
      PRODUCTLINE.PRODUCT_LINE, PRODUCTLINE.CODE,
      PRODUCT.PRODUCT_NAME, PRODUCT.PRODUCT_VENDOR,
      PRODUCT.QUANTITY_IN_STOCK)
    .from(PRODUCTLINE)
    .join(PRODUCT)
    .on(PRODUCTLINE.PRODUCT_LINE.eq(PRODUCT.PRODUCT_LINE))
    .fetchInto(FlatProductline.class);
//  .fetch(mapping(FlatProductline::new));
```

Alternatively, you can map UDTs and/or embeddable types. For instance, here is a user-defined POJO that fetches a String and an embeddable type containing a UDT. For the embeddable type, we relied on the jOOQ-generated POJO:

```
import jooq.generated.embeddables.pojos.ManagerStatus;

public class SimpleManagerStatus {
```

```
    private Long managerId;
    private ManagerStatus ms;

    // constructors, getters, setters, toString()
}
```

And, the jOOQ query is as follows:

```
List<SimpleManagerStatus> result =
       ctx.select(MANAGER.MANAGER_ID, MANAGER.MANAGER_STATUS)
    .from(MANAGER).fetchInto(SimpleManagerStatus.class);
```

More examples are available in the bundled code (that is, in the *PojoTypes* application, which is available for PostgreSQL). Next, let's talk about the different types of POJOs supported by jOOQ.

Types of POJOs

Besides the typical POJOs from the previous section, jOOQ also supports several other types of POJOs. For instance, it supports immutable POJOs.

Immutable POJOs

A user-defined immutable POJO can be written as follows:

```
public final class ImmutableCustomer {

    private final String customerName;
    private final YearMonth ym;

    // constructor and only getters
}
```

And a jOOQ query that maps to this POJO is shown next:

```
List<ImmutableCustomer> result = ctx.select(
      CUSTOMER.CUSTOMER_NAME,
      CUSTOMER.FIRST_BUY_DATE.coerce(YEARMONTH))
    .from(CUSTOMER)
    .fetchInto(ImmutableCustomer.class);
//  .fetch(mapping(ImmutableCustomer::new));
```

To work as expected, immutable POJOs require an exact match between the fetched fields and the POJO's fields (the constructor arguments). However, you can explicitly relax this match via `@ConstructorProperties` (`java.beans.ConstructorProperties`). Please check the bundled code (*Example 2.2*) for a meaningful example.

jOOQ can generate immutable POJOs on our behalf via the following configuration in the `<generate/>` tag:

```
<immutablePojos>true</immutablePojos>
```

Also, it can generate `@ConstructorProperties` via the following:

```
<constructorPropertiesAnnotationOnPojos>
   true
</constructorPropertiesAnnotationOnPojos>
```

In the bundled code, next to the other examples, you can also practice mapping UDTs and embeddable types via user-defined immutable POJOs.

POJOs decorated with @Column (jakarta.persistence.Column)

jOOQ can map a `Result` object to a POJO as follows:

```
public class JpaCustomer {

   @Column(name = "customer_name")
   public String cn;

   @Column(name = "first_buy_date")
   public YearMonth ym;
}
```

As you can see, jOOQ recognizes the `@Column` annotation and uses it as the primary source for mapping metainformation:

```
List<JpaCustomer> result = ctx.select(CUSTOMER.CUSTOMER_NAME,
       CUSTOMER.FIRST_BUY_DATE.coerce(YEARMONTH))
   .from(CUSTOMER).fetchInto(JpaCustomer.class);
```

jOOQ can generate such POJOs via the following configuration in `<generate/>`:

```
<jpaAnnotations>true</jpaAnnotations>
```

Check out more examples in the bundled code.

JDK 16 records

Consider the following JDK 16 record:

```
public record RecordDepartment(
   String name, Integer code, String[] topic) {}
```

And the jOOQ query is as follows:

```
List<RecordDepartment> result = ctx.select(
        DEPARTMENT.NAME, DEPARTMENT.CODE, DEPARTMENT.TOPIC)
              .from(DEPARTMENT)
              .fetchInto(RecordDepartment.class);
           // .fetch(mapping(RecordDepartment::new));
```

Alternatively, here is a user-defined JDK 16 record along with a UDT type:

```
public record RecordEvaluationCriteria(
   Integer communicationAbility, Integer ethics,
   Integer performance, Integer employeeInput) {}

public record RecordManager(
   String managerName, RecordEvaluationCriteria rec) {}
```

And the jOOQ query is as follows:

```
List<RecordManager> result = ctx.select(
       MANAGER.MANAGER_NAME, MANAGER.MANAGER_EVALUATION)
   .from(MANAGER).fetchInto(RecordManager.class);
```

Or you can use a user-defined JDK 16 record with an embeddable type (here, we are using the POJO generated by jOOQ for the embeddable type):

```
import jooq.generated.embeddables.pojos.OfficeFullAddress;

public record RecordOffice(
   String officecode, OfficeFullAddress ofa) {}
```

And here is the jOOQ query:

```
List<RecordOffice> result = ctx.select(
         OFFICE.OFFICE_CODE, OFFICE.OFFICE_FULL_ADDRESS)
   .from(OFFICE).fetchInto(RecordOffice.class);
```

jOOQ can generate JDK 16 records on our behalf via the following configuration in `<generate/>`:

```
<pojosAsJavaRecordClasses>true</pojosAsJavaRecordClasses>
```

In the bundled code, you can practice JDK 16 records for UDT, embeddable types, and more.

Interfaces and abstract classes

Finally, jOOQ can map a result into interfaces (abstract classes) known as "proxyable" types. You can find examples in the bundled code and in the jOOQ manual at `https://www.jooq.org/doc/latest/manual/sql-execution/fetching/pojos/`.

Moreover, jOOQ can generate interfaces on our behalf via this configuration in the `<generate/>` tag:

```
<interfaces>true</interfaces>
```

If POJOs are also generated, then they will implement these interfaces.

Useful configurations for POJOs

Among POJO's configurations, we can ask jOOQ to not generate the `toString()` method for the POJO via the `<pojosToString/>` flag, to not generate serializable POJOs (to not implement `Serializable`) via the `<serializablePojos/>` flag, and to generate fluent setters via the `<fluentSetters/>` flag. Moreover, besides POJOs for Java, we can ask jOOQ to generate POJOs for Kotlin via the `<pojosAsKotlinDataClasses/>` flag or for Scala via the `<pojosAsScalaCaseClasses/>` flag.

In addition, under the `<database/>` tag, we can force `LocalDateTime` into POJOs via `<dateAsTimestamp/>` and use unsigned types via `<unsignedTypes/>`.

The complete code is named *PojoTypes* (which is available for PostgreSQL (Maven/Gradle)). In the next section, let's talk about record mappers.

jOOQ record mappers

Sometimes, we need a custom mapping that cannot be achieved via the `fetchInto()` method, the `fetchMap()` method, the `fetchGroups()` method, or the `Records` utility. A simple approach relies on `Iterable.forEach(Consumer)`, as shown in the following mapping:

```
ctx.select(EMPLOYEE.FIRST_NAME,
           EMPLOYEE.LAST_NAME, EMPLOYEE.EMAIL)
```

```
.from(EMPLOYEE)
.forEach((Record3<String, String, String> record) -> {
   System.out.println("\n\nTo: "
    + record.getValue(EMPLOYEE.EMAIL));
   System.out.println("From: "
    + "hrdepartment@classicmodelcars.com");
   System.out.println("Body: \n   Dear, "
    + record.getValue(EMPLOYEE.FIRST_NAME)
    + " " + record.getValue(EMPLOYEE.LAST_NAME) + " ...");
});
```

You can check out this example for MySQL in *ForEachConsumer*.

However, especially for such cases, jOOQ provides a functional interface that allows us to express the custom mappings of a jOOQ result. In this context, we have `org.jooq.RecordMapper`, which returns the result produced after applying a custom mapping to the jOOQ result. For instance, let's consider a legacy POJO that was implemented via the `Builder` pattern and is named `LegacyCustomer`:

```
public final class LegacyCustomer {

   private final String customerName;
   private final String customerPhone;
   private final BigDecimal creditLimit;
   ...
   public static CustomerBuilder getBuilder(
                                String customerName) {
      return new LegacyCustomer.CustomerBuilder(customerName);
   }

   public static final class CustomerBuilder {
      ...
      public LegacyCustomer build() {
         return new LegacyCustomer(this);
      }
   }
   ...
}
```

Mapping a jOOQ result into `LegacyCustomer` can be done via a `RecordMapper` parameter, as follows:

```
List<LegacyCustomer> result
        = ctx.select(CUSTOMER.CUSTOMER_NAME, CUSTOMER.PHONE,
                     CUSTOMER.CREDIT_LIMIT)
    .from(CUSTOMER)
    .fetch((Record3<String, String, BigDecimal> record) -> {
       LegacyCustomer customer = LegacyCustomer.getBuilder(
          record.getValue(CUSTOMER.CUSTOMER_NAME))
          .customerPhone(record.getValue(CUSTOMER.PHONE))
          .creditLimit(record.getValue(CUSTOMER.CREDIT_LIMIT))
          .build();

       return customer;
});
```

This example is available in the bundled code, *RecordMapper* (which is available for PostgreSQL), next to other examples such as using a `RecordMapper` parameter to map a jOOQ result into a max-heap. Moreover, in *Chapter 18, jOOQ SPI (Providers and Listeners)*, you'll see how to configure record mappers via `RecordMapperProvider` so that jOOQ will automatically pick them up.

However, if you need more generic mapping algorithms, then we have to check out some third-party libraries that work with jOOQ. In the top three such libraries, we have ModelMapper, SimpleFlatMapper, and Orika Mapper.

It is beyond the scope of this book to deep dive into all these libraries. Therefore, I decided to go with the SimpleFlatMapper library (`https://simpleflatmapper.org/`). Let's assume the following one-to-many mapping:

```
public class SimpleProductLine {

   private String productLine;
   private String textDescription;
   private List<SimpleProduct> products;
   ...
}

public class SimpleProduct {

   private String productName;
```

```
    private String productVendor;
    private Short quantityInStock;
    ...
}
```

Essentially, SimpleFlatMapper can map a jOOQ result via `SelectQueryMapper`, as shown in the following self-explanatory example:

```
private final SelectQueryMapper<SimpleProductLine> sqMapper;
private final DSLContext ctx;

public ClassicModelsRepository(DSLContext ctx) {
    this.ctx = ctx;
    this.sqMapper = SelectQueryMapperFactory
       .newInstance().newMapper(SimpleProductLine.class);
}

public List<SimpleProductLine> findProductLineWithProducts() {

    List<SimpleProductLine> result = sqMapper.asList(
      ctx.select(PRODUCTLINE.PRODUCT_LINE,
          PRODUCTLINE.TEXT_DESCRIPTION,
          PRODUCT.PRODUCT_NAME, PRODUCT.PRODUCT_VENDOR,
          PRODUCT.QUANTITY_IN_STOCK)
         .from(PRODUCTLINE)
         .innerJoin(PRODUCT)
         .on(PRODUCTLINE.PRODUCT_LINE
            .eq(PRODUCT.PRODUCT_LINE))
         .orderBy(PRODUCTLINE.PRODUCT_LINE));

    return result;
}
```

In this code, SimpleFlatMapper maps the jOOQ result, so it acts directly on the jOOQ records. This code is available in the *SFMOneToManySQM* application (available for MySQL). However, as you can see in the *SFMOneToManyJM* application, this library can also take advantage of the fact that jOOQ allows us to manipulate the `ResultSet` object itself, so it can act directly on the `ResultSet` object via an API named `JdbcMapper`. This way, SimpleFlatMapper bypasses the jOOQ mapping to `Record`.

Moreover, the bundled code includes applications for mapping one-to-one and many-to-many relationships next to *SFMOneToManyTupleJM*, which is an application that combines SimpleFlatMapper and the jOOL `Tuple2` API to map a one-to-many relationship without using POJOs. For brevity, we cannot list this code in the book, so you need to reserve some time to explore it by yourself.

From another perspective, via the same `SelectQueryMapper` and `JdbcMapper` APIs, the SimpleFlatMapper library can co-work with jOOQ to map chained and/or nested `JOIN` statements. For instance, consider this model:

```
public class SimpleEmployee {

   private Long employeeNumber;
   private String firstName;
   private String lastName;

   private Set<SimpleCustomer> customers;
   private Set<SimpleSale> sales;
   ...
}

public class SimpleCustomer { private String customerName; … }
public class SimpleSale { private Float sale; … }
```

Using the SimpleFlatMapper and jOOQ combination, we can populate this model as follows:

```
this.sqMapper = …;

List<SimpleEmployee> result = sqMapper.asList(
   ctx.select(EMPLOYEE.EMPLOYEE_NUMBER, EMPLOYEE.FIRST_NAME,
       EMPLOYEE.LAST_NAME, CUSTOMER.CUSTOMER_NAME, SALE.SALE_)
      .from(EMPLOYEE)
      .leftOuterJoin(CUSTOMER)
      .on(CUSTOMER.SALES_REP_EMPLOYEE_NUMBER
        .eq(EMPLOYEE.EMPLOYEE_NUMBER))
      .leftOuterJoin(SALE)
      .on(EMPLOYEE.EMPLOYEE_NUMBER
        .eq(SALE.EMPLOYEE_NUMBER))
      .where(EMPLOYEE.OFFICE_CODE.eq(officeCode))
      .orderBy(EMPLOYEE.EMPLOYEE_NUMBER));
```

The complete code is named *SFMMultipleJoinsSQM*. The version of this code that uses `JdbcMapper` is named *SFMMultipleJoinsJM*. Moreover, in the bundled code, you can find an example of mapping a deep hierarchical JOIN of type (`EMPLOYEE` has `CUSTOMER` has `ORDER` has `ORDERDETAIL` has `PRODUCT`). This JOIN is also mapped in *SFMMultipleJoinsInnerLevelsTupleJM* using jOOL `Tuple2` and no POJOs. Anyway, even if such things work, I don't recommend you to do it in real applications. You better rely on the SQL/JSON/XML operators or MULTISET, as you'll do later.

Again, for brevity, we cannot list this code in the book, so you need to reserve some time to explore it by yourself. At this point, we have reached the climax of this chapter. It's time to beat the drums because the next section covers the outstanding mapping support of jOOQ SQL/JSON and SQL/XML.

The mighty SQL/JSON and SQL/XML support

Starting with jOOQ 3.14, we have support for mapping a result set to any kind of hierarchical/nested structure that can be shaped via JSON or XML into, practically, almost anything. For instance, if you develop a REST API, you can return JSON/XML data in the exact desired shape without mapping anything to your domain model.

As you probably know, most RDBMSs support SQL/JSON (standard or vendor-specific), and some of them support SQL/XML, too.

Handling SQL/JSON support

In a nutshell, for SQL/JSON, we can talk about the following operators that have a jOOQ implementation in the `org.jooq.impl.DSL` class:

- `JSON_OBJECT` (`DSL.jsonObject()`, `DSL.jsonEntry()`), `JSON_ARRAY` (`DSL.jsonArray()`), and `JSON_VALUE` (`DSL.jsonValue()`) to construct JSON data from values
- `JSON_ARRAYAGG` (`DSL.jsonArrayAgg()`) and `JSON_OBJECTAGG` (`DSL.jsonObjectAgg()`) to aggregate data into nested JSON documents
- `JSON_EXISTS` (`DSL.jsonExists()`) to query documents with the JSON path
- `JSON_TABLE` (`DSL.jsonTable()`) to transform JSON values into SQL tables
- SQL Server's `FOR JSON` syntax (including `ROOT`, `PATH`, `AUTO`, and `WITHOUT_ARRAY_WRAPPER`); the jOOQ commercial edition emulates the `FOR JSON` syntax for the databases that don't support it (in this book, you can see this for SQL Server and Oracle)

Let's see some introductory examples of these operators via the jOOQ DSL API.

Constructing and aggregating JSON data from the values

Constructing JSON data from values can be done via the `JSON_OBJECT` operator. This is implemented in jOOQ via different flavors of the `DSL.jsonObject()` method. For instance, you can map the `CUSTOMER.CUSTOMER_NAME` and `CUSTOMER.CREDIT_LIMIT` fields to an `org.jooq.JSON` object as follows:

```
Result<Record1<JSON>> result = ctx.select(jsonObject(
   key("customerName").value(CUSTOMER.CUSTOMER_NAME),
   key("creditLimit").value(CUSTOMER.CREDIT_LIMIT))
      .as("json_result"))
   .from(CUSTOMER).fetch();
```

In contrast to the `key().value()` construction, we can use `jsonObject(JSONEntry<?>... entries)`, as follows:

```
Result<Record1<JSON>> result = ctx.select(jsonObject(
   jsonEntry("customerName", CUSTOMER.CUSTOMER_NAME),
   jsonEntry("creditLimit", CUSTOMER.CREDIT_LIMIT))
      .as("json_result"))
   .from(CUSTOMER).fetch();
```

The returned `Result` object (remember that this is a wrapper of `java.util.List`) has a size equal to the number of fetched customers. Each `Record1` object wraps an `org.jooq.JSON` instance representing a customer name and credit limit. If we just want to format this `Result` object as a JSON, we can call the `formatJSON()` method (this will be presented in the next chapter). This will return a simple formatted representation such as the one here:

```
System.out.println(result.formatJSON());
```

```
{
 "fields": [{"name": "json_result", "type": "JSON"}],
 "records":
 [
  [{"creditLimit": 21000, "customerName": "Australian Home"}],
  [{"creditLimit": 21000, "customerName": "Joliyon"}],
  [{"creditLimit": 21000, "customerName": "Marquez Xioa"}]
  ...
```

```
  ]
}
```

However, this response is too verbose to send to the client. For instance, you might only need the "records" key. In such cases, we can rely on the formatJSON(JSONFormat) flavor as follows:

```
System.out.println(
   result.formatJSON(JSONFormat.DEFAULT_FOR_RECORDS));
```

```
[
 [{"creditLimit": 50000.00, "customerName":"GOLD"}],
 [{"creditLimit": null, "customerName": "Australian Home"}],
 [{"creditLimit": null, "customerName": "Joliyon"}],
 ...
]
```

Supposing that you just want to send the first JSON array, you can extract it from the Result object as result.get(0).value1().data():

```
result.get(0) // 0-first JSON, 1-second JSON, 2-third JSON …
.value1()     // this is the value from Record1, a JSON
.data()       // this is the data of the first JSON as String
```

```
{"creditLimit": 21000.00, "customerName": "Australian Home"}
```

However, perhaps you are planning to send all these JSONs as a List<String> to the client. Then, rely on fetchInto(String.class), which will return all of the JSONs as a List<String>. Note that each String is a JSON:

```
List<String> result = ctx.select(jsonObject(
      jsonEntry("customerName", CUSTOMER.CUSTOMER_NAME),
      jsonEntry("creditLimit", CUSTOMER.CREDIT_LIMIT))
         .as("json_result"))
   .from(CUSTOMER).fetchInto(String.class);
```

Also, you can send the response as a list of JSON arrays. Just wrap each JSON object into an array via jsonArray(), as follows:

```
List<String> result = ctx.select(jsonArray(jsonObject(
   jsonEntry("customerName", CUSTOMER.CUSTOMER_NAME),
   jsonEntry("creditLimit", CUSTOMER.CREDIT_LIMIT)))
```

```
        .as("json_result"))
    .from(CUSTOMER).fetchInto(String.class);
```

This time, the first JSON array (at index 0 in the list) is *[{"creditLimit": 21000.00, "customerName": "Australian Home"}]*, the second one (at index 1 in the list) is *[{"creditLimit": 21000, "customerName": "Joliyon"}]*, and so on.

However, it is more practical to aggregate all of these JSONs into a single array. This is possible via `jsonArrayAgg()`, which will return a single JSON array containing all of the fetched data:

```
String result = ctx.select(jsonArrayAgg(jsonObject(
      jsonEntry("customerName", CUSTOMER.CUSTOMER_NAME),
      jsonEntry("creditLimit", CUSTOMER.CREDIT_LIMIT)))
         .as("json_result"))
   .from(CUSTOMER).fetchSingleInto(String.class);
```

The aggregated JSON array is given here:

```
[
  {"creditLimit": 21000,"customerName": "Australian Home"},
  {"creditLimit": 21000,"customerName": "Joliyon"},
  ...
]
```

However, we can also aggregate the fetched data as a single JSON object that has `CUSTOMER_NAME` as the key and `CREDIT_LIMIT` as the value. This can be done via the `jsonObjectAgg()` method, as follows:

```
String result = ctx.select(jsonObjectAgg(
      CUSTOMER.CUSTOMER_NAME, CUSTOMER.CREDIT_LIMIT)
         .as("json_result"))
   .from(CUSTOMER).fetchSingleInto(String.class);
```

This time, the resulting JSON is as follows:

```
{
  "Joliyon": 21000,
  "Falafel 3": 21000,
  "Petit Auto": 79900,
…
}
```

If you are a SQL Server fan, then you know that fetching data as JSON can be done via the non-standard `FOR JSON` syntax. jOOQ supports this syntax via the `forJson()` API. It also supports clauses such as `ROOT` via `root()`, `PATH` via `path()`, `AUTO` via `auto()`, and `WITHOUT_ARRAY_WRAPPER` via `withoutArrayWrapper()`. Here is an example that produces nested results by using dot-separated column names via `PATH`:

```
Result<Record1<JSON>> result = ctx.select(
       CUSTOMER.CONTACT_FIRST_NAME, CUSTOMER.CREDIT_LIMIT,
       PAYMENT.INVOICE_AMOUNT.as("Payment.Amount"),
       PAYMENT.CACHING_DATE.as("Payment.CachingDate"))
    .from(CUSTOMER)
    .join(PAYMENT)
    .on(CUSTOMER.CUSTOMER_NUMBER.eq(PAYMENT.CUSTOMER_NUMBER))
    .orderBy(CUSTOMER.CREDIT_LIMIT).limit(5)
    .forJSON().path().root("customers")
    .fetch();
```

And here is an example of using `AUTO`, which automatically produces the output based on the structure of the `SELECT` statement:

```
Result<Record1<JSON>> result = ctx.select(
       CUSTOMER.CONTACT_FIRST_NAME, CUSTOMER.CREDIT_LIMIT,
       PAYMENT.INVOICE_AMOUNT, PAYMENT.CACHING_DATE)
   .from(CUSTOMER)
   .join(PAYMENT)
   .on(CUSTOMER.CUSTOMER_NUMBER.eq(PAYMENT.CUSTOMER_NUMBER))
   .orderBy(CUSTOMER.CREDIT_LIMIT).limit(5)
   .forJSON().auto().withoutArrayWrapper().fetch();
```

You can check out these examples in the bundled code for *SimpleJson* and get familiar with the produced JSONs. For now, let's talk about ordering and limiting the content of the resulting JSON when using SQL-standard JSON operators (for SQL Server's `FOR JSON` syntax, consider the previous two examples).

Using ORDER BY and LIMIT

When we don't use aggregation operators, ordering and limiting are quite similar to regular queries. For instance, you can order by `CUSTOMER_NAME` and limit the result to three JSONs as follows:

```
List<String> result = ctx.select(jsonObject(
      key("customerName").value(CUSTOMER.CUSTOMER_NAME),
```

```
key("creditLimit").value(CUSTOMER.CREDIT_LIMIT))
      .as("json_result"))
      .from(CUSTOMER)
   .orderBy(CUSTOMER.CUSTOMER_NAME).limit(3)
   .fetchInto(String.class);
```

On the other hand, when the aggregation operators (`jsonArrayAgg()` and `jsonObjectAgg()`) are involved, limiting should be done before the aggregation (for instance, in a subquery, `JOIN`, and more). Otherwise, this operation will be applied to the resulted aggregation itself, not to the aggregated data. During aggregation, ordering can be done before limiting, respectively. For instance, in the following example, the subquery orders the customers by `CUSTOMER_NAME` and limits the returned result to 3, while the aggregation orders this result by `CREDIT_LIMIT`:

```
String result = ctx.select(jsonArrayAgg(jsonObject(
      jsonEntry("customerName", field("customer_name")),
      jsonEntry("creditLimit", field("credit_limit"))))
        .orderBy(field("credit_limit")).as("json_result"))
   .from(select(CUSTOMER.CUSTOMER_NAME, CUSTOMER.CREDIT_LIMIT)
     .from(CUSTOMER).orderBy(CUSTOMER.CUSTOMER_NAME).limit(3))
   .fetchSingleInto(String.class);
```

The resulting aggregation is ordered by `CREDIT_LIMIT`:

```
[
  {"creditLimit": 0,"customerName": "American Souvenirs Inc"},
  {"creditLimit": 61100,"customerName": "Alpha Cognac"},
  {"creditLimit": 113000,"customerName": "Amica Models & Co."}
]
```

More examples are available in the bundled code for *SimpleJson*. Note that, in the applications that uses PostgreSQL and Oracle, you can see the SQL standard's `NULL ON NULL` (`nonOnNull()`) and `ABSENT ON NULL` (`absentOnNull()`) syntax at work. For now, let's query documents with the JSON path.

Querying JSON documents with the JSON path

Via `JSON_EXISTS` and `JSON_VALUE`, we can query and construct JSON documents that rely on the JSON path. In order to practice jOOQ's `jsonExists()` and `jsonValue()` queries, let's consider the `MANAGER.MANAGER_DETAIL` field, which stores data in JSON format. Please take a quick look at this JSON so that you can become familiar with its structure and content.

Now, selecting the `MANAGER.MANAGER_ID` and `MANAGER.MANAGER_NAME` fields of the managers that are also shareholders(with the `"shareholder"` key in JSON) can be done via `jsonExists()` and the JSON path, as follows:

```
Result<Record2<Long, String>> result = ctx.select(
        MANAGER.MANAGER_ID, MANAGER.MANAGER_NAME)
    .from(MANAGER)
    .where(jsonExists(MANAGER.MANAGER_DETAIL, "$.shareholder"))
    .fetch();
```

If the fetched JSON is constructed from JSON values, then we should rely on `jsonValue()` and the JSON path. For instance, fetching the cities of all managers can be done like this:

```
Result<Record1<JSON>> result = ctx.select(
      jsonValue(MANAGER.MANAGER_DETAIL, "$.address.city")
         .as("city"))
   .from(MANAGER).fetch();
```

Combining `jsonExists()` and `jsonValue()` allows us to query and construct JSON results from JSON documents. For instance, in PostgreSQL and Oracle, we can select the emails of the managers that had the role of *Principal Manager* by exploiting the JSON path:

```
Result<Record1<JSON>> result = ctx.select(
      jsonValue(MANAGER.MANAGER_DETAIL, "$.email").as("email"))
    .from(MANAGER)
    .where(jsonExists(MANAGER.MANAGER_DETAIL,
      "$[*] ? (@.projects[*].role == \"Principal Manager\")"))
    .fetch();
```

More examples are available in the bundled code, *SimpleJson*. Next, let's tackle `JSON_TABLE`.

Transforming JSON values into SQL tables

Transforming JSON values into SQL tables can be done via the `JSON_TABLE` operator, which, in jOOQ, is equivalent to the `jsonTable()` method. For instance, let's build a SQL table containing all projects of the development type via the `jsonTable(Field<JSON> json, Field<String> path)` flavor:

```
Result<Record> result = ctx.select(table("t").asterisk())
    .from(MANAGER, jsonTable(MANAGER.MANAGER_DETAIL,
```

```
                              val("$.projects[*]"))
   .column("id").forOrdinality()
   .column("name", VARCHAR).column("start", DATE)
   .column("end", DATE).column("type", VARCHAR)
   .column("role", VARCHAR).column("details", VARCHAR).as("t"))
   .where(field("type").eq("development")).fetch();
```

This query will produce a table, as follows:

```
+----+-------------------+----------+----------+-----------+-------------+--------------------+
|  id|name               |start     |end       |type       |role         |details             |
+----+-------------------+----------+----------+-----------+-------------+--------------------+
|   1|SBS (aka. Shared ... |2015-08-01|2019-11-21|development|Sr. Manager  |Develop a powerful ...|
|   2|RLCM (aka. Role  ... |2011-03-01|2015-07-01|development|Manager Leader|Increase income   ...|
|   2|Draft Management ... |2014-06-06|2016-08-02|development|Manager Leader|First e-management ...|
+----+-------------------+----------+----------+-----------+-------------+--------------------+
```

Figure 8.2 – The result of the previous query

Once you fetch a SQL table, you can think and act on it in the same way as any other database table. For brevity, I simply used `VARCHAR`, but it is better to specify a size in order to avoid defaulting to `VARCHAR(max)`.

More examples, including how to use `JSON_TABLE` with aggregates, `ORDER BY`, `LIMIT`, and how to transform back into JSON from a SQL table, are available in the bundled code, *SimpleJson*.

Handling relationships via SQL/JSON

The well-known one-to-one, one-to-many, and many-to-many relationships can be easily shaped via SQL/JSON support.

Mapping relationships to JSON

So, if by any chance you had a feeling that there is a shortcoming in jOOQ regarding mapping relationships, then you'll be very happy to see that a one-to-many relationship can be easily fetched directly into JSON as follows (in this case, we're looking at the relationship between `PRODUCTLINE` and `PRODUCT`):

```
Result<Record1<JSON>> result = ctx.select(jsonObject(
  key("productLine").value(PRODUCTLINE.PRODUCT_LINE),
  key("textDescription").value(PRODUCTLINE.TEXT_DESCRIPTION),
  key("products").value(select(jsonArrayAgg(
    jsonObject(key("productName").value(PRODUCT.PRODUCT_NAME),
     key("productVendor").value(PRODUCT.PRODUCT_VENDOR),
```

```
        key("quantityInStock").value(PRODUCT.QUANTITY_IN_STOCK)))
      .orderBy(PRODUCT.QUANTITY_IN_STOCK))
      .from(PRODUCT)
      .where(PRODUCTLINE.PRODUCT_LINE
        .eq(PRODUCT.PRODUCT_LINE)))))
        .from(PRODUCTLINE).orderBy(PRODUCTLINE.PRODUCT_LINE)
        .fetch();
```

As you can infer from the preceding code, expressing the one-to-one and many-to-many relationships is just a matter of juggling with the SQL/JSON operators. You can find these examples, including how to use `JOIN` instead of a `SELECT` subquery, in the bundled code for *JsonRelationships*.

If you think that `Result<Record1<JSON>>` is not ready to be sent to the client (for instance, via a REST controller), then decorate it a little bit more by aggregating all the product lines under a JSON array and relying on `fetchSingleInto()`:

```
String result = ctx.select(
   jsonArrayAgg(jsonObject(…))
   …
    .orderBy(PRODUCTLINE.PRODUCT_LINE))
    .from(PRODUCTLINE).fetchSingleInto(String.class);
```

In SQL Server, we can obtain a similar result via `forJson()`:

```
Result<Record1<JSON>> result = ctx.select(
 PRODUCTLINE.PRODUCT_LINE.as("productLine"),
 PRODUCTLINE.TEXT_DESCRIPTION.as("textDescription"),
 select(PRODUCT.PRODUCT_NAME.as("productName"),
  PRODUCT.PRODUCT_VENDOR.as("productVendor"),
  PRODUCT.QUANTITY_IN_STOCK.as("quantityInStock"))
 .from(PRODUCT)
 .where(PRODUCT.PRODUCT_LINE.eq(PRODUCTLINE.PRODUCT_LINE))
 .orderBy(PRODUCT.QUANTITY_IN_STOCK)
 .forJSON().path().asField("products"))
  .from(PRODUCTLINE)
  .orderBy(PRODUCTLINE.PRODUCT_LINE).forJSON().path().fetch();
```

Or we can obtain a `String` via `formatJSON(JSONformat)`:

```
String result = ctx.select(
  PRODUCTLINE.PRODUCT_LINE.as("productLine"),
```

```
...
.forJSON().path()
.fetch()
.formatJSON(JSONFormat.DEFAULT_FOR_RECORDS);
```

Both examples will produce a JSON, as follows (as you can see, altering the default JSON keys inferred from the field names can be done with aliases via `as("alias")`):

```
[
  {
    "productLine": "Classic Cars",
    "textDescription": "Attention car enthusiasts...",
    "products": [
      {
        "productName": "1968 Ford Mustang",
        "productVendor": "Autoart Studio Design",
        "quantityInStock": 68
      },
      {
        "productName": "1970 Chevy Chevelle SS 454",
        "productVendor": "Unimax Art Galleries",
        "quantityInStock": 1005
      }
      ...
    ]
  },
  {
    "productLine": "Motorcycles", ...
  }
]
```

In the bundled code for *JsonRelationships*, you can find a lot of examples to do with one-to-one, one-to-many, and many-to-many relationships. Moreover, you can check out several examples of how to map arrays and UDTs into JSON.

Mapping JSON relationships to POJOs

As you just saw, jOOQ can fetch and map a relationship directly into JSON. However, that's not all! jOOQ can go even further and map the resulted JSON to the domain model (POJOs). Yes, you read that right; as long as we have Gson, Jackson (Spring Boot has this by default), or JAXB in the classpath, jOOQ can automatically map the query results to our nested data structures. This is quite useful when you don't actually need the JSON itself – you can just rely on JSON to facilitate the nesting data structures and map them back to Java. For instance, let's assume the following domain model:

```
public class SimpleProductLine {

    private String productLine;
    private String textDescription;
    private List<SimpleProduct> products;
}

public class SimpleProduct {

    private String productName;
    private String productVendor;
    private Short quantityInStock;
}
```

Can we populate this model from jOOQ `Result` by just using jOOQ? Yes, we can do it via SQL/JSON support, as follows:

```
List<SimpleProductLine> result = ctx.select(jsonObject(
   key("productLine").value(PRODUCTLINE.PRODUCT_LINE),
   key("textDescription").value(PRODUCTLINE.TEXT_DESCRIPTION),
   key("products").value(select(jsonArrayAgg(jsonObject(
     key("productName").value(PRODUCT.PRODUCT_NAME),
     key("productVendor").value(PRODUCT.PRODUCT_VENDOR),
     key("quantityInStock").value(PRODUCT.QUANTITY_IN_STOCK)))
       .orderBy(PRODUCT.QUANTITY_IN_STOCK)).from(PRODUCT)
       .where(PRODUCTLINE.PRODUCT_LINE
          .eq(PRODUCT.PRODUCT_LINE)))))
   .from(PRODUCTLINE)
   .orderBy(PRODUCTLINE.PRODUCT_LINE)
   .fetchInto(SimpleProductLine.class);
```

That's so cool, right?! The same thing can be accomplished for one-to-one and many-to-many relationships, as you can see in the bundled code. All examples are available in *JsonRelationshipsInto*.

Mapping arbitrary models

If you think that what you've just seen is impressive, then get ready for more because jOOQ can fetch and map almost any kind of arbitrary model, not just the well-known 1:1,1:n, and n:n relationships. Let's consider the following three arbitrary models:

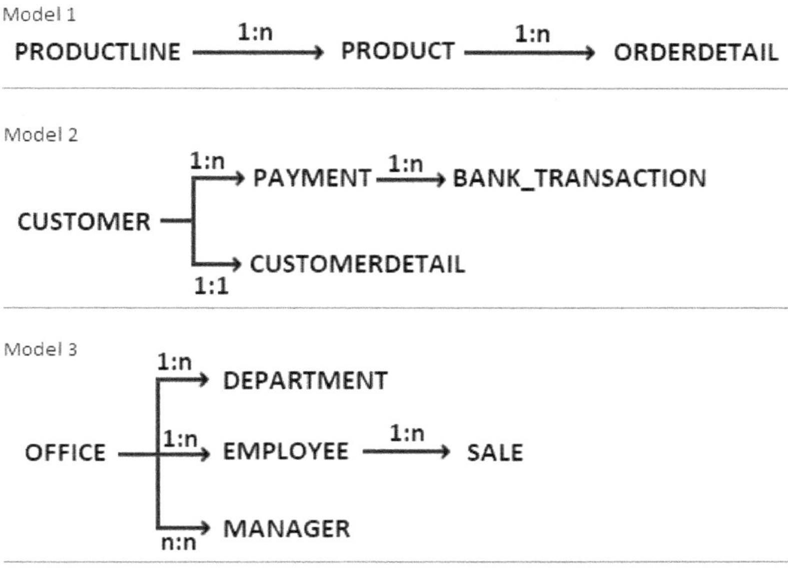

Figure 8.3 – Arbitrary domain models

Which one do you choose to be exemplified in the book? The second one (**Model 2**), of course! So, our goal is to write a query that returns a JSON that mirrors this model. For this, we rely on `jsonObject()` and `jsonArrayAgg()`, as follows:

```
Result<Record1<JSON>> result = ctx.select(jsonObject(
 jsonEntry("customerName", CUSTOMER.CUSTOMER_NAME),
 jsonEntry("creditLimit", CUSTOMER.CREDIT_LIMIT),
 jsonEntry("payments", select(jsonArrayAgg(jsonObject(
  jsonEntry("customerNumber", PAYMENT.CUSTOMER_NUMBER),
  jsonEntry("invoiceAmount", PAYMENT.INVOICE_AMOUNT),
  jsonEntry("cachingDate", PAYMENT.CACHING_DATE),
  jsonEntry("transactions", select(jsonArrayAgg(jsonObject(
    jsonEntry("bankName", BANK_TRANSACTION.BANK_NAME),
    jsonEntry("transferAmount",
              BANK_TRANSACTION.TRANSFER_AMOUNT)))
    .orderBy(BANK_TRANSACTION.TRANSFER_AMOUNT))
    .from(BANK_TRANSACTION)
```

```
    .where(BANK_TRANSACTION.CUSTOMER_NUMBER
      .eq(PAYMENT.CUSTOMER_NUMBER)
       .and(BANK_TRANSACTION.CHECK_NUMBER
         .eq(PAYMENT.CHECK_NUMBER))))))
    .orderBy(PAYMENT.CACHING_DATE))
    .from(PAYMENT)
    .where(PAYMENT.CUSTOMER_NUMBER
      .eq(CUSTOMER.CUSTOMER_NUMBER))),
  jsonEntry("details", select(
   jsonObject(jsonEntry("city", CUSTOMERDETAIL.CITY),
    jsonEntry("addressLineFirst",
          CUSTOMERDETAIL.ADDRESS_LINE_FIRST),
    jsonEntry("state", CUSTOMERDETAIL.STATE)))
   .from(CUSTOMERDETAIL)
   .where(CUSTOMERDETAIL.CUSTOMER_NUMBER
    .eq(CUSTOMER.CUSTOMER_NUMBER)))))
   .from(CUSTOMER).orderBy(CUSTOMER.CREDIT_LIMIT).fetch();
```

Lukas Eder states that:

"What I always like to mention in this regard is that there are no accidental Cartesian Products or costly de-duplication going on (as with JPA), because all the data is already nested correctly in SQL, and transferred optimally. This approach should be the first choice when nesting collections with SQL or producing JSON/XML for some frontend. Never use ordinary joins, which should be used only for flat results or aggregations."

Take your time to dissect this query and check out the bundled code to see the output. Moreover, in the bundled code, you can practice **Model 1** and **Model 3**, too. For each of these models, you have the JSON result and the corresponding mapping to POJOs. The application is named *NestedJson*.

I'm sure that, as a SQL Server fan, you are impatient to see the version of the previous query expressed via `forJson()`, so here it is:

```
Result<Record1<JSON>> result = ctx.select(
 CUSTOMER.CUSTOMER_NAME, CUSTOMER.CREDIT_LIMIT,
 select(PAYMENT.CUSTOMER_NUMBER, PAYMENT.INVOICE_AMOUNT,
  PAYMENT.CACHING_DATE,
   select(BANK_TRANSACTION.BANK_NAME,
          BANK_TRANSACTION.TRANSFER_AMOUNT)
   .from(BANK_TRANSACTION)
   .where(BANK_TRANSACTION.CUSTOMER_NUMBER
```

```
      .eq(PAYMENT.CUSTOMER_NUMBER)
      .and(BANK_TRANSACTION.CHECK_NUMBER
        .eq(PAYMENT.CHECK_NUMBER)))
    .orderBy(BANK_TRANSACTION.TRANSFER_AMOUNT)
    .forJSON().path().asField("transactions")).from(PAYMENT)
  .where(PAYMENT.CUSTOMER_NUMBER
    .eq(CUSTOMER.CUSTOMER_NUMBER))
  .orderBy(PAYMENT.CACHING_DATE)
  .forJSON().path().asField("payments"),
  select(CUSTOMERDETAIL.CITY,
   CUSTOMERDETAIL.ADDRESS_LINE_FIRST,CUSTOMERDETAIL.STATE)
  .from(CUSTOMERDETAIL)
   .where(CUSTOMERDETAIL.CUSTOMER_NUMBER
    .eq(CUSTOMER.CUSTOMER_NUMBER))
  .forJSON().path().asField("details")).from(CUSTOMER)
  .orderBy(CUSTOMER.CREDIT_LIMIT).forJSON().path().fetch();
```

Of course, you can check out these examples in the bundled code next to the examples for **Model 1** and **Model 3**.

> **Important Note**
>
> Next to SQL/JSON support, jOOQ also provides SQL/JSONB support. You can explicitly use JSONB via `org.jooq.JSONB` and the operators such as `jsonbObject()`, `jsonbArrayAgg()`, and `jsonbTable()`.

Now, it is time to talk about SQL/XML support.

Handling SQL/XML support

In a nutshell, for SQL/XML, we can talk about the following operators that have a jOOQ implementation in the `org.jooq.impl.DSL` class:

- `XMLELEMENT` (`DSL.xmlelement()`), `XMLATTRIBUTES` (`DSL.xmlattributes()`), `XMLFOREST` (`DSL.xmlforest()`), `XMLCONCAT` (`DSL.xmlconcat()`), and `XMLCOMMENT` (`DSL.xmlcomment()`) to construct XML data
- `XMLAGG` (`DSL.xmlagg()`) to aggregate data into nested XML documents
- `XMLEXISTS` (`DSL.xmlexists()`) and `XMLQUERY` (`DSL.xmlquery()`) to query XML documents with `XPath`

- `XMLPARSE` (`DSL.xmlparseContent()` and `DSL.xmlparseDocument()`) for parsing XML content and documents
- `XMLPI` (`DSL.xmlpi()`) for producing XML processing instructions
- `XMLTABLE` (`DSL.xmltable()`) to transform XML values into SQL tables

SQL Server's `FOR XML` syntax (including `ROOT`, `PATH`, `ELEMENTS`, `RAW`, and `AUTO`, and `EXPLICIT (jOOQ 3.17.x +)`) – jOOQ's commercial editions emulate the `FOR XML` syntax for databases that don't support it (in this book, you can practice this for SQL Server and Oracle).

Let's see some introductory examples of these operators via the jOOQ DSL API.

Constructing and aggregating XML data from values

A good start for constructing XML data from values relies on the `XMLELEMENT` operator. In jOOQ, `XMLELEMENT` is rendered via the `xmlelement()` method. For instance, the following snippet of code fetches the `CUSTOMER_NAME` field of each customer and uses it as the text of an XML element named `<name/>`:

```
Result<Record1<XML>> result = ctx.select(
      xmlelement("name", CUSTOMER.CUSTOMER_NAME))
   .from(CUSTOMER).fetch();
```

The returned `Result` has a size that is equal to the number of fetched customers. Each `Record1` wraps an `org.jooq.XML` instance representing a `<name/>` element. If we just want to format this `Result` as an XML, we can call the `formatXML()` method (this will be presented in the next chapter). This will return a simple formatted representation such as the one here:

```
<result xmlns="http:…">
<fields>
  <field name="xmlconcat" type="XML"/>
</fields>
<records>
  <record xmlns="http:…">
    <value field="xmlconcat">
      <name>Australian Home</name>
    </value>
  </record>
```

```
  <record xmlns="http:...">
    <value field="xmlconcat">
      <name>Joliyon</name>
    </value>
  </record>
...
```

However, this response is too verbose to send to the client. For instance, you might only need the `"records"` element. In such cases, we can rely on the `formatXML(XMLFormat)` flavor, as you'll see in the bundled code. Supposing that you want to just send the first `<name/>` element, you can extract it from the `Result` object as `result.get(0).value1().data()`:

```
result.get(0) // 0-first <name/>, 1-second <name/> ...
.value1()     // this is the value from Record1, a XML
.data()       // this is the data of the first XML as String
```

```
<name>Australian Home</name>
```

However, perhaps you are planning to send all of these `<name/>` tags as a `List<String>` to the client. Then, rely on `fetchInto(String.class)` to return all the `<name/>` elements as a `List<String>`. Note that each `String` is a `<name/>`:

```
List<String> result = ctx.select(
    xmlelement("name", CUSTOMER.CUSTOMER_NAME))
  .from(CUSTOMER).fetchInto(String.class);
```

Alternatively, it would be more practical to aggregate all these `<name/>` elements as a single `String`. This is possible via `xmlagg()`, which returns a single XML containing all of the fetched data (for convenience, let's aggregate everything under the `<names/>` tag):

```
String result = ctx.select(xmlelement("names", xmlagg(
      xmlelement("name", CUSTOMER.CUSTOMER_NAME))))
  .from(CUSTOMER).fetchSingleInto(String.class);
```

The aggregated XML is shown here:

```
<names>
  <name>Australian Home</name>
  <name>Joliyon</name>
  ...
</names>
```

What about adding attributes to the XML elements? This can be done via `xmlattributes()`, as shown in the following intuitive example:

```
Result<Record1<XML>> result = ctx.select(xmlelement("contact",
   xmlattributes(CUSTOMER.CONTACT_FIRST_NAME.as("firstName"),
   CUSTOMER.CONTACT_LAST_NAME.as("lastName"), CUSTOMER.PHONE)))
  .from(CUSTOMER).fetch();
```

The expected XML will look like this:

```
<contact firstName="Sart"
         lastName="Paoule" phone="40.11.2555"/>
```

A relatively useful XML operator is `xmlforest()`. This operator converts its parameters into XML and returns an XML fragment obtained by the concatenation of these converted arguments. Here is an example:

```
Result<Record1<XML>> result = ctx.select(
  xmlelement("allContacts", xmlagg(xmlelement("contact",
   xmlforest(CUSTOMER.CONTACT_FIRST_NAME.as("firstName"),
             CUSTOMER.CONTACT_LAST_NAME.as("lastName"),
             CUSTOMER.PHONE)))))
  .from(CUSTOMER).fetch();
```

The effect of `xmlforest()` can be seen in the resulting XML:

```
<allContacts>
      <contact>
           <firstName>Sart</firstName>
           <lastName>Paoule</lastName>
           <phone>40.11.2555</phone>
      </contact>
...
</allContacts>
```

If you are a SQL Server fan, then you know that fetching data as XML can be done via the non-standard `FOR XML` syntax. jOOQ supports this syntax via the `forXml()` API. It also supports clauses such as `ROOT` via `root()`, `PATH` via `path()`, `AUTO` via `auto()`, `RAW` via `raw()`, and `ELEMENTS` via `elements()`, as you can see in the following example:

```
Result<Record1<XML>> result = ctx.select(
         OFFICE.OFFICE_CODE, OFFICE.CITY, OFFICE.COUNTRY)
```

```
    .from(OFFICE)
    .forXML().path("office").elements().root("offices")
    .fetch();
```

The produced XML looks like this:

```
<offices>
     <office>
           <office_code>1</office_code>
           <city>San Francisco</city>
           <country>USA</country>
     </office>
     <office>
           <office_code>10</office_code>
     </office>
     <office>
           <office_code>11</office_code>
           <city>Paris</city>
           <country>France</country>
     </office>
...
</offices>
```

Note that missing tags (check the second `<office/>` instance, which does not have `<city/>` or `<country/>`) represent missing data.

As a side note, allow me to mention that jOOQ can also transform XML into an `org.w3c.dom.Document` by calling a flavor of `intoXML()` on `Record1<XML>`. Moreover, you'll love jOOX, or object-oriented XML (`https://github.com/jOOQ/jOOX`), which can be used to XSL transform or navigate the resulting XML document in a jQuery style.

I totally agree (sharing his enthusiasm) with Lukas Eder, who states:

"I don't know about you, but when I see these examples, I just want to write a huge application using jOOQ :) I mean, how else would anyone ever want to query databases and produce JSON or XML documents??"

You can see these examples (alongside many others) in the bundled code for *SimpleXml* and get familiar with the produced XMLs. For now, let's talk about how to order and limit the content of the resulting XML.

Using ORDER BY and LIMIT

When we don't use the `xmlagg()` aggregation operator, ordering and limiting is the same as for regular queries. For instance, you can order by `CUSTOMER_NAME` and limit the result to three XMLs as follows:

```
Result<Record1<XML>> result = ctx.select(
    xmlelement("name", CUSTOMER.CUSTOMER_NAME))
  .from(CUSTOMER)
  .orderBy(CUSTOMER.CUSTOMER_NAME).limit(3).fetch();
```

On the other hand, when the `xmlagg()` aggregation operator is used, then limiting should be done before the aggregation (for instance, in a subquery, `JOIN`, and more). Otherwise, this operation will be applied to the resulting aggregation itself. During aggregation, ordering can be done before limiting, respectively. For instance, in the following example, the subquery orders the customers by `CONTACT_LAST_NAME` and limits the returned results to 3, while the aggregation orders this result by `CONTACT_FIRST_NAME`:

```
String result = ctx.select(xmlelement("allContacts",
 xmlagg(xmlelement("contact",
   xmlforest(field("contact_first_name").as("firstName"),
    field("contact_last_name").as("lastName"),field("phone"))))
  .orderBy(field("contact_first_name"))))
  .from(select(CUSTOMER.CONTACT_FIRST_NAME,
               CUSTOMER.CONTACT_LAST_NAME, CUSTOMER.PHONE)
  .from(CUSTOMER).orderBy(CUSTOMER.CONTACT_LAST_NAME).limit(3))
  .fetchSingleInto(String.class);
```

The resulted aggregation is ordered by `CUSTOMER_FIRST_NAME`:

```
<allContacts>
     <contact>
           <firstName>Mel</firstName>
           <lastName>Andersen</lastName>
           <phone>030-0074555</phone>
     </contact>
     <contact>
           <firstName>Paolo</firstName>
           <lastName>Accorti</lastName>
           <phone>011-4988555</phone>
     </contact>
     <contact>
```

```
            <firstName>Raanan</firstName>
            <lastName>Altagar,G M</lastName>
            <phone>+ 972 9 959 8555</phone>
      </contact>
</allContacts>
```

More examples are available in the bundle code for *SimpleXml*. For now, let's learn how to query XML documents with XPath.

Querying XML documents with XPath

Querying XML documents can be done via the XPath expressions, and we can distinguish between queries that check for the existence of an element/attribute via `XMLEXISTS` (`xmlexists()`) and queries that fetches certain data from an XML document via `XMLQUERY` (`xmlquery()`). For instance, in `PRODUCTLINE`, we have a field named `HTML_DESCRIPTION` that holds the description of a product line in XML format. If a product line has a description, then this description starts with the `<productline/>` tag. So, fetching all product lines that have a description can be done via `xmlexists()`, as follows:

```
Result<Record1<String>> result =
   ctx.select(PRODUCTLINE.PRODUCT_LINE)
      .from(PRODUCTLINE)
      .where(xmlexists("/productline")
      .passing(PRODUCTLINE.HTML_DESCRIPTION)).fetch();
```

In `xmlexists("/productline").passing(…)`,`/productline` represents the XPath that should be searched, and the argument of the `passing()` method represents the XML document (or fragment) in which this XPath is searched.

On the other hand, the following snippet of code relies on `xmlquery()` to fetch an XML containing certain data from `HTML_DESCRIPTION`:

```
String result = ctx.select(xmlagg(
   xmlquery("productline/capacity/c[position()=last()]")
     .passing(PRODUCTLINE.HTML_DESCRIPTION)))
     .from(PRODUCTLINE).fetchSingleInto(String.class);
```

Of course, the argument of `passing()` can be an XML build from certain fields, too:

```
Result<Record1<XML>> result = ctx.select(
    xmlquery("//contact/phone").passing(
    xmlelement("allContacts", xmlagg(xmlelement("contact",
```

```
    xmlforest(CUSTOMER.CONTACT_FIRST_NAME.as("firstName"),
              CUSTOMER.CONTACT_LAST_NAME.as("lastName"),
              CUSTOMER.PHONE))))))
  .from(CUSTOMER).fetch();
```

This query fetches all the `<phone/>` tags from the given XML (for instance, `<phone>(26) 642-7555</phone>`). More examples are available in *SimpleXml*. Next, let's tackle `XMLTABLE`.

Transforming XML values into SQL tables

Transforming XML values into SQL tables can be done via the `XMLTABLE` operator, which, in jOOQ, is equivalent to `xmltable()`. For instance, let's build a SQL table containing the details of each product line extracted from `HTML_DESCRIPTION`:

```
Result<Record> result = ctx.select(table("t").asterisk())
   .from(PRODUCTLINE, xmltable("//productline/details")
     .passing(PRODUCTLINE.HTML_DESCRIPTION)
     .column("id").forOrdinality()
     .column("power", VARCHAR)
     .column("type", VARCHAR)
     .column("nr_of_lines", INTEGER).path("type/@nr_of_lines")
     .column("command", VARCHAR).path("type/@command")
     .as("t")).fetch();
```

This query will produce a table as follows:

```
+----+-------+----+-----------+-------+
|  id|power  |type|nr_of_lines|command|
+----+-------+----+-----------+-------+
|   1|5000kw |1   |          5|ERP    |
|   1|2000kw |2   |          3|OPS    |
|   1|20000kw|1   |          1|ERP    |
+----+-------+----+-----------+-------+
```

Figure 8.4 – The result of the previous query

Once you fetch a SQL table, you can think and act on it in the same way as any other database table. For brevity, I simply used `VARCHAR`, but it is better to specify a size in order to avoid defaulting to `VARCHAR(max)`.

More examples, including how to use `XMLTABLE` with aggregates, `ORDER BY`, `LIMIT`, and how to transform back into XML from a SQL table, are available in *SimpleXml*.

Handling relationships via SQL/XML

Handling the typical 1:1, 1:n, and n:n relationships can be done via jOOQ SQL/XML support. Let's go through a quick rundown of it.

Mapping relationships to XML

Most of the time, such relationships can be materialized into XML via a thoughtful combination of `xmlelement()`, `xmlagg()`, and `xmlforest()`. Since you are already familiar with the one-to-many relationship between `PRODUCTLINE` and `PRODUCT`, let's shape it into XML via SQL/XML support:

```
Result<Record1<XML>> result = ctx.select(
  xmlelement("productLine",
  xmlelement("productLine", PRODUCTLINE.PRODUCT_LINE),
  xmlelement("textDescription", PRODUCTLINE.TEXT_DESCRIPTION),
  xmlelement("products", field(select(xmlagg(
  xmlelement("product", xmlforest(
   PRODUCT.PRODUCT_NAME.as("productName"),
   PRODUCT.PRODUCT_VENDOR.as("productVendor"),
   PRODUCT.QUANTITY_IN_STOCK.as("quantityInStock")))))
 .from(PRODUCT)
 .where(PRODUCTLINE.PRODUCT_LINE.eq(PRODUCT.PRODUCT_LINE))))))
 .from(PRODUCTLINE)
 .orderBy(PRODUCTLINE.PRODUCT_LINE).fetch();
```

As you can infer from the preceding code, expressing the one-to-one and many-to-many relationships is just a matter of juggling with the SQL/XML operators. You can find these examples, including how to use `JOIN` instead of a `SELECT` subquery, in the bundled code for *XmlRelationships*.

If you think that `Result<Record1<XML>>` is not ready to be sent to the client (for instance, via a REST controller), then decorate it a little bit more by aggregating all the product lines under a XML element (root) and relying on `fetchSingleInto()`, as follows:

```
String result = ctx.select(
  xmlelement("productlines", xmlagg(
  xmlelement("productLine",
  ...
 .from(PRODUCTLINE).fetchSingleInto(String.class);
```

In SQL Server, we can obtain a similar result via `forXml()`:

```
Result<Record1<XML>> result = ctx.select(
  PRODUCTLINE.PRODUCT_LINE.as("productLine"),
  PRODUCTLINE.TEXT_DESCRIPTION.as("textDescription"),
  select(PRODUCT.PRODUCT_NAME.as("productName"),
    PRODUCT.PRODUCT_VENDOR.as("productVendor"),
    PRODUCT.QUANTITY_IN_STOCK.as("quantityInStock"))
      .from(PRODUCT)
      .where(PRODUCT.PRODUCT_LINE.eq(PRODUCTLINE.PRODUCT_LINE))
      .forXML().path().asField("products"))
  .from(PRODUCTLINE)
  .forXML().path("productline").root("productlines")
  .fetch();
```

Or we can obtain a `String` via `formatXML(XMLformat)`:

```
String result = ctx.select(
       PRODUCTLINE.PRODUCT_LINE.as("productLine"),
  ...
  .forXML().path("productline").root("productlines")
  .fetch()
  .formatXML(XMLFormat.DEFAULT_FOR_RECORDS);
```

Both examples will produce almost an identical XML, as follows (as you can see, altering the default XML tags inferred from the field names can be done with aliases via `as("alias")`):

```
<productlines>
  <productline>
    <productLine>Classic Cars</productLine>
    <textDescription>Attention ...</textDescription>
    <products>
      <product>
        <productName>1952 Alpine Renault 1300</productName>
        <productVendor>Classic Metal Creations</productVendor>
        <quantityInStock>7305</quantityInStock>
      </product>
      <product>
        <productName>1972 Alfa Romeo GTA</productName>
        <productVendor>Motor City Art Classics</productVendor>
```

```
      <quantityInStock>3252</quantityInStock>
    </product>
    ...
  </products>
</productline>
<productline>
  <productLine>Motorcycles</productLine>
  ...
```

You can check out these examples in the *XmlRelationships* application.

Mapping arbitrary nested models

jOOQ allows us to map arbitrarily nested models, not just the well-known 1:1,1:n, and n:n relationships, via SQL/XML support. Remember **Model 2** (see *Figure 8.3*)? Well, you already know how to fetch and map that model via SQL/JSON support, so this time, let's see how it can be done via SQL/XML:

```
Result<Record1<XML>> result = ctx.select(
 xmlelement("customer",
  xmlelement("customerName", CUSTOMER.CUSTOMER_NAME),
  xmlelement("creditLimit", CUSTOMER.CREDIT_LIMIT),
  xmlelement("payments", field(select(xmlagg(
  xmlelement("payment", // optional
   xmlforest(PAYMENT.CUSTOMER_NUMBER.as("customerNumber"),
    PAYMENT.INVOICE_AMOUNT.as("invoiceAmount"),
    PAYMENT.CACHING_DATE.as("cachingDate"),
    field(select(xmlagg(xmlelement("transaction", // optional
     xmlforest(BANK_TRANSACTION.BANK_NAME.as("bankName"),
     BANK_TRANSACTION.TRANSFER_AMOUNT.as("transferAmount")))))
 .from(BANK_TRANSACTION)
 .where(BANK_TRANSACTION.CUSTOMER_NUMBER
   .eq(PAYMENT.CUSTOMER_NUMBER)
     .and(BANK_TRANSACTION.CHECK_NUMBER
       .eq(PAYMENT.CHECK_NUMBER)))).as("transactions")))))
 .from(PAYMENT).where(PAYMENT.CUSTOMER_NUMBER
   .eq(CUSTOMER.CUSTOMER_NUMBER)))),
  xmlelement("details", field(select(xmlagg(
   xmlforest(CUSTOMERDETAIL.ADDRESS_LINE_FIRST
    .as("addressLineFirst"),
    CUSTOMERDETAIL.STATE.as("state"))))
```

```
  .from(CUSTOMERDETAIL)
  .where(CUSTOMERDETAIL.CUSTOMER_NUMBER
    .eq(CUSTOMER.CUSTOMER_NUMBER))))))
  .from(CUSTOMER).orderBy(CUSTOMER.CREDIT_LIMIT).fetch();
```

This is the power of example; there is not much else to say. Take your time to dissect this query, and check out the bundled code to see the output. Moreover, in the bundled code, you can see **Model 1** and **Model 3**, too. The application is named *NestedXml*.

As a SQL Server fan, you might be more interested in the previous query expressed via `forXML()`, so here it is:

```
Result<Record1<XML>> result = ctx.select(
 CUSTOMER.CUSTOMER_NAME, CUSTOMER.CREDIT_LIMIT,
 select(PAYMENT.CUSTOMER_NUMBER, PAYMENT.INVOICE_AMOUNT,
  PAYMENT.CACHING_DATE, select(BANK_TRANSACTION.BANK_NAME,
   BANK_TRANSACTION.TRANSFER_AMOUNT).from(BANK_TRANSACTION)
   .where(BANK_TRANSACTION.CUSTOMER_NUMBER
    .eq(PAYMENT.CUSTOMER_NUMBER)
     .and(BANK_TRANSACTION.CHECK_NUMBER
       .eq(PAYMENT.CHECK_NUMBER)))
   .orderBy(BANK_TRANSACTION.TRANSFER_AMOUNT)
   .forXML().path().asField("transactions")).from(PAYMENT)
   .where(PAYMENT.CUSTOMER_NUMBER
     .eq(CUSTOMER.CUSTOMER_NUMBER))
   .orderBy(PAYMENT.CACHING_DATE)
   .forXML().path().asField("payments"),
  select(CUSTOMERDETAIL.CITY,
   CUSTOMERDETAIL.ADDRESS_LINE_FIRST, CUSTOMERDETAIL.STATE)
   .from(CUSTOMERDETAIL)
   .where(CUSTOMERDETAIL.CUSTOMER_NUMBER
     .eq(CUSTOMER.CUSTOMER_NUMBER))
   .forXML().path().asField("details"))
   .from(CUSTOMER).orderBy(CUSTOMER.CREDIT_LIMIT)
   .forXML().path().fetch();
```

In the bundled code, *NestedXml*, you can practice many more examples that, for brevity reasons, couldn't be listed here. Remember that, especially for this chapter, I beat the drums. Now, it is time to bring in an entire orchestra and pay tribute to the coolest feature of jOOQ mapping. Ladies and gentlemen, allow me to introduce the MULTISET!

Nested collections via the astonishing MULTISET

The MULTISET value constructor (or MULTISET for short) is a SQL standard future that shapes nested subqueries (except scalar subqueries) into a single nested collection value. jOOQ 3.15+ provides marvelous and glorious support for MULTISET. It's marvelous because despite its tremendous power, it is quite easy (effortless) and intuitive to use via jOOQ, and it is glorious because it can produce any nested collection value of jOOQ `Record` or DTO (POJO/Java records) in a fully type-safe manner, with 0 reflections, no N+1 risks, no deduplications. This allows the database to perform nesting and to optimize the query execution plan.

Consider the well-known one-to-many relationship between `PRODUCTLINE` and `PRODUCT`. We can fetch and map this relationship via jOOQ's `<R extends Record> Field<Result<R>> multiset(Select<R> select)`, in jOOQ before 3.17.x, and `Field<Result<R>> multiset(TableLike<R> table)` starting with jOOQ 3.17.x as follows (later, we will refer to this example as *Exhibit A*):

```
var result = ctx.select(
        PRODUCTLINE.PRODUCT_LINE, PRODUCTLINE.TEXT_DESCRIPTION,
    multiset(
        select(PRODUCT.PRODUCT_NAME, PRODUCT.PRODUCT_VENDOR,
                PRODUCT.QUANTITY_IN_STOCK)
            .from(PRODUCT)
            .where(PRODUCTLINE.PRODUCT_LINE
                .eq(PRODUCT.PRODUCT_LINE))
    ).as("products")) // MULTISET ends here
   .from(PRODUCTLINE)
   .orderBy(PRODUCTLINE.PRODUCT_LINE)
   .fetch();
```

So, the usage is quite simple! The jOOQ `multiset()` constructor gets a `SELECT` statement as an argument(or, a table-like object, starting with jOOQ 3.17.x). Formally speaking, the result set of this `SELECT` statement represents a collection that will be nested in the outer collection (the result set produced by the outer `SELECT` statement). By nesting/mixing `multiset()` and `select()` (or `selectDistinct()`), we can achieve any level or shape/hierarchy of nested collections. Previously, we used the Java 10 `var` keyword as the type of result, but the real type is `Result<Record3<String, String, Result<Record3<String, String, Integer>>>>`. Of course, more nesting will produce a really hard-to-digest `Result` object, so using `var` is the recommended way to go. As you already intuited, `Result<Record3<String,`

`String, Integer>>` is produced by the `SELECT` statement from `multiset()`, while `Result<Record3<String, String, nested_result>>` is produced by the outer `SELECT` statement. The following diagram will help you to better understand this type:

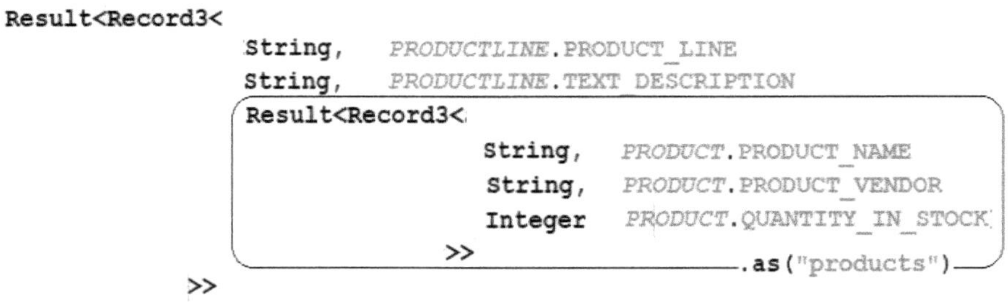

```
Result<Record3<
                 String,   PRODUCTLINE.PRODUCT_LINE
                 String,   PRODUCTLINE.TEXT_DESCRIPTION
                 Result<Record3<
                                String,   PRODUCT.PRODUCT_NAME
                                String,   PRODUCT.PRODUCT_VENDOR
                                Integer   PRODUCT.QUANTITY_IN_STOCK
                            >>                       .as("products")
              >>
```

Figure 8.5 – The type returned by the previous query

Since MULTISET has quite poor native support in databases, jOOQ has to emulate it via the SQL/JSON or SQL/XML operators. For instance, the previous query renders the following SQL in MySQL (check out how jOOQ uses `json_merge_preserve()` and `json_array()`):

```
SET @t = @@group_concat_max_len;
SET @@group_concat_max_len = 4294967295;

SELECT `classicmodels`.`productline`.`product_line`,
       `classicmodels`.`productline`.`text_description`,

  (SELECT coalesce(json_merge_preserve('[]', concat('[',
          group_concat(json_array(`v0`, `v1`, `v2`) separator
          ','), ']')), json_array())
   FROM
     (SELECT `classicmodels`.`product`.`product_name` AS `v0`,
         `classicmodels`.`product`.`product_vendor` AS `v1`,
         `classicmodels`.`product`.`quantity_in_stock` AS `v2`
      FROM `classicmodels`.`product`
      WHERE `classicmodels`.`productline`.`product_line` =
            `classicmodels`.`product`.`product_line`)
                 AS `t`) AS `products`
FROM `classicmodels`.`productline`
ORDER BY `classicmodels`.`productline`.`product_line`;

SET @@group_concat_max_len = @t;
```

At any moment, you can transform this collection of `Record` into plain JSON or XML via `formatJSON()`/`formatXML()`. However, allow me to take this opportunity to highlight that if all you want is to fetch a JSON/XML (since this is what your client needs), then it is better to use the SQL/JSON and SQL/XML operators directly (as you saw in the previous section) instead of passing through MULTISET. You can find examples in the bundled code, *MultisetRelationships*, alongside examples of how to use MULTISET for one-to-one and many-to-many relationships. In the example for many-to-many relationships, you can see how well the jOOQ type-safe implicit (one-to-one) join feature fits with MULTISET.

Remember **Model 2** (see *Figure 8.3*)? Well, you already know how to fetch and map that model via SQL/JSON and SQL/XML support, so let's see how to do it via MULTISET, too (later on, we will refer to this example as *Exhibit B*):

```
var result = ctx.select(
 CUSTOMER.CUSTOMER_NAME, CUSTOMER.CREDIT_LIMIT,
 multiset(select(PAYMENT.CUSTOMER_NUMBER,
                 PAYMENT.INVOICE_AMOUNT, PAYMENT.CACHING_DATE,
         multiset(select(BANK_TRANSACTION.BANK_NAME,
                         BANK_TRANSACTION.TRANSFER_AMOUNT)
                 .from(BANK_TRANSACTION)
                 .where(BANK_TRANSACTION.CUSTOMER_NUMBER
                  .eq(PAYMENT.CUSTOMER_NUMBER)
                  .and(BANK_TRANSACTION.CHECK_NUMBER
                   .eq(PAYMENT.CHECK_NUMBER)))
                 .orderBy(BANK_TRANSACTION.TRANSFER_AMOUNT)))
         .from(PAYMENT)
         .where(PAYMENT.CUSTOMER_NUMBER
          .eq(CUSTOMER.CUSTOMER_NUMBER))).as("payments"),
 multiset(select(CUSTOMERDETAIL.CITY,
                 CUSTOMERDETAIL.ADDRESS_LINE_FIRST,
                 CUSTOMERDETAIL.STATE)
         .from(CUSTOMERDETAIL)
         .where(CUSTOMERDETAIL.CUSTOMER_NUMBER
          .eq(CUSTOMER.CUSTOMER_NUMBER)))
           .as("customer_details"))
 .from(CUSTOMER)
 .orderBy(CUSTOMER.CREDIT_LIMIT.desc())
 .fetch();
```

This time, the returned type is quite verbose: `Result<Record4<String, BigDecimal, Result<Record4<Long, BigDecimal, LocalDateTime, Result<Record2<String, BigDecimal>>>>, Result<Record3<String, String, String>>>>`. The following diagram explains this:

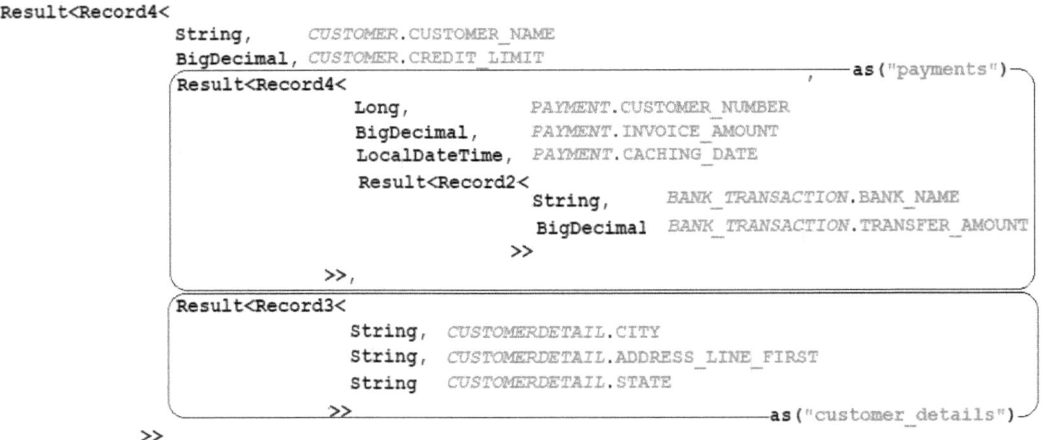

Figure 8.6 – The type returned by the previous query

You can find this example next to Model 1 and Model 3 in the application named *NestedMultiset*. Next, let's see how we can map MULTISET to DTO (for instance, POJO and Java 16 records).

Mapping MULTISET to DTO

Having the result of a MULTISET as a generic structural type is cool, but most probably, you'll love to have a `List` of POJO/Java records instead. For instance, if we think of *Exhibit A*, then you'll probably write the following Java records as the mapping model:

```
public record RecordProduct(String productName,
  String productVendor, Integer quantityInStock) {}
public record RecordProductLine(String productLine,
  String textDescription, List<RecordProduct> products) {}
```

So, you're expecting that `Result<Record3<String, String, Integer>>` will be fetched via MULTISET to be mapped to `List<RecordProduct>` and the whole query result to `List<RecordProductLine>`. The first part can be accomplished via the new ad hoc `Field.convertFrom()` converter, which was introduced in *Chapter 7, Types, Converters, and Bindings*. With `Field.convertFrom()`, we convert the given `Field<T>` (here, `Field<Result<Record3<String, String,`

`Integer>>>` is returned by `multiset()`) into a read-only `Field<U>` (here, `Field<List<RecordProduct>>`) for ad hoc usage:

```
Field<List<RecordProduct>> result = multiset(
  select(PRODUCT.PRODUCT_NAME, PRODUCT.PRODUCT_VENDOR,
         PRODUCT.QUANTITY_IN_STOCK)
 .from(PRODUCT)
 .where(PRODUCTLINE.PRODUCT_LINE.eq(PRODUCT.PRODUCT_LINE)))
  .as("products").convertFrom(r ->
                  r.map(mapping(RecordProduct::new)));
```

The `r` parameter from `r -> r.map(mapping(RecordProduct::new))` is `Result<Record3<String, String, Integer>>`, so this lambda can be seen as `Result<Record3<String, String, Integer>> -> RecordProduct`. The `r.map(…)` part is the `Result.map(RecordMapper<R, E>)` method. Finally, the `Records.mapping()` method (introduced earlier in this chapter) turns the constructor reference of the `Function3<String, String, Integer, RecordProduct>` type into a `RecordMapper` parameter, which is further used to turn a `Result<Record3<String, String, Integer>>` into a `List<RecordProduct>`. The resulting `Field<List<SimpleProduct>>` (which is like any other jOOQ `Field`) is now part of the outer `SELECT` next to `PRODUCTLINE.PRODUCT_LINE` (which is a `String`), and `PRODUCTLINE.TEXT_DESCRIPTION` (which is also a `String`).

So, our last mission is to convert the outer-most `Result3<String, String, List<RecordProduct>>` into `List<RecordProductLine>`. For this, we rely only on `mapping()`, as follows:

```
List<RecordProductLine> resultRecord = ctx.select(
  PRODUCTLINE.PRODUCT_LINE, PRODUCTLINE.TEXT_DESCRIPTION,
  multiset(
    select(PRODUCT.PRODUCT_NAME, PRODUCT.PRODUCT_VENDOR,
           PRODUCT.QUANTITY_IN_STOCK)
    .from(PRODUCT)
    .where(PRODUCTLINE.PRODUCT_LINE.eq(PRODUCT.PRODUCT_LINE)))
      .as("products").convertFrom(r ->
         r.map(Records.mapping(RecordProduct::new))))
    .from(PRODUCTLINE)
    .orderBy(PRODUCTLINE.PRODUCT_LINE)
    .fetch(mapping(RecordProductLine::new));
```

Done! Now, we can manipulate the `List<RecordProductLine>`. You can find this example in *MultisetRelationshipsInto*. By applying what we've learned here to the more complex *Exhibit B*, we obtain the following model:

```
public record RecordBank (
 String bankName, BigDecimal transferAmount) {}

public record RecordCustomerDetail(
 String city, String addressLineFirst, String state) {}

public record RecordPayment(
 Long customerNumber, BigDecimal invoiceAmount,
 LocalDateTime cachingDate, List<RecordBank> transactions) {}

public record RecordCustomer(String customerName,
 BigDecimal creditLimit, List<RecordPayment> payments,
 List<RecordCustomerDetail> details) {}
```

And the *Exhibit B* query is as follows:

```
List<RecordCustomer> resultRecord = ctx.select(
 CUSTOMER.CUSTOMER_NAME, CUSTOMER.CREDIT_LIMIT,
 multiset(select(PAYMENT.CUSTOMER_NUMBER,
                PAYMENT.INVOICE_AMOUNT, PAYMENT.CACHING_DATE,
  multiset(select(BANK_TRANSACTION.BANK_NAME,
                BANK_TRANSACTION.TRANSFER_AMOUNT)
           .from(BANK_TRANSACTION)
           .where(BANK_TRANSACTION.CUSTOMER_NUMBER
            .eq(PAYMENT.CUSTOMER_NUMBER)
             .and(BANK_TRANSACTION.CHECK_NUMBER
              .eq(PAYMENT.CHECK_NUMBER)))
           .orderBy(BANK_TRANSACTION.TRANSFER_AMOUNT))
           .convertFrom(r -> r.map(mapping(RecordBank::new))))
  .from(PAYMENT)
  .where(PAYMENT.CUSTOMER_NUMBER
   .eq(CUSTOMER.CUSTOMER_NUMBER))).as("payments")
  .convertFrom(r -> r.map(mapping(RecordPayment::new))),
 multiset(select(CUSTOMERDETAIL.CITY,
                CUSTOMERDETAIL.ADDRESS_LINE_FIRST,
                CUSTOMERDETAIL.STATE)
         .from(CUSTOMERDETAIL)
```

```
        .where(CUSTOMERDETAIL.CUSTOMER_NUMBER
         .eq(CUSTOMER.CUSTOMER_NUMBER)))
        .as("customer_details")
        .convertFrom(r ->
          r.map(mapping(RecordCustomerDetail::new))))
  .from(CUSTOMER)
  .orderBy(CUSTOMER.CREDIT_LIMIT.desc())
  .fetch(mapping(RecordCustomer::new));
```

This example, next to the examples for Model 1 and Model 3 from *Figure 8.3*, is available in *NestedMultiset*. Next, let's tackle the `MULTISET_AGG()` function.

The MULTISET_AGG() function

The jOOQ `MULTISET_AGG()` function is a synthetic aggregate function that can be used as an alternative to MULTISET. Its goal is to aggregate data into a nested collection represented as a jOOQ `Result` in a type-safe manner. The `MULTISET_AGG()` function is a convenient solution when we need to order by some aggregate value or create a `WHERE` statement based on result of not-deeply nested collection. For instance, the well-known one-to-many `PRODUCTLINE:PRODUCT` relationship can be aggregated as a nested collection as follows (the result type is `Result<Record3<String, String, Result<Record3<String, String, Integer>>>>`):

Starting with jOOQ 3.17.x, we can turn an expression of type `String`, `Name`, `Field`, and so in, into a multiset via `DSL.asMultiset()` methods. Check out the jOOQ documentation for more details.

```
var result = ctx.select(
  PRODUCTLINE.PRODUCT_LINE, PRODUCTLINE.TEXT_DESCRIPTION,
  multisetAgg(PRODUCT.PRODUCT_NAME, PRODUCT.PRODUCT_VENDOR,
              PRODUCT.QUANTITY_IN_STOCK).as("products"))
   .from(PRODUCTLINE)
   .join(PRODUCT)
   .on(PRODUCTLINE.PRODUCT_LINE.eq(PRODUCT.PRODUCT_LINE))
   .groupBy(PRODUCTLINE.PRODUCT_LINE,
            PRODUCTLINE.TEXT_DESCRIPTION)
   .orderBy(PRODUCTLINE.PRODUCT_LINE)
   .fetch();
```

This example is available, along with more examples, in *MultisetAggRelationships*.

Mapping the `Result` object to a DTO (for instance, POJO and Java 16 records) is accomplished by following the same principles as in the case of MULTISET:

```
List<RecordProductLine> resultRecord = ctx.select(
  PRODUCTLINE.PRODUCT_LINE, PRODUCTLINE.TEXT_DESCRIPTION,
  multisetAgg(PRODUCT.PRODUCT_NAME, PRODUCT.PRODUCT_VENDOR,
              PRODUCT.QUANTITY_IN_STOCK).as("products")
    .convertFrom(r -> r.map(mapping(RecordProduct::new))))
  .from(PRODUCTLINE)
  .join(PRODUCT)
  .on(PRODUCTLINE.PRODUCT_LINE.eq(PRODUCT.PRODUCT_LINE))
  .groupBy(PRODUCTLINE.PRODUCT_LINE,
           PRODUCTLINE.TEXT_DESCRIPTION)
  .orderBy(PRODUCTLINE.PRODUCT_LINE)
  .fetch(mapping(RecordProductLine::new));
```

This example is available alongside other examples in *MultisetAggRelationshipsInto*. Next, let's try to compare MULTISETs.

Comparing MULTISETs

MULTISETs can be used in predicates, too. Check out the following example:

```
ctx.select(count().as("equal_count"))
   .from(EMPLOYEE)
   .where(multiset(selectDistinct(SALE.FISCAL_YEAR)
                   .from(SALE)
                   .where(EMPLOYEE.EMPLOYEE_NUMBER
                      .eq(SALE.EMPLOYEE_NUMBER)))
        .eq(multiset(select(val(2003).union(select(val(2004))
                  .union(select(val(2007)))))))
   .fetch();
```

But when we can say that two MULTISETs are equal? Check out the following examples that are meant to clarify this:

```
// A
ctx.selectCount()
   .where(multiset(select(val("a"), val("b"), val("c")))
   .eq(multiset(select(val("a"), val("b"), val("c")))))
```

```
    .fetch();

// B
ctx.selectCount()
    .where(multiset(select(val("a"), val("b"), val("c")))
    .eq(multiset(select(val("a"), val("c"), val("b")))))
    .fetch();

// C
ctx.selectCount()
    .where(multiset(select(val("a")).union(select(val("b"))
                    .union(select(val("c")))))
    .eq(multiset(select(val("a")).union(select(val("b"))
                    .union(select(val("c"))))))
    .fetch();

// D
ctx.selectCount()
    .where(multiset(select(val("a")).union(select(val("b"))
                    .union(select(val("c")))))
    .eq(multiset(select(val("a")).union(select(val("b"))))))
    .fetch();
```

So, which of A, B, C, and D will return 1? The correct answer is A and C. This means that two MULTISETs are equal if they have the exactly same number of elements in the same order. The application is named *MultisetComparing*. Feel free to determine when a MULTISET X is greater/lesser/contained … than a MULTISET Y.

Also, don't forget to read `https://blog.jooq.org/jooq-3-15s-new-multiset-operator-will-change-how-you-think-about-sql/` and `https://blog.jooq.org/the-performance-of-various-to-many-nesting-algorithms/`. It looks as though jOOQ 3.17 will enrich MULTISET support with even more cool features, `https://twitter.com/JavaOOQ/status/1493261571103105030`, so stay tuned!

Moreover, since `MULTISET` and `MULTISET_AGG()` are such hot topics you should constantly update your skills from real scenarios exposed at `https://stackoverflow.com/search?q=%5Bjooq%5D+multiset`.

Next, let's talk about lazy fetching.

Lazy fetching

Hibernate JPA guy: So, how do you handle huge result sets in jOOQ?

jOOQ guy (me): jOOQ supports lazy fetching.

Hibernate JPA guy: And how do you manage `LazyInitializationException`?

jOOQ guy (me): For Hibernate JPA users that have just got here, I'd like to stress this right from the start – don't assume that jOOQ lazy fetching is related to or similar to Hibernate JPA lazy loading. jOOQ doesn't have and doesn't need a Persistence Context and doesn't rely on a `Session` object and proxy objects. Your code is not prone to any kind of lazy loading exceptions!

Then, what is jOOQ lazy fetching?

Well, most of the time, fetching the entire result set into memory is the best way to exploit your RDBMS (especially in web applications that face high traffic by optimizing small result sets and short transactions). However, there are cases (for instance, you might have a huge result set) when you'll like to fetch and process the result set in small chunks (for example, one by one). For such scenarios, jOOQ comes with the `org.jooq.Cursor` API. Practically, jOOQ holds a reference to an open result set and allows you to iterate (that is, load and process into memory) the result set via a number of methods such as `fetchNext()`, `fetchNextOptional()`, `fetchNextInto()`, and `fetchNextOptionalInto()`. However, to get a reference to an open result set, we have to call the `fetchLazy()` method, as shown in the following examples:

```
try (Cursor<CustomerRecord> cursor
    = ctx.selectFrom(CUSTOMER).fetchLazy()) {

   while (cursor.hasNext()) {
      CustomerRecord customer = cursor.fetchNext();
      System.out.println("Customer:\n" + customer);
   }
}
```

Notice that we are relying on the try-with-resources wrapping to ensure that the underlying result set is closed at the end of the iterating process. In this snippet of code, jOOQ fetches the records from the underlying result set into memory one by one via `fetchNext()`, but this doesn't mean that the JDBC driver does the same thing. JDBC drivers act differently across different database vendors and even across different versions of the same database. For instance, MySQL and PostgreSQL pre-fetches all records in a single database round-trip, SQL Server uses adaptive buffering (in the JDBC URL, we have `selectMethod = direct; responseBuffering = adaptive;`) and a default fetch size of 128 to avoid out-of-memory errors, and Oracle JDBC fetches a result set of 10 rows at a time from the database cursor (on the JDBC URL level, this can be altered via the `defaultRowPrefetch` property).

> **Important Note**
> Bear in mind that the fetch size is just a JDBC hint trying to instruct the driver about the number of rows to fetch in one go from the database. However, the JDBC driver is free to ignore this hint.

In jOOQ, configuring the fetch size can be done via `ResultQuery.fetchSize(int size)` or `Settings.withFetchSize(int size)`. jOOQ uses this configuration to set the underlying `Statement.setFetchSize(int size)` JDBC. Most JDBC drivers only apply this setting in certain contexts. For instance, MySQL should only apply this setting if we do the following:

- Set a forward-only result set (this can be set via jOOQ, `resultSetType()`).
- Set a concurrency read-only result set (via jOOQ, `resultSetConcurrency()`).

The fetch size is either set to `Integer.MIN_VALUE` for fetching records one by one or to the desired size while adding `useCursorFetch=true` to the JDBC URL for relying on cursor-based streaming.

Here is a snippet of code that takes advantage of these settings for MySQL:

```
try (Cursor<CustomerRecord> cursor = ctx.selectFrom(CUSTOMER)
   .resultSetType(ResultSet.TYPE_FORWARD_ONLY)
   .resultSetConcurrency(ResultSet.CONCUR_READ_ONLY)
   .fetchSize(Integer.MIN_VALUE).fetchLazy()) {

   while (cursor.hasNext()) {
      CustomerRecord customer = cursor.fetchNext();
      System.out.println("Customer:\n" + customer);
  }
}
```

The complete example is named *LazyFetching*.

On the other hand, PostgreSQL uses the fetch size if we do the following:

- Set forward-only result set (can be set via jOOQ, `resultSetType()`)
- Disable the *auto-commit* mode (in Spring Boot with the default Hikari CP connection pool, this can be done in `application.properties` via the following flag-property, `spring.datasource.hikari.auto-commit=false`, or in jOOQ via `<autoCommit>false</autoCommit>` in the `<jdbc/>` tag of your configuration)

So, the code for PostgreSQL can be as follows:

```
try ( Cursor<CustomerRecord> cursor = ctx.selectFrom(CUSTOMER)
  .resultSetType(ResultSet.TYPE_FORWARD_ONLY) // default
  .fetchSize(1).fetchLazy()) {

  while (cursor.hasNext()) {
    CustomerRecord customer = cursor.fetchNext();
    System.out.println("Customer:\n" + customer);
  }
}
```

Moreover, in PostgreSQL, the fetch size can be altered via `defaultRowFetchSize` and added to the JDBC URL. The complete example is also named *LazyFetching*.

For SQL Server and Oracle, we can rely on the default fetch size since both of them prevent out-of-memory errors. Nevertheless, enabling the fetch size in SQL Server is quite challenging while using the Microsoft JDBC driver (as in this book). It is much simpler if you rely on the jTDS driver.

Our examples for SQL Server and Oracle (*LazyFetching*) rely on the default fetching size; therefore, 128 for SQL Server and 10 for Oracle.

Finally, you can combine `ResultSet` and the Cursor API as follows:

```
ResultSet rs = ctx.selectFrom(CUSTOMER)
  .fetchLazy().resultSet();
Cursor<Record> cursor = ctx.fetchLazy(rs);
```

Additionally, you can do it like this:

```
Cursor<Record> result = ctx.fetchLazy(
   rs, CUSTOMER.CUSTOMER_NAME, CUSTOMER.CREDIT_LIMIT);
Cursor<Record> result = ctx.fetchLazy(
```

```
    rs, VARCHAR, DECIMAL);
Cursor<Record> result = ctx.fetchLazy(
    rs, String.class, BigDecimal.class);
```

Next, let's talk about lazy fetching via streaming.

Lazy featching via fetchStream()/fetchStreamInto()

In jOOQ, lazy fetching can also be achieved via `fetchStream()`/`fetchStreamInto()`. This method keeps an open JDBC result set internally and allows us to stream its content (that is, lazy fetching the result set into memory). For example, plain SQL can take advantage of `DSLContext.fetchStream()`, as follows:

```
try ( Stream<Record> stream
       = ctx.fetchStream("SELECT sale FROM sale")) {
 stream.filter(rs -> rs.getValue("sale", Double.class) > 5000)
       .forEach(System.out::println);
}
```

Or we can use the generated Java-based schema, as follows:

```
try ( Stream<SaleRecord> stream
       = ctx.selectFrom(SALE).fetchStream()) {
 stream.filter(rs -> rs.getValue(SALE.SALE_) > 5000)
       .forEach(System.out::println);
}
```

This code works in the same way as the next one, which uses `stream()`, not `fetchStream()`:

```
try ( Stream<SaleRecord> stream
       = ctx.selectFrom(SALE).stream()) {
 stream.filter(rs -> rs.getValue(SALE.SALE_) > 5000)
       .forEach(System.out::println);
}
```

However, pay attention as this code is not the same as the next one (the previous example uses `org.jooq.ResultQuery.stream()`, while the next example uses `java.util.Collection.stream()`):

```
ctx.selectFrom(SALE)
 .fetch()  // jOOQ fetches the whole result set into memory
```

```
          // and closes the database connection
 .stream() // stream over the in-memory result set
          // (no database connection is active)
 .filter(rs -> rs.getValue(SALE.SALE_) > 5000)
 .forEach(System.out::println);
```

Here, the `fetch()` method fetches the whole result set into memory and closes the database connection – this time, we don't need the try-with-resources wrapping since we are, essentially, streaming a list of records. Next, the `stream()` method opens a stream over the in-memory result set and no database connection is kept open. So, pay attention to how you write such snippets of code since you will be prone to accidental mistakes – for instance, you might need lazy fetching but accidentally add `fetch()`, or you might want eager fetching but accidentally forget to add `fetch()`.

Using org.jooq.ResultQuery.collect()

Sometimes, we need the stream pipeline to apply specific operations (for instance, `filter()`), and to collect the results, as shown in the following example:

```
try ( Stream<Record1<Double>> stream = ctx.select(SALE.SALE_)
      .from(SALE).fetchStream()) {  // jOOQ API ends here
 SimpleSale result = stream.filter( // Stream API starts here
   rs -> rs.getValue(SALE.SALE_) > 5000)
      .collect(Collectors.teeing(
        summingDouble(rs -> rs.getValue(SALE.SALE_)),
        mapping(rs -> rs.getValue(SALE.SALE_), toList()),
        SimpleSale::new));
}
```

However, if we don't actually need the stream pipeline (for instance, we don't need the `filter()` call or any other operation), and all we want is to lazily collect the result set, then it is pointless calling `fetchStream()`. But if we remove `fetchStream()`, how can we still collect in a lazy fashion? The answer is the jOOQ `collect()` method, which is available in `org.jooq.ResultQuery`. This method is very handy because it can handle resources internally and bypass the intermediate `Result` data structure. As you can see, there is no need to use try-with-resources after removing the `fetchStream()` call:

```
SimpleSale result = ctx.select(SALE.SALE_).from(SALE)
 .collect(Collectors.teeing( // org.jooq.ResultQuery.collect()
   summingDouble(rs -> rs.getValue(SALE.SALE_)),
   mapping(rs -> rs.getValue(SALE.SALE_), toList()),
   SimpleSale::new));
```

However, please bear in mind the following note.

> **Important Note**
>
> It is always a good practice to ensure that streaming is really needed. If your stream operations have SQL counterparts (for example, `filter()` can be replaced with a `WHERE` clause and `summingDouble()` can be replaced with the SQL's `SUM()` aggregate function), then go for the SQL. This will be much faster due to the significantly lower data transfer. So, always ask yourself: "*Can I translate this streaming operation into my SQL?*" If yes, then do it! If not, then go for streaming, as we will do in the following example.

Here is another example that lazy fetches groups. jOOQ doesn't fetch everything in memory thanks to `collect()`, and since we also set the fetch size, the JDBC driver fetches the result set in small chunks (here, a chunk has five records). The PostgreSQL version is as follows:

```
Map<Productline, List<Product>> result = ctx.select()
  .from(PRODUCTLINE).leftOuterJoin(PRODUCT)
  .on(PRODUCTLINE.PRODUCT_LINE.eq(PRODUCT.PRODUCT_LINE))
  .fetchSize(5) // Set the fetch size for JDBC driver
  .collect(Collectors.groupingBy(
     rs -> rs.into(Productline.class),
           Collectors.mapping(
     rs -> rs.into(Product.class), toList())));
```

The complete application is named *LazyFetchingWithStreams*. Next, let's talk about asynchronous fetching.

Asynchronous fetching

Whenever you consider that you need asynchronous fetching (for instance, a query takes too long to wait for it or multiple queries can run independently of each other (non-atomically)) you can rely on the jOOQ + `CompletableFuture` combination. For instance, the following asynchronous operation chains an `INSERT` statement, an `UPDATE` statement, and a `DELETE` statement using the `CompletableFuture` API and the threads obtained from the default `ForkJoinPool` API (if you are not familiar with this API, then you can consider purchasing the *Java Coding Problems* book from Packt, which dives deeper into this topic):

```
@Async
public CompletableFuture<Void> insertUpdateDeleteOrder() {
```

```
 return CompletableFuture.supplyAsync(() -> {
  return ctx.insertInto(ORDER)
   .values(null, LocalDate.of(2003, 2, 12),
    LocalDate.of(2003, 3, 1), LocalDate.of(2003, 2, 27),
    "Shipped", "New order inserted...", 363L, BigDecimal.ZERO)
   .returning().fetchOne();
  }).thenApply(order -> {
    order.setStatus("ON HOLD");
    order.setComments("Reverted to on hold ...");
    ctx.executeUpdate(order);

    return order.getOrderId();
  }).thenAccept(id -> ctx.deleteFrom(ORDER)
    .where(ORDER.ORDER_ID.eq(id)).execute());
}
```

This example is available for MySQL next to another one in the application named *SimpleAsync*.

You can exploit `CompletableFuture` and jOOQ, as demonstrated in the previous example. However, you can also rely on two jOOQ shortcuts, `fetchAsync()` and `executeAsync()`. For instance, let's suppose that we want to fetch managers (`MANAGER`), offices (`OFFICE`), and employees (`EMPLOYEE`) and serve them to the client in HTML format. Fetching managers, offices, and employees can be done asynchronously since these three queries are not dependent on each other. In this context, the jOOQ `fetchAsync()` method allows us to write the following three methods:

```
@Async
public CompletableFuture<String> fetchManagersAsync() {

   return ctx.select(MANAGER.MANAGER_ID, MANAGER.MANAGER_NAME)
   .from(MANAGER).fetchAsync()
   .thenApply(rs -> rs.formatHTML()).toCompletableFuture();
}

@Async
public CompletableFuture<String> fetchOfficesAsync() {

 return ctx.selectFrom(OFFICE).fetchAsync()
   .thenApply(rs -> rs.formatHTML()).toCompletableFuture();
```

```
}

@Async
public CompletableFuture<String> fetchEmployeesAsync() {

  return ctx.select(EMPLOYEE.OFFICE_CODE,
                    EMPLOYEE.JOB_TITLE, EMPLOYEE.SALARY)
 .from(EMPLOYEE).fetchAsync()
 .thenApply(rs -> rs.formatHTML()).toCompletableFuture();
}
```

Next, we wait for these three asynchronous methods to complete via the `CompletableFuture.allOf()` method:

```
public String fetchCompanyAsync() {

 CompletableFuture<String>[] fetchedCf
                          = new CompletableFuture[]{
  classicModelsRepository.fetchManagersAsync(),
  classicModelsRepository.fetchOfficesAsync(),
  classicModelsRepository.fetchEmployeesAsync()};

  // Wait until they are all done
  CompletableFuture<Void> allFetchedCf
                          = CompletableFuture.allOf(fetchedCf);
  allFetchedCf.join();

  // collect the final result
  return allFetchedCf.thenApply(r -> {
   StringBuilder result = new StringBuilder();

   for (CompletableFuture<String> cf : fetchedCf) {
    result.append(cf.join());
   }

   return result.toString();
   }).join();
}
```

The `String` returned by this method (for instance, from a REST controller) represents a piece of HTML produced by jOOQ via the `formatHTML()` method. Curious about what this HTML looks like? Then, simply run the *FetchAsync* application under MySQL and use the provided controller to fetch the data in a browser. You might also like to practice the *ExecuteAsync* (which is available for MySQL) application that uses the jOOQ `executeAsync()` method as an example.

Lukas Eder mentions that:

"Perhaps worth mentioning that there's an ExecutorProvider SPI that allows for routing these async executions elsewhere when the default ForkJoinPool is not the correct place? jOOQ's own CompletionStage implementations also make sure that everything is always executed on the Executor provided by ExecutorProvider, unlike the JDK APIs, which always defaults back to the ForkJoinPool again (unless that has changed, recently)."

Next, let's tackle reactive fetching.

Reactive fetching

Reactive fetching refers to the use of a reactive API in combination with jOOQ. Since you are using Spring Boot, there is a big chance that you are already familiar with the Project Reactor reactive library (`https://projectreactor.io/`) or the `Mono` and `Flux` APIs. So, without going into further detail, let's take an example of combining `Flux` and jOOQ in a controller:

```
@GetMapping(value = "/employees",
            produces = MediaType.TEXT_EVENT_STREAM_VALUE)
public Flux<String> fetchEmployees() {

 return Flux.from(
   ctx.select(EMPLOYEE.FIRST_NAME, EMPLOYEE.LAST_NAME,
              EMPLOYEE.JOB_TITLE, EMPLOYEE.SALARY)
      .from(EMPLOYEE))
      .map(e -> e.formatHTML())
        .delayElements(Duration.ofSeconds(2))
        .share();
}
```

So, jOOQ is responsible for fetching some data from `EMPLOYEE`, and `Flux` is responsible for publishing the fetched data. You can practice this example in *SimpleReactive* (which is available for MySQL).

What about a more complex example? One of the important architectures that can be applied to mitigate data loss in a streaming pipeline is **Hybrid Message Logging** (**HML**). Imagine a streaming pipeline for meetup RSVPs. In order to ensure that we don't lose any RSVPs, we can rely on **Receiver-Based Message Logging** (**RBML**) to write every received RSVP to stable storage (for instance, PostgreSQL) before any action is performed on it. Moreover, we can rely on **Sender-Based Message Logging** (**SBML**) to write each RSVP in the stable storage right before we send it further on in the pipeline (for example, to a message queuing). This is the RSVP that was processed by the application business logic, so it might not the same as the received RSVP. The following diagram represents data flowing through an HML implementation:

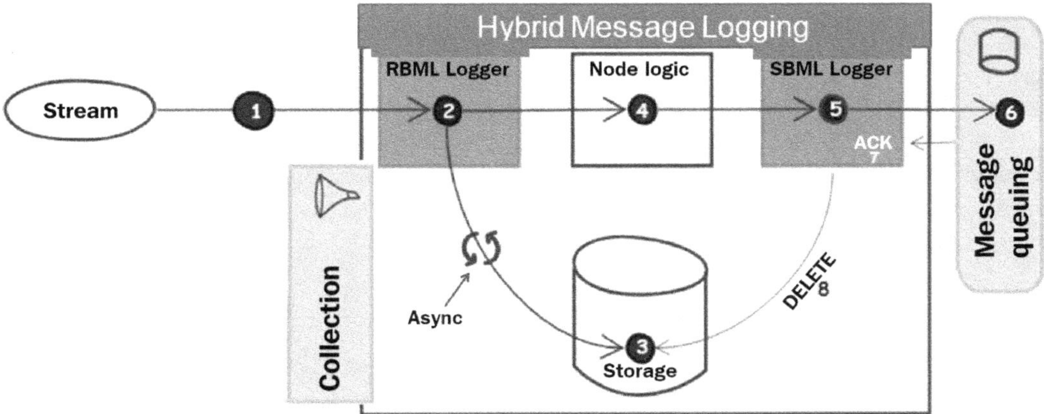

Figure 8.7 – Data flow through HML

Based on the preceding diagram, we can implement the processing and recovery of data asynchronously. For instance, the RBML part can be expressed in jOOQ as follows:

```
public void insertRsvps(String message) {

 Flux<RsvpDocument> fluxInsertRsvp =
  Flux.from(ctx.insertInto(RSVP_DOCUMENT)
   .columns(RSVP_DOCUMENT.ID, RSVP_DOCUMENT.RSVP,
              RSVP_DOCUMENT.STATUS)
   .values((long) Instant.now().getNano(), message, "PENDING")
   .returningResult(RSVP_DOCUMENT.ID, RSVP_DOCUMENT.RSVP,
                 RSVP_DOCUMENT.STATUS))
   .map(rsvp -> new RsvpDocument(rsvp.value1(), rsvp.value2(),
           rsvp.value3()));
```

```
    processRsvp(fluxInsertRsvp);
}
```

On the other hand, the SBML part can be expressed as follows:

```
private void recoverRsvps() {
 Flux<RsvpDocument> fluxFindAllRsvps = Flux.from(
  ctx.select(RSVP_DOCUMENT.ID, RSVP_DOCUMENT.RSVP,
             RSVP_DOCUMENT.STATUS)
     .from(RSVP_DOCUMENT))
     .map(rsvp -> new RsvpDocument(rsvp.value1(),
            rsvp.value2(), rsvp.value3()));

 processRsvp(fluxFindAllRsvps);
}
```

What about deleting or updating an RSVP? For the complete code, check out the *HML* application, which is available for PostgreSQL.

Summary

This was a big chapter that covered one of the most powerful capabilities of jOOQ, fetching and mapping data. As you learned, jOOQ supports a wide range of approaches for fetching and mapping data, from simple fetching to record mappers, to the fancy SQL/JSON and SQL/XML, to the marvelous and glorious MULTISET support, and finally, to lazy, asynchronous, and reactive fetching. In the next chapter, we will talk about how to batch and bulk data.

Part 3: jOOQ and More Queries

In this part, we go a step further and tackle some hot topics, such as identifiers, batching, bulking, pagination, optimistic locking, and HTTP long conversations.

By the end of this part, you will know how to use the jOOQ capabilities for implementing the aforementioned topics.

This part contains the following chapters:

9
CRUD, Transactions, and Locking

In this chapter, we'll cover a must-know mix of fundamental notions about CRUD operations, transactions, and locking. These three topics are heavily exploited in almost any database application. In a common scenario, an application has a significant number of CRUD operations that are executed in explicitly demarcated logical transactions and, in certain cases, they also need to explicitly control the concurrent access to data to prevent **race conditions**, **lost updates**, and other **SQL phenomena** (or SQL anomalies).

In this chapter, we will cover the following topics:

- CRUD
- Navigating (updatable) records
- Transactions
- Locking

Let's get started!

Technical requirements

The code for this chapter can be found on GitHub at `https://github.com/PacktPublishing/jOOQ-Masterclass/tree/master/Chapter09`.

CRUD

Besides the awesome DSL-fluent API for expressing complex SQL, jOOQ can be used to express *everyday* SQL operations as well. These are known as **CRUD** operations (**Create** (`INSERT`), **Read** (`SELECT`), **Update** (`UPDATE`), and **Delete** (`DELETE`)), and jOOQ facilitates them via a dedicated API that involves `UpdatableRecord` types. In other words, the jOOQ Code Generator generates a `UpdatableRecord` (a record that can be fetched and stored again in the database) for each table that has a primary key (not just a simple unique key!). Tables without a primary key (`org.jooq.TableRecord`) are rightly considered non-updatable by jOOQ. You can easily recognize a jOOQ `UpdatableRecord` because it has to extend the `UpdatableRecordImpl` class (simply inspect your generated records from `jooq.generated.tables.records`). Next, jOOQ exposes a CRUD API that allows you to operate directly on these updatable records instead of writing DSL-fluent queries (which fits better for complex queries that involve more than one table).

If you need a quick reminder about jOOQ records, please check out *Chapter 3, jOOQ Core Concepts*.

> **Important Note**
>
> The jOOQ CRUD API fits like a glove for **normalized** databases, so for tables that have primary keys (simple or composed) or unique lifespans, a primary key is inserted only once into a table. Once it's been inserted, it cannot be changed or re-inserted after being deleted.
>
> However, as you know, jOOQ tries to greet you in any case you have, so if you need updatable primary keys, then rely on `Settings.updatablePrimaryKeys()`.

The jOOQ CRUD API facilitates several operations, including insert (`insert()`), update (`update()`), delete (`delete()`), merge (`merge()`) and the handy store (`store()`). Besides these operations, we have the well-known `selectFrom()`, which is useful for reading (`SELECT`) from a single table directly into a `Result` of updatable records.

However, before we look at several CRUD examples, is important to know about a set of methods that can influence the behavior of updatable records and CRUD operations. These methods are `attach()`, `detach()`, `original()`, `changed()`, `reset()`, and `refresh()`.

Attaching/detaching updatable records

Roughly, jOOQ updatable records are just Java objects that can live independently of the database and can be manipulated in memory. So long as an updatable record doesn't need to interact with the database, it can remain in the **detached** state. In this state, the updatable record doesn't hold any reference to the database or to the connection that created it. However, before interacting with the database (insert, delete, update, and so on), an updatable record must be **attached** to a valid `Configuration` that, among other things, has the coordinates to connect to the database that this updatable record will interact with. After an updatable record has been attached, it will remain like this for as long as the corresponding `Configuration` lives or until it is explicitly detached by calling `detach()`.

When we fetch updatable records from the database, jOOQ will automatically attach them to the currently used `Configuration` and implicitly to the involved database connection. This connection may be used internally for subsequent interactions with the fetched updatable records with the database.

> **Important Note**
>
> To be precise, jOOQ doesn't hold references to a JDBC `Connection` but to `ConnectionProvider` from `Configuration`. In terms of transactions or connection pooling, this might be relevant. For instance, if we are using a Spring `TransactionAwareDataSourceProxy`, an attached record can be fetched in one transaction and stored in another transparently.

Consider the following read operation:

```
SaleRecord sr = ctx.selectFrom(SALE)
   .where(SALE.SALE_ID.eq(1L)).fetchSingle();
```

The `sr` record was automatically attached by jOOQ to the `Configuration` part of `ctx`. Next, we can successfully execute other operations that interact with the database, such as `sr.update()`, `sr.delete()`, and so on.

Now, let's consider a brand-new record that's been created by the client, as follows:

```
SaleRecord srNew = new SaleRecord(...);
```

Such records are not automatically attached to any existent `Configuration`. They haven't been fetched from the database, so jOOQ will justifiably expect that you'll explicitly/manually attach them to a `Configuration` whenever this is necessary (for instance, before calling `sr.insert()`). This can be done by explicitly calling the `DSLContext.attach()` or `UpdatableRecord.attach()` methods, as follows:

```
ctx.attach(srNew);
srNew.attach(ctx.configuration());
```

However, to avoid the `attach()` explicit call, we can rely on the `DSLContext.newRecord()` alternative. Since `DSLContext` contains the `Configuration` part, jOOQ will do the attachment automatically. So, here, you should use the following code snippet (if you want to populate the record from a POJO, then use the `newRecord(Table<R> table, Object o)` flavor):

```
SaleRecord srNew = ctx.newRecord(SALE);
srNew.setFiscalYear(…);
...
```

Once `srNew` has been attached, we can execute operations that interact with the database. Attempting to execute such operations on a detached updatable record will lead to an exception that states `org.jooq.exception.DetachedException`: *Cannot execute query. No Connection configured.*

An updatable record can be explicitly detached with `UpdatableRecord.detach()`:

```
srNew.detach(); // equivalent to srNew.attach(null);
```

While an updatable record is serializable, the underlying `Connection` (or `DataSource`) of `Configuration` is non-serializable. Nevertheless, you don't have to detach records before serialization. The internals of `DefaultConfiguration` ensure that anything that isn't `Serializable` (for instance, `DefaultConnectionProvider`) isn't serialized. Re-attaching will still be necessary after de-serialization, though. Support for serialization on jOOQ radar will eventually be deprecated: `https://github.com/jOOQ/jOOQ/issues/2359`.

From this, don't conclude that serializing/de-serializing records is a day-to-day task. Most of the time, records are used to populate views (for instance, via Thymeleaf) or they are exported as CSV, JSON, HTML, and so on via jOOQ support, as you'll see in the next chapter. None of these actions requires an explicit detach.

What's an original (updatable) record?

Every (updatable) record holds a reference to its *current* values and its *original* values.

If the record was fetched from the database, then the fetched values represent the *original* and the *current* values at the same time. Next, the *current* values can be modified in memory, while the *original* values remain in place. For example, let's assume that the following query fetches the `SALE.FISCAL_YEAR` field, which is *2005* and that, after fetching it, we set it to *2002*:

```
SaleRecord sr = ctx.selectFrom(SALE)
   .where(SALE.SALE_ID.eq(1L)).fetchSingle();

// or, a little bit more concise
SaleRecord sr = ctx.fetchSingle(SALE, SALE.SALE_ID.eq(1L));

sr.setFiscalYear(2002);
```

At this point, the *original* value of the fiscal year is *2005* and the *current* value is *2002*. After inserting/updating a record, the *current* values that have been inserted/updated become the *original* values. For instance, after updating `sr`, the *original* value of the fiscal year becomes *2002*, just like the *current* value. This way, `sr` mirrors the latest state of the database. So, after an update, the *original* values reflect only what has been sent to the database by default. Trigger-generated values are not fetched back by default. For that to work, `Settings.returnAllOnUpdatableRecord()` is required.

In the case of a new record, the *original* values are always `null` and remain like this until the record is inserted or updated.

Whenever we need the *original* values, we can call `Record.original()`. Without arguments, the `original()` method returns a brand-new record that's been populated with the *original* values. If `sr` is attached when calling `original()`, then `srOriginal` is attached as well; otherwise, it is detached:

```
SaleRecord srOriginal = sr.original();
```

By specifying an argument as a `Field`, `Name`, `String`, or integer (index), we can extract the *original* value of a certain field. Here is a type-safe and a non-type-safe approach:

```
int fiscalYear = sr.original(SALE.FISCAL_YEAR);
int fiscalYear = (int) sr.original("fiscal_year");
```

Having the *current* and *original* values of a record in your hands can be useful for making decisions after comparing these values between them or with other values, for creating side-by-side views of original data and current data, and so on. In the bundled code, called *OriginalRecords* (available for MySQL), you can see an example of rendering a side-by-side view for a `PRODUCT` via Thymeleaf. The relevant Thymeleaf code is as follows:

```
<tr>
  <td> Product Buy Price:</td>
  <td th:text = "${product.original('buy_price')}"/></td>
</tr>
<tr>
  <td> Product MSRP:</td>
  <td th:text = "${product.original('msrp')}"/></td>
</tr>
```

You'll see something similar to the following:

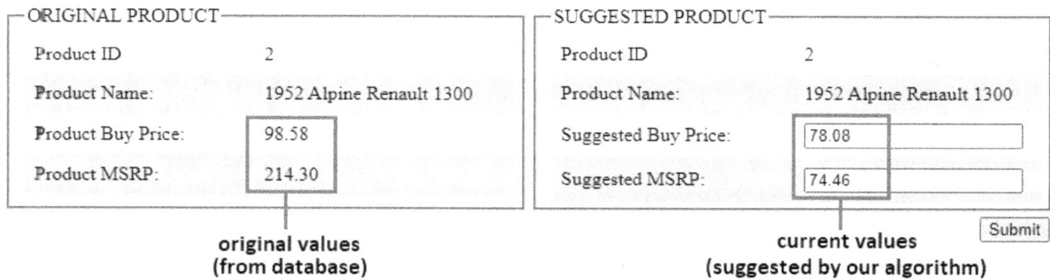

Figure 9.1 – Side-by-side view

Next, let's focus on marking (updatable) records as changed/unchanged.

Marking (updatable) records as changed/unchanged

A record whose *original* values are exactly the same as the *current* values is considered *unchanged*. By calling the `Record.changed()` flag method, we can find out if a record is considered as *changed* or *unchanged* by jOOQ. For instance, a record that has been fetched from the database and never modified is *unchanged*:

```
SaleRecord sr = ctx.selectFrom(SALE)
   .where(SALE.SALE_ID.eq(1L)).fetchSingle();
sr.changed(); // false

sr.setFiscalYear(2005);
sr.changed(); // true
```

Even if we set the same fiscal year that was fetched from the database (*2005*), this record is marked as *changed*. On the other hand, a brand-new record is considered *changed*:

```
SaleRecord sr = new SaleRecord(null, 2021, 453.2, 1504L, ...);
sr.changed(); // true

SaleRecord sr = new SaleRecord();
sr.setFiscalYear(2021);
sr.setSale(4500.25);
...
sr.changed(); // true
```

Notice that `changed()` operates at the field level since it's a `BitSet` that's the same length as the number of record fields. In other words, each field of a record has the `changed()` flag method. All we have to do is pass the field as an argument of `changed()` as a `Field`, `Name`, `String`, or `int` (representing the index):

```
boolean changed = sr.changed(SALE.FISCAL_YEAR);
boolean changed = sr.changed("fiscal_year");
```

Whenever we attempt to insert or update a record, jOOQ inspects the *changed* flags to determine which fields should be part of the generated query. This is great, because by rendering only the *changed* fields, jOOQ allows *default* values (specified via `CREATE TABLE` DDL statements) to be set for the omitted fields. If none of the fields were *changed* then jOOQ can prevent an insert/update from being executed (we'll cover this in more detail later in this chapter).

However, we can enforce/suppress the execution of such statements by explicitly turning on (or off) the *changed* flag. We can do so by marking all the fields as *changed*/*unchanged*:

```
sr.changed(true/false);
```

We can also do the same by marking only certain fields:

```
sr.changed(SALE.FISCAL_YEAR, true/false);
sr.changed("fiscal_year", true/false);
```

Pay attention to how you juggle these flags since is quite easy to mess things up and render some unfortunate DML statements. Next, let's talk about resetting records.

Resetting an (updatable) record

By resetting a record, we **reset** all the **current** values to the **original** values and all the **changed** flags to `false`. This can be accomplished without interacting with the database. We can reset all the fields by using `Record.reset()`:

```
sr.reset();
```

We can reset only certain fields by using `reset(Field/Name/String/int)`:

```
sr.reset(SALE.FISCAL_YEAR);
sr.reset("fiscal_year");
```

You'll see this method at work in the forthcoming sections. Finally, let's talk about how to refresh a record.

Refreshing an updatable record

By refreshing an updatable record, we synchronize the record's original values with the database's latest state, and we revert the current values in case they had been set. Practically, `UpdatableRecord.refresh()` is materialized in a `SELECT` round trip that loads the data in the original values of the record and sets all the changed flags to `false`. Refreshing all the fields of the record can be done as follows:

```
sr.refresh();
```

You can partially refresh the updatable record using `refresh(Field<?>... fields)` or `refresh(Collection<? extends Field<?>> fields)`:

```
sr.refresh(SALE.FISCAL_YEAR, SALE.SALE_);
```

In this case, `SELECT` is rendered to fetch only the specified fields. As you'll see shortly, this `refresh()` method is quite useful when mixed with optimistic locking.

You can find these examples in *SimpleCRUDRecords* (available for MySQL and PostgreSQL). Next, let's talk about inserting updatable records.

Inserting updatable records

Just a quick note to be sure that this section is not misunderstood: `insert()` can be called on any `TableRecord`, not just the `UpdatableRecord` ones; no primary key is necessary for `insert()`.

Now, inserting updatable records can be done via `UpdatableRecord.insert()` and its overloads. Typically, we insert brand-new records or records that have been loaded from POJOs (for instance, via the `from(POJO)` or `newRecord(Table<R> table, Object o)` methods) that are considered and treated by jOOQ as new records. Here is a simple and classical example of creating and inserting an updatable record:

```
SaleRecord sr = ctx.newRecord(SALE);
sr.setFiscalYear(2021);
...
sr.insert();
```

However, if you create it like so, then you need to explicitly call `attach()`:

```
SaleRecord sr = new SaleRecord();
sr.setFiscalYear(2021);
...
ctx.attach(sr);
sr.insert();
```

jOOQ generates and executes the `INSERT` statement. Moreover, by default, jOOQ tries to load any generated keys (the `IDENTITY/SEQUENCE` values, which are supported by most databases) from the database and turn them back into records that conform to the following note.

> **Important Note**
>
> The JDBC `getGeneratedKeys()` approach is only used when there's no better approach. In Db2, Firebird, MariaDB, Oracle, PostgreSQL, and, soon, H2, there is native `RETURNING` or `<data change delta table>` support that favors single round trips. Sometimes, if `getGeneratedKeys()` isn't supported, or only poorly, then an extra round trip may be needed with an extra `SELECT` statement. This is particularly true for many dialects when `Settings.returnAllOnUpdatableRecord()` is active.

So, calling `sr.getSaleId()` after `INSERT` will return the database-generated primary key.

Sometimes, you just need to re-insert the data that's contained in an updatable record. By default, since the record is *unchanged* after being inserted (`changed()` returns `false`), executing another `insert()` renders an `INSERT` of defaults (`INSERT INTO sale VALUES (default, default, ...)`). If the `CREATE TABLE` DDL doesn't provide default values for all the rendered defaults, then this will result in an error; that is, *Field 'foo' doesn't have a default value.* But, as you can see in the bundled code, this behavior can be controlled via `withInsertUnchangedRecords(false)`. Setting this flag to `false` will suppress any attempt to execute an `INSERT` of defaults.

To insert the same data without creating a new record, you can manually mark the record fields as changed (notice that we mark the primary key as *unchanged*, so it is omitted from the generated `INSERT` to avoid a duplicate key error):

```
sr.changed(true);
sr.changed(SALE.SALE_ID, false);
sr.insert();
```

Of course, if you want to re-insert only certain fields, then mark only those fields as *changed*. On the other hand, if you want to re-insert the data and create a new record as well, then rely on the `UpdatableRecord.copy()` method. The `copy()` method is quite handy because it duplicates this record in memory, marks all fields as *changed*, and doesn't copy the primary key or any other main unique key:

```
SaleRecord srCopy = sr.copy();
srCopy.insert();

// or, shortly
sr.copy().insert();
```

More examples, including inserting without returning the generated primary key and inserting and returning all fields, can be found in the bundled code, *SimpleCRUDRecords* (available for MySQL and PostgreSQL). Next, let's focus on updating records.

Updating updatable records (this sounds funny)

Typically, an updatable record is fetched from the database and is changed in memory. Subsequently, these changes are propagated to the database by calling `UpdatableRecord.update()` or its flavors, which allows us to nominate the fields that should be updated. Let's fetch a record and change it:

```
SaleRecord sr = ctx.selectFrom(SALE)
   .where(SALE.SALE_ID.eq(1L)).fetchSingle();

sr.setFiscalYear(2000);
sr.setSale(1111.25);
```

If we print `sr` at the console, then the result will look similar to the following:

sale_id	fiscal_year	sale	employee_number	hot	rate	vat	fiscal_month	revenue_growth	trend
1	*2000	*1111.25	1370	0	{null}	{null}	1	0.0	CONSTANT

Figure 9.2 – Displaying the modified record on the console

Notice that `fiscal_year` and `sale` are marked with an asterisk (`*`) by jOOQ. This asterisk highlights the fields that have been changed and that will participate in the following `UPDATE`:

```
sr.update();
```

The rendered SQL in MySQL dialect is as follows (the rendered `UPDATE` relies on the primary key):

```
UPDATE `classicmodels`.`sale`
SET `classicmodels`.`sale`.`fiscal_year` = 2000,
    `classicmodels`.`sale`.`sale` = 1111.25
WHERE `classicmodels`.`sale`.`sale_id` = 1
```

By default, updating a record that is already up to date has no effect. Of course, if you rely on `changed()` to mark all/some fields as *changed*, then you force jOOQ to execute the corresponding `UPDATE`. You can practice forcing an update via `Settings.withUpdateUnchangedRecords(UpdateUnchangedRecords)` in the bundled code.

Deleting updatable records

A fetched updatable record can be deleted via `UpdatableRecord.delete()`:

```
SaleRecord sr = ctx.selectFrom(SALE)
   .where(SALE.SALE_ID.eq(5L)).fetchSingle();

sr.delete();

// MySQL rendered SQL
DELETE FROM `classicmodels`.`sale` WHERE `classicmodels`
  .`sale`.`sale_id` = 5
```

As you can see, the rendered `DELETE` relies on the primary key (or main unique key). After deletion, all the fields of the deleted record are automatically marked as *changed*, so you can easily insert it again by calling `insert()`.

Merging updatable records

Whenever we want to execute a `MERGE` statement for an updatable record (brand new or fetched from the database), we can call `UpdatableRecord.merge()`. In this case, jOOQ renders an `INSERT ... ON DUPLICATE KEY UPDATE` (this is emulated, depending on the used dialect), so it delegates the task of choosing between `INSERT` and `UPDATE` to the database. Here is an example:

```
SaleRecord srNew = ctx.newRecord(SALE);
srNew.setFiscalYear(2000);
...
srNew.merge();
```

In this case, `srNew` will be inserted. Here is another example:

```
SaleRecord srFetched = ctx.selectFrom(SALE)
   .where(SALE.SALE_ID.eq(1L)).fetchSingle();
srFetched.setFiscalYear(2005);
...
srFetched.merge();
```

Here, `srFetched` will be updated based on the primary key. Practically, jOOQ will render an SQL that updates the row, regardless of which (unique) key value is already present.

Storing updatable records

Storing an updatable record can be done by calling the `UpdatableRecord.store()` method. This method results in an `INSERT` or an `UPDATE`, depending on the primary key's state. The decision of rendering an `INSERT` or an `UPDATE` is made by jOOQ, not by the database, as in the case of `MERGE`.

Typically, calling `store()` for new updatable records results in an `INSERT`:

```
SaleRecord srNew = ctx.newRecord(SALE);
srNew.setFiscalYear(2000);
...
srNew.store(); // jOOQ render an INSERT
```

If the updatable record was fetched from the database and its primary key was not changed, then jOOQ will render an `UPDATE`:

```
SaleRecord srFetched = ctx.selectFrom(SALE)
   .where(SALE.SALE_ID.eq(5L)).fetchSingle();
srFetched.setFiscalYear(2005);
srFetched.changed(SALE.SALE_, true);
...
srFetched.store(); // jOOQ render an UPDATE
```

If the updatable record was fetched from the database and its primary key was changed, then jOOQ will render an `INSERT`:

```
srFetched.setSaleId(…);
srFetched.store(); // jOOQ render an INSERT
```

However, we can still force an `UPDATE` of the primary key via `withUpdatablePrimaryKeys(true)`:

```
DSLContext derivedCtx = ctx.configuration().derive(
  new Settings().withUpdatablePrimaryKeys(true)).dsl();

SaleRecord sr = derivedCtx.selectFrom(SALE)
   .where(SALE.SALE_ID.eq(7L)).fetchSingle();

sr.setSaleId(...);
sr.store(); // jOOQ render an UPDATE of primary key
```

However, as Lukas Eder shared: *"I think it's worth mentioning that updating primary keys is very much against all principles of normalization. It was introduced for those cases where users have very good reasons to do so, and those reasons are very rare (usually data migrations or fixing broken data, but even then, they're probably more likely to use SQL statements than updatable records)."*

You can see these examples in *SimpleCRUDRecords* (available for MySQL and PostgreSQL).

On the other hand, if you prefer to work with POJOs and jOOQ's DAO, then you'll like to check out the examples from *SimpleDaoCRUDRecords* (available for MySQL and PostgreSQL). These examples relies on DAO's `insert()`, `update()`, `delete()`, and `merge()`. Moreover, you'll see the `withReturnRecordToPojo()` setting at work. Next, let's focus on using updatable records in web applications.

Using updatable records in HTTP conversations

jOOQ s updatable records can be used in web applications or, in other words, in conversations that span across requests over the stateless HTTP protocol. Next, we'll develop several Spring Boot samples that are meant to highlight what we've learned so far.

Using insert(), update(), and delete()

Let's try to build a Spring Boot sample application that uses updatable records and `insert()`, `update()`, and `delete()`. While relying on the Spring MVC design pattern, let's consider the following scenario: our main goal is to provide a list of all bank transactions (`BANK_TRANSACTION`) that belong to the same payment (`PAYMENT`) of a certain customer. The user should be able to insert a new bank transaction and delete or modify an existing one.

Listing all bank transactions

The page that displays the bank transactions should look as follows (*transactions.html*):

All bank transactions

Id	Bank Name	Bank IBAN	Transfer amount		Status	
6	Optimus Bank	8V34VN5U435334	4588.36	...	UNAUTHORIZED	Modify
7	Optimus Bank	8V34VN5U435334	8987.36	...	UNAUTHORIZED	Modify
8	Optimus Bank	8V34VN5U435334	2544.36	...	FAILED	Modify
9	Optimus Bank	8V34VN5U435334	5312.23	...	RETRIED	Modify

Insert new bank transaction

Figure 9.3 – All bank transactions of a certain payment

Let's start from the controller endpoint, which should be accessed to produce the output shown in the preceding screenshot:

```
@GetMapping("/transactions")
public String loadAllBankTransactionOfCertainPayment(
            SessionStatus sessionStatus, Model model) {

  sessionStatus.setComplete();
  model.addAttribute(ALL_BANK_TRANSACTION_ATTR,
    classicModelsService
```

```
        .loadAllBankTransactionOfCertainPayment());

    return "transactions";
}
```

In the highlighted code, we call the service that's responsible for accessing the repository that executes the query for fetching all the transactions for a certain payment. This query is quite simple (of course, in reality, you won't hardcode the values of `CUSTOMER_NUMBER` and `CHECK_NUMBER` – these can represent something such as the login payment credentials):

```
public Result<BankTransactionRecord>
      fetchAllBankTransactionOfCertainPayment() {

   return ctx.selectFrom(BANK_TRANSACTION)
      .where(BANK_TRANSACTION.CUSTOMER_NUMBER.eq(333L)
      .and(BANK_TRANSACTION.CHECK_NUMBER.eq("NF959653")))
      .fetch();
}
```

Next, the fetched `Result<BankTransactionRecord>` is returned in the controller endpoint listed previously and stored in the model (Spring Boot's `Model`) as a request attribute named `all` (`ALL_BANK_TRANSACTION_ATTR = "all"`). To render the page that's returned from this controller endpoint (*transactions.html*) we can rely on the popular Thymeleaf template engine (of course, you can use any other template engine):

```
<tr th:each="t : ${all}">
 <td><span th:text="${t.transactionId}">ID</span></td>
 <td><span th:text="${t.bankName}">Bank Name</span></td>
 ...
 <td><span th:text="${t.status}">Status</span></td>
</tr>
```

From the returned page (*transactions.html*), we can choose to insert a new transaction or modify an existing one.

Inserting a new bank transaction

Inserting a new bank transaction can be done by rendering the link like so:

```
<a href="/newbanktransaction">Insert new bank transaction</a>
```

This link reaches a controller endpoint that looks like this:

```
@GetMapping("/newbanktransaction")
public String newBankTransaction(Model model) {

  model.addAttribute(NEW_BANK_TRANSACTION_ATTR,
                           new BankTransactionRecord());

  return "newtransaction";
}
```

So, this controller endpoint creates a new `BankTransactionRecord` that is stored in the model via the `NEW_BANK_TRANSACTION_ATTR` request attribute. The returned page, *newtransaction.html*, is rendered like so:

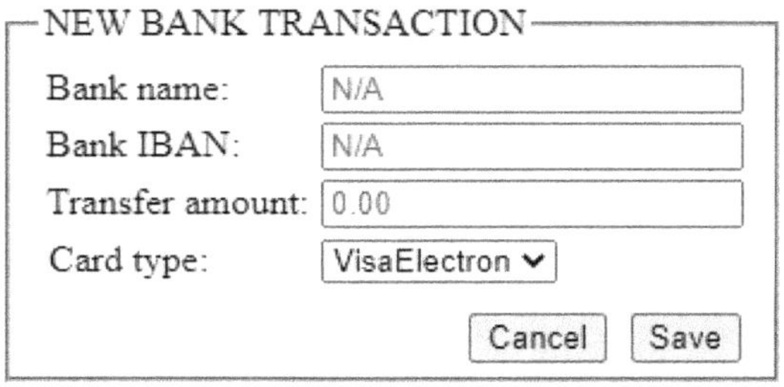

Figure 9.4 – Creating a new bank transaction

Pressing the *Save* button triggers a `POST` request that reaches the following controller endpoint (`/new`):

```
@PostMapping("/new")
public String newBankTransaction(
  @ModelAttribute BankTransactionRecord btr,
  RedirectAttributes redirectAttributes) {

  classicModelsService.newBankTransaction(btr);
  redirectAttributes.addFlashAttribute(
    INSERT_DELETE_OR_UPDATE_BANK_TRANSACTION_ATTR, btr);
```

```
  return "redirect:success";
}
```

So, Spring Boot populates the `btr` record with the submitted data, and we insert it into the database (before inserting it, in the service method (not listed here), we associate this new transaction with the corresponding payment via `btr.setCustomerNumber()` and `btr.setCheckNumber()`):

```
@Transactional
public int newBankTransaction(BankTransactionRecord btr) {

  ctx.attach(btr);

  return btr.insert();
}
```

Since this is a new bank transaction, we must attach it before inserting it.

Updating a bank transaction

Let's consider that updating a bank transaction implies a four-step wizard. In the case of simple wizards, we can use a single `<form/>` that is submitted at the last step of the wizard. However, in the case of dynamic wizards, we must use one `<form/>` per panel since we must submit the data at each step to decide which is going to be the next panel and what it will contain. So, in such cases, we must implement a long HTTP conversation that's capable of storing the user data while navigating back and forth between the panels. Commonly, this is done by storing data via the client's HTTP session.

Let's keep it as simple as possible and assume that the four-step wizard looks as follows:

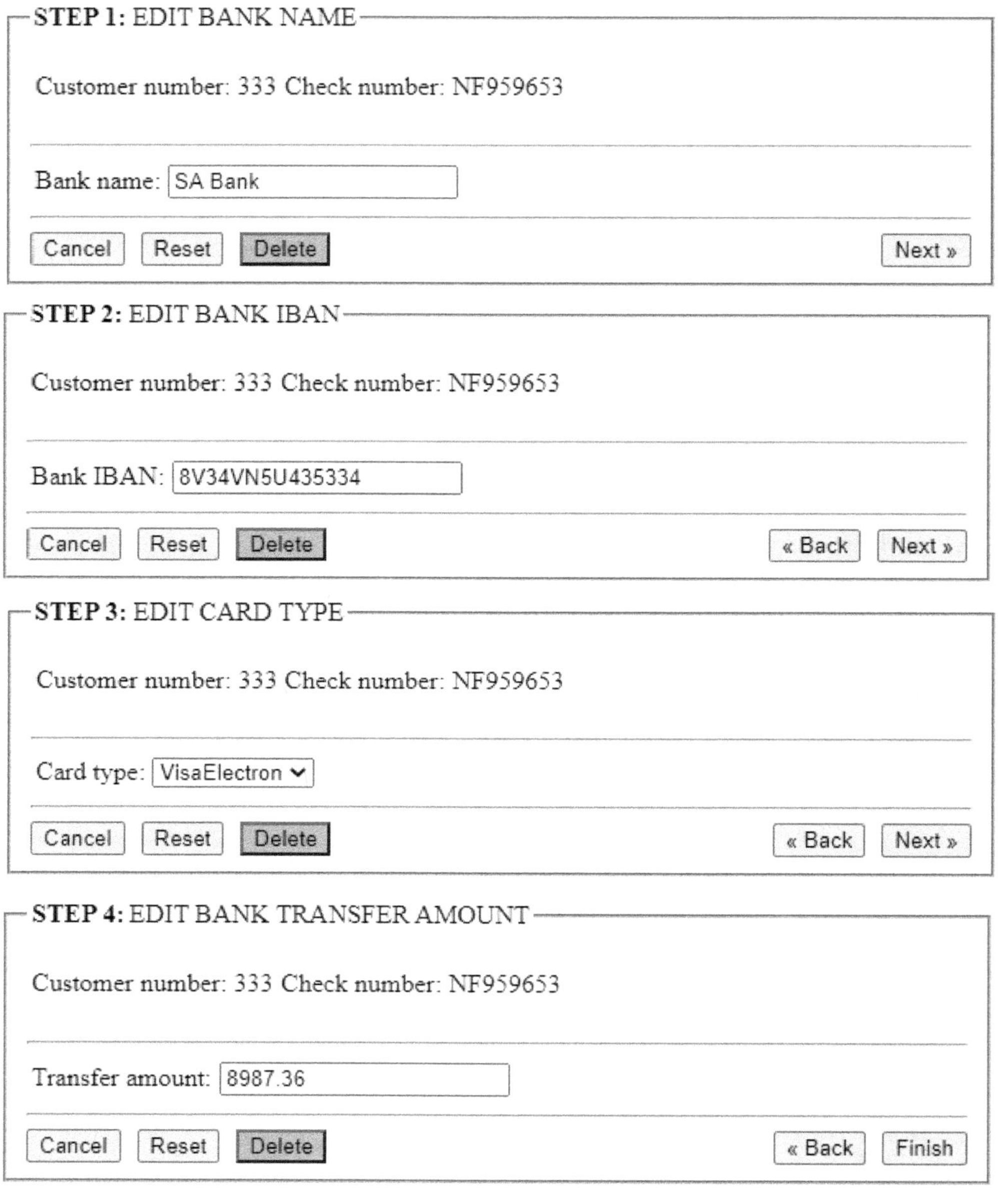

Figure 9.5 – Four-step wizard

Before entering this wizard, we must click on the *Modify* link that corresponds to the bank transaction that we plan to edit. This will hit the following controller endpoint while sending the transaction ID:

```
@GetMapping("/editbankname/{id}")
public String loadBankTransaction(
       @PathVariable(name = "id") Long id, Model model) {

  model.addAttribute(BANK_TRANSACTION_ATTR,
    classicModelsService.loadBankTransaction(id));

  return "redirect:/editbankname";
}
```

`BankTransactionRecord` can be fetched via the following repository method:

```
public BankTransactionRecord fetchBankTransaction(Long id) {

  return ctx.selectFrom(BANK_TRANSACTION)
     .where(BANK_TRANSACTION.TRANSACTION_ID.eq(id))
     .fetchSingle();
}
```

Since this is the transaction that should live across our wizard panels, we must store it in the model via the session attribute, `BANK_TRANSACTION_ATTR = "bt"`. Next, we must return the first panel of the wizard.

Edit bank name

Once we've edited the bank name, we must click on *Next* to submit the data. This reaches the following controller endpoint:

```
@PostMapping("/name")
public String editBankName(
   @ModelAttribute(BANK_TRANSACTION_ATTR)
                  BankTransactionRecord btr) {

   return "redirect:editbankiban";
}
```

Here, we just allow Spring Boot to synchronize the `btr` session record with the submitted data. Next, we must return the second panel.

Edit IBAN

Once we've edited the bank name, we must edit the IBAN and click *Next* (we can also click *Back* and edit the bank name again). After editing the IBAN, the submitted data hits the controller endpoint:

```
@PostMapping("/iban")
public String editBankIban(
    @ModelAttribute(BANK_TRANSACTION_ATTR)
        BankTransactionRecord btr) {

    return "redirect:editcardtype";
}
```

Again, we allow Spring Boot to synchronize the `btr` session record with the submitted data. Next, we must return the third panel.

Edit the card type

Once we've edited the bank's IBAN, we must choose the card type and click *Next* (we can also click *Back* and edit the bank IBAN again). After choosing the card type, the submitted data hits the controller endpoint:

```
@PostMapping("/cardtype")
public String editCardType(
    @ModelAttribute(BANK_TRANSACTION_ATTR)
        BankTransactionRecord btr) {

    return "redirect:editbanktransfer";
}
```

Again, we allow Spring Boot to synchronize the `btr` session record with the submitted data. Next, we must return the last panel.

Edit the transferred amount

Finally, we must edit the transferred amount and submit it to the controller endpoint:

```
@PostMapping("/transfer")
public String updateBankTransfer(
  @ModelAttribute(BANK_TRANSACTION_ATTR)
    BankTransactionRecord btr, SessionStatus sessionStatus,
```

```
    RedirectAttributes redirectAttributes) {

  classicModelsService.updateBankTransaction(btr);
  redirectAttributes.addFlashAttribute(
    INSERT_DELETE_OR_UPDATE_BANK_TRANSACTION_ATTR, btr);

  sessionStatus.setComplete();

  return "redirect:success";
}
```

The UPDATE method is used in a repository method, as follows:

```
@Transactional
public int updateBankTransaction(BankTransactionRecord btr) {

  return btr.update();
}
```

Finally, we must clean up the HTTP session to remove the updatable record.

Resetting the wizard data

It is a common feature of any wizard to provide a *Reset* button for reverting the data from the current panel or from the entire wizard to the latest saved data. Our *Reset* button is relying on jOOQ's `reset()` method to reset the wizard (all three panels):

```
@GetMapping("/reset/{page}")
    public String reset(
    @PathVariable(name = "page") String page, Model model) {

  if (model.containsAttribute(BANK_TRANSACTION_ATTR)) {
   ((BankTransactionRecord) model.getAttribute(
     BANK_TRANSACTION_ATTR)).reset();
  }

  return "redirect:/" + page;
}
```

Of course, you can use `reset(Field/Name/String/int)` to implement a reset per panel feature. Finally, let's delete a bank transaction.

Deleting a bank transaction

As shown in *Figure 9.5*, each panel of our wizard contains a *Delete* button, which allows us to delete this bank transaction. The code for its controller endpoint is as follows:

```
@GetMapping("/delete")
public String deleteBankTransaction(
  SessionStatus sessionStatus, Model model,
          RedirectAttributes redirectAttributes) {
  ...
  BankTransactionRecord btr = (BankTransactionRecord)
    model.getAttribute(BANK_TRANSACTION_ATTR);
  classicModelsService.deleteBankTransaction(btr);
  sessionStatus.setComplete();
  ...
}
```

And the DELETE is rendered by a call of `delete()` in the following repository method:

```
@Transactional
public int deleteBankTransaction(BankTransactionRecord btr) {

  return btr.delete();
}
```

The complete code is called *CRUDRecords*. If you prefer to use POJOs and jOOQ's DAO, then check out *DaoCRUDRecords* and the REST version (for Postman, ARC, and so on), which is called *DaoCRUDRESTRecords*. These three applications are available for MySQL. For brevity, we skipped any validation and error handling code.

Using merge() versus store()

Let's consider the following scenario: we have loaded and displayed the payments (`PAYMENT`) of a certain customer (for instance, `PAYMENT.CUSTOMER_NUMBER.eq(103L)`). The user should be able to insert new payments for this customer or update the amount of an existing payment. To solve this task, we have two approaches that are almost the same. These are shown in the following screenshot:

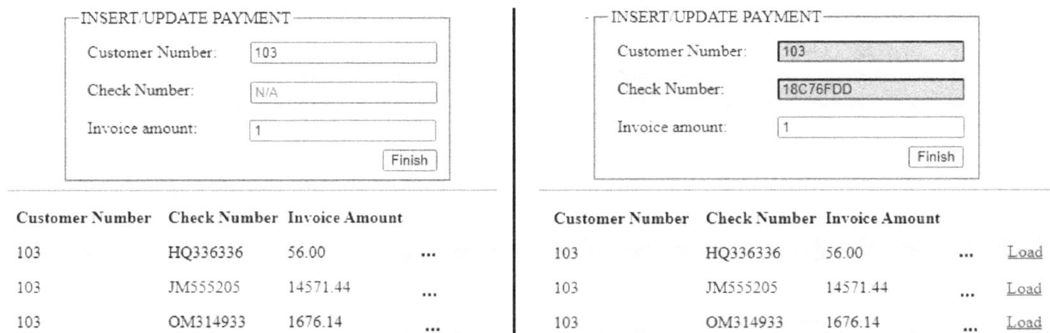

Figure 9.6 – Insert/update payment

Regarding the design on the left-hand side, to insert a new payment, we can simply type a new (unique) *Check Number* (for instance, received via SMS) and the corresponding *Invoice Amount*. To update the *Invoice Amount* of an existing payment, we must type its current *Check Number* from the bottom table (for instance, to update the second payment from the table, we must type the *Check Number JM555205*).

Regarding the right-hand side design, to insert a new payment, we just type the *Invoice Amount*; the *Check Number* is auto-generated and pre-filled by the application. However, to update the *Invoice Amount* of an existing payment, we must load the payment first via the corresponding *Load* link in the bottom table. This will fetch the corresponding payment from the database so that we can type in the new amount value and update it.

Implementing the left-hand side design via merge()

Let's focus on the left-hand side design. After the user submits the payment form, Spring Boot creates a new `PaymentRecord` and populates it with the submitted data. Next, based on the submitted *Check Number*, we must determine if this is a new payment or an update of an existing payment to execute an `INSERT` or an `UPDATE`. So, it is time for `merge()` to do its job and render an SQL that delegates the task of choosing between `INSERT` and `UPDATE` to the database:

```
@PostMapping("/merge")
public String mergePayment(PaymentRecord pr) {

  classicModelsService.mergePayment(pr);

  return "redirect:payments";
}
```

```
@Transactional
public int mergePayment(PaymentRecord pr) {

   ctx.attach(pr);

   return pr.merge();
}
```

That's all the important code! Notice that, before merging, we need to attach the relevant `PaymentRecord`. Remember that Spring Boot has created this record, so it is not attached to any `Configuration`.

Check out the complete application for this code, which is called *MergeRecords*. If you prefer to use POJOs and jOOQ's DAO, then check out *DaoMergeRecords*. Both applications are available for MySQL.

Implementing the right-hand side design via store()

If we wish to implement the right-hand side design, then we must start by preparing a brand-new `PaymentRecord` (for instance, we must generate the *Check Number*) and storing it via an HTTP session attribute (`PAYMENT_ATTR`). This `PaymentRecord` is returned to the user. However, if the user wants to update the *Invoice Amount* of an existing payment, then they have the option to click on the corresponding *Load* link in the bottom table. The following query can be used to fetch the relevant `RecordPayment`:

```
public PaymentRecord fetchPayment(Long nr, String ch) {

  return ctx.selectFrom(PAYMENT)
    .where(row(PAYMENT.CUSTOMER_NUMBER, PAYMENT.CHECK_NUMBER)
    .eq(row(nr, ch)))
    .fetchSingle();
}
```

The fetched `PaymentRecord` overrides the one from the HTTP session and is returned to the user. When the user submits the data, Spring Boot synchronizes the `PaymentRecord` value that's stored in `PAYMENT_ATTR` (which can be the new `PaymentRecord` or the fetched `PaymentRecord`) with the submitted data. This time, we can let jOOQ choose between `INSERT` and `UPDATE` via `store()` since this method distinguishes between a new `PaymentRecord` and a fetched `PaymentRecord` and acts accordingly:

```
@PostMapping("/store")
public String storePayment(SessionStatus sessionStatus,
```

```
          @ModelAttribute(PAYMENT_ATTR) PaymentRecord pr) {

   pr.setCachingDate(LocalDateTime.now());
   classicModelsService.storePayment(pr);
   sessionStatus.setComplete();

   return "redirect:payments";
}

@Transactional
public int storePayment(PaymentRecord pr) {

  ctx.attach(pr);

  return pr.store();
}
```

The application that uses `store()` is named *StoreRecords* (available for MySQL). Now, let's move on and talk about navigating (updatable) records.

Navigating (updatable) records

jOOQ exposes several navigation methods that can be used for attached (updatable) records only (`TableRecord` and `UpdatableRecord`). To use these methods, please consider the following note.

> **Important Note**
> While these methods are very convenient/appealing, they are also a big N+1 risk. `UpdatableRecord` is great for CRUD, but if you aren't using CRUD, then you shouldn't use `UpdatableRecord`. It's better to project only the columns you need and try to use joins or other SQL utilities to fetch data from multiple tables.

These methods navigate based on the foreign key references. For instance, with an attached `DepartmentRecord`, we can navigate its parent (`OFFICE`) via `fetchParent(ForeignKey<R, O> key)`, as shown in the following example:

```
public OfficeRecord fetchOfficeOfDepartment(
     DepartmentRecord dr) {
```

```
  return dr.fetchParent(Keys.DEPARTMENT_OFFICE_FK);
  // or, Keys.DEPARTMENT_OFFICE_FK.fetchParent(dr);
}
```

The `Keys.DEPARTMENT_OFFICE_FK` foreign key was generated by the jOOQ Code Generator based on our `CREATE TABLE` DDL. In terms of MySQL dialect, jOOQ renders the following SQL:

```
SELECT
  `classicmodels`.`office`.`office_code`,
  ...
  `classicmodels`.`office`.`location`
FROM
  `classicmodels`.`office`
WHERE
  `classicmodels`.`office`.`office_code` in (?)
```

You can also fetch `Table<OfficeRecord>` via `parent()`:

```
Table<OfficeRecord> tor =
   dr.parent(Keys.DEPARTMENT_OFFICE_FK);
Table<OfficeRecord> tor =
   Keys.DEPARTMENT_OFFICE_FK.parent(dr);
```

Next, with an attached `OfficeRecord`, we can fetch the employees (`EMPLOYEE`) via `fetchChildren(ForeignKey<O,R> key)`, as follows:

```
public Result<EmployeeRecord>
         fetchEmployeesOfOffice(OfficeRecord or) {

  return or.fetchChildren(Keys.EMPLOYEE_OFFICE_FK);
  // or, Keys.EMPLOYEE_OFFICE_FK.fetchChildren(or);
}
```

This time, the SQL that's rendered for the MySQL dialect is as follows:

```
SELECT
  `classicmodels`.`employee`.`employee_number`,
  ...
  `classicmodels`.`employee`.`monthly_bonus`
FROM
```

```
  `classicmodels`.`employee`
WHERE
  `classicmodels`.`employee`.`office_code` in (?)
```

You can also fetch `Table<OfficeRecord>` via `children()` (using `children()` is often preferable to `fetchChildren()` because it encourages writing queries rather than navigating `UpdatableRecord` directly):

```
Table<EmployeeRecord> ter =
   or.children(Keys.EMPLOYEE_OFFICE_FK);
Table<EmployeeRecord> ter =
   Keys.EMPLOYEE_OFFICE_FK.children(or);
```

Next, we can reuse `fetchChildren()` to fetch the customers (`CUSTOMER`) of a certain employee (`EmployeeRecord`). This will result in every `CustomerRecord` of that `EmployeeRecord`. Finally, with an attached `CustomerRecord`, we can fetch its details (`CUSTOMERDETAIL`) via `fetchChild(ForeignKey<O, R> key)`, as follows:

```
public CustomerdetailRecord
      fetchCustomerdetailOfCustomer(CustomerRecord cr) {

  return cr.fetchChild(Keys.CUSTOMERDETAIL_CUSTOMER_FK);
}
```

The rendered SQL for the MySQL dialect is as follows:

```
SELECT
  `classicmodels`.`customerdetail`.`customer_number`,
  ...
  `classicmodels`.`customerdetail`.`country`
FROM
  `classicmodels`.`customerdetail`
WHERE
  `classicmodels`.`customerdetail`.`customer_number` in (?)
```

In the bundled code (*NavigationRecords*, which is available for MySQL), you can see all these methods collaborating to obtain something similar to the following:

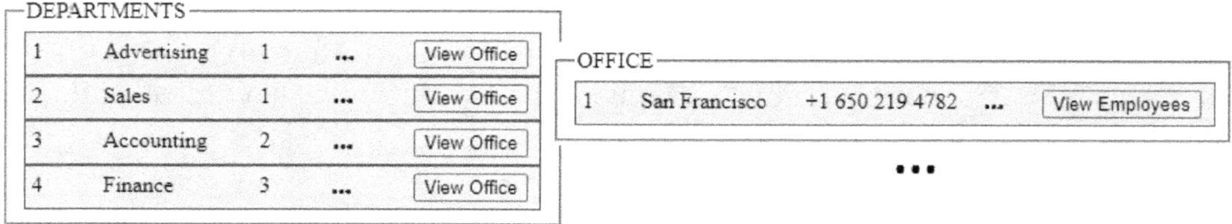

Figure 9.7 – Navigating between records

These methods are also quite handy for looping the parent/children of a record and taking some action. Here is an example of using `fetchParent()` to fetch the `EmployeeRecord` details of each `SaleRecord` that has less than *2,000* sales:

```
for (SaleRecord sale : ctx.fetch(SALE, SALE.SALE_.lt(2000d))){

  if ("Sales Rep".equals(sale.fetchParent(
          Keys.SALE_EMPLOYEE_FK).getJobTitle())) {
    sale.delete();
  }
}
```

In the previous example, each call of `fetchParent()` executes a separate `SELECT`, which is far away from being a good choice. However, an interesting method that's helpful in this case is `fetchParents()`, which can fetch all the parents of a list of records in a single `SELECT`. This means that we can rewrite the previous query like so:

```
List<SaleRecord> sales
    = ctx.fetch(SALE, SALE.SALE_.lt(2000d));
List<EmployeeRecord> employees
    = Keys.SALE_EMPLOYEE_FK.fetchParents(sales);

for (SaleRecord sale : sales) {
  for (EmployeeRecord employee : employees) {
    if (Objects.equals(sale.getEmployeeNumber(),
        employee.getEmployeeNumber()) && "Sales Rep".equals(
          employee.getJobTitle())) {
        sale.delete();
        break;
```

```
        }
    }
}
```

If you need Table<EmployeeRecord>, then use parents():

```
Table<EmployeeRecord> employeesTable
   = Keys.SALE_EMPLOYEE_FK.parents(sales);
```

> **Important Note**
>
> Note that these kinds of loops are really bad from a performance perspective!
>
> If there's no business logic in the client, it should be a single DELETE statement with a semi-join (for instance, an IN predicate). So, don't take these loop examples at face value. I know that this approach feels easier but I strongly recommend avoiding it. Don't implement such loops all over the application and then complain about jOOQ being slow, just like when people complain about Hibernate being slow when these navigational loops are simply wrong.
>
> The only reason why anyone should ever process data row by row is that each row requires very complex business logic that can't be expressed in SQL or otherwise pushed into the database. People get this wrong in all languages, including PL/SQL. They loop over rows because it's convenient and they prefer 3GLs over SQL-the-4GL, and then they run queries on a row-by-row basis because they can. So, to justify the previous loops, we need to at least add some businessLogicHere(saleRecord) method calls to hint at the row-by-row approach being necessary in this particular case.

You can find these examples in *NavigationParentsRecords* (available for MySQL). Next, let's focus on using explicit transactions and jOOQ queries.

Transactions

Among other benefits, transactions give us the ACID properties. We can distinguish between *read-only* and *read-write* transactions, different isolation levels, different propagation strategies, and so on. While Spring Boot supports a comprehensive transactional API (Spring TX) that's commonly used via @Transactional and TransactionTemplate, jOOQ comes with a simple transaction API (and an org.jooq.TransactionProvider SPI) that fits perfectly in the context of fluent style.

The following diagram highlights the main implementations of this SPI:

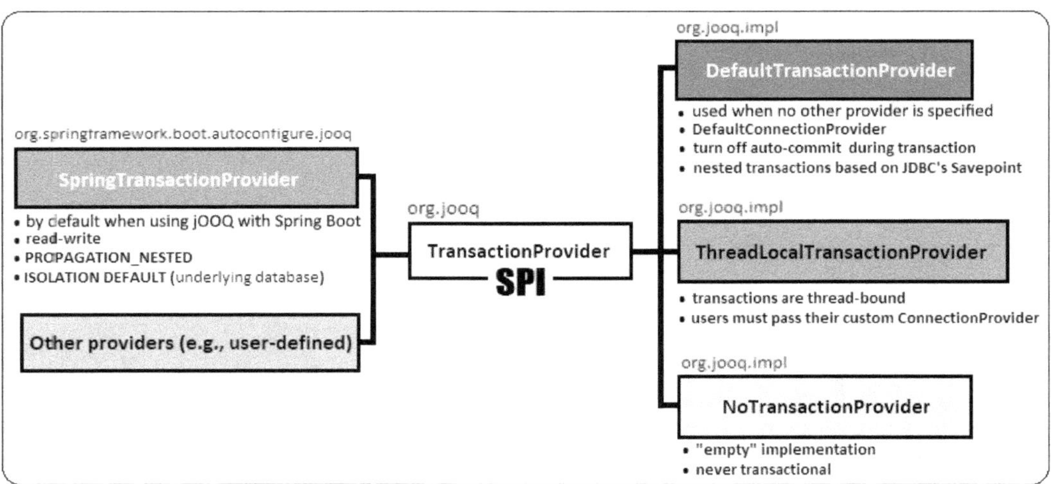

Figure 9.8 – jOOQ transaction providers

Starting with jOOQ 3.17 we have support for transactions in R2DBC as well. So, jOOQ 3.17 come with support for reactive transactions.

Mainly, the jOOQ API for blocking transactions can be used like so:

```
ctx.transaction(configuration -> {
  DSL.using(configuration)...
  // or, configuration.dsl()...
}

var result = ctx.transactionResult(configuration -> {
  return DSL.using(configuration)...
  // or, return configuration.dsl()...
}
```

Here, we have `transaction(TransactionalRunnable)`, which returns `void` and `transactionResult(TransactionalCallable)` for returning a result. The former wraps the transactional code in jOOQ's `org.jooq.TransactionalRunnable` functional interface, while the latter wraps the transactional code in jOOQ's `org.jooq.TransactionalCallable` functional interface.

> **Important Note**
>
> Pay attention to the fact that, inside a jOOQ transaction, you must use `DSLContext` that's been obtained from the given (**derived**) `configuration`, not `ctx` (the injected `DSLContext`).

SpringTransactionProvider

In terms of Spring Boot's context, jOOQ delegates the task of handling the transactions (*begin*, *commit*, and *rollback*) to `SpringTransactionProvider`, an implementation of the `org.jooq.TransactionProvider` SPI that's meant to allow Spring Transaction to be used with JOOQ. By default, you'll get a *read-write* transaction with no name (`null`) whose propagation is set to `PROPAGATION_NESTED`, and the isolation level is set to the default isolation level of the underlying database; that is, `ISOLATION_DEFAULT`.

If you ever want to decouple `SpringTransactionProvider` (for instance, to avoid potential incompatibilities between Spring Boot and jOOQ), then use the following code:

```
// affects ctx
ctx.configuration().set((TransactionProvider) null);

// create derived DSLContext
ctx.configuration().derive((TransactionProvider) null).dsl();
```

Once you've decoupled `SpringTransactionProvider`, jOOQ will execute the transaction via the jOOQ's `DefaultTransactionProvider` and `DefaultConnectionProvider` with *auto-commit* mode set to `false` (if it was `true` before the transaction, then jOOQ will restore it after the transaction). `DefaultTransactionProvider` supports nested transactions that have been implemented via JDBC's `java.sql.Savepoint`. In *Chapter 18, jOOQ SPI (Providers and Listeners)*, you'll learn how to implement a `TransactionProvider`, but for now, let's look at some examples of jOOQ transactions. Let's start from a simple transaction that highlights the commit/rollback:

```
ctx.transaction(configuration -> {
  DSL.using(configuration).delete(SALE)
    .where(SALE.SALE_ID.eq(1L)).execute();

  DSL.using(configuration).insertInto(TOKEN)
    .set(TOKEN.SALE_ID, 1L).set(TOKEN.AMOUNT, 1000d)
    .execute();

  // at this point transaction should commit, but the error
```

```
  // caused by the previous INSERT will lead to rollback
});
```

If you want to handle/prevent rollbacks, then you can wrap the transactional code in a `try-catch` block and act as you consider; if you want to do some work (for instance, do some cleanup work) and roll back, then just throw the exception at the end of the `catch` block. Otherwise, by catching `RuntimeException`, we can prevent a rollback from occurring if something went wrong while executing the SQL statements from jOOQ:

```
ctx.transaction(configuration -> {

  try {
      // same DMLs as in the previous example
  } catch (RuntimeException e) {
    System.out.println("I've decided that this error
                        doesn't require rollback ...");
  }
});
```

jOOQ nested transactions look like Matrioska dolls. We nest the transactional code by nesting calls of `transaction()`/`transactionResult()`. Here, the transactions will be automatically demarcated by jOOQ with savepoints. Of course, no one is prevented from extracting these lambdas into methods and composing them as higher-order functions, just like you can compose Spring-annotated transactional methods.

Here is an example of nesting two jOOQ transactions:

```
public void nestedJOOQTransaction() {

  ctx.transaction(outer -> {

    DSL.using(outer).delete(SALE) // or, outer.dsl()
       .where(SALE.SALE_ID.eq(2L)).execute();

    // savepoint created
    DSL.using(outer).transaction(inner -> {
      DSL.using(inner).insertInto(TOKEN) // or, inner.dsl()
        .set(TOKEN.SALE_ID, 1L)
        .set(TOKEN.AMOUNT, 1000d).execute();
    });
  });
}
```

By default, if something goes wrong in one of the transactions, then the subsequent transactions (inner transactions) will not be executed and all the outer transactions will be rolled back. But sometimes, we may want to roll back only the current transaction and not affect the outer transactions, as shown in the following example:

```
ctx.transaction(outer -> {
  try {
    DSL.using(outer).delete(SALE)
        .where(SALE.SALE_ID.eq(1L)).execute();

         // savepoint created
         try {
           DSL.using(outer)
               .transaction(inner -> {
                 DSL.using(inner).insertInto(TOKEN)
                    .set(TOKEN.SALE_ID, 1L)
                    .set(TOKEN.AMOUNT, 1000d).execute();
                 });
          } catch (RuntimeException e) { throw e; }
  } catch (RuntimeException e) {
    System.out.println("I've decided that this error doesn't
         require rollback of the outer transaction ...");
    // throw e; // rollback
  }
});
```

You can check out these examples in *JOOQTransaction* (available for MySQL).

ThreadLocalTransactionProvider

Another jOOQ built-in transaction provider is `ThreadLocalTransactionProvider`. This provider implements thread-bound transaction semantics. In other words, a transaction and its associated `Connection` will never leave the thread that started the transaction.

An important requirement of `ThreadLocalTransactionProvider` is that we must pass a custom `ConnectionProvider` implementation directly to this provider instead of passing it to `Configuration`. We can write our own `CustomProvider` or rely on a jOOQ built-one such as `MockConnectionProvider` (for tests), `DefaultConnectionProvider`, `DataSourceConnectionProvider`, or `NoConnectionProvider`.

For instance, if we choose `DataSourceConnectionProvider`, then, in a Spring Boot application, we can use a `DataSource` (for instance, `HikariDataSource`) that has already been prepared by Spring Boot, as follows:

```
@Configuration
public class JooqConfig {

  @Bean
  @ConditionalOnMissingBean(org.jooq.Configuration.class)
  public DefaultConfiguration jooqConfiguration(
          JooqProperties properties, DataSource ds) {

    final DefaultConfiguration defaultConfig =
      new DefaultConfiguration();
    final ConnectionProvider cp =
      new DataSourceConnectionProvider(ds);

    defaultConfig
     .set(properties.determineSqlDialect(ds))
     .set(new ThreadLocalTransactionProvider(cp, true));

    /* or, as a derived configuration
    final org.jooq.Configuration derivedConfig = defaultConfig
      .derive(properties.determineSqlDialect(ds))
      .derive(new ThreadLocalTransactionProvider(cp, true));
    */

    return defaultConfig;
  }
}
```

Alternatively, if you are using Spring Boot 2.5.0+, then you can profit from the `DefaultConfigurationCustomizer` functional interface. This interface defines a method called `customize(DefaultConfiguration configuration)`, which is a handy way to customize jOOQ's `DefaultConfiguration`:

```
@Configuration
public class JooqConfig
  implements DefaultConfigurationCustomizer {
```

```
  private final DataSource ds;

  public JooqConfig(DataSource ds) {
    this.ds = ds;
  }

  @Override
  public void customize(DefaultConfiguration configuration) {

    configuration.set(new ThreadLocalTransactionProvider(
            new DataSourceConnectionProvider(ds), true));
  }
}
```

Done! Now, we can inject the `DSLContext` information that's been built by Spring Boot based on our `Configuration` and take advantage of thread-bound transaction semantics, which is usually exactly what Spring uses. You can check out an example by looking at *ThreadLocalTransactionProvider{1,2}*, which is available for MySQL.

Next, let's talk about jOOQ asynchronous transactions.

jOOQ asynchronous transactions

While `transaction()` and `transactionResult()` are synchronous, jOOQ also has `transactionAsync()` and `transactionResultAsync()`, which can be used to shape asynchronous transactions. Here are two asynchronous transactions that act independently of each other – they run in concurrent threads. The first one commits, while the second one rolls back:

```
// this transaction commits
@Async
public CompletableFuture<Integer>
             executeFirstJOOQTransaction() {

  return ctx.transactionResultAsync(configuration -> {

    int result = 0;

    result += DSL.using(configuration).insertInto(TOKEN)
      .set(TOKEN.SALE_ID, 1L).set(TOKEN.AMOUNT, 500d)
      .execute();
```

```
    result += DSL.using(configuration).insertInto(TOKEN)
      .set(TOKEN.SALE_ID, 1L).set(TOKEN.AMOUNT, 1000d)
      .execute();

    return result;
  }).toCompletableFuture();
}

// this transaction is roll backed
@Async
public CompletableFuture<Integer>
          executeSecondJOOQTransaction() {

  return ctx.transactionResultAsync(configuration -> {

    int result = 0;

    result += DSL.using(configuration).delete(SALE)
      .where(SALE.SALE_ID.eq(2L)).execute();

    result += DSL.using(configuration).insertInto(TOKEN)
      .set(TOKEN.SALE_ID, 2L).set(TOKEN.AMOUNT, 1000d)
      .execute();

    return result;
  }).toCompletableFuture();
}
```

If you don't want to rely on the default `Executor` (`ForkJoinPool.commonPool()`), then use `transactionAsync(Executor exctr, TransactionalRunnable tr)` or `transactionResultAsync(Executor exctr, TransactionalCallable<T> tc)`, respectively. But unlike what `CompletableFuture` does, jOOQ will remember `Executor` in its `CompletionStage` implementation so that it doesn't have to be provided on every ensuing asynchronous call.

However, asynchronous transactions work very badly with Spring, which usually assumes thread-bound transactions. Go to `https://github.com/spring-projects/spring-boot/issues/24049` to see a discussion about this.

Check out the complete code in *JOOQTransactionAsync* (available for MySQL). Next, let's look at some examples of using/choosing `@Transactional` or the jOOQ transaction API.

@Transactional versus the jOOQ transaction API

Right off the bat, I wish to enforce an important note (most of you probably already know and respect these statements, but a quick reminder is always welcome).

> **Important Note**
>
> A non-transactional-context refers to a context with no explicit transaction boundaries, *not* to a context with no physical database transaction. All database statements are executed in the context of a physical database transaction.
>
> Without specifying the explicit boundaries of the transaction (via `@Transactional`, `TransactionTemplate`, the jOOQ transaction API, and so on), jOOQ *may* use a different database connection for each statement. Whether or not jOOQ uses a different connection per statement is defined by `ConnectionProvider`. This statement is true for `DataSourceConnectionProvider` (and even then, it depends on `DataSource`) but false for `DefaultConnectionProvider`. In the worst-case scenario, this means that the statements that define a *logical transaction* don't benefit from ACID and they are prone to lead to *race condition* bugs and SQL *phenomena*. Each statement is executed in a separate transaction (*auto-commit* mode), which may result in a high database connection acquisition request rate, which is not good! On medium/large applications, reducing the database connection acquisition request rate next to short transactions will sustain performance since your application will be ready to face high traffic (a high number of concurrent requests).
>
> Never combine `@Transactional/TransactionTemplate` and the jOOQ transaction API to solve a task in common (the same is true for Java/Jakarta EE transactions, of course). This may lead to unexpected behaviors. So long as Spring transactions and jOOQ transactions are not interleaved, it is safe to use them in the same application.

The best way to use Spring transactions *only* consists of annotating your repository/service class with `@Transactional(readOnly=true)` and explicitly setting `@Transactional` only on methods that should be allowed to execute write statements. However, if the same repository/service uses jOOQ transactions as well, then you should explicitly annotate each method, not the class itself. This way, you avoid inheriting `@Transactional(readOnly=true)` in methods that explicitly use jOOQ transactions.

Now, let's consider several examples that are meant to reveal the best practices for using transactions. Let's start with the following snippet of code:

```
public void fetchWithNoTransaction() {

   ctx.selectFrom(SALE).fetchAny();
   ctx.selectFrom(TOKEN).fetchAny();
}
```

This method runs in a non-transactional context and executes two read statements. Each read is executed by the database in a separate physical transaction that requires a separate database connection. Keep in mind that this may not be true, depending on `ConnectionProvider`. Relying on `@Transactional(readOnly=true)` is much better:

```
@Transactional(readOnly=true)
public void fetchWithTransaction() {

   ctx.selectFrom(SALE).fetchAny();
   ctx.selectFrom(TOKEN).fetchAny();
}
```

This time, a single database connection and a single transaction are used. `readOnly` come with a bunch of advantages, including that your team members cannot accidentally add write statements (such attempt result in an error), read-only transactions can be optimized at the database level (this is database vendor-specific), you must explicitly set the transaction isolation level as expected, and so on.

Moreover, having no transaction and setting auto-commit to `true` only makes sense if you execute a single read-only SQL statement, but it doesn't lead to any significant benefit. Therefore, even in such cases, it's better to rely on explicit (declarative) transactions.

However, if you consider that the `readOnly=true` flag isn't needed, then the following code can be executed in a jOOQ transaction as well (by default, this is a read-write transaction):

```
public void fetchWithTransaction() {

  ctx.transaction(configuration -> {

    DSL.using(configuration).selectFrom(SALE).fetchAny();
    DSL.using(configuration).selectFrom(TOKEN).fetchAny();
```

```
    // Implicit commit executed here
  });
}
```

Notice that, exactly like Spring's `TransactionTemplate` (which can be used as well), the jOOQ transaction can strictly demarcate the transactional code. In other words, the `@Transactional` annotation acquires the database connection and starts the transaction immediately when entering the method. Then, it commits the transaction at the end of the method. This means that the potentially non-transactional code of a `@Transactional` method (the code that shapes business logic that doesn't need to run in a transaction) still runs inside the current transaction, which can lead to a long-running transaction. On the other hand, jOOQ transactions (just like `TransactionTemplate`) allow us to isolate and orchestrate the transactional code to run in transactions and the rest of the code outside of transactions. Let's look at a scenario where using a jOOQ transaction (or `TransactionTemplate`) is a better choice than using `@Transactional`:

```
@Transactional
public void fetchAndStreamWithTransactional() {

  ctx.update(EMPLOYEE).set(EMPLOYEE.SALARY,
    EMPLOYEE.SALARY.plus(1000)).execute();

  ctx.selectFrom(EMPLOYEE)
      .fetch() // jOOQ fetches the whole result set into memory
               // via the connection opened by @Transactional
      .stream()// stream over the in-memory result set
               // (database connection is active)
      .map()   // ... more time-consuming pipeline operations
               // holds the transaction open
      .forEach(System.out::println);
}
```

In this case, jOOQ fetches the whole result set into memory via the connection that's been opened by `@Transactional`. This means that the streaming operations (for instance, `map()`) don't need transactions, but Spring will close this transaction at the end of the method. This can result in a potentially long-running transaction. While we can avoid this issue by splitting the code into separate methods, we can also rely on a jOOQ transaction (or `TransactionTemplate`):

```
public void fetchAndStreamWithJOOQTransaction() {

  Result<EmployeeRecord> result =
```

```
    ctx.transactionResult(configuration -> {

      DSL.using(configuration).update(EMPLOYEE)
         .set(EMPLOYEE.SALARY, EMPLOYEE.SALARY.plus(1000))
         .execute();

      return DSL.using(configuration).selectFrom(EMPLOYEE)
               .fetch();
  });

  result.stream() // stream over the in-memory result set
                  // (database connection is closed)
        .map()    // ... more time-consuming pipeline
                  // operations, but the transaction is closed
        .forEach(System.out::println);
}
```

This is much better because we've removed the streaming operations from the transaction.

In terms of executing one or more DML operations in a method, it should be annotated with `@Transactional`, explicitly use the jOOQ transaction API, or use `TransactionTemplate` to demarcate the transactional code. Otherwise, Spring Boot will report an `SQLException`: *Connection is read-only. Queries leading to data modification are not allowed.* You can see such an example next to the previous examples in *SpringBootTransactional* (available for MySQL).

It is a well-known shortcoming of Spring transactions that `@Transactional` is ignored if it is added to a `private`, `protected`, or package-protected method or to a method that's been defined in the same class as where it is invoked. By default, `@Transactional` only works on `public` methods that should be added to classes and are different from where they are invoked. However, this issue can easily be avoided by using the jOOQ transaction API or `TransactionTemplate`, which don't suffer from these issues. You can explore some examples by looking at the *JOOQTransactionNotIgnored* application (available for MySQL).

A strong argument for choosing the Spring transactions for our jOOQ queries is that we can benefit from Spring transactions' isolation levels and propagation strategies. In the bundled code, you can find a suite of seven applications – one for each of the seven propagation levels supported by Spring transactions – that exemplifies the usage of jOOQ queries and Spring transaction propagations. These applications are called *Propagation{Foo}* and are available for MySQL.

In conclusion, jOOQ queries can be used in the following circumstances:

- Only with Spring transactions (you can take advantage of Spring transactions' features at full capacity)
- Only with jOOQ transactions (in the context of Spring Boot, you'll get read-write, nested transactions that rely on the database's isolation level)
- By combining them without interleaving Spring with jOOQ transactions to accomplish common tasks (in other words, once you open a Spring transaction, ensure that any subsequent inner transaction is a Spring one as well. If you open a jOOQ transaction, then ensure that any subsequent inner transaction is a jOOQ one.)

Hooking reactive transactions

As mentioned earlier, starting with jOOQ 3.17, we can take advantage of reactive transactions or transactions in R2DBC. Reactive transactions are easy to use via `Publisher`, as they have the same semantics as JDBC's blocking APIs Here is an example of how to write a nested reactive transaction:

```
Flux<?> flux = Flux.from(
 ctx.transactionPublisher(outer -> Flux.from(
  DSL.using(outer).delete(SALE) // or, outer.dsl()
                .where(SALE.SALE_ID.eq(2L)))
     .thenMany(Flux.from(
     DSL.using(outer).transactionPublisher( // or, outer.dsl()
      inner -> Flux.from(
      DSL.using(inner).insertInto(TOKEN) // or, inner.dsl()
                     .set(TOKEN.SALE_ID, 1L)
                     .set(TOKEN.AMOUNT, 1000d)
      )))
  )));

flux.subscribe();
```

Being in Spring Boot, this example relies on Project Reactor (`https://projectreactor.io/`), but you can use any other reactive library. More examples are available in the bundled code for MySQL, *jOOQReactiveTransaction*.

Locking

Locking is used to orchestrate concurrent access to data to prevent *race condition* threads, deadlocks, *lost updates*, and other SQL *phenomena*.

Among the most popular locking mechanisms, we have **optimistic** and **pessimistic** *locking*. As you'll see shortly, jOOQ supports both of them for CRUD operations. So, let's start with *optimistic locking*.

Optimistic locking overview

Optimistic locking is commonly related to the **lost updates** SQL phenomena, so let's quickly overview this anomaly.

A *lost update* is a popular anomaly that can seriously affect data integrity. A transaction reads a record and uses this information to make business decisions (for instance, decisions that may lead to that record being modified) without being aware that, in the meantime, a concurrent transaction has modified that record and committed it. When the first transaction commits, it is unaware of the lost update. This may cause data integrity issues (for example, the inventory can report a negative value, a certain payment can be lost, and so on).

Consider the scenario shown in the following diagram:

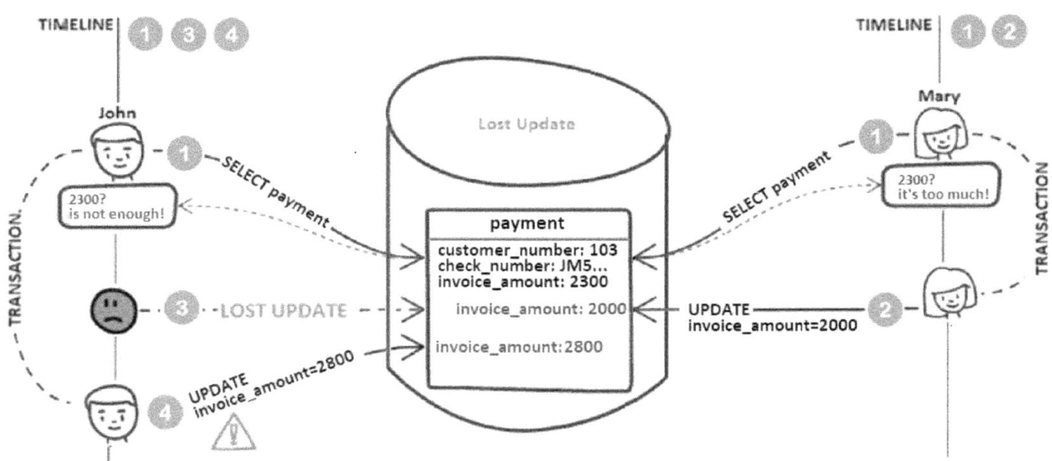

Figure 9.9 – Lost update phenomena

If we take this scenario step-by-step, then the following occurs:

1. John and Mary fetch the invoice amount (*2,300*) of the same payment.
2. Mary considers that the current invoice amount is too much, so she updates the amount from *2,300* to *2,000*.
3. John's transaction is not aware of Mary's update.
4. John considers that the current invoice amount is not enough, so he updates the amount to *2,800*, without being aware of Mary's decision.

This anomaly affects the **Read Committed** isolation level and can be avoided by setting the **Repeatable Read** or **Serializable** isolation level. For the **Repeatable Read** isolation level without **Multi-Version Concurrency Control** (**MVCC**), the database uses shared locks to reject other transactions' attempts to modify an already fetched record.

However, in the presence of MVCC databases, there is no need for locks since we can use the application-level optimistic locking mechanism. Typically, application-level optimistic locking starts by adding an integer field (typically named `version`) to the corresponding table(s). By default, this field is 0, and each `UPDATE` attempts to increment it by 1, as shown in the following diagram (this is also known as *versioned* optimistic locking):

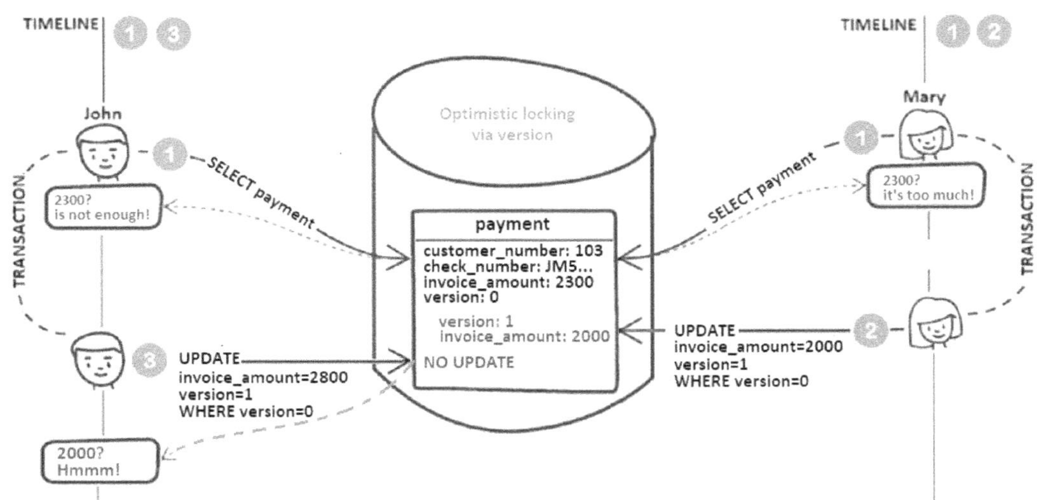

Figure 9.10 – Versioned optimistic locking (via numeric field)

This time, John is aware of Mary's decision, so he can make further decisions based on this information. In application-level optimistic locking, the application is responsible for handling the `version` field. The application should set the `version` value and append the proper `WHERE` clause to the executed `UPDATE/DELETE` to check the `version` value against the database. Moreover, if no `UPDATE/DELETE` has happened because

`WHERE version=?` failed, then the application is responsible for signaling this behavior, meaning that the corresponding transaction contains *stale* data. Commonly, it does this by throwing a meaningful exception. As you'll see next, jOOQ aligns with this behavior.

With long conversations that span several (HTTP) requests, besides the application-level optimistic locking mechanism, you must keep the old data snapshots (for instance, jOOQ updatable records). In web applications, they can be stored in the HTTP session.

jOOQ optimistic locking

By default, the jOOQ optimistic locking mechanism for CRUD operations is disabled. It can be enabled via the `withExecuteWithOptimisticLocking()` setting, as shown here:

```
@Configuration
public class JooqSetting {

  @Bean
  public Settings jooqSettings() {
    return new Settings()
      .withExecuteWithOptimisticLocking(true);
  }
}
```

Of course, you can toggle this setting locally by using a **derived** `Configuration` as well.

jOOQ optimistic locking via SELECT ... FOR UPDATE

At this point, without further settings, jOOQ will intercept any CRUD `UPDATE`/`DELETE` (executed explicitly via `update()`/`delete()` or generated via `merge()`/`store()`), and will attempt to determine if the record contains *stale* data. For this, jOOQ acquires a database-exclusive read/write lock for the involved data via `SELECT ... FOR UPDATE`, which is practically done via pessimistic locking. Next, jOOQ compares the fetched data with the data to be updated/deleted. Practically, the fetched data is compared against the *original* values of the record to be updated/deleted. If the record data is not *stale*, then jOOQ executes `UPDATE`/`DELETE` against the database; otherwise, it will throw a specific `org.jooq.exception.DataChangedException`.

For instance, before updating the amount of a payment (`PAYMENT.INVOICE_AMOUNT`), jOOQ will execute the following `SELECT` (MySQL dialect):

```
SELECT
  `classicmodels`.`payment`.`customer_number`,
```

```
   ...
FROM `classicmodels`.`payment` WHERE
  (`classicmodels`.`payment`.`customer_number` = ? AND
  `classicmodels`.`payment`.`check_number` = ?) FOR UPDATE
```

It's quite easy to enable this type of optimistic locking, but it has two main shortcomings: it uses exclusive locks and is applied to all CRUD `DELETE`/`UPDATE`, which means it's also applied to all tables.

However, jOOQ also supports optimistic locking via the TIMESTAMP or VERSION fields. This kind of implementation is more popular, so let's look at this next.

jOOQ optimistic locking via the TIMESTAMP/VERSION fields

We already know from the previous section that jOOQ optimistic locking is enabled via the `withExecuteWithOptimisticLocking(true)` flag. Next, we must add a field of the `TIMESTAMP` type (for TIMESTAMP optimistic locking) or the `INT` type (for VERSION optimistic locking) for the corresponding table(s). For instance, let's add the `PAYMENT` table (for the MySQL dialect):

```
CREATE TABLE `payment` (
  `customer_number` Bigint NOT NULL,
  `check_number` Varchar(50) NOT NULL,
  ...
  `version` INT NOT NULL DEFAULT 0,
  `modified` TIMESTAMP NOT NULL DEFAULT NOW(),
  CONSTRAINT `payment_pk`
    PRIMARY KEY (`customer_number`,`check_number`),
  ...
) ENGINE=InnoDB DEFAULT CHARSET=latin1;
```

Of course, you don't have to add both of them! Decide what type of optimistic locking you need and add the corresponding field. Next, we must inform the jOOQ Code Generator about the fact that these fields should be used for optimistic locking. We can do so programmatically or declaratively. For Maven applications, you can do this via `<recordVersionFields/>` or `recordTimestampFields/>`, respectively:

```
<database>
  <!-- numeric column for versioned optimistic locking -->
  <recordVersionFields>version</recordVersionFields>

  <!-- timestamp column for versioned optimistic locking -->
```

```
  <recordTimestampFields>modified</recordTimestampFields>
</database>
```

For Gradle, please look at the bundled code.

At this point, jOOQ optimistic locking based on the TIMESTAMP/VERSION fields is ready to be used. Two jOOQ flags are useful for controlling (enable/disable) optimistic locking based on the TIMESTAMP/VERSION fields. These two flags are set to `true` by default (`withUpdateRecordVersion()` and `withUpdateRecordTimestamp()`), so you don't have to enable them explicitly. However, you can use them to disable this type of optimistic locking.

Nevertheless, at this point, there is an important aspect that you should be aware of. So far, jOOQ uses optimistic locking based on the TIMESTAMP/VERSION fields for any record of the `PaymentRecord` type, that is updated/deleted, but it still employees `SELECT ... FOR UPDATE` for the rest of the records that execute CRUD `UPDATE`/`DELETE` statements. If this is not what you need, then you can explicitly enable the jOOQ flag setting (disabled by default); that is, `withExecuteWithOptimisticLockingExcludeUnversioned()`. For instance, you can instruct jOOQ to use only optimistic locking based on the TIMESTAMP/VERSION fields, like this:

```
@Bean // VERSION field (numeric)
public Settings jooqSettings() {
  return new Settings()
    .withUpdateRecordVersion(true) // it can be omitted
    .withExecuteWithOptimisticLocking(true)
    .withExecuteWithOptimisticLockingExcludeUnversioned(true);
}

@Bean // TIMESTAMP field (timestamp)
public Settings jooqSettings() {
  return new Settings()
    .withUpdateRecordTimestamp(true) // it can be omitted
    .withExecuteWithOptimisticLocking(true)
    .withExecuteWithOptimisticLockingExcludeUnversioned(true);
}
```

So, if we group all these settings into a logical diagram, we can obtain something:

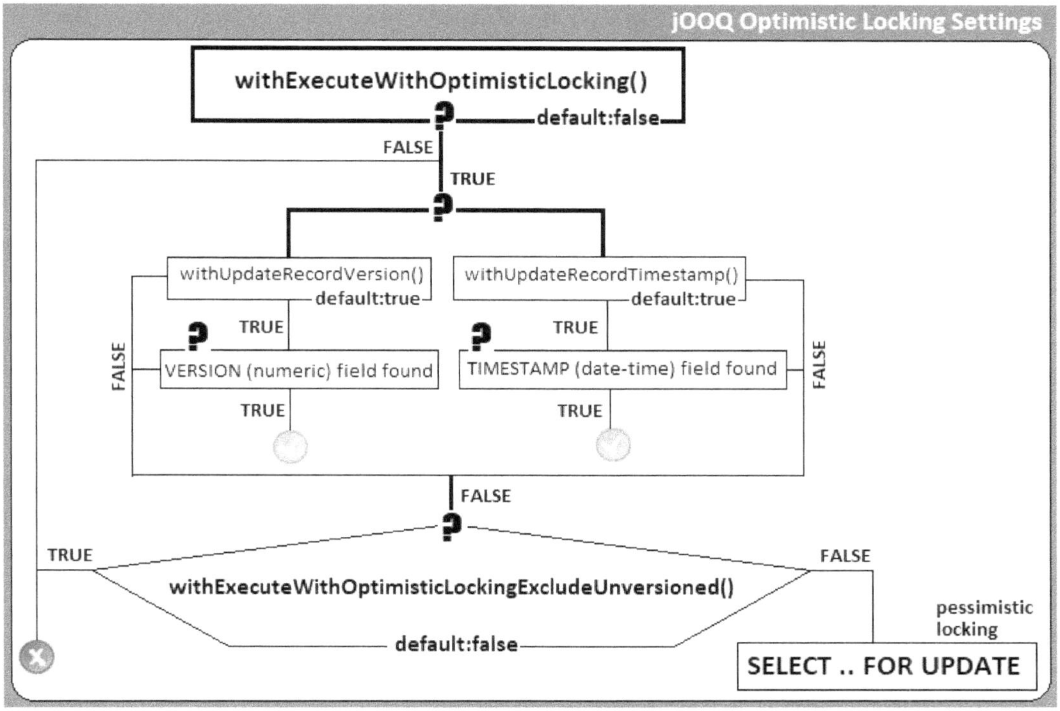

Figure 9.11 – jOOQ optimistic locking settings

If you can choose between versions and timestamps, then go for versions. jOOQ has to support timestamps too if a legacy system uses them, or for quick wins, but with timestamps, there's always the risk of a lack of precision. For instance, when two updates happen in a very short time, the timestamps may still be the same. This can't happen with versions.

Next, let's try to apply the jOOQ optimist locking based on the VERSION field to the *StoreRecords* application (the application that's available for MySQL that uses `store()` was created in the *Using merge() versus store()* section).

Let's look at some code

I assume that you are already familiar with the *StoreRecords* code that shapes the following scenario: we must load and display the payments (`PAYMENT`) of a certain customer (for instance, `PAYMENT.CUSTOMER_NUMBER.eq(103L)`). The user should be able to insert new payments for this customer or update the amount of an existing payment via a user interface, as shown in the following screenshot:

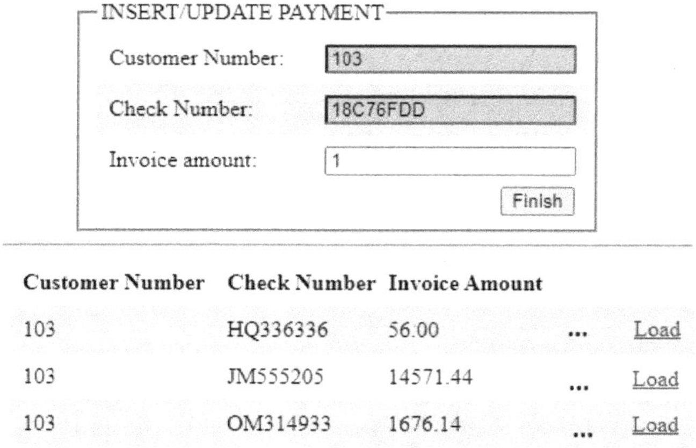

INSERT/UPDATE PAYMENT

Customer Number: 103

Check Number: 18C76FDD

Invoice amount: 1

Finish

Customer Number	Check Number	Invoice Amount		
103	HQ336336	56.00	...	Load
103	JM555205	14571.44	...	Load
103	OM314933	1676.14	...	Load

Figure 9.12 – INSERT/UPDATE PAYMENT

Behind the scenes, we use `store()`:

```
@Transactional
public int storePayment(PaymentRecord pr) {

  ctx.attach(pr);

  return pr.store();
}
```

Here, if two concurrent transactions update the same payment, then our code is prone to the *lost updates* anomaly, so we must engage in optimistic locking.

So far, we've already added the `version` field to `PAYMENT`:

```
CREATE TABLE `payment` (
   ...
   `version` INT NOT NULL DEFAULT 0,
   ...
}
```

We have also added the settings for enabling jOOQ optimistic locking based on the VERSION field, so we have set the following:

```
<database>
  <recordVersionFields>version</recordVersionFields>
</database>
```

We have also set the following:

```
@Bean
public Settings jooqSettings() {
 return new Settings()
   .withExecuteWithOptimisticLocking(true)
   .withExecuteWithOptimisticLockingExcludeUnversioned(true);
}
```

So far, so good! From the perspective of optimistic locking, the interesting part starts when we call the `store()` method. If we attempt to store a new `PaymentRecord`, then `store()` will produce an `INSERT` statement that is not affected by optimistic locking. However, if this `PaymentRecord` needs to be updated, then optimistic locking will enrich the `WHERE` clause of the generated `UPDATE` (the same goes for `DELETE`) with an explicit check of the `version` number, as shown in the following MySQL `UPDATE`:

```
UPDATE
  `classicmodels`.`payment`
SET
  `classicmodels`.`payment`.`invoice_amount` = ?,
  `classicmodels`.`payment`.`version` = ?
WHERE
  (
    `classicmodels`.`payment`.`customer_number` = ?
    and `classicmodels`.`payment`.`check_number` = ?
    and `classicmodels`.`payment`.`version` = ?
  )
```

If the `version` number from the database doesn't match the version number from the `WHERE` clause, then this record contains *stale* data (another transaction has modified this data). This will lead to a jOOQ `DataChangedException` that can be handled in our controller endpoint, as shown here:

```
@PostMapping("/store")
public String storePayment(SessionStatus sessionStatus,
  RedirectAttributes redirectAttributes,
  @ModelAttribute(PAYMENT_ATTR) PaymentRecord pr,
  BindingResult bindingResult) {

  if (!bindingResult.hasErrors()) {
    try {
```

```
        classicModelsService.storePayment(pr);
        sessionStatus.setComplete();
    } catch (org.jooq.exception.DataChangedException e) {
        bindingResult.reject("",
          "Another user updated the data.");
    }
  }

  if (bindingResult.hasErrors()) {
      redirectAttributes.addFlashAttribute(
        BINDING_RESULT, bindingResult);
  }

  return "redirect:payments";
}
```

So, if `DataChangedException` occurs, then we must add a global error in `BindingResult` that contains the message *Another user updated the data.* This message will be rendered via Thymeleaf next to a button labeled *Refresh*, as shown in the following screenshot:

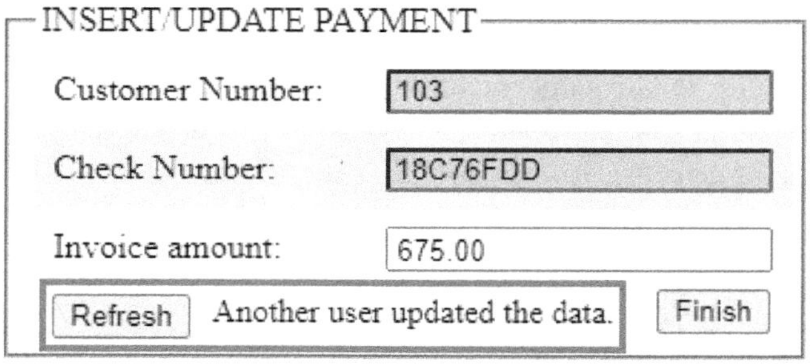

Figure 9.13 – Signaling stale data to the user

Remember jOOQ's `refresh()` method? This is the perfect place to highlight its usability because the user should see the latest state of the record. This is the perfect job for `refresh()`. In this case, the `reset()` method doesn't help since `reset()` restores the record to its in-memory original values, which is a different thing. So, let's use `refresh()` to execute a `SELECT` that will fetch the latest state of this `PaymentRecord`:

```
@GetMapping(value = "/refresh")
public String refreshPayment(Model model) {
```

```
  if (model.containsAttribute(PAYMENT_ATTR)) {
          classicModelsService.refreshPayment(
      (PaymentRecord) model.getAttribute(PAYMENT_ATTR));
  }

  return "redirect:payments";
}

public void refreshPayment(PaymentRecord pr) {

  pr.refresh();
}
```

After refreshing, the user sees the data that was updated earlier by the concurrent transaction and can decide whether they wish to continue with their update. To reproduce this scenario, follow these steps:

1. Launch two browsers (mainly, two users or HTTP sessions).
2. In both, use the *Load* link to fetch the same payment.
3. For user A, update the invoice amount and click *Finish*. This should successfully update the payment.
4. For user B, update the invoice amount and click *Finish*. Since user A already updated this payment, you should see the message shown in the preceding screenshot.
5. Click *Refresh*. Now, you should see the invoice amount that was set by user A.
6. For user B, try to update again. This time, it will work as expected.

In a conclusion, if an explicit `UPDATE/DELETE` or `UPDATE` resulted from calling `store()`, jOOQ VERSION/TIMESTAMP optimistic locking will enrich the `WHERE` clause of the generated `UPDATE/DELETE` with an explicit check of the numeric values of timestamp fields. In the case of calling `merge()`, an explicit `INSERT` or `UPDATE` is rendered, depending on whether the numeric/timestamp values are present or not in the record.

The complete code for this example can be found in *OLVersionStoreRecords*. The alternative, which is for using a TIMESTAMP field, can be found in *OLTimestampStoreRecords*. Finally, the `SELECT ... FOR UPDATE` solution can be found in *OLPLStoreRecords*. All of them are available for MySQL.

Retrying a failed transaction

Let's consider that our scenario gets updated. If a transaction didn't update a payment with an invoice amount larger than the current amount, then this transaction should be retried without user interaction (so, in this case, we don't care about the lost updates). Otherwise, the user should see the current amount and perform the update from the interface (there will be no *Refresh* button since the refresh should be done automatically).

But how can we retry a failed transaction from the application? In Spring Boot, this is equivalent to executing the failed `@Transactional storePayment(PaymentRecord pr)` method again, which can be done via Spring Retry. Once you've added Spring Retry (see the bundled code), you must adjust the `storePayment(PaymentRecord pr)` method, as shown here:

```
@Transactional
@Retryable(
  value = org.jooq.exception.DataChangedException.class,
  maxAttempts = 2, backoff = @Backoff(delay = 100))
public int storePayment(PaymentRecord pr) {

  int stored = 0;

  try {
      ctx.attach(pr);
      stored = pr.store();
  } catch (org.jooq.exception.DataChangedException e) {

    BigDecimal invoiceAmount = pr.getInvoiceAmount();
    pr.refresh();

    if (invoiceAmount.doubleValue() >
            pr.getInvoiceAmount().doubleValue()) {
      pr.setInvoiceAmount(invoiceAmount);
      throw e;
    }

    throw new OptimisticLockingRetryFailed(e.getMessage());
  }

  return stored;
}
```

So, this time, we catch `DataChangedException` and analyze the *current* value of the invoice amount against the refreshed record (the latest state from the database). If the current amount is larger than the fetched amount, then we set it in place of the fetched amount and throw the caught `DataChangedException`. This will trigger the Spring Retry mechanism, will should retry this transaction. Otherwise, we must throw a custom `OptimisticLockingRetryFailed` exception, which will lead to an explicit message for the user. You can practice this example in *OLRetryVersionStoreRecords* (available for MySQL).

Pessimistic locking overview

Pessimistic locking is about locking rows (or cells) via exclusive/shared locks until the transaction that acquired these locks finishes its tasks. Depending on the lock's strength, other transactions may partially interact with these rows/cells or they will have to abort or wait for the resource to become available (lock-free). From the well-known `SELECT … FOR UPDATE` (exclusive read/write lock for rows (record lock)) and `SELECT … FOR UPDATE OF` (exclusive read/write locks for cells specific to Oracle) to `SELECT … FOR UPDATE NOWAIT` and `SELECT … FOR UPDATE WAIT n` (also specific to Oracle), to the more relaxed `SELECT ... FOR UPDATE SKIP LOCKED`, `SELECT … FOR SHARE`, and PostgreSQL-specific `SELECT … FOR NO KEY UPDATE` and `SELECT … FOR KEY SHARE`, jOOQ supports them all.

jOOQ pessimistic locking

As we mentioned in the *Pessimistic locking overview* section, jOOQ supports a significant number of locks of the `SELECT … FOR FOO` type. For instance, we can explicitly call `SELECT … FOR UPDATE` via `forUpdate()`:

```
ctx.selectFrom(PRODUCTLINE)
    .where(PRODUCTLINE.PRODUCT_LINE.eq("Classic Cars"))
    .forUpdate()
    .fetchSingle();
```

If transaction A executes this statement, then it locks the corresponding rows. The other transaction, transaction B, must wait for transaction A to release this exclusive lock before performing its tasks on the same resource. Check out this scenario in the *ForUpdate* application (available for MySQL) – pay attention that this application results in an exception: `MySQLTransactionRollbackException`: *Lock wait timeout exceeded; try restarting transaction*. Also, check out *ForUpdateForeignKey* (available for PostgreSQL) – this example highlights the effect of `FOR UPDATE` on foreign keys that's caused by the fact this lock affects the referenced rows from other tables as well, not just the rows from the current table.

So, remaining in the same context, `SELECT ... FOR UPDATE` locks all the selected rows across all the involved tables (listed in the `FROM` clause, joined, and so on). If table X and table Y are involved in such a case, then `SELECT ... FOR UPDATE` locks the rows of both tables, even if transaction A affects only rows from table X. On the other hand, transaction B needs to acquire locks from table Y, but it cannot do so until transaction A releases the locks on tables X and Y.

For such scenarios, Oracle has `SELECT ... FOR UPDATE OF`, which allows us to nominate the columns that should be locked. In this case, Oracle only acquires locks on the rows of the table(s) that have the column name listed in `FOR UPDATE OF`. For instance, the following statements only lock rows from `PRODUCTLINE`, even if the `PRODUCT` table is also involved:

```
ctx.select(PRODUCTLINE.PRODUCT_LINE, PRODUCTLINE.CODE,
     PRODUCT.PRODUCT_LINE, PRODUCT.PRODUCT_NAME,
     PRODUCT.PRODUCT_SCALE)
   .from(PRODUCTLINE).join(PRODUCT).onKey()
   // lock only rows from PRODUCTLINE
   .forUpdate().of(PRODUCTLINE.PRODUCT_LINE, PRODUCTLINE.CODE)
   .fetch();
```

Since the `PRODUCT` table isn't locked, another statement can obtain locks on its rows:

```
ctx.select(PRODUCTLINE.PRODUCT_LINE, PRODUCTLINE.CODE,
           PRODUCT.PRODUCT_LINE, PRODUCT.PRODUCT_NAME,
           PRODUCT.PRODUCT_SCALE)
   .from(PRODUCTLINE).join(PRODUCT).onKey()
   // lock only rows from PRODUCT
   .forUpdate().of(PRODUCT.PRODUCT_NAME)
   .fetch();
```

If we remove `.of(PRODUCT.PRODUCT_NAME)`, then this statement will attempt to lock rows from `PRODUCTLINE` as well, so it will have to wait for the lock to release on that table. You can check out this example by going to the *ForUpdateOf* application (available for Oracle).

If a transaction were to acquire a lock or fail immediately, then we should use `SELECT ... FOR UPDATE NOWAIT`:

```
ctx.selectFrom(PRODUCT)
   .forUpdate()
   .noWait() // acquire the lock or fails immediately
   .fetch();
```

However, if the transaction needs to wait for a fixed amount of time, then we must rely on the `SELECT ... FOR UPDATE WAIT n` lock (Oracle), where `n` is the time to wait, given in seconds:

```
ctx.selectFrom(PRODUCT)
    .forUpdate()
    .wait(15)
    .fetch();
```

You can check out an example in *ForUpdateWait* (available for Oracle). As you'll see, transaction A acquires a lock immediately, while transaction B waits for a certain amount of time before acquiring a lock on the same resource. If this time expires before transaction A releases the lock, then you'll get an error stating *ORA-30006: resource busy; acquire with WAIT timeout expired.*

Let's consider the following scenario: to provide a high-quality description of products, we have reviewers that analyze each product and write a proper description. Since this is a concurrent process on the `PRODUCT` table, the challenge consists of coordinating the reviewers so that they don't review the same product at the same time. To pick a product for review, the reviewer should skip the products that have already been reviewed (`PRODUCT.PRODUCT_DESCRIPTION.eq("PENDING")`) and the products that are currently in review. This is what we call a concurrent table-based queue (also known as job queues or batch queues).

This is a job for `SKIP LOCKED`. This SQL option is available in many databases (Oracle, MySQL 8, PostgreSQL 9.5, and so on) and it instructs the database to skip the locked rows and to lock the rows that have not been locked previously:

```
Result<ProductRecord> products = ctx.selectFrom(PRODUCT)
   .where(PRODUCT.PRODUCT_DESCRIPTION.eq("PENDING"))
   .orderBy(PRODUCT.PRODUCT_ID).limit(3)
   .forUpdate()
   .skipLocked()
   .fetch();
```

If transaction A executes this statement, then it may lock the `PENDING` products with IDs *1*, *2*, and *3*. While transaction A holds this lock, transaction B executes the same statement and will lock the `PENDING` products with IDs *4*, *5*, and *6*. You can see this scenario in *ForUpdateSkipLocked* (available for MySQL).

A weaker form of `SELECT ... FOR UPDATE` is the `SELECT ... FOR SHARE` query. This ensures referential integrity when inserting child records for a parent. For instance, transaction A executes the following:

```
SaleRecord sr = ctx.selectFrom(SALE)
   .where(SALE.SALE_ID.eq(2L))
   .forShare()
   .fetchSingle();

ctx.insertInto(TOKEN)
   .set(TOKEN.SALE_ID, sr.getSaleId())
   .set(TOKEN.AMOUNT, 1200.5)
   .execute();
```

However, transaction B cannot `UPDATE` if transaction A holds the `SHARE` lock:

```
ctx.update(SALE)
   .set(SALE.SALE_, SALE.SALE_.plus(1000))
   .where(SALE.SALE_ID.eq(2L))
   .execute();
```

Also, transaction C cannot `DELETE`:

```
ctx.delete(SALE)
   .where(SALE.SALE_ID.eq(2L))
   .execute();
```

You can check out this example in *ForShare* (available for PostgreSQL).

Starting with version 9.3, PostgreSQL supports two more locking clauses: `SELECT ... FOR NO KEY UPDATE` and `SELECT ... FOR KEY SHARE`. The former acts similarly to the `FOR UPDATE` locking clause but it does not block `SELECT ... FOR KEY SHARE`. For instance, transaction A uses `SELECT ... FOR NO KEY UPDATE`:

```
SaleRecord sr = ctx.selectFrom(SALE)
   .where(SALE.SALE_ID.eq(2L))
   .forNoKeyUpdate()
   .fetchSingle();

ctx.insertInto(TOKEN)
   .set(TOKEN.SALE_ID, sr.getSaleId())
```

```
    .set(TOKEN.AMOUNT, 1200.5)
    .execute();
```

Even if transaction A holds a lock on this resource, transaction B can acquire a `SELECT … FOR KEY SHARE`:

```
ctx.selectFrom(SALE)
    .where(SALE.SALE_ID.eq(2L))
    .forKeyShare()
    .fetchSingle();
```

However, transaction C cannot acquire a `SELECT … FOR SHARE` on this resource if transaction A doesn't release its lock:

```
ctx.selectFrom(SALE)
    .where(SALE.SALE_ID.eq(2L))
    .forShare()
    .fetchSingle();
```

You can check out this example in *ForNoKeyUpdate* (available for PostgreSQL).

Finally, `SELECT … FOR KEY SHARE` is the weakest lock. For instance, transaction A acquires the following type of lock:

```
SaleRecord sr = ctx.selectFrom(SALE)
                   .where(SALE.SALE_ID.eq(2L))
                   .forKeyShare()
                   .fetchSingle();

ctx.insertInto(TOKEN)
    .set(TOKEN.SALE_ID, sr.getSaleId())
    .set(TOKEN.AMOUNT, 1200.5)
    .execute();
```

While transaction A holds this lock, transaction B can perform updates if it doesn't attempt to update `SALE_ID`:

```
ctx.update(SALE)
    .set(SALE.SALE_, SALE.SALE_.plus(1000))
    .where(SALE.SALE_ID.eq(2L))
    .execute();
```

Transaction B will have to wait for transaction A to release the lock since it attempts to update SALE_ID:

```
ctx.update(SALE)
   .set(SALE.SALE_ID, SALE.SALE_ID.plus(50))
   .where(SALE.SALE_ID.eq(2L))
   .execute();
```

Finally, transaction C cannot DELETE if transaction A holds the KEY SHARE lock:

```
ctx.delete(SALE)
   .where(SALE.SALE_ID.eq(2L))
   .execute();
```

You can see this example in *ForNoKeyUpdate* (available for PostgreSQL).

Deadlocks

Deadlocks are not specific to databases – they can occur in any scenario involving a concurrency environment (concurrency control) and they mainly define a situation when two processes cannot advance because they are waiting for each other to finish (release the lock). In the case of a database, a classical deadlock can be represented like so:

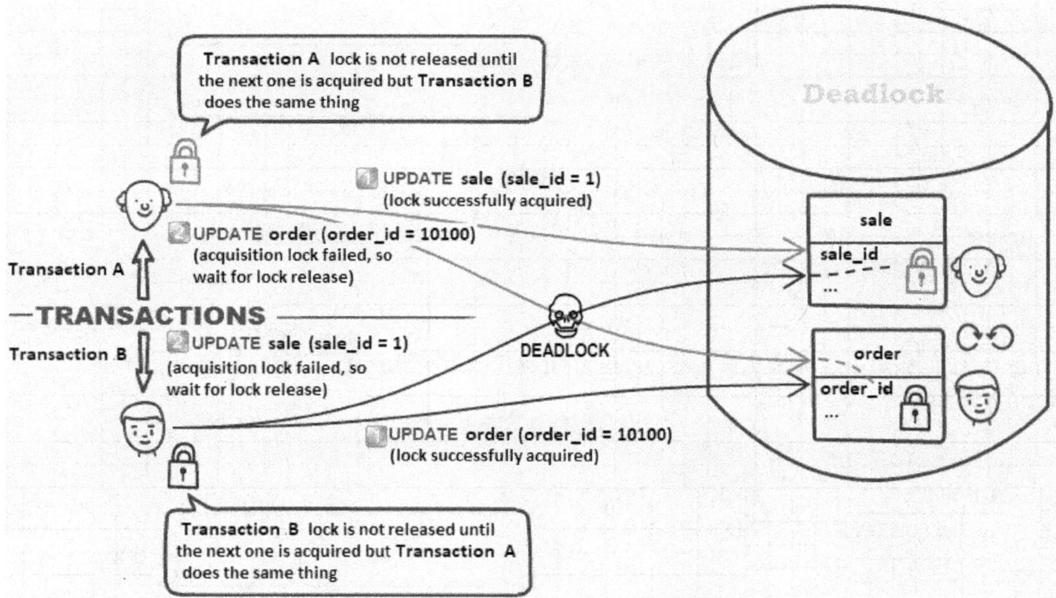

Figure 9.14 – A classical case of a deadlock

Here, we have two transactions that don't use explicit locks (the database itself relies on locks since it detects whether a transaction has attempted to modify data). Transaction A has acquired a lock on the `SALE` resource, and it doesn't release it until it manages to acquire another lock on the `ORDER` resource, which is currently locked by transaction B. At the same time, transaction B holds a lock on the `ORDER` resource, and it doesn't release it until it manages to acquire a lock on the `SALE` resource, which is currently locked by transaction A. You can see this scenario exemplified in *Deadlock* (available for MySQL).

However, using explicit locks doesn't mean that deadlocks cannot happen anymore. For instance, in *DeadlockShare* (available for MySQL), you can see explicit usage of `SELECT ... FOR SHARE` that causes a deadlock. It is very important to understand what each type of lock does and what other locks are permitted (if any) while a certain lock is present. The following table covers the common locks:

Locks	UPDATE	NO KEY UPDATE	SHARE	KEY SHARE
UPDATE	✗	✗	✗	✗
NO KEY UPDATE	✗	✗		
SHARE	✗	✗		
KEY SHARE	✗			

Figure 9.15 – Lock acquisition permissions

Databases automatically scan transactions to discover deadlocks (or the so-called lock-wait cycles). When a deadlock occurs, the database attempts to fix it by aborting one of the transactions. This releases the lock and allows the other transaction to progress. In this context, always rely on `NOWAIT` or explicit short timeouts to avoid deadlocks. While the database can recover after a deadlock, it can only do so after the timeout (if any). So, a long timeout means keeping a database connection busy for a long time, and this is a performance penalty. Moreover, locking too much data may affect scalability.

Summary

I'm glad that you've come this far and that we've managed to cover the three main topics of this chapter – CRUD, transactions, and locking. At this point, you should be familiar with jOOQ UpdatableRecords and how they work in the context of CRUD operations. Among other things, we've learnt about cool stuff such as the must-know `attach()`/`detach()`, the handy `original()` and `reset()`, and the fancy `store()` and `refresh()` operations. After that, we learned how to handle transactions in the context of Spring Boot and jOOQ APIs before tackling optimistic and pessimistic locking in jOOQ.

In the next chapter, we'll learn how to batching, bulking and loading files into the database via jOOQ. We'll also do single-thread and multi-thread batching.

10
Exporting, Batching, Bulking, and Loading

Manipulating large amounts of data requires serious skills (know-how and programming skills) in exporting, batching, bulking, and loading data. Each of these areas requires a significant amount of code and a lot of time to be implemented and tested against real datasets. Fortunately, jOOQ provides comprehensive APIs that cover all these operations and expose them in a fluent style, while hiding the implementation details. In this context, our agenda includes the following:

- Exporting data in text, JSON, XML, CSV, charts, and `INSERT` statements
- Batching `INSERT`, `UPDATE`, `DELETE`, `MERGE`, and `Record`
- Bulking queries
- Loading JSON, CSV, arrays, and `Record`

Let's get started!

Technical requirements

The code for this chapter can be found on GitHub at `https://github.com/PacktPublishing/jOOQ-Masterclass/tree/master/Chapter10`.

Exporting data

Exporting (or formatting) data is achievable via the `org.jooq.Formattable` API. jOOQ exposes a suite of `format()` and `formatFoo()` methods that can be used to format `Result` and `Cursor` (remember `fetchLazy()` from *Chapter 8, Fetching and Mapping*) as text, JSON, XML, CSV, XML, charts, and `INSERT` statements. As you can see in the documentation, all these methods come in different flavors capable of exporting data into a string or a file via the Java `OutputStream` or `Writer` APIs.

Exporting as text

I'm sure that you have already seen in your console output something similar to the following:

```
+------------+----------+----------------------------------------+
|product_line|product_id|product_name                            |
+------------+----------+----------------------------------------+
|Vintage Cars|        80|1936 Mercedes Benz 500k Roadster        |
|Vintage Cars|        29|1932 Model A Ford J-Coupe               |
|Vintage Cars|        33|1928 Mercedes-Benz SSK                  |
|Vintage Cars|        61|1939 Chevrolet Deluxe Coupe             |
|Vintage Cars|        64|1938 Cadillac V-16 Presidential Limousine|
+------------+----------+----------------------------------------+
|...record(s) truncated...
110
```

Figure 10.1 – Tabular text data

This textual tabular representation can be achieved via the `format()` method. A flavor of this method takes an integer argument representing the maximum number of records to include in the formatted result (by default, jOOQ logs just the first five records of the result formatted via jOOQ's text export, but we can easily format and log all the `fetch` records as `result.format(result.size())`. But, if you need a fine-tuning of this output, then jOOQ has a dedicated immutable class named `TXTFormat` with a lot of intuitive options available in the documentation. Using this class in conjunction with exporting the resulting text into a file named `result.txt` via `format(Writer writer, TXTFormat format)` can be done as shown in the following example:

```
try (BufferedWriter bw = Files.newBufferedWriter(
  Paths.get("result.txt"), StandardCharsets.UTF_8,
  StandardOpenOption.CREATE_NEW, StandardOpenOption.WRITE)) {
```

```
 ctx.select(PRODUCTLINE.PRODUCT_LINE, PRODUCT.PRODUCT_ID,
            PRODUCT.PRODUCT_NAME)
    .from(PRODUCTLINE)
    .join(PRODUCT).onKey()
    .fetch()
    .format(bw, new TXTFormat().maxRows(25).minColWidth(20));
} catch (IOException ex) { // handle exception }
```

You can see this example in the bundled code, `Format` (available for MySQL and PostgreSQL), next to other examples.

Exporting JSON

Exporting `Result/Cursor` as JSON can be done via `formatJSON()` and its overloads. Without arguments, `formatJSON()` produces a JSON containing two main arrays: an array named `"fields"`, representing a header (as you'll see later, this can be useful for importing the JSON into the database), and an array named `"records"`, which wraps the fetched data. Here is such an output:

```
{
 "fields": [
  {"schema": "public", "table": "productline", "name":
   "product_line", "type": "VARCHAR"},
  {"schema": "public", "table": "product", "name":
   "product_id", "type": "BIGINT"},
  {"schema": "public", "table": "product", "name":
   "product_name", "type": "VARCHAR"}
 ],
 "records": [
  ["Vintage Cars", 80, "1936 Mercedes Benz 500k Roadster"],
  ["Vintage Cars", 29, "1932 Model A Ford J-Coupe"],
  ...
 ]
}
```

So, this JSON can be obtained via the `formatJSON()` method without arguments, or via `formatJSON(JSONFormat.DEFAULT_FOR_RESULTS)`. If we want to render only the `"records"` array and avoid rendering the header represented by the `"fields"` array, then we can rely on `formatJSON(JSONFormat.DEFAULT_FOR_RECORDS)`.

This produces something as shown here (as you'll see later, this can also be imported back into the database):

```
[
  ["Vintage Cars", 80, "1936 Mercedes Benz 500k Roadster"],
  ["Vintage Cars", 29, "1932 Model A Ford J-Coupe"],
  ...
]
```

`DEFAULT_FOR_RESULTS` and `DEFAULT_FOR_RECORDS` are two statics of the immutable `org.jooq.JSONFormat` used to fine-tune JSON imports/exports. When these statics are not enough, we can instantiate `JSONFormat` and fluently append a suite of intuitive options such as the ones from this example (check all the available options in the jOOQ documentation):

```
JSONFormat jsonFormat = new JSONFormat()
   .indent(4)        // defaults to 2
   .header(false)    // default to true
   .newline("\r")    // "\n" is default
   .recordFormat(
      JSONFormat.RecordFormat.OBJECT); // defaults to ARRAY
```

Further, let's use `jsonFormat` in the context of exporting a JSON into a file via `formatJSON(Writer writer, JSONFormat format)`:

```
try ( BufferedWriter bw = Files.newBufferedWriter(
  Paths.get("resultObject.json"), StandardCharsets.UTF_8,
  StandardOpenOption.CREATE_NEW, StandardOpenOption.WRITE)) {
  ctx.select(PRODUCTLINE.PRODUCT_LINE, PRODUCT.PRODUCT_ID,
             PRODUCT.PRODUCT_NAME)
     .from(PRODUCTLINE)
     .join(PRODUCT).onKey()
     .fetch()
     .formatJSON(bw, jsonFormat);
} catch (IOException ex) { // handle exception }
```

The resulting JSON looks like this (also importable into the database):

```
[
  {
    "product_line": "Vintage Cars",
    "product_id": 80,
```

```
    "product_name": "1936 Mercedes Benz 500k Roadster"
  },
...
]
```

If we fetch a single Record (so, not Result/Cursor, via fetchAny(), for instance), then formatJSON() will return an array containing only the data, as in this sample of fetching Record3<String, Long, String>:

```
["Classic Cars",2,"1952 Alpine Renault 1300"]
```

But, if we explicitly mention JSONFormat.RecordFormat.OBJECT, then this becomes the following:

```
{"product_line":"Classic Cars","product_id":2,
 "product_name":"1952 Alpine Renault 1300"}
```

You can check out this example in the bundled code, *Format* (available for MySQL and PostgreSQL), next to other examples including formatting a UDT, an array type, and an embeddable type as JSON.

Export XML

Exporting Result/Cursor as XML can be done via formatXML() and its overloads. Without arguments, formatXML() produces an XML containing two main elements: an element named <fields/>, representing a header, and an element named <records/>, which wraps the fetched data. Here is such an output:

```
<result xmlns="http:...">
<fields>
  <field schema="public" table="productline"
           name="product_line" type="VARCHAR"/>
  <field schema="public" table="product"
           name="product_id" type="BIGINT"/>
  <field schema="public" table="product"
           name="product_name" type="VARCHAR"/>
</fields>
<records>
  <record xmlns="http:...">
    <value field="product_line">Vintage Cars</value>
    <value field="product_id">80</value>
    <value field="product_name">1936 Mercedes Benz ...</value>
```

```
  </record>
  ...
</records>
</result>
```

The jOOQ code that produced this output is as follows:

```
ctx.select(PRODUCTLINE.PRODUCT_LINE,
           PRODUCT.PRODUCT_ID, PRODUCT.PRODUCT_NAME)
   .from(PRODUCTLINE)
   .join(PRODUCT).onKey()
   .fetch()
   .formatXML();
```

So, this XML can be obtained via the `formatXML()` method without arguments or via `formatXML(XMLFormat.DEFAULT_FOR_RESULTS)`. If we want to keep only the `<records/>` element and avoid rendering the `<fields/>` element, then use `formatJXML(XMLFormat.DEFAULT_FOR_RECORDS)`. This is an output sample:

```
<result>
 <record>
  <value field="product_line">Vintage Cars</value>
  <value field="product_id">80</value>
  <value field="product_name">1936 Mercedes Benz ...</value>
 </record>
...
</result>
```

`DEFAULT_FOR_RESULTS` and `DEFAULT_FOR_RECORDS` are two statics of the immutable `org.jooq.XMLFormat`, used to fine-tune XML imports/exports. Besides these, we can instantiate `XMLFormat` and fluently append a suite of intuitive options. For instance, the previous snippets of XML are rendered based on the default record format, `XMLFormat.RecordFormat.VALUE_ELEMENTS_WITH_FIELD_ATTRIBUTE`; notice the `<value/>` element and the `field` attribute. But, using `XMLFormat`, we can go for two other options: `VALUE_ELEMENTS` and `COLUMN_NAME_ELEMENTS`. The former formats the records using just the `<value/>` element as follows:

```
<record xmlns="http:...">
  <value>Vintage Cars</value>
  <value>29</value>
  <value>1932 Model A Ford J-Coupe</value>
</record>
```

COLUMN_NAME_ELEMENTS uses the column names as elements. Let's use this setting next to header(false) to format the MANAGER.MANAGER_EVALUATION UDT (available in the PostgreSQL schema):

```
ctx.select(MANAGER.MANAGER_ID, MANAGER.MANAGER_EVALUATION)
   .from(MANAGER)
   .fetch()
   .formatXML(new XMLFormat()
    .header(false)
    .recordFormat(XMLFormat.RecordFormat.COLUMN_NAME_ELEMENTS))
```

The resulting XML looks like this:

```
<record xmlns="http...">
 <manager_id>1</manager_id>
 <manager_evaluation>
  <record xmlns="http...">
    <communication_ability>67</communication_ability>
    <ethics>34</ethics>
    <performance>33</performance>
    <employee_input>66</employee_input>
  </record>
 </manager_evaluation>
</record>
```

If we fetch a single Record (so, no Result/Cursor via fetchAny(), for instance) then formatXML() will return an XML containing only the data, as in this sample of fetching Record3<String, Long, String>:

```
<record>
  <value field="product_line">Classic Cars</value>
  <value field="product_id">2</value>
  <value field="product_name">1952 Alpine Renault 1300</value>
</record>
```

Of course, you can alter this default output via XMLFormat. For instance, let's consider that we have this record:

```
<Record3<String, Long, String> oneResult = …;
```

And, let's apply `RecordFormat.COLUMN_NAME_ELEMENTS`:

```
String xml = oneResult.formatXML(new XMLFormat().recordFormat(
             XMLFormat.RecordFormat.COLUMN_NAME_ELEMENTS));
```

The rendered XML is as follows:

```
<record xmlns="http://...">
  <product_line>Classic Cars</product_line>
  <product_id>2</product_id>
  <product_name>1952 Alpine Renault 1300</product_name>
</record>
```

Consider this example next to others (including exporting XML into a file) in the bundled code, `Format` (available for MySQL and PostgreSQL).

Exporting HTML

Exporting `Result`/`Cursor` as HTML can be done via `formatHTML()` and its overloads. By default, jOOQ attempts to wrap the fetched data in a simple HTML table, therefore, expect to see tags such as `<table/>`, `<th/>`, and `<td/>` in the resultant HTML. For instance, formatting the `MANAGER.MANAGER_EVALUATION` UDT (available in the PostgreSQL schema) can be done as follows:

```
ctx.select(MANAGER.MANAGER_NAME, MANAGER.MANAGER_EVALUATION)
   .from(MANAGER)
   .fetch()
   .formatHTML();
```

The resultant HTML looks like this:

```
<table>
 <thead>
  <tr>
   <th>manager_name</th>
   <th>manager_evaluation</th>
  </tr>
 </thead>
 <tbody>
  <tr>
   <td>Joana Nimar</td>
```

```
    <td>(67, 34, 33, 66)</td>
  </tr>
  ...
```

Notice that the value of MANAGER_EVALUATION, *(67, 34, 33, 66)*, is wrapped in a <td/> tag. But, maybe you'd like to obtain something like this:

```
<h1>Joana Nimar</h1>
<table>
 <thead>
  <tr>
   <th>communication_ability</th>
   <th>ethics</th>
   <th>performance</th>
   <th>employee_input</th>
  </tr>
 </thead>
 <tbody>
  <tr>
   <td>67</td>
   <td>34</td>
   <td>33</td>
   <td>66</td>
  </tr>
 </tbody>
</table>
```

We can obtain this HTML by decorating our query as follows:

```
ctx.select(MANAGER.MANAGER_NAME, MANAGER.MANAGER_EVALUATION)
   .from(MANAGER)
   .fetch()
   .stream()
   .map(e -> "<h1>".concat(e.value1().concat("</h1>"))
                   .concat(e.value2().formatHTML()))
   .collect(joining("<br />"))
```

Check out more examples in the bundled code, *Format* (available for MySQL and PostgreSQL).

Exporting CSV

Exporting `Result/Cursor` as CSV can be done via `formatCSV()` and its overloads. By default, jOOQ renders a CSV file as the one here:

```
city,country,dep_id,dep_name
Bucharest,"","",""
Campina,Romania,3,Accounting
Campina,Romania,14,IT
...
```

Among the handy overloads, we have `formatCSV(boolean header, char delimiter, String nullString)`. Via this method, we can specify whether the CSV header should be rendered (by default, `true`), the record's delimiter (by default, a *comma*), and a string for representing `NULL` values (by default, `""`). Next to this method, we also have a suite of combinations of these arguments such as `formatCSV(char delimiter, String nullString)`, `formatCSV(char delimiter)`, and `formatCSV(boolean header, char delimiter)`. Here is an example that renders the header (default) and uses `TAB` as a delimiter and `"N/A"` for representing `NULL` values:

```
ctx.select(OFFICE.CITY, OFFICE.COUNTRY,
           DEPARTMENT.DEPARTMENT_ID.as("dep_id"),
           DEPARTMENT.NAME.as("dep_name"))
   .from(OFFICE).leftJoin(DEPARTMENT).onKey().fetch()
   .formatCSV('\t', "N/A");
```

The resulting CSV looks like this:

```
City        country     dep_id      dep_name
Bucharest   N/A         N/A         N/A
Campina     Romania     3           Accounting
...
Hamburg     Germany     N/A         N/A
London      UK          N/A         N/A
NYC         USA         4           Finance
...
Paris       France      2           Sales
```

Whenever we need more options, we can rely on the immutable `CSVFormat`. Here is an example of using `CSVFormat` and exporting the result in a file:

```
try (BufferedWriter bw = Files.newBufferedWriter(
  Paths.get("result.csv"), StandardCharsets.UTF_8,
```

```
  StandardOpenOption.CREATE_NEW, StandardOpenOption.WRITE)) {
  ctx.select(OFFICE.CITY, OFFICE.COUNTRY,
             DEPARTMENT.DEPARTMENT_ID, DEPARTMENT.NAME)
    .from(OFFICE).leftJoin(DEPARTMENT).onKey()
    .fetch()
     .formatCSV(bw, new CSVFormat()
        .delimiter("|").nullString("{null}"));
} catch (IOException ex) { // handle exception }
```

The complete code next to other examples is available in the bundled code, *Format* (available for MySQL and PostgreSQL).

Exporting a chart

Exporting `Result/Cursor` as a chart may result in something as observed in this figure:

```
168.8|               ccccccccccccccc
162.4|               ccccccccccccccc
156.1|               ccccccccccccccc
149.8|               aaaaaaaaaaaaaaa
143.5|               aaaaaaaaaaaaaaa
137.2|               aaaaaaaaaaaaaaa
130.9|               aaaaaaaaaaaaaaa
124.6|cccccccccccccccaaaaaaaaaaaaaaa
118.3|cccccccccccccccaaaaaaaaaaaaaaa
  112|bbbbbbbbbbbbbbbaaaaaaaaaaaaaaa
105.7|aaaaaaaaaaaaaaaaaaaaaaaaaaaaaa
 99.3|aaaaaaaaaaaaaaaaaaaaaaaaaaaaaa
   93|aaaaaaaaaaaaaaaaaaaaaaaaaaaaaa
 86.7|aaaaaaaaaaaaaaaaaaaaaaaaaaaaaa
 80.4|aaaaaaaaaaaaaaaaaaaaaaaaaaaaaa
 74.1|aaaaaaaaaaaaaaaaaaaaaaaaaaaaaa
 67.8|aaaaaaaaaaaaaaaaaaaaaaaaaaaaaa
 61.5|aaaaaaaaaaaaaaaaaaaaaaaaaaaaaa
 55.2|aaaaaaaaaaaaaaaaaaaaaaaaaaaaaa
 48.9|aaaaaaaaaaaaaaaaaaaaaaaaaaaaaa
 42.6|aaaaaaaaaaaaaaaaaaaaaaaaaaaaaa               ccccccccccccccc
 36.2|aaaaaaaaaaaaaaaaaaaaaaaaaaaaaa               aaaaaaaaaaaaaaabbbbbbbbbbbbbbb
 29.9|aaaaaaaaaaaaaaaaaaaaaaaaaaaaaacccccccccccccccaaaaaaaaaaaaaaaaaaaaaaaaaaaaaa
-----+---------------------------------------------------------------------------
     |      29             33             61             64             80
```

Figure 10.2 – jOOQ chart sample

This is an area chart containing three graphs: *a*, *b*, and *c*. Graph a represents `PRODUCT.BUY_PRICE`, graph *b* represents `PRODUCT.MSRP`, and graph *c* represents `avg(ORDERDETAIL.PRICE_EACH)`. While this chart can be displayed on the console, it can be exported to a file as shown here:

```
try (BufferedWriter bw = Files.newBufferedWriter(
 Paths.get("result2Chart.txt"), StandardCharsets.UTF_8,
 StandardOpenOption.CREATE_NEW, StandardOpenOption.WRITE)) {
 ctx.select(PRODUCT.PRODUCT_ID, PRODUCT.BUY_PRICE,
            field("avg_price"), PRODUCT.MSRP)
   .from(PRODUCT, lateral(select(
         avg(ORDERDETAIL.PRICE_EACH).as("avg_price"))
        .from(ORDERDETAIL)
        .where(PRODUCT.PRODUCT_ID.eq(ORDERDETAIL.PRODUCT_ID))))
     .limit(5).fetch()
     .formatChart(bw, cf);
} catch (IOException ex) { // handle exception }
```

Obviously, the chart is obtained via the `formatChart()` method. More precisely, in this example, via `formatChart(Writer writer, ChartFormat format)`. The `ChartFormat` class is immutable and contains a suite of options for customizing the chart. While you can check all of them in the jOOQ documentation, here is the `cf` used in this example:

```
DecimalFormat decimalFormat = new DecimalFormat("#.#");
ChartFormat cf = new ChartFormat()
 .showLegends(true, true)      // show legends
 .display(ChartFormat.Display.DEFAULT) // or,
                               // HUNDRED_PERCENT_STACKED
 .categoryAsText(true)         // category as text
 .type(ChartFormat.Type.AREA)  // area chart type
 .shades('a', 'b', 'c')        // shades of PRODUCT.BUY_PRICE,
                               // PRODUCT.MSRP,
                               // avg(ORDERDETAIL.PRICE_EACH)
 .values(1, 2, 3)              // value source column numbers
 .numericFormat(decimalFormat);// numeric format
```

The complete code next to other examples is available in the bundled code in the application named *Format* (available for MySQL and PostgreSQL).

Exporting INSERT statements

jOOQ can export Result/Cursor as INSERT statements via the formatInsert() method and its overloads. By default, if the first record is TableRecord, then formatInsert() uses the first record's TableRecord.getTable() method to generate INSERT statements into this table, otherwise, it generates INSERT statements into UNKNOWN_TABLE. In both cases, jOOQ calls the Result.fields() method to determine the column names.

Here is an example that exports the generated INSERT statements into a file on disk. The INSERT statements are generated into a database table named product_stats specified via formatInsert(Writer writer, Table<?> table, Field<?>... fields):

```
try (BufferedWriter bw = Files.newBufferedWriter(
  Paths.get("resultInserts.txt"), StandardCharsets.UTF_8,
  StandardOpenOption.CREATE_NEW, StandardOpenOption.WRITE)) {
  ctx.select(PRODUCT.PRODUCT_ID, PRODUCT.BUY_PRICE,
             field("avg_price"), PRODUCT.MSRP)
     .from(PRODUCT, lateral(select(
       avg(ORDERDETAIL.PRICE_EACH).as("avg_price"))
       .from(ORDERDETAIL)
       .where(PRODUCT.PRODUCT_ID.eq(ORDERDETAIL
         .PRODUCT_ID))))
     .limit(5)
     .fetch()
     .formatInsert(bw, table("product_stats"));
} catch (IOException ex) { // handle exception }
```

A generated INSERT statement looks like the following:

```
INSERT INTO product_stats VALUES (29, 108.06, 114.23, 127.13);
```

The complete code next to other examples, including exporting INSERT statements for UDT, JSON, array, and embeddable types, is available in the bundled code, Format (available for MySQL and PostgreSQL). Next, let's talk about batching.

Batching

Batching can be the perfect solution for avoiding performance penalties caused by a significant number of separate database/network round trips representing inserts, deletes, updates, merges, and so on. For instance, without batching, having 1,000 inserts requires

1,000 separate round trips, while employing batching with a batch size of 30 will result in 34 separate round trips. The more inserts (statements) we have, the more helpful batching is.

Batching via DSLContext.batch()

The `DSLContext` class exposes a suite of `batch()` methods that allow us to execute a set of queries in batch mode. So, we have the following `batch()` methods:

```
BatchBindStep batch(String sql)
BatchBindStep batch(Query query)

Batch batch(String... queries)
Batch batch(Query... queries)

Batch batch(Queries queries)
Batch batch(Collection<? extends Query> queries)

Batch batch(String sql, Object[]... bindings)
Batch batch(Query query, Object[]... bindings)
```

Behind the scenes, jOOQ implements these methods via JDBC's `addBatch()`. Each query is accumulated in the batch via `addBatch()`, and in the end, it calls the JDBC `executeBatch()` method to send the batch to the database.

For instance, let's assume that we need to batch a set of `INSERT` statements into the `SALE` table. If you have a Hibernate (JPA) background, then you know that this kind of batch will not work because the `SALE` table has an auto-incremented primary key, and Hibernate will automatically disable/prevent insert batching. But, jOOQ doesn't have such issues, so batching a set of inserts into a table having an auto-incremented primary key can be done via `batch(Query... queries)`, as follows:

```
int[] result = ctx.batch(
  ctx.insertInto(SALE, SALE.FISCAL_YEAR, SALE.EMPLOYEE_NUMBER,
        SALE.SALE_, SALE.FISCAL_MONTH, SALE.REVENUE_GROWTH)
     .values(2005, 1370L, 1282.64, 1, 0.0),
  ctx.insertInto(SALE, SALE.FISCAL_YEAR, SALE.EMPLOYEE_NUMBER,
          SALE.SALE_, SALE.FISCAL_MONTH, SALE.REVENUE_GROWTH)
     .values(2004, 1370L, 3938.24, 1, 0.0),
  ...
 ).execute();
```

The returned array contains the number of affected rows per `INSERT` statement (in this case, `[1, 1, 1, …]`). While executing several queries without bind values can be done as you just saw, jOOQ allows us to execute one query several times with bind values as follows:

```
int[] result = ctx.batch(
  ctx.insertInto(SALE, SALE.FISCAL_YEAR,SALE.EMPLOYEE_NUMBER,
        SALE.SALE_, SALE.FISCAL_MONTH, SALE.REVENUE_GROWTH)
     .values((Integer) null, null, null, null, null))
     .bind(2005, 1370L, 1282.64, 1, 0.0)
     .bind(2004, 1370L, 3938.24, 1, 0.0)
     ...
 .execute();
```

Notice that you will have to provide *dummy* bind values for the original query, and this is commonly achieved via `null` values, as in this example. jOOQ generates a single query (`PreparedStatement`) with placeholders (`?`) and will loop the bind values to populate the batch. Whenever you see that `int[]` contains a negative value (for instance, `-2`) it means that the affected row count value couldn't be determined by JDBC.

In most cases, JDBC prepared statements are better, so, whenever possible, jOOQ relies on `PreparedStatement` (`www.jooq.org/doc/latest/manual/sql-execution/statement-type/`). But, we can easily switch to static statements (`java.sql.Statement`) via `setStatementType()` or `withStatementType()` as in the following example (you can also apply this globally via `@Bean`):

```
int[] result = ctx.configuration().derive(new
 Settings().withStatementType(StatementType.STATIC_STATEMENT))
 .dsl().batch(
    ctx.insertInto(SALE, SALE.FISCAL_YEAR,
          SALE.EMPLOYEE_NUMBER, SALE.SALE_, SALE.FISCAL_MONTH,
          SALE.REVENUE_GROWTH)
       .values((Integer) null, null, null, null, null))
       .bind(2005, 1370L, 1282.64, 1, 0.0)
       .bind(2004, 1370L, 3938.24, 1, 0.0)
   ...
   .execute();
```

This time, the bind values will be automatically inlined into a static batch query. This is the same as the first examples from this section, which use `batch(Query... queries)`.

Obviously, using binding values is also useful for inserting (updating, deleting, and so on) a collection of objects. For instance, consider the following list of `SimpleSale` (POJO):

```
List<SimpleSale> sales = List.of(
   new SimpleSale(2005, 1370L, 1282.64, 1, 0.0),
   new SimpleSale(2004, 1370L, 3938.24, 1, 0.0),
   new SimpleSale(2004, 1370L, 4676.14, 1, 0.0));
```

First, we define the proper `BatchBindStep` containing one `INSERT` (it could be `UPDATE`, `DELETE`, and so on, as well):

```
BatchBindStep batch = ctx.batch(ctx.insertInto(SALE,
      SALE.FISCAL_YEAR, SALE.EMPLOYEE_NUMBER, SALE.SALE_,
      SALE.FISCAL_MONTH, SALE.REVENUE_GROWTH)
   .values((Integer) null, null, null, null, null));
```

Second, we bind the values and execute the batch:

```
sales.forEach(s -> batch.bind(s.getFiscalYear(),
    s.getEmployeeNumber(), s.getSale(),
    s.getFiscalMonth(),s.getRevenueGrowth()));
batch.execute();
```

You can find these examples in the bundled code, `BatchInserts`, next to examples for batching updates, `BatchUpdates`, and deletes, `BatchDeletes`, as well. But, we can also combine all these kinds of statements in a single `batch()` method, as follows:

```
int[] result = ctx.batch(
 ctx.insertInto(SALE, SALE.FISCAL_YEAR,SALE.EMPLOYEE_NUMBER,
      SALE.SALE_, SALE.FISCAL_MONTH, SALE.REVENUE_GROWTH)
    .values(2005, 1370L, 1282.64, 1, 0.0),
 ctx.insertInto(SALE, SALE.FISCAL_YEAR, SALE.EMPLOYEE_NUMBER,
  SALE.SALE_, SALE.FISCAL_MONTH, SALE.REVENUE_GROWTH)
   .values(2004, 1370L, 3938.24, 1, 0.0),
 ...
 ctx.update(EMPLOYEE).set(EMPLOYEE.SALARY,
   EMPLOYEE.SALARY.plus(1_000))
    .where(EMPLOYEE.SALARY.between(100_000, 120_000)),
 ctx.update(EMPLOYEE).set(EMPLOYEE.SALARY,
   EMPLOYEE.SALARY.plus(5_000))
    .where(EMPLOYEE.SALARY.between(65_000, 80_000)),
 ...
```

```
ctx.deleteFrom(BANK_TRANSACTION)
    .where(BANK_TRANSACTION.TRANSACTION_ID.eq(1)),
ctx.deleteFrom(BANK_TRANSACTION)
    .where(BANK_TRANSACTION.TRANSACTION_ID.eq(2)),
...
).execute();
```

While using `batch()` methods, jOOQ will always preserve your order of statements and will send all these statements in a single batch (round trip) to the database. This example is available in an application named *CombineBatchStatements*.

During the batch preparation, the statements are accumulated in memory, so you have to pay attention to avoid memory issues such as OOMs. You can easily emulate a batch size by calling the jOOQ batch in a `for` loop that limits the number of statements to a certain value. You can execute all batches in a single transaction (in case of an issue, roll back all batches) or execute each batch in a separate transaction (in case of an issue, roll back only the last batch). You can see these approaches in the bundled code, `EmulateBatchSize`.

While a synchronous batch ends up with an `execute()` call, an asynchronous batch ends up with an `executeAsync()` call. For example, consider the application named *AsyncBatch*. Next, let's talk about batching records.

Batching records

Batching records is another story. The jOOQ API for batching records relies on a set of dedicated methods per statement type as follows:

- `INSERT`: `batchInsert()` follows `TableRecord.insert()` semantics
- `UPDATE`: `batchUpdate()` follows `UpdatableRecord.update()` semantics
- `DELETE`: `batchDelete()` follows `UpdatableRecord.delete()` semantics
- `MERGE`: `batchMerge()` follows `UpdatableRecord.merge()` semantics
- `INSERT/UPDATE`: `batchStore()` follows `UpdatableRecord.store()` semantics

Next, we'll cover each of these statements but before that, let's point out an important aspect. By default, all these methods create batch operations for executing a certain type of query with bind values. jOOQ preserves the order of the records as long as the records generate the same SQL with bind variables, otherwise, the order is changed to group together the records that share the same SQL with bind variables. So, in the best-case scenario, when all records generate the same SQL with bind variables, there will be a single batch operation, while in the worst-case scenario, the number of records will be equal to the number of batch operations. In short, the number of batch operations that will be executed is equal to the number of distinct rendered SQL statements.

If we switch from the default `PreparedStatement` to a static `Statement` (`StatementType.STATIC_STATEMENT`), then the record values are inlined. This time, there will be just one batch operation and the order of records is preserved exactly. Obviously, this is preferable when the order of records must be preserved and/or the batch is very large, and rearranging the records can be time-consuming and results in a significant number of batch operations.

Batch records insert, update, and delete

Let's consider the following set of `Record`:

```
SaleRecord sr1 = new SaleRecord(…, 2005, 1223.23, 1370L, …);
SaleRecord sr2 = new SaleRecord(…, 2004, 5483.33, 1166L, …);
SaleRecord sr3 = new SaleRecord(…, 2005, 9022.21, 1370L, …);
```

Inserting these records in batch can be done as follows:

```
int[] result = ctx.batchInsert(sr3, sr1, sr2).execute();
```

In this case, these records are inserted in a single batch operation since the generated SQL with bind variables is the same for `sr1` to `sr3`. Moreover, the batch preserves the order of records as given (`sr3`, `sr1`, and `sr2`). If we want to update, and respectively to delete these records, then we replace `batchInsert()` with `batchUpdate()`, and, respectively, `batchDelete()`. You can also have these records in a collection and pass that collection to `batchInsert()`, as in this example:

```
List<SaleRecord> sales = List.of(sr3, sr1, sr2);
int[] result = ctx.batchInsert(sales).execute();
```

Next, let's consider a mix of records:

```
SaleRecord sr1 = new SaleRecord(…);
SaleRecord sr2 = new SaleRecord(…);
BankTransactionRecord bt1 = new BankTransactionRecord(…);
SaleRecord sr3 = new SaleRecord(…);
SaleRecord sr4 = new SaleRecord(…);
BankTransactionRecord bt2 = new BankTransactionRecord(…);
```

Calling `batchInsert(bt1, sr1, sr2, bt2, sr4, sr3)` is executed in two batch operations, one for `SaleRecord` and one for `BankTransactionRecord`. jOOQ will group `SaleRecord` (`sr1`, `sr2`, `sr3`, and `sr4`) in one batch operation and `BankTransactionRecord` (`bt1` and `bt2`) in another batch operation, so the order of records in not preserved (or, is partially preserved) since (`bt1`, `sr1`, `sr2`, `bt2`, `sr4`, and `sr3`) may become ((`bt1` and `bt2`), (`sr1`, `sr2`, `sr4`, and `sr3`)).

Finally, let's consider these records:

```
SaleRecord sr1 = new SaleRecord();
sr1.setFiscalYear(2005);
sr1.setSale(1223.23);
sr1.setEmployeeNumber(1370L);
sr1.setTre"d("UP");
sr1.setFiscalMonth(1);
sr1.setRevenueGrowth(0.0);

SaleRecord sr2 = new SaleRecord();
sr2.setFiscalYear(2005);
sr2.setSale(9022.21);
sr2.setFiscalMonth(1);
sr2.setRevenueGrowth(0.0);

SaleRecord sr3 = new SaleRecord();
sr3.setFiscalYear(2003);
sr3.setSale(8002.22);
sr3.setEmployeeNumber(1504L);
sr3.setFiscalMonth(1);
sr3.setRevenueGrowth(0.0);
```

If we execute `batchInsert(sr3, sr2, sr1)`, then there will be three batch operations, since `sr1`, `sr2`, and `sr3` produce three SQLs that will different bind variables. The order of records is preserved as `sr3`, `sr2`, and `sr1`. The same flow applies for `batchUpdate()` and `batchDelete()`.

Any of these examples can take advantage of JDBC static statements by simply adding the `STATIC_STATEMENT` setting as follows:

```
int[] result = ctx.configuration().derive(new Settings()
  .withStatementType(StatementType.STATIC_STATEMENT))
  .dsl().batchInsert/Update/…(…).execute();
```

You can practice these examples in *BatchInserts*, *BatchUpdates*, and *BatchDeletes*.

Batch merge

As you already know from the bullet list from the *Batching records* section, `batchMerge()` is useful for executing batches of `MERGE` statements. Mainly, `batchMerge()` conforms to the `UpdatableRecord.merge()` semantics covered in *Chapter 9, CRUD, Transactions, and Locking*.

In other words, `batchMerge()` renders the synthetic `INSERT ... ON DUPLICATE KEY UPDATE` statement emulated depending on dialect; in MySQL, via `INSERT ... ON DUPLICATE KEY UPDATE`, in PostgreSQL, via `INSERT ... ON CONFLICT`, and in SQL Server and Oracle, via `MERGE INTO`. Practically, `batchMerge()` renders an `INSERT ... ON DUPLICATE KEY UPDATE` statement independent of the fact that the record has been previously fetched from the database or is created now. The number of distinct rendered SQL statements gives us the number of batches. So, by default (which means default settings, default changed flags, and no optimistic locking), jOOQ renders a query that delegates to the database the decision between insert and update based on the primary key uniqueness. Let's consider the following records:

```
SaleRecord sr1 = new SaleRecord(1L, 2005, 1223.23, ...);
SaleRecord sr2 = new SaleRecord(2L, 2004, 543.33, ...);
SaleRecord sr3 = new SaleRecord(9999L, 2003, 8002.22, ...);
```

We execute a merge in batch as shown here:

```
int[] result = ctx.batchMerge(sr1, sr2, sr3).execute();
```

For instance, in PostgreSQL, the render SQL is as follows:

```
INSERT INTO "public"."sale" ("sale_id",
                             "fiscal_year", ..., "trend")
  VALUES (?, ?, ..., ?) ON CONFLICT ("sale_id") DO
  UPDATE SET "sale_id" = ?,
             "fiscal_year" = ?, ..., "trend" = ?
  WHERE "public"."sale"."sale_id" = ?
```

Because `sr1` (having primary key *1*) and `sr2` (having primary key *2*) already exist in the `SALE` table, the database will decide to update them, while `sr3` (having primary key *9999*) will be inserted, since it doesn't exist in the database. There will be just one batch since the generated SQL with bind variables is the same for all `SaleRecord`. The order of records is preserved. More examples are available in `BatchMerges`.

Batch store

`batchStore()` is useful for executing `INSERT` or `UPDATE` statements in the batch. Mainly, `batchStore()` conforms to `UpdatableRecord.store()`, which was covered in the previous chapter. So, unlike `batchMerge()`, which delegates the decision of choosing between update or insert to the database, `batchStore()` allows jOOQ to decide whether INSERT or UPDATE should be rendered by analyzing the state of the primary key's value.

For instance, let's rely on defaults (which means default settings, default changed flags, and no optimistic locking), the following two records are used for executing in a batch store:

```
SaleRecord sr1 = new SaleRecord(9999L,
                                2005, 1223.23, 1370L, ...);

SaleRecord sr2 = ctx.selectFrom(SALE)
   .where(SALE.SALE_ID.eq(1L)).fetchOne();
sr2.setFiscalYear(2006);

int[] result = ctx.batchStore(sr1, sr2).execute();
```

Since `sr1` is a brand-new `SaleRecord`, it will result in `INSERT`. On the other hand, `sr2` was fetched from the database and it was updated, so it will result in `UPDATE`. Obviously, the generated SQL statements are not the same, therefore, there will be two batch operations and the order will be preserved as `sr1` and `sr2`.

Here is another example that updates `SaleRecord` and adds a few more:

```
Result<SaleRecord> sales = ctx.selectFrom(SALE).fetch();

 // update all sales
 sales.forEach(sale -> { sale.setTrend("UP"); });

 // add more new sales
 sales.add(new SaleRecord(...));
 sales.add(new SaleRecord(...));
 ...

 int[] result = ctx.batchStore(sales)
  .execute();
```

We have two batch operations: a batch that contains all updates needed to update the fetched `SaleRecord` and a batch that contains all inserts needed to insert the new `SaleRecord`.

In the bundled code, you can find more examples that couldn't be listed here because they are large, so take your time to practice examples from `BatchStores`. This was the last topic of this section. Next, let's talk about the *batched connection* API.

Batched connection

Besides the batching capabilities covered so far, jOOQ also comes with an API named `org.jooq.tools.jdbc.BatchedConnection`. Its main purpose is to buffer already existing jOOQ/JDBC statements and execute them in batches without requiring us to change the SQL strings or the order of execution. We can use `BatchedConnection` explicitly or indirectly via `DSLContext.batched(BatchedRunnable runnable)` or `DSLContext.batchedResult(BatchedCallable<T> callable)`. The difference between them consists of the fact that the former returns `void` and the latter returns `T`.

For instance, let's assume that we have a method (service) that produces a lot of `INSERT` and `UPDATE` statements:

```
void insertsAndUpdates(Configuration c) {

   DSLContext ctxLocal = c.dsl();

   ctxLocal.insertInto(…).execute();
   …
   ctxLocal.update(…).execute();
   …
}
```

To improve the performance of this method, we can simply add batch-collecting code via `DSLContext.batched()`, as here:

```
public void batchedInsertsAndUpdates() {

   ctx.batched(this::insertsAndUpdates);
}
```

Of course, if `INSERT` statements are produced by an `inserts(Configuration c)` method and `UPDATE` statements by another method, `updates(Configuration c)`, then both of them should be collected:

```
public void batchedInsertsAndUpdates() {

   ctx.batched((Configuration c) -> {
```

```
      inserts(c);
      updates(c);
   });
}
```

Moreover, this API can be used for batching jOOQ records as well. Here is a sample:

```
ctx.batched(c -> {

  Result<SaleRecord> records = c.dsl().selectFrom(SALE)
     .limit(5).fetch();

  records.forEach(record -> {
    record.setTrend("CONSTANT");
    ...
    record.store();
  });
});
```

Or, here is another example:

```
List<SaleRecord> sales = List.of(
  new SaleRecord(...), new SaleRecord(...), ...
);

ctx.batched(c -> {

  for (SaleRecord sale : sales) {
    c.dsl().insertInto(SALE)
           .set(sale)
           .onDuplicateKeyUpdate()
           .set(SALE.SALE_, sale.getSale())
           .execute();
  }
}); // batching is happening here
```

Notice that jOOQ will preserve exactly your order of statements and this order may affect the number of batch operations. Read carefully the following note, since it is very important to have it in your mind while working with this API.

> **Important Note**
>
> jOOQ automatically creates a new batch every time it detects that:
>
> - The SQL string changes (even whitespace is considered a change).
>
> - A query produces results (for instance, SELECT); such queries are not part of the batch.
>
> - A static statement occurs after a prepared statement (or vice versa).
>
> - A JDBC interaction is invoked (transaction committed, connection closed, and so on).
>
> - The batch size threshold is reached.

As an important limitation, notice that the affected row count value will be reported always by the JDBC `PreparedStatement.executeUpdate()` as 0.

Notice that the last bullet from the previous note refers to a batch size threshold. Well, this API can take advantage of `Settings.batchSize()`, which sets the maximum batch statement size as here:

```
@Bean
public Settings jooqSettings() {
   return new Settings().withBatchSize(30);
}
```

Moreover, if we rely on `BatchedConnection` explicitly, then we can wrap the JDBC connection and specify the batch size as an argument via the `BatchedConnection (Connection delegate, int batchSize)` constructor as follows (here, the batch size is set to 2; consider reading the comments):

```
try ( BatchedConnection conn = new BatchedConnection(
 DriverManager.getConnection(
     "jdbc:mysql://localhost:3306/classicmodels",
     "root", "root"), 2)) {

  try ( PreparedStatement stmt = conn.prepareStatement(
    "insert into `classicmodels`.`sale` (`fiscal_year`,
    `employee_number`, `sale`, `fiscal_month`,
    `revenue_growth`) " + "values (?, ?, ?, ?, ?);")) {

   // the next 2 statements will become the first batch
   stmt.setInt(1, 2004);
   stmt.setLong(2, 1166L);
```

```
    stmt.setDouble(3, 543.33);
    stmt.setInt(4, 1);
    stmt.setDouble(5, 0.0);
    stmt.executeUpdate();

    stmt.setInt(1, 2005);
    stmt.setLong(2, 1370L);
    stmt.setDouble(3, 9022.20);
    stmt.setInt(4, 1);
    stmt.setDouble(5, 0.0);
    stmt.executeUpdate();

    // reached batch limit so this is the second batch
    stmt.setInt(1, 2003);
    stmt.setLong(2, 1166L);
    stmt.setDouble(3, 3213.0);
    stmt.setInt(4, 1);
    stmt.setDouble(5, 0.0);
    stmt.executeUpdate();
   }

   // since the following SQL string is different,
   // next statements represents the third batch
   try ( PreparedStatement stmt = conn.prepareStatement(
    "insert into `classicmodels`.`sale` (`fiscal_year`,
     `employee_number`, `sale`, `fiscal_month`,
       `revenue_growth`,`trend`) "
       + "values (?, ?, ?, ?, ?, ?);")) {

      stmt.setInt(1, 2004);
      stmt.setLong(2, 1166L);
      stmt.setDouble(3, 4541.35);
      stmt.setInt(4, 1);
      stmt.setDouble(5, 0.0);
      stmt.setString(6, "UP");
      stmt.executeUpdate();

      stmt.setInt(1, 2005);
      stmt.setLong(2, 1370L);
      stmt.setDouble(3, 1282.64);
```

```
    stmt.setInt(4, 1);
    stmt.setDouble(5, 0.0);
    stmt.setString(6, "DOWN");
    stmt.executeUpdate();
  }
} catch (SQLException ex) { … }
```

Moreover, `BatchedConnection` implements `java.sql.Connection`, so you can use the entire arsenal of `Connection` methods, including methods for shaping the behavior of transactions. More examples are available in `Batched`.

Next, let's tackle two special cases encountered in PostgreSQL and SQL Server.

Batching and fetching sequences in PostgreSQL/Oracle

As you know, PostgreSQL/Oracle can rely on sequences for providing primary keys (and other unique values). For instance, our PostgreSQL `employee` table uses the following sequence for producing sequence values for `employee_number`:

```
CREATE TABLE "employee" (
  "employee_number" BIGINT NOT NULL,
  ...
  CONSTRAINT "employee_pk" PRIMARY KEY ("employee_number"),
  ...
);

CREATE SEQUENCE "employee_seq" START 100000 INCREMENT 10
        MINVALUE 100000 MAXVALUE 10000000
        OWNED BY "employee"."employee_number";
```

But, in the context of batching, fetching the `employee` primary keys from the application requires a database round trip (`SELECT`) for each primary key. Obviously, it is a performance penalty to have a batch of *n* `INSERT` statements and execute *n* round trips (`SELECT` statements) just to fetch their primary keys. Fortunately, jOOQ leverages at least two solutions. One of them is to inline sequence references in SQL statements (the `EMPLOYEE_SEQ.nextval()` call):

```
int[] result = ctx.batch(
  ctx.insertInto(EMPLOYEE, EMPLOYEE.EMPLOYEE_NUMBER,
                 EMPLOYEE.LAST_NAME, ...)
     .values(EMPLOYEE_SEQ.nextval(), val("Lionel"), ...),
  ctx.insertInto(EMPLOYEE, EMPLOYEE.EMPLOYEE_NUMBER,
```

```
                    EMPLOYEE.LAST_NAME...)
      .values(EMPLOYEE_SEQ.nextval(), val("Ion"), ...),
   ...
 ).execute();
```

Another approach is to pre-fetch a number of *n* primary keys via `SELECT`:

```
var ids = ctx.fetch(EMPLOYEE_SEQ.nextvals(n));
```

Then, use these primary keys in batch (notice the `ids.get(n).value1()` call):

```
int[] result = ctx.batch(
  ctx.insertInto(EMPLOYEE, EMPLOYEE.EMPLOYEE_NUMBER,
                 EMPLOYEE.LAST_NAME, ...)
     .values(ids.get(0).value1(), "Lionel", ...),
  ctx.insertInto(EMPLOYEE, EMPLOYEE.EMPLOYEE_NUMBER,
                 EMPLOYEE.LAST_NAME, ...)
     .values(ids.get(1).value1(), "Ion", ...),
  ...
).execute();
```

Both of these examples rely on the `public static final EMPLOYEE_SEQ` field or, more precisely, on `jooq.generated.Sequences.EMPLOYEE_SEQ`. Mainly, the jOOQ Code Generator will generate a sequence object per database sequence and each such object has access to methods such as `nextval()`, `currval()`, `nextvals(int n)`, and others, which will be covered in *Chapter 11, jOOQ Keys*.

Of course, if you rely on an auto-generated sequence from `(BIG)SERIAL` or on a sequence associated as default (for example, in the `sale` table, we have a sequence associated to `sale_id` as `DEFAULT NEXTVAL ('sale_seq')`), then the simplest way to batch is to omit the primary key field in statements, and the database will do the rest. The previous examples, along with many more, are available in *BatchInserts* for PostgreSQL.

SQL Server IDENTITY columns and explicit values

Inserting explicit values for the SQL Server `IDENTITY` columns results in the error *Cannot insert explicit value for identity column in table 'table_name' when* `IDENTITY_INSERT` *is set to* `OFF`. Bypassing this error can be done by setting `IDENTITY_INSERT` to `ON` before `INSERT`. In the context of batching, this can be done as shown here:

```
int[] result = ctx.batch(
  ctx.query("SET IDENTITY_INSERT [sale] ON"),
```

```
  ctx.insertInto(SALE, SALE.SALE_ID, SALE.FISCAL_YEAR, …)
     .values(1L, 2005, …),
  ctx.insertInto(SALE, SALE.SALE_ID, SALE.FISCAL_YEAR, …)
     .values(2L, 2004, …),
   ...
  ctx.query("SET IDENTITY_INSERT [sale] OFF")
).execute();
```

You can find this example in `BatchInserts` for SQL Server. Next, let's talk about bulking.

Bulking

Writing bulk queries in jOOQ is just a matter of using the jOOQ DSL API. For instance, a bulk insert SQL looks like this:

```
INSERT IGNORE INTO `classicmodels`.`order` (
  `order_date`, `required_date`, `shipped_date`,
  `status`, `comments`, `customer_number`, `amount`)
VALUES (?, ?, ?, ?, ?, ?, ?), (?, ?, ?, ?, ?, ?, ?),
       (?, ?, ?, ?, ?, ?, ?)
```

This can be expressed in jOOQ by chaining the `values()` call:

```
ctx.insertInto(ORDER)
   .columns(ORDER.ORDER_DATE, ORDER.REQUIRED_DATE,
           ORDER.SHIPPED_DATE, ORDER.STATUS,
           ORDER.COMMENTS,ORDER.CUSTOMER_NUMBER,
           ORDER.AMOUNT)
    .values(LocalDate.of(2004,10,22), LocalDate.of(2004,10,23),
     LocalDate.of(2004,10,23", "Shipped",
     "New order inserted...", 363L, BigDecimal.valueOf(322.59))
    .values(LocalDate.of(2003,12,2), LocalDate.of(2003,1,3),
     LocalDate.of(2003,2,26), "Resolved",
     "Important order ...", 128L, BigDecimal.valueOf(455.33))
    ...
    .onDuplicateKeyIgnore() // onDuplicateKeyUpdate().set(...)
   .execute()
```

Or, you can use a bulk update SQL as follows:

```
update `classicmodels`.`sale`
set
  `classicmodels`.`sale`.`sale` = case when
  `classicmodels`.`sale`.`employee_number` = ? then (
    `classicmodels`.`sale`.`sale` + ?
  ) when `classicmodels`.`sale`.`employee_number` = ? then (
    `classicmodels`.`sale`.`sale` + ?
  ) when `classicmodels`.`sale`.`employee_number` = ? then (
    `classicmodels`.`sale`.`sale` + ?
  ) end
where
  `classicmodels`.`sale`.`employee_number` in (?, ?, ?)
```

It can be expressed in jOOQ as follows:

```
ctx.update(SALE).set(SALE.SALE_,
case_()
 .when(SALE.EMPLOYEE_NUMBER.eq(1370L), SALE.SALE_.plus(100))
 .when(SALE.EMPLOYEE_NUMBER.eq(1504L), SALE.SALE_.plus(500))
 .when(SALE.EMPLOYEE_NUMBER.eq(1166L), SALE.SALE_.plus(1000)))
.where(SALE.EMPLOYEE_NUMBER.in(1370L, 1504L, 1166L))
.execute();
```

More examples are available in `Bulk` for MySQL. Next, let's talk about the *Loader* API, which has built-in bulk support.

Loading (the Loader API)

Whenever we need to load (import) our database tables with data coming from different sources (CSV, JSON, and so on), we can rely on the jOOQ Loader API (`org.jooq.Loader`). This is a fluent API that allows us to smoothly tackle the most important challenges, such as handling duplicate keys, bulking, batching, committing, and error handling.

The Loader API syntax

Typically, we have a file containing the data to be imported in a common format such as CSV or JSON, and we customize the Loader API general syntax to fit our needs:

```
ctx.loadInto(TARGET_TABLE)
   .[options]
   .[source and source to target mapping]
   .[listeners]
   .[execution and error handling]
```

While `TARGET_TABLE` is obviously the table in which the data should be imported, let's see what options we have.

Options

We can mainly distinguish between three types of options that can be used for customizing the import process: options for handling duplicate keys, throttling options, and options for handling failures (errors). The following diagram highlights each category of options and the valid paths that can be used for chaining these options fluently:

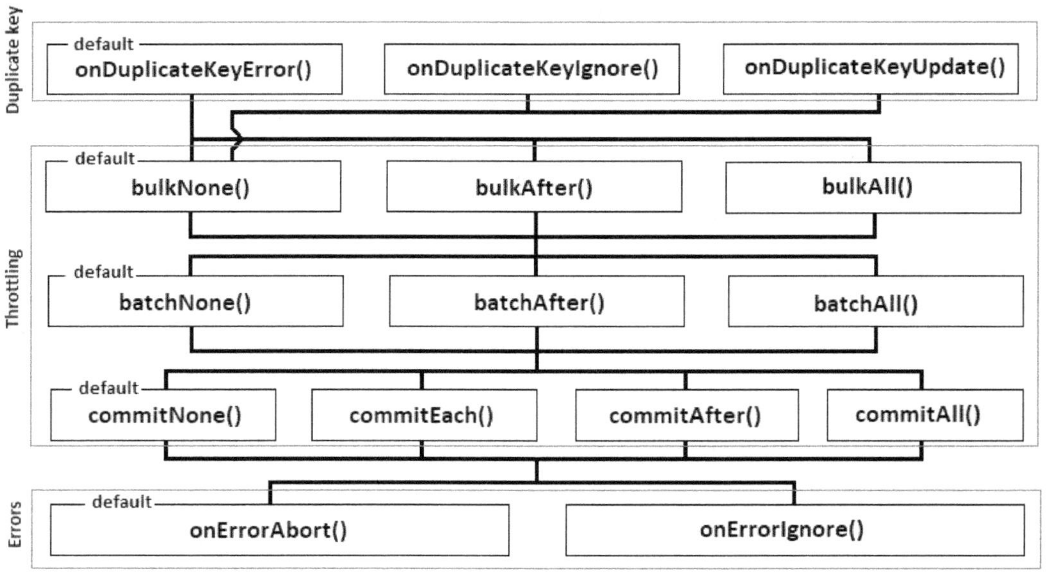

Figure 10.3 – The Loader API options

Let's explore each of these categories, starting with the one for tackling duplicate keys.

Duplicate keys options

A duplicate key occurs when a unique key exists in the table and we attempt to import a record having the same key. By unique key, jOOQ means any unique key, not only primary keys.

So, handling duplicate keys can be done via `onDuplicateKeyError()`, which is the default, or via `onDuplicateKeyIgnore()` or `onDuplicateKeyUpdate()`. The default behavior throws an exception if there are any duplicate keys.

By explicitly using `onDuplicateKeyIgnore()`, we instruct jOOQ to skip any duplicate key without throwing an exception (this is the synthetic `ON DUPLICATE KEY IGNORE` clause, which can be emulated by jOOQ depending on dialect). We can instruct jOOQ to execute `UPDATE` instead of `INSERT` via `onDuplicateKeyUpdate()` (this is the synthetic `ON DUPLICATE KEY UPDATE` clause, which can be emulated by jOOQ depending on dialect).

Throttling options

There are three throttling options that can be used to fine-tune the import. These options refer to bulking, batching, and committing. jOOQ allows us to explicitly use any combination of these options or to rely on the following defaults: no bulking, batching, and committing.

Bulking can be set via `bulkNone()` (which is the default and means that no bulking will be used), `bulkAfter(int rows)` (which allows us to specify how many rows will be inserted in one bulk via a multi-row `INSERT` (`insert into ... (...) values (?, ?, ?,...), (?, ?, ?,...), (?, ?, ?,...), ...`), and `bulkAll()` (which attempts to create one bulk from the entire source of data).

As you can see from *Figure 10.3*, `bulkNone()` is the only one that can be chained after all options used for handling duplicate values. The `bulkAfter()` and `bulkAll()` methods can be chained only after `onDuplicateKeyError()`. Moreover, `bulkNone()`, `bulkAfter()`, and `bulkAll()` are mutually exclusive.

Batching can be avoided via the default `batchNone()`, or it can be explicitly set via `batchAfter(int bulk)` or `batchAll()`. Explicitly specifying the number of bulk statements that should be sent to the server as a single JDBC batch statement can be accomplished via `batchAfter(int bulk)`. On the other hand, sending a single batch containing all bulks can be done via `batchAll()`. If bulking is not used (`bulkNone()`) then it is as if each row represents a bulk, so, for instance, `batchAfter(3)` means to create batches of three rows each.

As you can see from *Figure 10.3*, `batchNone()`, `batchAfter()`, and `batchAll()` are mutually exclusive.

Finally, committing data to the database can be controlled via four dedicated methods. By default, `commitNone()` leaves committing and rolling back operations up to client code (for instance, via `commitNone()`, we can allow Spring Boot to handle commit and rollback). But, if we want to commit after a certain number of batches, then we have to use `commitAfter(int batches)` or the handy `commitEach()` method, which is equivalent to `commitAfter(1)`. And, if we decide to commit all batches at once, then we need `commitAll()`. If batching is not used (relying on `batchNone()`), then it is as if each batch is a bulk, (for instance, `commitAfter(3)` means to commit after every three bulks). If bulking is not used either (relying on `bulkNone()`), then it is as if each bulk is a row (for instance, `commitAfter(3)` means to commit after every three rows).

As you can see from *Figure 10.3*, `commitNone()`, `commitAfter()`, `commitEach()`, and `commitAll()` are mutually exclusive.

Error options

Attempting to manipulate (import) large amounts of data is a process quite prone to errors. While some of the errors are fatal and should stop the importing process, others can be safely ignored or postponed to be resolved after import. In the case of fatal errors, the Loader API relies on a method named `onErrorAbort()`. If an error occurs, then the Loader API stops the import process. On the other hand, we have `onErrorIgnore()`, which instructs the Loader API to skip any insert that caused an error and try to execute the next one.

Special cases

While finding the optimal combination of these options is a matter of benchmarking, there are several things that you should know, as follows:

- If there are no unique keys in our table then `onDuplicateKeyUpdate()` acts exactly as `onDuplicateKeyIgnore()`.
- If `bulkAll()` + `commitEach()` or `bulkAll()` + `commitAfter()` is used, then jOOQ forces the usage of `commitAll()`.
- If `batchAll()` + `commitEach()` or `batchAll()` + `commitAfter()` is used, then jOOQ forces the usage of `commitAll()`.

Next, let's quickly cover the supported sources of data.

Importing data sources

Providing the source of data can be accomplished via dedicated methods that are specific to the supported different data types. For instance, if the data source is a CSV file, then we rely on the `loadCSV()` method; if it is a JSON file, then we rely on the `loadJSON()` method; and if it is an XML file, then we rely on `loadXML()`. Moreover, we can import arrays via `loadArrays()` and jOOQ `Records` via `loadRecords()`.

The `loadCSV()`, `loadJSON()`, and `loadXML()` methods come in 10+ flavors that allow us to load data from `String`, `File`, `InputStream`, and `Reader`. On the other hand, `loadArrays()` and `loadRecords()` allow us to load data from an array, `Iterable`, `Iterator`, or `Stream`.

Listeners

The Loader API comes with import listeners to be chained for keeping track of import progress. We mainly have `onRowStart(LoaderRowListener listener)` and `onRowEnd(LoaderRowListener listener)`. The former specifies a listener to be invoked before processing the current row, while the latter specifies a listener to be invoked after processing the current row. `LoaderRowListener` is a functional interface.

Execution and error handling

After the Loader API is executed, we have access to meaningful feedback that is available through the returned `org.jooq.Loader`. For instance, we can find out the number of executed bulks/batches via the `executed()` method, the number of processed rows via the `processed()` method, the number of stored rows (`INSERT`/`UPDATE`) via the `stored()` method, the number of ignored rows (caused by errors or duplicate keys) via the `ignored()` method, and the potential errors via the `errors()` method as `List<LoaderError>`. As you'll see in the next section of examples, `LoaderError` contains details about the errors (if any).

Examples of using the Loader API

Finally, after all this theory, it is time to see some examples of loading CSV, JSON, `Record`, and arrays. All these examples are executed and dissected in the context of Spring Boot `@Transactional`. Feel free to practice them under the jOOQ transactional context by simply removing `@Transactional` and wrapping the code as follows:

```
ctx.transaction(configuration -> {
   // Loader API code
   configuration.dsl()…
});
```

So, let's start by loading some CSV.

Loading CSV

Loading CSV is accomplished via the `loadCSV()` method. Let's start with a simple example based on the following typical CSV file (`in.csv`):

```
sale_id,fiscal_year,sale,employee_number,…,trend
1,2003,5282.64,1370,0,…,UP
2,2004,1938.24,1370,0,…,UP
3,2004,1676.14,1370,0,…,DOWN
…
```

Obviously, this data should be imported in the `sale` table, so `TARGET_TABLE` (`Table<R>`) that should be passed to `loadInto()` is `SALE`. Pointing jOOQ to this file is accomplished via the `loadCSV()` method as follows:

```
ctx.loadInto(SALE)
   .loadCSV(Paths.get("data", "csv", "in.csv").toFile(),
            StandardCharsets.UTF_8)
   .fieldsCorresponding()
   .execute();
```

This code relies on the default options. Notice the call of the `fieldsCorresponding()` method. This method signals to jOOQ that all input fields having a corresponding field in `SALE` (with the same name) should be loaded. Practically, in this case, all fields from the CSV file have a correspondent in the `SALE` table, so all of them will be imported.

But, obviously, this is not always the case. Maybe we want to load only a subset of scattered fields. In such cases, simply pass *dummy* nulls for the field indexes (positions) that shouldn't be loaded (this is an index/position-based field mapping). This time, let's collect the number of processed rows as well via `processed()`:

```
int processed = ctx.loadInto(SALE)
 .loadCSV(Paths.get("data", "csv", "in.csv").toFile(),
        StandardCharsets.UTF_8)
 .fields(null, SALE.FISCAL_YEAR, SALE.SALE_,
        null, null, null, null, SALE.FISCAL_MONTH,
        SALE.REVENUE_GROWTH,SALE.TREND)
 .execute()
 .processed();
```

This code loads from CSV only SALE.FISCAL_YEAR, SALE.SALE_, SALE.FISCAL_MONTH, SALE.REVENUE_GROWTH, and SALE.TREND. Notice that we've used the fields() method instead of fieldsCorresponding(), since fields() allows us to keep only the desired fields and skip the rest. A sample of the resultant INSERT (in MySQL dialect) looks like this:

```
INSERT INTO `classicmodels`.`sale` (`fiscal_year`, `sale`,
               `fiscal_month`, `revenue_growth`, `trend`)
VALUES (2005, 5243.1, 1, 0.0, 'DOWN')
```

While this CSV file is a typical one (first line header, data separated by a comma, and so on), sometimes we have to deal with CSV files that are quite customized, such as the following one:

```
1|2003|5282.64|1370|0|{null}|{null}|1|0.0|*UP*
2|2004|1938.24|1370|0|{null}|{null}|1|0.0|*UP*
3|2004|1676.14|1370|0|{null}|{null}|1|0.0|*DOWN*
...
```

This CSV file contains the same data as the previous one expect that there is no header line, the data separator is |, the quote mark is *, and the null values are represented as {null}. Loading this CSV file into SALE requires the following code:

```
List<LoaderError> errors = ctx.loadInto(SALE)
 .loadCSV(Paths.get("data", "csv", "in.csv").toFile(),
       StandardCharsets.UTF_8)
 .fields(SALE.SALE_ID, SALE.FISCAL_YEAR, SALE.SALE_,
         SALE.EMPLOYEE_NUMBER, SALE.HOT, SALE.RATE, SALE.VAT,
         SALE.FISCAL_MONTH, SALE.REVENUE_GROWTH, SALE.TREND)
 .ignoreRows(0)
 .separator('|').nullString("{null}").quote('*')
 .execute()
 .errors();
```

First of all, since there is no header, we rely on fields() to explicitly specify the list of fields (SALE_ID is mapped to index 1 in CSV, FISCAL_YEAR to index 2, and so on). Next, we call ignoreRows(0); by default, jOOQ skips the first line, which is considered the header of the CSV file, but since there is no header in this case, we have to instruct jOOQ to take into account the first line as a line containing data. Obviously, this method is useful for skipping *n* rows as well. Taking it a step further, we call separator(), nullString(), and quote() to override the defaults. Finally, we call errors() and collect potential errors in List<LoaderError>. This is an optional step and is not related to this particular example. In the bundled code (*LoadCSV* for MySQL), you can see how to loop this list and extract valuable information about what happened during the loading process. Moreover, you'll see more examples of loading CSV files. Next, let's explore more examples for loading a JSON file.

Loading JSON

Loading JSON is done via the `loadJSON()` method. Let's start with a JSON file:

```
{
  "fields": [
    {
      "schema": "classicmodels",
      "table": "sale",
      "name": "sale_id",
      "type": "BIGINT"
    },
    ...
  ],
  "records": [
    [1, 2003, 5282.64, 1370, 0, null, null, 1, 0.0, "UP"],
    [2, 2004, 1938.24, 1370, 0, null, null, 1, 0.0, "UP"],
    ...
  ]
}
```

This JSON file was previously exported via `formatJSON()`. Notice the `"fields"` header, which is useful for loading this file into the `SALE` table via the mapping provided by the `fieldsCorresponding()` method. Without a header, the `fieldsCorresponding()` method cannot produce the expected results since the input fields are missing. But, if we rely on the `fields()` method, then we can list the desired fields (all or a subset of them) and count on index-based mapping without worrying about the presence or absence of the `"fields"` header. Moreover, this time, let's add an `onRowEnd()` listener as well:

```
ctx.loadInto(SALE)
   .loadJSON(Paths.get("data", "json", "in.json").toFile(),
       StandardCharsets.UTF_8)
   .fields(null, SALE.FISCAL_YEAR, SALE.SALE_, null, null,
       null, null, SALE.FISCAL_MONTH, SALE.REVENUE_GROWTH,
       SALE.TREND)
   .onRowEnd(ll -> {
```

```
    System.out.println("Processed row: "
       + Arrays.toString(ll.row()));
    System.out.format("Executed: %d, ignored: %d, processed:
       %d, stored: %d\n", ll.executed(), ll.ignored(),
       ll.processed(), ll.stored());
    })
  .execute();
```

After each row is processed you'll see an output in the log as shown here:

```
Processed row: [28, 2005, 5243.1, 1504, …, DOWN]
Executed: 28, ignored: 0, processed: 28, stored: 28
```

But, let's look at a JSON file without the "`fields`" header, as follows:

```
[
    {
      "fiscal_month": 1,
      "revenue_growth": 0.0,
      "hot": 0,
      "vat": null,
      "rate": null,
      "sale": 5282.64013671875,
      "trend": "UP",
      "sale_id": 1,
      "fiscal_year": 2003,
      "employee_number": 1370
    },
    {
      …
    },
…
```

This kind of JSON can be loaded via `fieldsCorresponding()` or via `fields()`. Since the field names are available as JSON keys, the `fieldsCorresponding()` method maps them correctly. Using `fields()` should be done by keeping in mind the order of keys in this JSON. So, `"fiscal_month"` is on index `1`, `"revenue_growth"` on index `2`, and so on. Here is an example that loads only `"fiscal_month"`, `"revenue_growth"`, `"sale"`, `"fiscal_year"`, and `"employee_number"`:

```
int processed = ctx.loadInto(SALE)
  .loadJSON(Paths.get("data", "json", "in.json").toFile(),
```

```
        StandardCharsets.UTF_8)
  .fields(SALE.FISCAL_MONTH, SALE.REVENUE_GROWTH,
        null, null, null, SALE.SALE_, null, null,
        SALE.FISCAL_YEAR, SALE.EMPLOYEE_NUMBER)
  .execute()
  .processed();
```

But, sometimes, the missing data is in JSON itself, as here:

```
[
    {
      "sale_id": 1,
      "fiscal_year": 2003,
      "sale": 5282.64
      "fiscal_month": 1,
      "revenue_growth": 0.0
    },
...
```

Here is another example:

```
[
  [
    1,
    2003,
    5282.64,
    1,
    0.0
  ],
...
```

This time, in both cases, we must rely on `fields()`, as here:

```
ctx.loadInto(SALE)
   .loadJSON(Paths.get("data", "json", "in.json").toFile(),
      StandardCharsets.UTF_8)
   .fields(SALE.SALE_ID, SALE.FISCAL_YEAR, SALE.SALE_,
     SALE.FISCAL_MONTH, SALE.REVENUE_GROWTH)
   .execute();
```

Next, let's assume that we have a JSON file that should be imported into the database using batches of size `2` (rows), so we need `batchAfter(2)`. The commit (as in all the previous examples) will be accomplished by Spring Boot via `@Transactional`:

```
@Transactional
public void loadJSON () {

 int executed = ctx.loadInto(SALE)
  .batchAfter(2)// each *batch* has 2 rows
  .commitNone() // this is default, so it can be omitted
  .loadJSON(Paths.get("data", "json", "in.json").toFile(),
     StandardCharsets.UTF_8)
  .fieldsCorresponding()
  .execute()
  .executed();
}
```

Since `commitNone()` is the default behavior, it could be omitted. Essentially, `commitNone()` allows `@Transactional` to handle the commit/rollback actions. By default, `@Transactional` commits the transaction at the end of the annotated method. If something goes wrong, the entire payload (all batches) is rolled back. But, if you remove `@Transactional`, then `auto-commit =true` goes into action. This commits after each batch (so, after every two rows). If something goes wrong, then there is no rollback action, but the loading process is aborted immediately since we rely on the default settings, `onDuplicateKeyError()` and `onErrorAbort()`. If we remove `@Transactional` and set `auto-commit` to `false` (`spring.datasource.hikari.auto-commit=false`), then nothing commits.

This example returns the number of executed batches via `executed()`. For instance, if there are 36 rows processed with `batchAfter(2)`, then `executed()` returns 18.

Next, let's consider a JSON file that contains duplicate keys. Every time a duplicate key is found, the Loader API should skip it, and, in the end, it should report the number of ignored rows. Moreover, the Loader API should commit after each batch of three rows:

```
int ignored = ctx.loadInto(SALE)
 .onDuplicateKeyIgnore()
 .batchAfter(3) // each *batch* has 3 rows
 .commitEach()  // commit each batch
 .loadJSON(Paths.get("data", "json", "in.json").toFile(),
     StandardCharsets.UTF_8)
 .fieldsCorresponding()
```

```
 .execute()
 .ignored();
```

If you want to execute `UPDATE` instead of ignoring duplicate keys, just replace `onDuplicateKeyIgnore()` with `onDuplicateKeyUpdate()`.

Finally, let's import a JSON using `bulkAfter(2)`, `batchAfter(3)`, and `commitAfter(3)`. In other words, each bulk has two rows, and each batch has three bulks. Therefore, six rows commit after three batches, that is nine bulks, so after 18 rows, you get the following:

```
int inserted = ctx.loadInto(SALE)
 .bulkAfter(2)    // each *bulk* has 2 rows
 .batchAfter(3)   // each *batch* has 3 *bulks*, so 6 rows
 .commitAfter(3) // commit after 3 *batches*, so after 9
                  // *bulks*, so after 18 rows
 .loadJSON(Paths.get("data", "json", "in.json").toFile(),
      StandardCharsets.UTF_8)
 .fieldsCorresponding()
 .execute()
 .stored();
```

If something goes wrong, the last uncommitted batch is rolled back without affecting the already committed batches. More examples are available in the bundled code, `LoadJSON`, for MySQL.

Loading records

Loading jOOQ `Record` via the Loader API is a straightforward process accomplished via the `loadRecords()` method. Let's consider the following set of records:

```
Result<SaleRecord> result1 = ...;

Result<Record3<Integer, Double, String>> result2 = ...;

Record3<Integer, Double, String>[] result3 = ...;

SaleRecord r1 = new SaleRecord(1L, ...);
SaleRecord r2 = new SaleRecord(2L, ...);
SaleRecord r3 = new SaleRecord(3L, ...);
```

Loading them can be done as follows:

```
ctx.loadInto(SALE)
   .loadRecords(result1)
   .fields(null, SALE.FISCAL_YEAR, SALE.SALE_,
           SALE.EMPLOYEE_NUMBER, SALE.HOT, SALE.RATE, SALE.VAT,
           SALE.FISCAL_MONTH, SALE.REVENUE_GROWTH, SALE.TREND)
   .execute();

ctx.loadInto(SALE).loadRecords(result2/result3)
   .fieldsCorresponding()
   .execute();

ctx.loadInto(SALE).loadRecords(r1, r2, r3)
   .fieldsCorresponding()
   .execute();
```

Let's look at loading the following map of `Record`:

```
Map<CustomerRecord, CustomerdetailRecord> result = …;
```

So, `CustomerRecord` should be loaded in `CUSTOMER`, and `CustomerdetailRecord` should be loaded in `CUSTOMERDETAIL`. For this, we can use `Map.keySet()` and `Map.values()` as follows:

```
ctx.loadInto(CUSTOMER)
   .onDuplicateKeyIgnore()
   .loadRecords(result.keySet())
   .fieldsCorresponding()
   .execute();

ctx.loadInto(CUSTOMERDETAIL)
   .onDuplicateKeyIgnore()
   .loadRecords(result.values())
   .fieldsCorresponding()
   .execute();
```

More examples are available in the bundled code, `LoadRecords`, for MySQL.

Loading arrays

Loading arrays is accomplished via the `loadArrays()` method. Let's consider the following array containing data that should be loaded into the `SALE` table:

```
Object[][] result = ctx.selectFrom(…).fetchArrays();
```

Loading this array can be done as follows:

```
ctx.loadInto(SALE)
   .loadArrays(Arrays.stream(result)) // Arrays.asList(result)
   .fields(null, SALE.FISCAL_YEAR, SALE.SALE_,
           SALE.EMPLOYEE_NUMBER, SALE.HOT, SALE.RATE, SALE.VAT,
           SALE.FISCAL_MONTH, SALE.REVENUE_GROWTH, SALE.TREND)
   .execute();
```

Here is another example that relies on `loadArrays(Object[]... os)`:

```
int executed = ctx.loadInto(SALE)
   .onDuplicateKeyIgnore()
   .batchAfter(2)
   .commitEach()
   .loadArrays(
     new Object[]{1, 2005, 582.64, 1370, 0,… , "UP"},
     new Object[]{2, 2005, 138.24, 1370, 0,… , "DOWN"},
     new Object[]{3, 2005, 176.14, 1370, 0,… , "DOWN"})
   .fields(SALE.SALE_ID, SALE.FISCAL_YEAR, SALE.SALE_,
           SALE.EMPLOYEE_NUMBER, SALE.HOT, SALE.RATE,
           SALE.VAT, SALE.FISCAL_MONTH, SALE.REVENUE_GROWTH,
           SALE.TREND)
   .execute()
   .ignored();
```

You can check out these examples next to others not listed here in the bundled code, `LoadArrays`, for MySQL. It is time to summarize this chapter.

Summary

In this chapter, we've covered four important topics: exporting, batching, bulking, and loading. As you saw, jOOQ comes with dedicated APIs for accomplishing each of these tasks that require a lot of complex code under the hood. Frequently, jOOQ simplifies the complexity (as usual) and allows us to focus on what we have to do and less on how we do it. For instance, it is amazing to see that it takes seconds to write a snippet of code for loading a CSV or a JSON file into the database while having fluent and smooth support for error handling control, diagnosis output, bulking, batching, and committing control.

In the next chapter, we will cover the jOOQ keys.

11
jOOQ Keys

Choosing the proper type of keys for our tables has a significant benefit on our queries. jOOQ sustains this statement by supporting a wide range of keys, from the well-known unique and primary keys to the fancy embedded and synthetic/surrogate keys. The most commonly used synthetic identifiers (or surrogate identifiers) are numerical or UUIDs. In comparison with natural keys, surrogate identifiers don't have a meaning or a correspondent in the real world. A surrogate identifier can be generated by a Numerical Sequence Generator (for instance, an identity or sequence) or by a Pseudorandom Number Generator (for instance, a GUID or UUID). Moreover, let me use this context to recall that in clustered environments, most relational databases rely on *numerical sequences* and different offsets per node to avoid the risk of conflicts. Use *numerical sequences* instead of UUIDs because they require less memory than UUIDs (a UUID requires 16 bytes, while `BIGINT` requires 8 bytes and `INTEGER` 4 bytes) and the index usage is more performant. Moreover, since UUIDs are not sequential, they introduce performance penalties at a clustered indexes level. More precisely, we will discuss an issue known as *index fragmentation*, which is caused by the fact that UUIDs are random. Some databases (for instance, MySQL 8.0) come with significant improvements in mitigating UUID performance penalties (there are three new functions – `UUID_TO_BIN`, `BIN_TO_UUID`, and `IS_UUID`) while other databases are still prone to these issues. As Rick James highlights, *"If you cannot avoid UUIDs (which would be my first recommendation) then..."* It is recommended to read his article (`http://mysql.rjweb.org/doc.php/uuid`) for a deeper understanding of the main issues and potential solutions.

For now, let's get back to our chapter, which will cover the following topics:

- Fetching a database-generated primary key
- Suppressing a primary key return on updatable records
- Updating a primary key of an updatable record
- Using database sequences
- Inserting a SQL Server IDENTITY
- Fetching the Oracle ROWID pseudo-column
- Comparing composite primary keys
- Working with embedded keys
- Working with jOOQ synthetic objects
- Overriding primary keys

Let's get started!

Technical requirements

The code for this chapter can be found on GitHub at `https://github.com/PacktPublishing/jOOQ-Masterclass/tree/master/Chapter11`.

Fetching the database-generated primary key

A typical scenario consists of fetching a database-generated (identity) primary key after an `INSERT` operation is executed via the `insertInto()` method or the updatable record's `insert()` method. If you are using `insertInto()` (`DSL.insertInto()` or `DSLContext.insertInto()`), the database-generated primary key can be obtained via the `returningResult()`/`returning()` methods. For instance, the identity primary key of `SALE` is shaped in MySQL via `AUTO_INCREMENT`, in SQL Server via `IDENTITY`, and for historic reasons (because both now support standard SQL `IDENTITY` columns), in PostgreSQL and Oracle via database sequences. In all these cases, the generated identity primary key of `SALE` can be fetched as here (`SALE.SALE_ID`):

```
long insertedId = ctx.insertInto(SALE, SALE.FISCAL_YEAR,
    SALE.SALE_, SALE.EMPLOYEE_NUMBER, SALE.FISCAL_MONTH,
    SALE.REVENUE_GROWTH)
  .values(2004, 2311.42, 1370L, 1, 0.0)
  .returningResult(SALE.SALE_ID)
```

```
.fetchOneInto(long.class);
// .fetchOne(); to fetch Record1<Long>
```

Alternatively, a convenient approach relies on the `getIdentity()` method, as shown here:

```
.returningResult(SALE.getIdentity().getField())
```

However, this approach is useful when your table has a single identity column; otherwise, it is better to explicitly list the identities that should be returned. However, don't get me wrong here – even if some databases (for example, PostgreSQL) support multiple identities, that is quite an unusual approach, which personally I don't like to use, but I'll cover it in this chapter. Also, check this tweet to get more details: `https://twitter.com/lukaseder/status/1205046981833482240`.

Now, the `insertedId` variable holds the database-generated primary key as a `Record1<Long>`. Getting the `long` value can be done via `fetchOne().value1()` or directly via `.fetchOneInto(long.class)`. The same practice is apparent for a bulk insert (a multi-record insert). This time, the generated primary keys are stored in `Result<Record1<Long>>` or `List<Long>`:

```
List<Long> insertedIds = ctx.insertInto(SALE,
    SALE.FISCAL_YEAR,SALE.SALE_, SALE.EMPLOYEE_NUMBER,
    SALE.FISCAL_MONTH, SALE.REVENUE_GROWTH)
 .values(2004, 2311.42, 1370L, 1, 0.0)
 .values(2003, 900.21, 1504L, 1, 0.0)
 .values(2005, 1232.2, 1166L, 1, 0.0)
 .returningResult(SALE.getIdentity().getField())
 // or, .returningResult(SALE.SALE_ID)
 .collect(intoList());
 // or, .fetchInto(Long.class);
```

For a special case when we cannot provide an identity, jOOQ allows us to use the handy `lastID()` method:

```
ctx.insertInto(SALE, SALE.FISCAL_YEAR, SALE.SALE_,
               SALE.EMPLOYEE_NUMBER, SALE.FISCAL_MONTH,
               SALE.REVENUE_GROWTH)
   .values(2002, 5411.42, 1504L, 1, 0.0)
   .execute();
```

```
//meanwhile, a concurrent transaction can sneak a INSERT

var lastId = ctx.lastID();
```

However, the `lastID()` method has at least two shortcomings that deserve our attention. In a concurrent transactional environment (for instance, a web application), there is no guarantee that the returned value belongs to the previous `INSERT` statement, since a concurrent transaction can sneak another `INSERT` between our `INSERT` and the `lastID()` call. In such a case, the returned value belongs to the `INSERT` statement executed by the concurrent transaction. In addition, `lastID()` is not quite useful in the case of bulk inserts, since it returns only the last-generated primary key (but maybe this is exactly what you need).

If you are inserting an updatable record, jOOQ will automatically return the generated identity primary key and populate the updatable record field, as shown here:

```
SaleRecord sr = ctx.newRecord(SALE);
sr.setFiscalYear(2021);
...

sr.insert();

// here you can call sr.getSaleId()
```

After insert, calling `sr.getSaleId()` returns the primary key generated by the database for this record. The same thing can be accomplished via jOOQ's DAO while inserting a POJO:

```
private final SaleRepository saleRepository; // injected DAO

Sale s = new Sale(); // jooq.generated.tables.pojos.Sale
s.setFiscalYear(2020);
...
saleRepository.insert(s);

// here you can call s.getSaleId()
```

This time, jOOQ set the generated primary key in the inserted POJO. You can find these examples in the *Keys* bundled code.

Suppressing a primary key return on updatable records

In the previous section, you saw that jOOQ automatically fetches and sets the generated primary key for updatable records. Suppressing this action can be done via the `withReturnIdentityOnUpdatableRecord()` flag setting. In some dialects, a database round trip (the `lastID()` style) can be prevented, so this is mostly a performance feature. By default, this flag is `true`, but if we explicitly set it to `false`, then jOOQ will no longer attempt to fetch the generated primary key:

```
DSLContext derivedCtx = ctx.configuration().derive(
 new Settings().withReturnIdentityOnUpdatableRecord(false))
              .dsl();

 SaleRecord sr = derivedCtx.newRecord(SALE);
 sr.setFiscalYear(2021);
 ...

 sr.insert();
```

This time, calling `sr.getSaleId()` returns `null`.

Updating a primary key of an updatable record

As a good practice, a primary key should never be updated anyway. But, who am I to judge?!

By default, calling the `store()` method after changing (to a non-null value) the primary key of an updatable record previously loaded via jOOQ causes an `INSERT` statement to be executed. However, we can force jOOQ to generate and execute an `UPDATE` of the primary key via the `withUpdatablePrimaryKeys()` flag setting:

```
DSLContext derivedCtx = ctx.configuration().derive(
  new Settings().withUpdatablePrimaryKeys(true)).dsl();

 SaleRecord sr = derivedCtx.selectFrom(SALE)
    .where(SALE.SALE_ID.eq(2L))
    .fetchSingle();

 sr.setSaleId(new_primary_key);

 sr.store(); // UPDATE primary key
```

Of course, we can also update the primary key via an explicit UPDATE, and if you really have to do it, then go for this instead of a jOOQ flag:

```
ctx.update(SALE)
   .set(SALE.SALE_ID, sr.getSaleId() + 1)
   .where(SALE.SALE_ID.eq(sr.getSaleId()))
   .execute();
```

You can find these examples in the *Keys* bundled code.

Using database sequences

To yield sequential numbers, databases such as PostgreSQL, SQL Server, and Oracle rely on sequences. A database sequence lives independently from tables – it can be associated with the primary key and non-primary key columns, it can be auto-generated (as in the case of PostgreSQL (BIG)SERIAL), it can be used across multiple tables, it can have independent permissions, it can have cycles, it can increment values in its own transactions to guarantee uniqueness across transactions using it, we can explicitly alter its values by setting minimum, maximum, increment, and current values, and so on.

For instance, let's consider the following sequence (employee_seq), defined in our PostgreSQL schema for the employee.employee_number primary key:

```
CREATE SEQUENCE "employee_seq" START 100000 INCREMENT 10
       MINVALUE 100000 MAXVALUE 10000000
       OWNED BY "employee"."employee_number";

CREATE TABLE "employee" (
  "employee_number" BIGINT NOT NULL,
  ...
);
```

The employee_seq sequence doesn't produce sequence values automatically on your insertions, so the application must explicitly manipulate it. On the other hand, the sale_seq sequence produces sequence values automatically on your insertions, and it looks like the following code block (you'll get an automatic value when the SALE_ID column is omitted from the INSERT statement or DEFAULT or DEFAULT VALUES is used; when users set SALE_ID to NULL explicitly, there's going to be a constraint violation error):

```
CREATE SEQUENCE "sale_seq" START 1000000;

CREATE TABLE "sale" (
```

```
  "sale_id" BIGINT NOT NULL DEFAULT NEXTVAL ('"sale_seq"'),
  ...
);
```

For each such sequence, the jOOQ Code Generator produces an `org.jooq.Sequence` instance in `Sequences` (take your time to check the `jooq.generated.Sequences` class). For `employee_seq`, we get this:

```
public static final Sequence<Long> EMPLOYEE_SEQ =
  Internal.createSequence("employee_seq", Public.PUBLIC,
   SQLDataType.BIGINT.nullable(false), 100000, 10, 100000,
    10000000, false, null);
```

The jOOQ API exposes several methods for obtaining information about a sequence. Among them, we have the following suggested methods (you can find out more in the jOOQ documentation):

```
String name = EMPLOYEE_SEQ.getName();
Field<Long> start = EMPLOYEE_SEQ.getStartWith();
Field<Long> min = EMPLOYEE_SEQ.getMinvalue();
Field<Long> max = EMPLOYEE_SEQ.getMaxvalue();
Field<Long> inc = EMPLOYEE_SEQ.getIncrementBy();
```

Besides these methods, we have three more that are very useful in daily tasks – `currval()`, `nextval()`, and `nextvals()`. The first one (`currval()`) attempts to return the *current* value in the sequence. This can be obtained in a `SELECT` statement:

```
long cr = ctx.fetchValue(EMPLOYEE_SEQ.currval());

long cr = ctx.select(EMPLOYEE_SEQ.currval())
  .fetchSingle().value1();

long cr = ctx.select(EMPLOYEE_SEQ.currval())
  .fetchSingleInto(Long.class); // or, fetchOneInto()
```

The second one, `nextval()`, attempts to return the *next* value in the sequence. It can be used as follows:

```
long nv = ctx.fetchValue(EMPLOYEE_SEQ.nextval());

long nv = ctx.select(EMPLOYEE_SEQ.nextval())
```

```
  .fetchSingle().value1();

long nv = ctx.select(EMPLOYEE_SEQ.nextval())
  .fetchSingleInto(Long.class); // or, fetchOneInto()
```

And here is a `SELECT` statement that fetches both, the current and the next value:

```
Record2<Long, Long> vals = ctx.fetchSingle(
           EMPLOYEE_SEQ.nextval(), EMPLOYEE_SEQ.currval());

Record2<Long, Long> vals = ctx.select(EMPLOYEE_SEQ.nextval(),
           EMPLOYEE_SEQ.currval()) .fetchSingle();
```

A potential issue of using sequences consists of selecting `currval()` from the sequence before *initializing* it within your session by selecting `nextval()` for it. Commonly, when you are in such a scenario, you'll get an explicit error that mentions that `currval()` is not yet defined in this session (for instance, in Oracle, this is ORA-08002). By executing `INSERT` or calling `nextval()` (for instance, in `SELECT` as the previous one), you'll initialize `currval()` as well.

If the sequence can produce values automatically then the best way to insert a new record is to simply omit the primary key field. Since `sale_seq` can produce values automatically, an `INSERT` can be like this:

```
ctx.insertInto(SALE, SALE.FISCAL_YEAR, SALE.SALE_,
               SALE.EMPLOYEE_NUMBER, SALE.FISCAL_MONTH,
               SALE.REVENUE_GROWTH)
   .values(2005, 1370L, 1282.641, 1, 0.0)
   .execute();
```

The database will use `sale_seq` to assign a value to the `SALE_ID` field (the primary key of `SALE`). This is like using any other type of identity associated with a primary key.

> **Important Note**
>
> There is no need to explicitly call the `currval()` or `nextval()` method as long as you don't have a specific case that requires a certain sequence value from a sequence that is auto-generated (for example, from `(BIG)SERIAL`) or set as default (for example, as `NOT NULL DEFAULT NEXTVAL ("'sale_seq'")`). Simply omit the primary key field (or whatever field uses the sequence) and let the database generate it.

However, if the sequence cannot automatically produce values (for instance, `employee_seq`), then an `INSERT` statement must rely on an explicit call of the `nextval()` method:

```
ctx.insertInto(EMPLOYEE, EMPLOYEE.EMPLOYEE_NUMBER,
               EMPLOYEE.LAST_NAME, EMPLOYEE.FIRST_NAME, ...)
   .values(EMPLOYEE_SEQ.nextval(),
      val("Lionel"), val("Andre"), ...)
   .execute();
```

Pay attention to how you interpret and use the `currval()` and `nextval()` methods. Once you fetch a sequence value via `nextval()` (for instance, via `SELECT`), you can safely use it later in subsequent queries (`INSERT`) because the database will not give this value to other (concurrent) transactions. So, `nextval()` is safe to be used by multiple concurrent transactions. On the other hand, in the case of `currval()`, you have to be aware of some aspects. Check this code:

```
ctx.insertInto(SALE, SALE.FISCAL_YEAR, SALE.SALE_,
               SALE.EMPLOYEE_NUMBER, SALE.FISCAL_MONTH,
               SALE.REVENUE_GROWTH)
   .values(2020, 900.25, 1611L, 1, 0.0)
   .execute();

// another transaction can INSERT and currval() is modified

long cr = ctx.fetchValue(SALE_SEQ.currval());
```

So, between the previous `INSERT` and `SELECT` of the current value, another transaction can execute `INSERT`, and `currval()` is modified/incremented (generally speaking, another transaction performs an action that updates the current value). This means that there is no guarantee that `cr` holds the value of `SALE_ID` of our `INSERT` (`SALE_ID` and `cr` can be different). If all we need is to get `SALE_ID` of our `INSERT`, then the best approach is to rely on `returningResult(SALE.SALE_ID)`, as you saw in the *Fetching a database-generated primary key* section.

Obviously, attempting to use the fetched `currval()` in subsequent `UPDATE`, `DELETE`, and so on statements falls under the same statement. For instance, there is no guarantee that the following `UPDATE` will update our previous `INSERT`:

```
ctx.update(SALE)
   .set(SALE.FISCAL_YEAR, 2005)
   .where(SALE.SALE_ID.eq(cr))
   .execute();
```

Another approach that should be avoided in a concurrent transactional environment is the following:

```
ctx.deleteFrom(SALE)
   .where(SALE.SALE_ID.eq(ctx.fetchValue(SALE_SEQ.currval())))
   .execute();
```

Even if this looks like a single query statement, it is not. This is materialized in a `SELECT` of the current value followed by a `DELETE`. Between these two statements, a concurrent transaction can still perform an `INSERT` that alters the current value (or, generally speaking, any kind of action that modifies/advances a sequence and returns a new value). Also, pay attention to these kinds of queries:

```
ctx.deleteFrom(SALE)
   .where(SALE.SALE_ID.eq(SALE_SEQ.currval()))
   .execute();
```

This renders a single `DELETE`, as shown here (the PostgreSQL dialect):

```
DELETE FROM "public"."sale" WHERE
  "public"."sale"."sale_id" = currval('"public"."sale_seq"')
```

This time, you definitely refer to the latest current value, whatever it is. For instance, this may result in deleting the latest inserted record (not necessarily by us), or it may hit a current value that is not associated with any record yet.

Furthermore, performing multi-inserts or batch inserts can take advantage of inlined `nextval()` references or pre-fetch a certain number of values via `nextvals()`:

```
List<Long> ids1 = ctx.fetchValues(EMPLOYEE_SEQ.nextvals(10));

List<Long> ids2 = ctx.fetch(EMPLOYEE_SEQ
   .nextvals(10)).into(Long.class);

List<Record1<Long>> ids3 = ctx.fetch(
   EMPLOYEE_SEQ.nextvals(10));
```

At this point, `ids1`, `ids2`, and `ids3` hold in memory 10 values that can be used in subsequent queries. Until we exhaust these values, there is no need to fetch others. This way, we reduce the number of database round trips. Here is an example of a multi-insert:

```
for (int i = 0; i < ids.size(); i++) {
 ctx.insertInto(EMPLOYEE, EMPLOYEE.EMPLOYEE_NUMBER,
```

```
                  EMPLOYEE.LAST_NAME...)
     .values(ids1.get(i), "Lionel", ...)
     .execute();
}
```

The pre-fetched values can be used to pre-set IDs of `Record` as well:

```
EmployeeRecord er = new EmployeeRecord(ids1.get(0),
  // or, ids2.get(0).value1(),
  "Lionel", ...);
```

You can find these examples in the *Keys* bundled code.

Inserting a SQL Server IDENTITY

This is not the first time in this book that we have talked about inserting SQL Server `IDENTITY` values, but let's consider this section a must-have for this chapter. The problem consists of the fact that SQL Server doesn't allow us to specify an explicit value for an `IDENTITY` field as the `PRODUCT` primary key:

```
CREATE TABLE [product] (
  [product_id] BIGINT NOT NULL IDENTITY,
  ...
);
```

In other words, the following `INSERT` statement will cause the following error – *Cannot insert explicit value for identity column in table 'product' when IDENTITY_INSERT is set to OFF*:

```
ctx.insertInto(PRODUCT, PRODUCT.PRODUCT_ID,
               PRODUCT.PRODUCT_LINE, PRODUCT.CODE,
               PRODUCT.PRODUCT_NAME)
   .values(5555L, "Classic Cars", 599302L, "Super TX Audi")
   .onDuplicateKeyIgnore();
```

So, the solution to this error is contained in the message. We have to set `IDENTITY_INSERT` to `ON`. However, this should be done in the SQL Server *current session context*. In other words, we have to issue the settings of `IDENTITY_INSERT` and the actual `INSERT` statements in the same batch, as shown here:

```
Query q1 = ctx.query("SET IDENTITY_INSERT [product] ON");
Query q2 = ctx.insertInto(PRODUCT, PRODUCT.PRODUCT_ID,
```

```
    PRODUCT.PRODUCT_LINE, PRODUCT.CODE, PRODUCT.PRODUCT_NAME)
  .values(5555L, "Classic Cars", 599302L, "Super TX Audi")
  .onDuplicateKeyIgnore(); // this will lead to a MERGE
Query q3 = ctx.query("SET IDENTITY_INSERT [product] OFF");

ctx.batch(q1, q2, q3).execute();
```

This time, there is no issue with inserting it into the `IDENTITY` column. You can find these examples in the *Keys* (for SQL Server) bundled code.

Fetching the Oracle ROWID pseudo-column

If you are a fan of the Oracle database, then it is impossible not to have heard about the ROWID pseudo-column. However, as a quick reminder, the ROWID pseudo-column is associated with each row by Oracle, and its main goal is to return the address of the row. The information contained by ROWID can be used to locate a certain row. In jOOQ, we can refer to ROWID via the `rowid()` method.

For instance, the following statement inserts a new `SALE` and fetches the generated primary key and the ROWID:

```
ctx.insertInto(SALE, SALE.FISCAL_YEAR, SALE.SALE_,
               SALE.EMPLOYEE_NUMBER, SALE.FISCAL_MONTH,
               SALE.REVENUE_GROWTH)
   .values(2004, 2311.42, 1370L, 1, 0.0)
   .returningResult(SALE.SALE_ID, rowid())
   .fetchOne();
```

The `rowid()` method returns a `String`, representing the value of ROWID (for instance, *AAAVO3AABAAAZzBABE*). We can use the ROWID for subsequent queries, such as locating a record:

```
String rowid = ...;
var result = ctx.selectFrom(SALE)
   .where(rowid().eq(rowid))
   .fetch();
```

However, as Lukas Eder shared: *"ROWIDs are not guaranteed to remain stable, so clients should never keep them around for long (for instance, outside of a transaction). But they can be useful to identify a row in a table without a primary key (for instance, a logging table)."*

In the bundled code, *Keys* (for Oracle), you can also see an example of using `rowid()` in the `SELECT`, `UPDATE`, and `DELETE` statements.

Comparing composite primary keys

By definition, a composite primary key involves two or more columns that should uniquely identify a record. A composite primary key is usually a natural key (even if it is composed of references to surrogate keys) and can often be preferable to surrogate keys in relationship tables: `https://blog.jooq.org/2019/03/26/the-cost-of-useless-surrogate-keys-in-relationship-tables/`. This means that predicates based on composite keys must contain all the involved columns. For instance, the `PRODUCTLINE` table has a composite key as (`PRODUCT_LINE`, `CODE`), and we can write a predicate for fetching a certain record by chaining the fields of the composite key via `and()`, as follows:

```
var result = ctx.selectFrom(PRODUCTLINE)
  .where(PRODUCTLINE.PRODUCT_LINE.eq("Classic Cars")
    .and(PRODUCTLINE.CODE.eq(599302L)))
  .fetchSingle();
```

Alternatively, we can separate fields from values using `row()` (the `eq()` method doesn't require an explicit `row()` constructor, so use it as you like):

```
var result = ctx.selectFrom(PRODUCTLINE)
   .where(row(PRODUCTLINE.PRODUCT_LINE, PRODUCTLINE.CODE)
     .eq(row("Classic Cars", 599302L)))
   .fetchSingle();
```

Using `row()` is also useful in conjunction with `in()`, `notIn()`, and so on:

```
result = ctx.selectFrom(PRODUCTLINE)
   .where(row(PRODUCTLINE.PRODUCT_LINE, PRODUCTLINE.CODE)
   .in(row("Classic Cars", 599302L),
       row("Trains", 123333L),
       row("Motorcycles", 599302L)))
```

Practically, in all these examples (available in *Keys*), you have to ensure that you don't forget any column of the composite key. This may become a struggle for composite keys containing more than two fields and/or in cases where the predicates involve more related conditions, and it is difficult to visually isolate the composite key fields.

A better approach is to employ embedded keys.

Working with embedded keys

As part of the embeddable types introduced in *Chapter 7, Types, Converters, and Bindings*, we have jOOQ-embedded keys. An embedded key is materialized by the jOOQ Code Generator into the implementation of the jOOQ `org.jooq.EmbeddableRecord` interface and a handy POJO class. An embedded key extends the default implementation of the `org.jooq.EmbeddableRecord` interface, which is `org.jooq.impl.EmbeddableRecordImpl`.

We can define embedded keys for primary and unique keys. Practically, we indicate to jOOQ the primary/unique keys that should become embedded keys, and jOOQ will generate the corresponding artifacts for each primary/unique key, as well as for each foreign key referencing these primary/unique keys. Roughly, embedded keys mirror the primary/unique keys and the corresponding foreign keys in Java classes.

However, in order to employ embedded keys, we need the following configuration:

```
// Maven and standalone
<database>
  ...
  <embeddablePrimaryKeys>.*</embeddablePrimaryKeys>
  <embeddableUniqueKeys>.*</embeddableUniqueKeys>
</database>

// Gradle
database {
  ...
  embeddablePrimaryKeys = '.*'
  embeddableUniqueKeys = '.*'
}

// programmatic
.withEmbeddablePrimaryKeys(".*")
.withEmbeddableUniqueKeys(".*")
```

Most probably, you'll not rely on a `.*` regular expression, since you'll not want to transform all your primary/unique keys into embedded keys. For instance, you may prefer to use embedded keys for composite keys only, so you have to use the proper regular expression for your case. Speaking about composite keys, how about creating an embedded key for the composite key of `PRODUCTLINE` (introduced in the previous section)?

```
CREATE TABLE [productline] (
  [product_line] VARCHAR(50) NOT NULL,
```

```
  [code] BIGINT NOT NULL,
  ...

  CONSTRAINT [productline_pk]
    PRIMARY KEY ([product_line],[code])
);
```

Indicate to jOOQ that we are interested in the (`product_line`, `code`) primary key via `<embeddablePrimaryKeys>productline_pk</embeddablePrimaryKeys>`, where `productline_pk` represents the name of the constraint that defines our composite primary key (if you want to list multiple constraints/primary keys, then use | as a separator).

> **Important Note**
>
> As a rule of thumb, it's always a good idea to explicitly name your constraints. This way, you never have to bother with dealing with vendor-specific generated names and potential issues. If you are not convinced that you should always name your constraints, then I suggest you read this meaningful article: `https://blog.jooq.org/how-to-quickly-rename-all-primary-keys-in-oracle/`.
>
> However, notice that MySQL ignores the constraint names on the primary key and defaults all to `PRIMARY`. In such a case, you cannot refer to a composite primary key via the name of its constraint but instead as `KEY_tablename_PRIMARY`. For instance, instead of `productline_pk`, use `KEY_productline_PRIMARY`.

At this point, jOOQ is ready to generate the classes for this embedded key, but let's take another action and customize the names of these classes. At this point, jOOQ relies on the default matcher strategy, so the names will be `ProductlinePkRecord.java` and `ProductlinePk.java`. But, we prefer `EmbeddedProductlinePkRecord.java` and `EmbeddedProductlinePk.java` respectively. As you already know, whenever we talk about renaming jOOQ things, we can rely on a configurative/programmatic matcher strategy and regular expressions (note that the `(?i:...)` directive is a thing to render the expression case-insensitive). In this case, we have the following:

```
<strategy>
 <matchers>
  <embeddables>
   <embeddable>
    <expression>.*_pk</expression>
    <recordClass>
```

```
    <expression>Embedded_$0_Record</expression>
    <transform>PASCAL</transform>
   </recordClass>
   <pojoClass>
    <expression>Embedded_$0</expression>
    <transform>PASCAL</transform>
   </pojoClass>
  </embeddable>
 </embeddables>
 </matchers>
</strategy>
```

Okay, so far, so good! At this point, the jOOQ Code Generator is ready to materialize our embedded key in `EmbeddedProductlinePkRecord.java` and `EmbeddedProductlinePk.java`. Also, jOOQ generates the `PRODUCTLINE_PK` field in the `Productline` class (see `jooq.generated.tables.Productline`), representing the embedded primary key.

Moreover, the jOOQ Code Generator searches the foreign keys referencing our composite key, and it should find the following two:

```
CREATE TABLE [product] (
  [product_line] VARCHAR(50) DEFAULT NULL,
  [code] BIGINT NOT NULL,
  ...
  CONSTRAINT [product_productline_fk]
    FOREIGN KEY ([product_line],[code])
    REFERENCES [productline] ([product_line],[code])
);

CREATE TABLE [productlinedetail] (
  [product_line] VARCHAR(50) NOT NULL,
  [code] BIGINT NOT NULL,
  ...
  CONSTRAINT [productlinedetail_productline_fk]
    FOREIGN KEY ([product_line],[code])
    REFERENCES [productline] ([product_line],[code])
);
```

For the `product_productline_fk` and `productlinedetail_productline_fk` constraints (of our foreign keys), jOOQ generates the `PRODUCT_PRODUCTLINE_FK` field in the `Product` class (see `jooq.generated.tables.Product`) and the `PRODUCTLINEDETAIL_PRODUCTLINE_FK` field in the `Productlinedetail` class (see `jooq.generated.tables.Productlinedetail`).

Now, let's practice! For instance, let's assume that we want to fetch the composite primary key of `PRODUCTLINE` and the creation date. Most probably, without using the embedded key, our `SELECT` statement will be something like this:

```
var result = ctx.select(PRODUCTLINE.PRODUCT_LINE,
           PRODUCTLINE.CODE, PRODUCTLINE.CREATED_ON) ...
```

We know that `PRODUCT_LINE` and `CODE` form our composite key. However, for someone who is not very familiar with our schema, it will be more convenient and less risky to rely on the `PRODUCTLINE_PK` embedded key and write this:

```
// Result<Record2<EmbeddedProductlinePkRecord, LocalDate>>
var result = ctx.select(PRODUCTLINE.PRODUCTLINE_PK,
                        PRODUCTLINE.CREATED_ON)...
```

Obviously, this is less verbose and much more expressive. There is no risk of forgetting a field of the composite key or mixing composite key fields with other fields (which just increases confusion), and we can add/remove a column from the composite key without modifying this code. Once we rerun the Code Generator, jOOQ will shape `PRODUCTLINE_PK` accordingly.

We can access data via getters, as shown here:

```
// '.get(0)' returns the first
// Record2<EmbeddedProductlinePkRecord, LocalDate>,
// while '.value1()' returns the EmbeddedProductlinePkRecord
result.get(0).value1().getProductLine()
result.get(0).value1().getCode()
```

Moreover, since the embedded key takes advantage of a generated POJO as well, we can fetch the composite key directly in the POJO. Look at how cool this is:

```
List<EmbeddedProductlinePk> result =
   ctx.select(PRODUCTLINE.PRODUCTLINE_PK)
      .from(PRODUCTLINE)
      .where(PRODUCTLINE.IMAGE.isNull())
      .fetchInto(EmbeddedProductlinePk.class);
```

The `EmbeddedProductlinePk` POJO exposes getters and setters to access the parts of the embedded composite key.

> **Important Note**
> Embedded keys are the embeddable types most prone to overlapping. By default, jOOQ tries to elegantly solve each overlapping case to our benefit, but when the ambiguity cannot be clarified, jOOQ will log such cases, and it's your job to act accordingly.

Let's go further and see other examples. For instance, searching a composite key in a certain collection of composite keys can be done, as shown here:

```
var result = ctx.selectFrom(PRODUCTLINE)
  .where(PRODUCTLINE.PRODUCTLINE_PK.in(
    new EmbeddedProductlinePkRecord("Classic Cars", 599302L),
    new EmbeddedProductlinePkRecord("Vintage Cars", 223113L)))
  .fetch();
```

Alternatively, joining `PRODUCTLINE` and `PRODUCT` can be done, as shown here (both the primary and foreign keys produce the primary key record):

```
var result = ctx.select(PRODUCTLINE.PRODUCTLINE_PK,
              PRODUCT.PRODUCT_ID, PRODUCT.PRODUCT_NAME)
   .from(PRODUCTLINE)
   .join(PRODUCT)
   .on(PRODUCTLINE.PRODUCTLINE_PK.eq(
         PRODUCT.PRODUCT_PRODUCTLINE_FK))
   .fetch();
```

Again, the code is less verbose and more expressive. However, more importantly, there is no risk of forgetting a column of the composite key in the join predicate. In addition, since both primary and foreign keys produce the primary key record, the predicate is valid only if we rely on matching primary/foreign key columns. This goes beyond type checking, since there is no risk of comparing wrong fields (for instance, fields that don't belong to the composite key but have the same type as the fields of the composite key).

As Lukas Eder mentioned: *"The type checking aspect is also interesting for single-column key types. With embeddable types, column types become "semantic," and what would otherwise be two compatible Field<Long> columns no longer are compatible. So, specifically in the case of JOIN predicates, it will no longer be possible to accidentally compare the wrong columns in* *on(). This could even help detect a forgotten foreign key constraint."* (`https://twitter.com/anghelleonard/status/1499751304532533251`)

This is a good opportunity to reflect on your favorite way to express the `JOIN` predicate with composite keys in jOOQ. The following figure summarizes several approaches, including a simple `and()`, using `row()`, an implicit join, a synthetic `onKey()`, and embedded keys:

A

```
var result = ctx.select(PRODUCTLINE.CREATED_ON,
      PRODUCT.PRODUCT_NAME, PRODUCT.BUY_PRICE)
    .from(PRODUCT)
    .leftOuterJoin(PRODUCTLINE)
    .on(PRODUCTLINE.PRODUCT_LINE.eq(PRODUCT.PRODUCT_LINE)
          .and(PRODUCTLINE.CODE.eq(PRODUCT.CODE)))
    .fetch();
```

using AND

B

```
var result = ctx.select(PRODUCTLINE.CREATED_ON,
      PRODUCT.PRODUCT_NAME, PRODUCT.BUY_PRICE)
    .from(PRODUCT)
    .leftOuterJoin(PRODUCTLINE)
    .on(row(PRODUCTLINE.PRODUCT_LINE, PRODUCTLINE.CODE)
          .eq(row(PRODUCT.PRODUCT_LINE, PRODUCT.CODE)))
    .fetch();
```

using row()

C

```
var result = ctx.select(PRODUCT.productline().CREATED_ON,
      PRODUCT.PRODUCT_NAME, PRODUCT.BUY_PRICE)
    .from(PRODUCT)
    .fetch();
```

using implicit join

D

```
var result = ctx.select(PRODUCTLINE.CREATED_ON,
      PRODUCT.PRODUCT_NAME, PRODUCT.BUY_PRICE)
    .from(PRODUCT)
    .leftOuterJoin(PRODUCTLINE)
    .onKey()
    .fetch();
```

synthetic onKey()

E

```
var result = ctx.select(PRODUCTLINE.CREATED_ON,
      PRODUCT.PRODUCT_NAME, PRODUCT.BUY_PRICE)
    .from(PRODUCT)
    .leftOuterJoin(PRODUCTLINE)
    .on(PRODUCTLINE.PRODUCTLINE_PK
       .eq(PRODUCT.PRODUCT_PRODUCTLINE_FK))
    .fetch();
```

enbedded keys

Figure 11.1 – The JOIN predicate with composite keys

How about updating/deleting/inserting an embedded key? Well, these examples speak for themselves:

```
EmbeddedProductlinePkRecord pk = new
    EmbeddedProductlinePkRecord("Turbo Jets", 908844L);

ctx.update(PRODUCTLINE)
   .set(PRODUCTLINE.TEXT_DESCRIPTION, "Not available")
   .where(PRODUCTLINE.PRODUCTLINE_PK.eq(pk))
   .execute();

ctx.deleteFrom(PRODUCTLINE)
   .where(PRODUCTLINE.PRODUCTLINE_PK.eq(pk))
   .execute();

ctx.insertInto(PRODUCTLINE, PRODUCTLINE.PRODUCTLINE_PK,
               PRODUCTLINE.TEXT_DESCRIPTION)
   .values(pk, "Some cool turbo engines")
   .execute();
```

Practice these examples in *EmbeddedCompositeKeys* (for SQL Server and Oracle). Alternatively, if you prefer to start with embedded keys for simple primary keys, then you can check out the *EmbeddedSimpleKeys* application (for SQL Server and Oracle). Next, let's talk about jOOQ synthetic objects.

Working with jOOQ synthetic objects

jOOQ *synthetic objects* is a powerful and exciting feature introduced in version 3.14 that reveals its full usability with database (updatable) views, databases that you cannot but want to alter, and legacy databases that have some *missing parts*. By *missing parts*, we mean identities, primary keys, unique keys, and foreign keys that simply don't exist, or do exist but are not enabled or reported by the database (and are not present in the database metadata). The jOOQ Code Generator can tackle this aspect by producing synthetic objects that emulate these *missing parts*. Let's adopt the learning by example technique to see how synthetic objects work.

Synthetic primary/foreign keys

Let's consider that we have the following two database views (in PostgreSQL):

```
CREATE OR REPLACE VIEW "customer_master" AS
SELECT "customerdetail"."city",
       "customerdetail"."country",
       "customerdetail"."state",
       "customerdetail"."postal_code",
       ...
FROM "customer"
JOIN "customerdetail"
  ON "customerdetail"."customer_number" =
     "customer"."customer_number"
WHERE "customer"."first_buy_date" IS NOT NULL;

CREATE OR REPLACE VIEW "office_master" AS
SELECT "office"."city",
       "office"."country",
       "office"."state",
       "office"."postal_code",
       ...
FROM "office"
WHERE "office"."city" IS NOT NULL;
```

Exactly as in the case of regular tables, jOOQ generates the corresponding records, tables, and POJOs for these views, so you'll have `CustomerMasterRecord` (a non-updatable record because the view is non-updatable) and `OfficeMasterRecord` (an updatable record because the view is updatable) in `jooq.generated.tables.records`, and `CustomerMaster` and `OfficeMaster` in `jooq.generated.tables` and `jooq.generated.tables.pojos` respectively.

Next, let's indulgently assume that a triad (`country`, `state`, and `city`) uniquely identifies a customer and an office, and we want to find customers that are in the same area as an office. For this, we can write `LEFT JOIN`, as shown in the following:

```
ctx.select(CUSTOMER_MASTER.CUSTOMER_NAME,
           CUSTOMER_MASTER.CREDIT_LIMIT,
           CUSTOMER_MASTER.CITY.as("customer_city"),
           OFFICE_MASTER.CITY.as("office_city"),
           OFFICE_MASTER.PHONE)
```

```
    .from(CUSTOMER_MASTER)
    .leftOuterJoin(OFFICE_MASTER)
    .on(row(CUSTOMER_MASTER.COUNTRY, CUSTOMER_MASTER.STATE,
            CUSTOMER_MASTER.CITY)
      .eq(row(OFFICE_MASTER.COUNTRY, OFFICE_MASTER.STATE,
              OFFICE_MASTER.CITY)))
    .orderBy(CUSTOMER_MASTER.CUSTOMER_NAME)
    .fetch();
```

Look at the `JOIN` statement's predicate! It is verbose and prone to mistakes. Moreover, if we modify (for instance, rename or remove) any of the columns involved in this predicate, then we have to adjust this predicate as well. However, there is nothing we can do, since a database view doesn't support primary/foreign keys, right? Actually, here is exactly where synthetic keys enter the scene. If jOOQ were able to give us a composite synthetic primary key for `OFFICE_MASTER` and a synthetic foreign key for `CUSTOMER_MASTER` referencing the `OFFICE_MASTER` synthetic primary key, then we could simplify and reduce the risk of mistakes in our `JOIN`. Practically, we could express our `JOIN` as an implicit `JOIN` or via `onKey()` exactly as in the case of regular tables.

However, remember that we said to *indulgently assume the uniqueness*. Note that we don't even need to make an assumption of uniqueness for the natural key (`country`, `state`, and `city`). Synthetic **primary keys/unique keys (PK/UK)** can even be used to enable some cool features for things that aren't actually candidate keys, or even unique. For example, there may be hundreds of reports that calculate stuff based on this "location relationship," and normalizing is not possible because this is a data warehouse, and so on.

Going further, jOOQ synthetic keys are shaped at the configuration level. For Maven and standalone configuration, we need the following intuitive snippet of code that defines the `office_master_pk` synthetic composite primary key and the `office_master_fk` synthetic foreign key (you should have no problem understanding this code by simply following the tag's name and its content in the context of previous database views):

```
<database>
 ...
 <syntheticObjects>
  <primaryKeys>
   <primaryKey>
    <name>office_master_pk</name>
    <tables>office_master</tables>
    <fields>
     <field>country</field>
     <field>state</field>
```

```
      <field>city</field>
     </fields>
    </primaryKey>
   </primaryKeys>
   <foreignKeys>
    <foreignKey>
     <name>office_master_fk</name>
     <tables>customer_master</tables>
     <fields>
      <field>country</field>
      <field>state</field>
      <field>city</field>
     </fields>
     <referencedTable>office_master</referencedTable>
     <referencedFields>
      <field>country</field>
      <field>state</field>
      <field>city</field>
     </referencedFields>
    </foreignKey>
   </foreignKeys>
  </syntheticObjects>
</database>
```

You can find the guidance for Gradle and the programmatic approach (which, in jOOQ style, is very intuitive as well) in the jOOQ manual.

Now, after running the jOOQ Code Generator, our `JOIN` can take advantage of the generated synthetic keys and be simplified via the synthetic `onKey()`, introduced in *Chapter 6*, *Tackling Different Kinds of JOIN Statements*. So, now we can write this:

```
ctx.select(CUSTOMER_MASTER.CUSTOMER_NAME,
           CUSTOMER_MASTER.CREDIT_LIMIT,
           CUSTOMER_MASTER.CITY.as("customer_city"),
           OFFICE_MASTER.CITY.as("office_city"),
           OFFICE_MASTER.PHONE)
   .from(CUSTOMER_MASTER)
   .leftOuterJoin(OFFICE_MASTER)
   .onKey()
   .orderBy(CUSTOMER_MASTER.CUSTOMER_NAME)
   .fetch();
```

In comparison to the previous approach, this is less verbose, less prone to mistakes, and robust against subsequent modification of the columns involved with the synthetic key. Of course, you can use `onKey()` to write `INNER JOIN` and `RIGHT JOIN` statements and so on. However, without synthetic keys, the usage of `onKey()` leads to `DataAccessException` – *No matching Key found between tables ["classicmodels"."customer_master"] and ["classicmodels"."office_master"].*

Even if `onKey()` works just fine, you'll most probably find synthetic **foreign keys (FKs)** even more powerful for implicit joins between views. Unlike `onKey()`, which can lead to ambiguities in complex `JOIN` graphs (or even in simple ones), implicit joins are always non-ambiguous.

So, sticking to `LEFT JOIN`, the previous `JOIN` can be simplified and reinforced even more by adopting an implicit `JOIN`:

```
ctx.select(CUSTOMER_MASTER.CUSTOMER_NAME,
        CUSTOMER_MASTER.CREDIT_LIMIT,
        CUSTOMER_MASTER.CITY.as("customer_city"),
        CUSTOMER_MASTER.officeMaster().CITY.as("office_city"),
        CUSTOMER_MASTER.officeMaster().PHONE)
    .from(CUSTOMER_MASTER)
    .orderBy(CUSTOMER_MASTER.CUSTOMER_NAME)
    .fetch();
```

So cool! There are no explicit columns in the join predicate, and we can modify the composite key without risks! Once we run the jOOQ Code Generator to reflect the changes, this code will work out of the box.

However, the implicit join example here might lead to a peculiar weirdness. Since this is a synthetic foreign key, and the synthetic primary key isn't actually/truly unique (we've just indulgently assumed the uniqueness), projecting an implicit join path means that we might get a Cartesian Product just from the projection, which is very surprising in SQL. A projection should never affect the cardinality of the result, but here we are... Perhaps this is a good opportunity to explore the `UNIQUE()` predicate to check whether their "candidate" key is actually unique: `https://www.jooq.org/doc/latest/manual/sql-building/conditional-expressions/unique-predicate/`.

You can practice this example in *SyntheticPkKeysImplicitJoin*.

Embedded keys for synthetic keys

Next, let's assume that we want to fetch some data from the `OFFICE_MASTER` table, based on a given set of `country`, `state`, and `city` triads. At this point, we can write this:

```
ctx.select(OFFICE_MASTER.OFFICE_CODE, OFFICE_MASTER.PHONE)
   .from(OFFICE_MASTER)
   .where(row(OFFICE_MASTER.COUNTRY, OFFICE_MASTER.STATE,
          OFFICE_MASTER.CITY).in(
      row("USA", "MA", "Boston"),
      row("USA", "CA", "San Francisco")))
   .fetch();
```

However, we know that (`country`, `state`, `city`) is actually our synthetic key. This means that if we define an embedded key for this synthetic key, then we should take advantage of embedded keys, exactly as we saw earlier in the *Working with embedded keys* section. Since the synthetic key name is `office_master_pk`, the embedded keys resume to this:

```
<embeddablePrimaryKeys>
 office_master_pk
</embeddablePrimaryKeys>
```

Rerun the jOOQ Code Generator to generate the jOOQ artifacts corresponding to this embedded key, `OfficeMasterPkRecord`, and the `OfficeMasterPk` POJO. This time, we can rewrite our query, as shown here:

```
ctx.select(OFFICE_MASTER.OFFICE_CODE, OFFICE_MASTER.PHONE)
   .from(OFFICE_MASTER)
   .where(OFFICE_MASTER.OFFICE_MASTER_PK.in(
     new OfficeMasterPkRecord("USA", "MA", "Boston"),
     new OfficeMasterPkRecord("USA", "CA", "San Francisco")))
   .fetch();
```

Alternatively, maybe we want to fetch an embedded key value in the `OfficeMasterPk` POJO:

```
List<OfficeMasterPk> result =
   ctx.select(OFFICE_MASTER.OFFICE_MASTER_PK)
      .from(OFFICE_MASTER)
      .where(OFFICE_MASTER.OFFICE_CODE.eq("1"))
      .fetchInto(OfficeMasterPk.class);
```

How about a JOIN, using explicitly OFFICE_MASTER_PK and OFFICE_MASTER_FK?

```
ctx.select(CUSTOMER_MASTER.CUSTOMER_NAME, ...)
   .from(CUSTOMER_MASTER)
   .innerJoin(OFFICE_MASTER)
     .on(OFFICE_MASTER.OFFICE_MASTER_PK
       .eq(CUSTOMER_MASTER.OFFICE_MASTER_FK))
   .orderBy(CUSTOMER_MASTER.CUSTOMER_NAME)
   .fetch();
```

Alternatively, maybe an update that has a predicate based on the embedded key:

```
ctx.update(OFFICE_MASTER)
   .set(OFFICE_MASTER.PHONE, "+16179821809")
   .where(OFFICE_MASTER.OFFICE_MASTER_PK.eq(
       new OfficeMasterPkRecord("USA", "MA", "Boston")))
   .execute();
```

You can practice these examples in *EmbeddedSyntheticKeys* for PostgreSQL.

Using navigation methods

Furthermore, if we inspect the generated jooq.generated.Keys, we notice the following generated keys for OFFICE_MASTER and CUSTOMER_MASTER:

```
UniqueKey<OfficeMasterRecord> OFFICE_MASTER_PK = ...
ForeignKey<CustomerMasterRecord, OfficeMasterRecord>
             CUSTOMER_MASTER__OFFICE_MASTER_FK = ...
```

These keys are quite useful in conjunction with jOOQ navigation methods – fetchParent(), fetchChildren(), fetchChild(), and so on. These methods were introduced in *Chapter 9, CRUD, Transactions, and Locking*, and here are two examples of using them to navigate our views:

```
CustomerMasterRecord cmr = ctx.selectFrom(CUSTOMER_MASTER)
 .where(CUSTOMER_MASTER.CUSTOMER_NAME
  .eq("Classic Legends Inc.")).fetchSingle();

OfficeMasterRecord parent = cmr.fetchParent(
 Keys.CUSTOMER_MASTER__OFFICE_MASTER_FK);

List<CustomerMasterRecord> children =
 parent.fetchChildren(Keys.CUSTOMER_MASTER__OFFICE_MASTER_FK);
```

You can practice these examples in *SyntheticPkKeysNavigation* for PostgreSQL.

Synthetic unique keys

In the previous section, we used a composite synthetic primary key built on the triad `country`, `state`, and `city`. However, if we look carefully, we notice that both views select `postal_code` as well. Since we don't have two offices in the same city, we can consider that `postal_code` (which has `CONSTRAINT "office_postal_code_uk" UNIQUE ("postal_code")` in the `office` table) is a unique key for `office_master` (of course, in reality, you have to pay attention to such assumptions; maybe the best way to represent an address is via `BLOB`, but let's continue with what we have). This means that we can use a synthetic unique key as well. By simply replacing the `<primaryKeys/>` tag with the `<uniqueKeys/>` tag, as shown here, we set up `postal_code` as a synthetic unique key:

```
<syntheticObjects>
 <uniqueKeys>
  <uniqueKey>
   <name>office_master_uk</name>
   <tables>office_master</tables>
   <fields>
    <field>postal_code</field>
   </fields>
  </uniqueKey>
 </uniqueKeys>
 <foreignKeys>
  <foreignKey>
   <name>customer_office_master_fk</name>
   <tables>customer_master</tables>
   <fields>
    <field>postal_code</field>
   </fields>
   <referencedTable>office_master</referencedTable>
   <referencedFields>
    <field>postal_code</field>
   </referencedFields>
  </foreignKey>
 </foreignKeys>
</syntheticObjects>
```

The good news is that our `JOIN` statements that rely on synthetic keys will work out of the box, even if we switched from a composite synthetic primary key to a simple synthetic unique key. The bundled code is *SyntheticUniqueKeysImplicitJoin* for PostgreSQL.

Synthetic identities

As you saw earlier, jOOQ can fetch an identity primary key after executing an insert (via `insertInto()...returningResult(pk)` or inserting an updatable record). However, not all identity columns must be primary keys as well. For instance, our `PRODUCT` table from PostgreSQL has two identity columns – one is also the primary key (`PRODUCT_ID`), while the second one is just a simple identity column (`PRODUCT_UID`):

```
CREATE TABLE "product" (
  "product_id" BIGINT
               NOT NULL DEFAULT NEXTVAL ('"product_seq"'),
  ...
  "product_id" BIGINT GENERATED BY DEFAULT AS IDENTITY
                      (START WITH 10 INCREMENT BY 10),
  CONSTRAINT "product"pk" PRIMARY KEY ("product_id"),
  ...
) ;
```

Fetching both identities via `insertInto() ... returningResult(pk)` can be done quite easily:

```
var result = ctx.insertInto(PRODUCT)
  .set(PRODUCT.PRODUCT_LIN", "Vintage Cars")
  .set(PRODUCT.CODE, 223113L)
  .set(PRODUCT.PRODUCT_NAME, "Rolls-Royce Dawn Drophead")
  .returningResult(PRODUCT.PRODUCT_ID, PRODUCT.PRODUCT_UID)
  .fetch();

result.get(0).value1(); // valid primary key (PRODUCT_ID)
result.get(0).value2(); // valid identity key (PRODUCT_UID)
```

There is no surprise here, since `returningResult()` instructs jOOQ to return all columns enlisted as an argument. However, inserting a record represents a more interesting case:

```
ProductRecord pr = ctx.newRecord(PRODUCT);
pr.setProductLine("Classic Cars");
pr.setCode(599302L);
```

```
pr.setProductName("1967 Chevrolet Camaro RS");

pr.insert();

pr.getProductId();  // valid primary key (PRODUCT_ID)
pr.getProductUid(); // valid identity key (PRODUCT_UID) WOW!
```

That's cool! Besides the identity primary key, jOOQ has also populated the record with the database-generated `PRODUCT_UID`. So, as long as the database reports a column as being an identity, jOOQ can detect it and act accordingly.

Okay, let's next focus on our Oracle schema that defines the `PRODUCT` table, like this:

```
CREATE TABLE product (
  product_id NUMBER(10) DEFAULT product_seq.nextval
                      NOT NULL,
  ...
  product_uid NUMBER(10) DEFAULT product_uid_seq.nextval
                      NOT NULL,
  CONSTRAINT product_pk PRIMARY KEY (product_id),
  ...
);

CREATE SEQUENCE product_seq START WITH 1000000 INCREMENT BY 1;
CREATE SEQUENCE product_uid_seq START WITH 10 INCREMENT BY 10;
```

In this scenario, `insertInto()` … `returningResult()` works as excepted, but after inserting a `ProductRecord`, we get back only the identity primary key (`PRODUCT_ID`), while calling `getProductUid()` will return `null`. In other words, jOOQ detected only `PRODUCT_ID` as being a primary key column, while `PRODUCT_UID` was not reported by the database as being an identity column. However, here is where the jOOQ synthetic identities come to the rescue. Synthetic identities allow us to configure jOOQ to treat as formal identities those columns that are not reported by the database as being identities. In this particular case, `PRODUCT_UID` falls under this umbrella, so here is the jOOQ-expected configuration for Maven (and standalone):

```
<syntheticObjects>
 <identities>
  <identity>
   <tables>product</tables>
   <fields>product_uid</fields>
```

```
  </identity>
 </identities>
</syntheticObjects>
```

If you have multiple tables/identities, then enlist them, separated by | as regular expressions. This time, after running the Code Generator and inserting a new `ProductRecord`, jOOQ fetches both `PRODUCT_ID` (check it via `getProductId()`) and `PRODUCT_UID` (check it via `getProductUid()`, which should return a valid integer). Moreover, this works for Oracle versions where formal identity columns are emulated using sequences and triggers (prior to Oracle 12c). So, another cool feature of jOOQ has been revealed.

The bundled code samples are *DetectIdentity* (for PostgreSQL) and *SyntheticIdentity* (for Oracle).

Hooking computed columns

A *computed column* is a column that cannot be written to. Its value is computed from a given *expression*. When a column is computed on read (for instance, in `SELECT` statements) it is known as a `VIRTUAL` column (in DDL, such columns appear roughly expressed as ... `GENERATED ALWAYS AS <expression> VIRTUAL`). Typically, `VIRTUAL` columns don't exist/appear in the database schema. On the other hand, a column that is computed on write (for instance, in `INSERT`, `UPDATE`, `MERGE` statements) is known as a `STORED` column (in DDL, some common syntax is `... GENERATED ALWAYS AS <expression> STORED`). Such columns exist/appear in your database schema.

Server side computed columns

In this context, jOOQ 3.16 added support for *server side computed columns*. jOOQ 's Code Generator is capable of detecting server side computed columns and marking them as *read-only* (`https://www.jooq.org/doc/latest/manual/code-generation/codegen-advanced/codegen-config-database/codegen-database-readonly-columns/`). In other words, for your convenience, such columns are automatically excluded from DML statements and occur only in `SELECT` statements. However, jOOQ allows us to fine tune read-only columns via a bunch of settings available at `https://www.jooq.org/doc/latest/manual/sql-building/column-expressions/readonly-columns/`. Moreover, jOOQ supports *synthetic read-only columns*, which are recognized by jOOQ if we configure them via `<readonlyColumns/>`, `<readonlyColumn/>` tags (Maven). You can explore this very exciting topic in jOOQ documentation at `https://www.jooq.org/doc/latest/manual/code-generation/codegen-advanced/codegen-config-`

`database/codegen-database-synthetic-objects/codegen-database-synthetic-readonly-columns/`, but for now, let's get back to the computed columns topic.

So, not all dialects support server side computed columns or expressions based on scalar subqueries (even correlated ones), or implicit joins. jOOQ 3.17 comes with a powerful feature that covers these limitations, and this feature is known as *client side computed columns*.

Client side computed columns

Check out the following configuration of a client side computed column (`VIRTUAL` alike):

```
<database>
 …
 <!-- Prepare the synthetic keys -->
 <syntheticObjects>
  <columns>
   <column>
    <name>REFUND_AMOUNT</name>
    <tables>BANK_TRANSACTION</tables>
    <type>DECIMAL(10,2)</type>
   </column>
  </columns>
 </syntheticObjects>

 <!-- Now tell the code generator
      how to compute the values -->
 <forcedTypes>
  <forcedType>
   <generator>
    ctx -> payment().INVOICE_AMOUNT.minus(
           DSL.sum(TRANSFER_AMOUNT))
   </generator>
   <includeExpression>REFUND_AMOUNT</includeExpression>
  </forcedType>
 </forcedTypes>
</database>
```

Because the forced type matches a synthetic column (`REFUND_AMOUNT`), jOOQ semantics stands for a `VIRTUAL` computed column. So, the column does not exist in the database schema, but the computation (here, an implicit join, but correlated subqueries is also a supported option) will be automatically present in all of your `SELECT` statements containing this column. In the bundled code available for SQL Server, *SyntheticComputedColumns*, you can see a query sample that uses the virtual column, `BANK_TRANSACTION.REFUND_AMOUNT`.

Now, check this out:

```
<database>
 <!-- Tell the code generator how to
      compute the values for an existing column -->
 <forcedTypes>
  <forcedType>
   <generator>
     ctx -> DSL.concat(OFFICE.COUNTRY, DSL.inline(", "),
            OFFICE.STATE, DSL.inline(", "), OFFICE.CITY)
   </generator>
   <includeExpression>
    CFFICE.ADDRESS_LINE_FIRST
   </includeExpression>
  </forcedType>
 </forcedTypes>
</database>
```

This time, the forced type matches an actual column (`OFFICE.ADDRESS_LINE_FIRST`), so jOOQ applies the semantics of a `STORED` computed column. In other words, the DML statements will be transformed to reflect the correct computation of the value, which will be written to your schema. You can check out an example in the bundled code, *StoredComputedColumns*, available for SQL Server. Moreover, if you are able to, take the time to read this great article: `https://blog.jooq.org/create-dynamic-views-with-jooq-3-17s-new-virtual-client-side-computed-columns/`.

Overriding primary keys

Let's consider the following schema fragment (from PostgreSQL):

```
CREATE TABLE "customer" (
  "customer_number" BIGINT NOT NULL
                         DEFAULT NEXTVAL ('"customer_seq"'),
  "customer_name" VARCHAR(50) NOT NULL,
```

```
  ...
  CONSTRAINT "customer_pk" PRIMARY KEY ("customer_number"),
  CONSTRAINT "customer_name_uk" UNIQUE ("customer_name")
  ...
);

CREATE TABLE "department" (
  "department_id" SERIAL NOT NULL,
  "code" INT NOT NULL,
  ...
  CONSTRAINT "department_pk" PRIMARY KEY ("department_id"),
  CONSTRAINT "department_code_uk" UNIQUE ("code")
  ...
);
```

The following is an example of updating a `CUSTOMER`:

```
CustomerRecord cr = ctx.selectFrom(CUSTOMER)
  .where(CUSTOMER.CUSTOMER_NAME.eq("Mini Gifts ..."))
  .fetchSingle();

cr.setPhone("4159009544");
cr.store();
```

Here is an example of updating a `DEPARTMENT`:

```
DepartmentRecord dr = ctx.selectFrom(DEPARTMENT)
  .where(DEPARTMENT.DEPARTMENT_ID.eq(1))
  .fetchSingle();

dr.setTopic(new String[] {"promotion", "market", "research"});
dr.store();
```

From *Chapter 9, CRUD, Transaction, and Locking*, we know how `store()` works; therefore, we know that the generated SQLs will rely on the `CUSTOMER` primary key and the `DEPARTMENT` primary key (the same behavior applies to `update()`, `merge()`, `delete()`, and `refresh()`). For instance, `cr.store()` executes the following `UPDATE`:

```
UPDATE "public"."customer" SET "phone" = ?
WHERE "public"."customer"."customer_number" = ?
```

Since CUSTOMER_NUMBER is the primary key of CUSTOMER, jOOQ uses it for appending the predicate to this UPDATE.

On the other hand, dr.store() executes this UPDATE:

```
UPDATE "public"."department" SET "topic" = ?::text[]
WHERE ("public"."department"."name" = ?
       AND "public"."department"."phone" = ?)
```

Something doesn't look right here, since our schema reveals that the primary key of DEPARTMENT is DEPARTMENT_ID, so why does jOOQ use here a composite predicate containing DEPARTMENT_NAME and DEPARTMENT_PHONE? This may look confusing, but the answer is quite simple. We actually defined a synthetic primary key (department_name and department_phone), which we reveal here:

```
<syntheticObjects>
 <primaryKeys>
  <primaryKey>
   <name>synthetic_department_pk</name>
   <tables>department</tables>
   <fields>
    <field>name</field>
    <field>phone</field>
   </fields>
  </primaryKey>
 </primaryKeys>
</syntheticObjects>
```

That's cool! So, jOOQ has used the synthetic key in place of the schema primary key. We can say that we overrode the scheme's primary key with a synthetic key.

Let's do it again! For instance, let's suppose that we want to instruct jOOQ to use the customer_name unique key for cr.store() and the code unique key for dr.store(). This means that we need the following configuration:

```
<syntheticObjects>
 <primaryKeys>
  <primaryKey>
   <name>synthetic_customer_name</name>
   <tables>customer</tables>
   <fields>
    <field>customer_name</field>
   </fields>
```

```
  </primaryKey>
  <primaryKey>
   <name>synthetic_department_code</name>
   <tables>department</tables>
   <fields>
    <field>code</field>
   </fields>
  </primaryKey>
 </primaryKeys>
</syntheticObjects>
```

This configuration overrides the schema defaults, and the generated SQL becomes the following:

```
UPDATE "public"."customer" SET "phone" = ?
WHERE "public"."customer"."customer_name" = ?

UPDATE "public"."department" SET "topic" = ?::text[]
WHERE "public"."department"."code" = ?
```

Cool, right?! The complete example is named *OverridePkKeys* for PostgreSQL.

Summary

I hope you enjoyed this short but comprehensive chapter about jOOQ keys. The examples from this chapter covered popular aspects of dealing with different kinds of keys, from unique/primary keys to jOOQ-embedded and synthetic keys. I really hope that you don't stop at these examples and get curious to deep dive into these amazing jOOQ features – for instance, an interesting topic that deserves your attention is read-only columns: `https://www.jooq.org/doc/dev/manual/sql-building/column-expressions/readonly-columns/`.

In the next chapter, we will tackle pagination and dynamic queries.

12
Pagination and Dynamic Queries

In this chapter, we'll talk about pagination and dynamic queries in jOOQ, two topics that work hand in hand in a wide range of applications to paginate and filter lists of products, items, images, posts, articles, and so on.

In this chapter, we will be covering the following topics:

- jOOQ offset pagination
- jOOQ keyset pagination
- Writing dynamic queries
- Infinite scrolling and dynamic filters

Let's get started!

Technical requirements

The code for this chapter can be found on GitHub at `https://github.com/PacktPublishing/jOOQ-Masterclass/tree/master/Chapter12`.

Offset and keyset pagination

Offset and keyset pagination (or seek, as Markus Winand calls it) represent two well-known techniques for paginating data while fetching it from the database. Offset pagination is quite popular because Spring Boot (more precisely, Spring Data Commons) provides two default implementations for it, via the `Page` and `Slice` APIs. Therefore, in terms of productivity, it's very convenient to rely on these implementations. However, in terms of performance, while your project evolves and data keeps accumulating, relying on offset pagination may lead to serious performance degradations. Nevertheless, as you'll see soon, jOOQ can help you to sweeten the situation.

Conversely, keyset pagination is a technique that sustains high performance, being faster and more stable than offset pagination. Keyset pagination really shines on paginating large datasets and infinite scrolling, while it leads to almost the same performance as offset pagination, especially for relatively small datasets. However, can you guarantee that the amount of data will not grow over time (sometimes, quite fast)? If yes, then using offset pagination should be okay. Otherwise, it is better to prevent this well-known performance issue right from the start and rely on keyset pagination. Don't think that this is premature optimization; think of it as the capability to make the right decisions depending on the business case you are modeling. And, as you'll see soon, jOOQ makes keyset pagination usage child's play.

Index scanning in offset and keyset

Dealing with offset pagination means you can ignore the performance penalty induced by throwing away *n* records before reaching the desired offset. A larger *n* leads to a significant performance penalty that equally affects both the `Page` and `Slice` APIs. Another penalty is the extra `SELECT COUNT` needed to count the total number of records. This extra `SELECT COUNT` is specific to the `Page` API only, so it doesn't affect the `Slice` API. Basically, this is the main difference between the `Page` and `Slice` APIs; the former contains the total number of records (useful for computing the total number of pages), while the latter can only tell whether there is at least one more page available or this is the last page.

Lukas Eder has a very nice observation here: *"Another way I tend to think about this is: What's the business value of being able to jump to page 2712? What does that page number even mean? Shouldn't the search be refined with better filters instead? On the other hand, jumping to the *next* page from any page is a very common requirement."*

An index scan in offset will traverse the range of indexes from the beginning to the specified offset. Basically, the offset represents the number of records that must be skipped before including them in the result set. So, the offset approach will traverse the already shown records, as shown in the following figure:

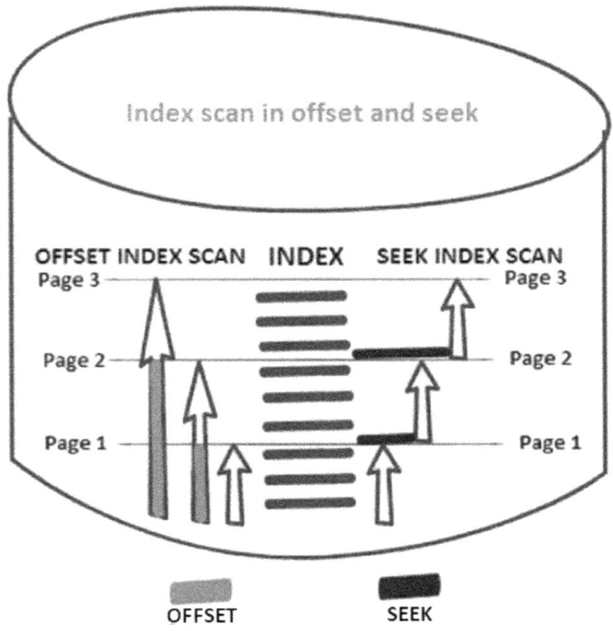

Figure 12.1 – An index scan in offset and keyset pagination

On the other hand, the index scan in the keyset pagination will traverse only the required values, starting with the last previous value (it skips the values until the last value previously fetched). In the keyset, the performance remains approximately constant in relation to the increase of the table records.

> **Important Note**
>
> An important reference and compelling argument against using offset pagination is mentioned on the *USE THE INDEX, LUKE!* website (`https://use-the-index-luke.com/no-offset`). I strongly suggest you take some time and watch this great presentation by Markus Winand (`https://www.slideshare.net/MarkusWinand/p2d2-pagination-done-the-postgresql-way?ref=https://use-the-index-luke.com/no-offset`), which covers important topics for tuning pagination-SQL, such as using indexes and row values (supported in PostgreSQL) in offset and keyset pagination.

Okay, after this summary of offset versus keyset pagination, let's see how jOOQ can mimic and even improve the default Spring Boot offset pagination implementation.

jOOQ offset pagination

Spring Boot implementation of offset pagination can be easily shaped via `LIMIT ... OFFSET` (or `OFFSET ... FETCH`) and `SELECT COUNT`. For instance, if we assume that a client gives us a page number (via the `page` argument) and `size` (the number of products to be displayed on `page`), then the following jOOQ query mimics the default Spring Boot pagination behavior for the `PRODUCT` table:

```
long total = ctx.fetchCount(PRODUCT);

List<Product> result = ctx.selectFrom(PRODUCT)
   .orderBy(PRODUCT.PRODUCT_ID)
   .limit(size)
   .offset(size * page)
   .fetchInto(Product.class);
```

If the client (for instance, the browser) expects as a response a serialization of the classical `Page<Product>` (`org.springframework.data.domain.Page`), then you can simply produce it, as shown here:

```
Page<Product> pageOfProduct = new PageImpl(result,
  PageRequest.of(page, size, Sort.by(Sort.Direction.ASC,
    PRODUCT.PRODUCT_ID.getName())), total);
```

However, instead of executing two `SELECT` statements, we can use the `COUNT()` window function to obtain the same result but with a single `SELECT`:

```
Map<Integer, List<Product>> result = ctx.select(
    PRODUCT.asterisk(), count().over().as("total"))
  .from(PRODUCT)
  .orderBy(PRODUCT.PRODUCT_ID)
  .limit(size)
  .offset(size * page)
  .fetchGroups(field("total", Integer.class), Product.class);
```

This is already better than the default Spring Boot implementation. When you return this `Map<Integer, List<Product>>` to the client, you can return a `Page<Product>` as well:

```
Page<Product> pageOfProduct
  = new PageImpl(result.values().iterator().next(),
    PageRequest.of(page, size, Sort.by(Sort.Direction.ASC,
       PRODUCT.PRODUCT_ID.getName())),
         result.entrySet().iterator().next().getKey());
```

Most probably, you'll prefer `Page` because it contains a set of metadata, such as the total number of records if we hadn't paginated (`totalElements`), the current page we're on (`pageNumber`), the actual page size (`pageSize`), the actual offsets of the returned rows (`offset`), and whether we are on the last page (`last`). You can find the previous examples in the bundled code as *PaginationCountOver*.

However, as Lukas Eder highlights in this article (`https://blog.jooq.org/2021/03/11/calculating-pagination-metadata-without-extra-roundtrips-in-sql/`), all this metadata can be obtained in a single SQL query; therefore, there is no need to create a `Page` object to have them available. In the bundled code (*PaginationMetadata*) you can practice Lukas's dynamic query via a REST controller.

jOOQ keyset pagination

Keyset (or seek) pagination doesn't have a default implementation in Spring Boot, but this shouldn't stop you from using it. Simply start by choosing a table's column that should act as the latest visited record/row (for instance, the id column), and use this column in the `WHERE` and `ORDER BY` clauses. The idioms relying on the ID column are as follows (sorting by multiple columns follows this same idea):

```
SELECT ... FROM ...
WHERE id < {last_seen_id}
ORDER BY id DESC
LIMIT {how_many_rows_to_fetch}

SELECT ... FROM ...
WHERE id > {last_seen_id}
ORDER BY id ASC
LIMIT {how_many_rows_to_fetch}
```

Or, like this:

```
SELECT ... FROM ...
WHERE ... AND id < {last_seen_id}
ORDER BY id DESC
LIMIT {how_many_rows_to_fetch}

SELECT ... FROM ...
WHERE ... AND id > {last_seen_id}
ORDER BY id ASC
LIMIT {how_many_rows_to_fetch}
```

Based on the experience gained so far, expressing these queries in jOOQ should be a piece of cake. For instance, let's apply the first idiom to the `PRODUCT` table via `PRODUCT_ID`:

```
List<Product> result = ctx.selectFrom(PRODUCT)
   .where(PRODUCT.PRODUCT_ID.lt(productId))
   .orderBy(PRODUCT.PRODUCT_ID.desc())
   .limit(size)
   .fetchInto(Product.class);
```

In MySQL, the rendered SQL is (where `productId` = `20` and `size` = `5`) as follows:

```
SELECT `classicmodels`.`product`.`product_id`,
       `classicmodels`.`product`.`product_name`,
       ...
FROM `classicmodels`.`product`
WHERE `classicmodels`.`product`.`product_id` < 20
ORDER BY `classicmodels`.`product`.`product_id` DESC
LIMIT 5
```

This was easy! You can practice this case in *KeysetPagination*.

However, keyset pagination becomes a little bit trickier if the `WHERE` clause becomes more complicated. Fortunately, jOOQ saves us from this scenario via a synthetic clause named `SEEK`. Let's dive into it!

The jOOQ SEEK clause

The jOOQ synthetic SEEK clause simplifies the implementation of keyset pagination. Among its major advantages, the SEEK clause is type-safe and is capable of generating/emulating the correct/expected WHERE clause (including the emulation of row value expressions).

For instance, the previous keyset pagination example can be expressed using the SEEK clause, as shown here (productId is provided by the client):

```
List<Product> result = ctx.selectFrom(PRODUCT)
  .orderBy(PRODUCT.PRODUCT_ID)
  .seek(productId)
  .limit(size)
  .fetchInto(Product.class);
```

Note that there is no explicit WHERE clause. jOOQ will generate it on our behalf, based on the seek() arguments. While this example may not look so impressive, let's consider another one. This time, let's paginate EMPLOYEE using the employee's office code and salary:

```
List<Employee> result = ctx.selectFrom(EMPLOYEE)
  .orderBy(EMPLOYEE.OFFICE_CODE, EMPLOYEE.SALARY.desc())
  .seek(officeCode, salary)
  .limit(size)
  .fetchInto(Employee.class);
```

Both officeCode and salary are provided by the client, and they land into the following generated SQL sample (where officeCode = 1, salary = 75000, and size = 10):

```
SELECT `classicmodels`.`employee`.`employee_number`,
  ...
FROM `classicmodels`.`employee`
WHERE (`classicmodels`.`employee`.`office_code` > '1'
   OR (`classicmodels`.`employee`.`office_code` = '1'
     AND `classicmodels`.`employee`.`salary` < 75000))
ORDER BY `classicmodels`.`employee`.`office_code`,
         `classicmodels`.`employee`.`salary` DESC
LIMIT 10
```

Check out the generated `WHERE` clause! I am pretty sure that you don't want to get your hands dirty and explicitly write this clause. How about the following example?

```
List<Orderdetail> result = ctx.selectFrom(ORDERDETAIL)
  .orderBy(ORDERDETAIL.ORDER_ID, ORDERDETAIL.PRODUCT_ID.desc(),
          ORDERDETAIL.QUANTITY_ORDERED.desc())
  .seek(orderId, productId, quantityOrdered)
  .limit(size)
  .fetchInto(Orderdetail.class);
```

And the following code is a sample of the generated SQL (where `orderId` = `10100`, `productId` = `23`, `quantityOrdered` = `30`, and `size` = `10`):

```
SELECT `classicmodels`.`orderdetail`.`orderdetail_id`,
      ...
FROM `classicmodels`.`orderdetail`
WHERE (`classicmodels`.`orderdetail`.`order_id` > 10100
  OR (`classicmodels`.`orderdetail`.`order_id` = 10100
  AND `classicmodels`.`orderdetail`.`product_id` < 23)
  OR (`classicmodels`.`orderdetail`.`order_id` = 10100
  AND `classicmodels`.`orderdetail`.`product_id` = 23
  AND `classicmodels`.`orderdetail`.`quantity_ordered` < 30))
ORDER BY `classicmodels`.`orderdetail`.`order_id`,
         `classicmodels`.`orderdetail`.`product_id` DESC,
         `classicmodels`.`orderdetail`.`quantity_ordered` DESC
LIMIT 10
```

After this example, I think is obvious that you should opt for the `SEEK` clause and let jOOQ do its job! Look, you can even do this:

```
List<Product> result = ctx.selectFrom(PRODUCT)
  .orderBy(PRODUCT.BUY_PRICE, PRODUCT.PRODUCT_ID)
  .seek(PRODUCT.MSRP.minus(PRODUCT.MSRP.mul(0.35)),
      val(productId))
  .limit(size)
  .fetchInto(Product.class);
```

You can practice these examples in *SeekClausePagination*, next to the other examples, including using jOOQ-embedded keys as arguments of the `SEEK` clause.

Implementing infinite scroll

Infinite scroll is a classical usage of keyset pagination and is gaining popularity these days. For instance, let's assume that we plan to obtain something, as shown in this figure:

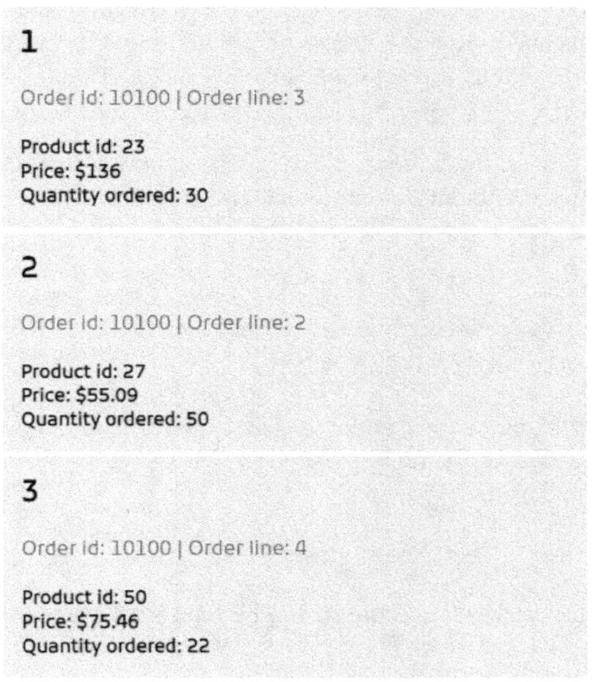

Figure 12.2 – An infinite scroll

So, we want an infinite scroll over the `ORDERDETAIL` table. At each scroll, we fetch the next *n* records via the `SEEK` clause:

```
public List<Orderdetail> fetchOrderdetailPageAsc(
         long orderdetailId, int size) {

  List<Orderdetail> result = ctx.selectFrom(ORDERDETAIL)
    .orderBy(ORDERDETAIL.ORDERDETAIL_ID)
    .seek(orderdetailId)
```

```
      .limit(size)
      .fetchInto(Orderdetail.class);

   return result;
}
```

This method gets the last visited `ORDERDETAIL_ID` and the number of records to fetch (`size`), and it returns a list of `jooq.generated.tables.pojos.Orderdetail`, which will be serialized in JSON format via a Spring Boot REST controller endpoint defined as `@GetMapping("/orderdetail/{orderdetailId}/{size}")`.

On the client side, we rely on the JavaScript Fetch API (of course, you can use XMLHttpRequest, jQuery, AngularJS, Vue, React, and so on) to execute an HTTP `GET` request, as shown here:

```
const postResponse
   = await fetch('/orderdetail/${start}/${size}');
const data = await postResponse.json();
```

For fetching exactly three records, we replace `${size}` with `3`. Moreover, the `${start}` placeholder should be replaced by the last visited `ORDERDETAIL_ID`, so the `start` variable can be computed as the following:

```
start = data[size-1].orderdetailId;
```

While scrolling, your browser will execute an HTTP request at every three records, as shown here:

```
http://localhost:8080/orderdetail/0/3
http://localhost:8080/orderdetail/3/3
http://localhost:8080/orderdetail/6/3
...
```

You can check out this example in *SeekInfiniteScroll*. Next, let's see an approach for paginating `JOIN` statements.

Paginating JOINs via DENSE_RANK()

Let's assume that we want to paginate offices (`OFFICE`) with employees (`EMPLOYEE`). If we apply a classical offset or keyset pagination to the `JOIN` between `OFFICE` and `EMPLOYEE`, then the result is prone to be truncated. Therefore, an office can be fetched with only a subset of its employees. For instance, while we think of a result page of size 3 as containing three offices with all their employees, we instead get a single office with three employees (even if this office has more employees). The following figure reveals what we expect versus what we get from a page of size 3 (offices):

Figure 12.3 – Join pagination

In order to obtain a result set like the one on the left-hand side of the preceding figure, we can rely on the `DENSE_RANK()` window function, which assigns a sequential number to different values of `a` within each group `b`, as shown in the following query:

```
Map<Office, List<Employee>> result = ctx.select().from(
  select(OFFICE.OFFICE_CODE, OFFICE...,
         EMPLOYEE.FIRST_NAME, EMPLOYEE...,
         denseRank().over().orderBy(
             OFFICE.OFFICE_CODE, OFFICE.CITY).as("rank"))
  .from(OFFICE)
  .join(EMPLOYEE)
  .on(OFFICE.OFFICE_CODE.eq(EMPLOYEE.OFFICE_CODE)).asTable("t"))
  .where(field(name("t", "rank")).between(start, end))
  .fetchGroups(Office.class, Employee.class);
```

The `start` and `end` variables represent the range of offices set via `DENSE_RANK()`. The following figure should clarify this aspect where `start` = `1` and `end` = `3` (the next page of three offices is between `start` = `4` and `end` = `6`):

office_code	rank	city	country	territory	state	first_name	last_name	employee_number	job_title	salary
1	1	San Francisco	USA	NA	CA	Gerard	Hernandez	1370	Sales Rep	65000
1	1	San Francisco	USA	NA	CA	Pamela	Castillo	1401	Sales Rep	55000
1	1	San Francisco	USA	NA	CA	Martin	Gerard	1702	Sales Rep	50000
1	1	San Francisco	USA	NA	CA	Mami	Nishi	1621	Sales Rep	55000
1	1	San Francisco	USA	NA	CA	Yoshimi	Kato	1625	Sales Rep	60000
1	1	San Francisco	USA	NA	CA	William	Patterson	1088	Sales Manager (AP...	80000
2	2	Boston	USA	NA	MA	Andy	Fixter	1611	Sales Rep	50000
2	2	Boston	USA	NA	MA	Peter	Marsh	1612	Sales Rep	55000
3	3	NYC	USA	NA	NY	Tom	King	1619	Sales Rep	60000
3	3	NYC	USA	NA	NY	Larry	Bott	1501	Sales Rep	50000

Figure 12.4 – The DENSE_RANK() effect

Here is a more compact version of the previous query, using the `QUALIFY` clause:

```
Map<Office, List<Employee>> result =
  ctx.select(OFFICE.OFFICE_CODE, OFFICE...,
             EMPLOYEE.FIRST_NAME, EMPLOYEE...)
     .from(OFFICE)
     .join(EMPLOYEE)
     .on(OFFICE.OFFICE_CODE.eq(EMPLOYEE.OFFICE_CODE))
     .qualify(denseRank().over()
        .orderBy(OFFICE.OFFICE_CODE, OFFICE.CITY)
        .between(start, end))
     .fetchGroups(Office.class, Employee.class);
```

You can check out these examples (offset, keyset, and `DENSE_RANK()` queries) via three REST controller endpoints in *DenseRankPagination* for MySQL. In all these cases, the returned `Map<Office, List<Employee>>` is serialized to JSON.

Paginating database views via ROW_NUMBER()

Let's consider the following figure, representing a snapshot of a database view named `PRODUCT_MASTER` (it could be a regular table as well):

product_line	product_name	product_scale
Classic Cars	1956 Porsche 356A Coupe	1:18
Classic Cars	1992 Porsche Cayenne Turbo S...	1:24
Classic Cars	1961 Chevrolet Impala	1:18
Classic Cars	1982 Camaro Z28	1:18
Motorcycles	1969 Harley Davidson Ultimate ...	1:10
Motorcycles	1996 Moto Guzzi 1100i	1:10
Motorcycles	2003 Harley-Davidson Eagle Dr...	1:10
Motorcycles	2002 Suzuki XREO	1:12
Motorcycles	1936 Harley Davidson El Knuckl...	1:18
Motorcycles	1957 Vespa GS150	1:18
Motorcycles	1997 BMW R 1100 S	1:24

Figure 12.5 – The PRODUCT_MASTER database view

Next, we want to paginate this view by PRODUCT_LINE (first column), so we have to take into account the fact that PRODUCT_LINE contains duplicates. While this is not an issue for offset pagination, it could produce weird results for keyset pagination relying only on PRODUCT_LINE and LIMIT (or counterparts) clauses. We can eliminate this issue by using the (PRODUCT_LINE, PRODUCT_NAME) combo in ORDER BY and WHERE predicate. This will work as expected because PRODUCT_NAME contains unique values.

However, let's try another approach, relying on the ROW_NUMBER() window function. This function assigns a database temporary sequence of values to rows. More precisely, the ROW_NUMBER() window function produces a sequence of values that starts from value 1 with an increment of 1. This is a temporary sequence of values (non-persistent) that is calculated dynamically at query execution time.

Based on ROW_NUMBER(), we can visualize PRODUCT_MASTER, as shown in this figure:

product_line	product_name	product_scale	rowNum
Classic Cars	1956 Porsche 356A Coupe	1:18	1
Classic Cars	1992 Porsche Cayenne Turbo S...	1:24	2
Classic Cars	1961 Chevrolet Impala	1:18	3
Classic Cars	1982 Camaro Z28	1:18	4
Motorcycles	1969 Harley Davidson Ultimate ...	1:10	5
Motorcycles	1996 Moto Guzzi 1100i	1:10	6
Motorcycles	2003 Harley-Davidson Eagle Dr...	1:10	7
Motorcycles	2002 Suzuki XREO	1:12	8
Motorcycles	1936 Harley Davidson El Knuckl...	1:18	9
Motorcycles	1957 Vespa GS150	1:18	10
Motorcycles	1997 BMW R 1100 S	1:24	11

Figure 12.6 – The ROW_NUMBER() effect

In this context, expressing pagination can be done as follows:

```
var result = ctx.select().from(
    select(PRODUCT_MASTER.PRODUCT_LINE,
      PRODUCT_MASTER.PRODUCT_NAME, PRODUCT_MASTER.PRODUCT_SCALE,
        rowNumber().over().orderBy(
          PRODUCT_MASTER.PRODUCT_LINE).as("rowNum"))
  .from(PRODUCT_MASTER).asTable("t"))
  .where(field(name("t", "rowNum")).between(start, end))
  .fetchInto(ProductMaster.class);
```

Alternatively, we can make it more compact via the `QUALIFY` clause:

```
var result = ctx.select(PRODUCT_MASTER.PRODUCT_LINE,
      PRODUCT_MASTER.PRODUCT_NAME, PRODUCT_MASTER.PRODUCT_SCALE)
  .from(PRODUCT_MASTER)
  .qualify(rowNumber().over()
     .orderBy(PRODUCT_MASTER.PRODUCT_LINE).between(start, end))
  .fetchInto(ProductMaster.class);
```

Fetching the first page of size 5 can be done via `start` = `1` and `end` = `5`. Fetching the next page of size 5 can be done via `start` = `6` and `end` = `10`. The complete example is available in the bundler code as *RowNumberPagination* for MySQL.

Okay, that's enough about pagination. Next, let's tackle dynamic queries (filters).

Writing dynamic queries

Commonly, a dynamic query contains no or some fixed parts and some other parts that can be appended at runtime to form a query that corresponds to a certain scenario or use case.

> **Important Note**
>
> In jOOQ, even when they look like static queries (due to jOOQ's API design), every SQL is dynamic; therefore, it can be broken up into query parts that can be fluently glued back in any valid jOOQ query. We already have covered this aspect in *Chapter 3, jOOQ Core Concepts*, in the *Understanding the jOOQ fluent API* section.

Dynamically creating an SQL statement on the fly is one of the favorite topics of jOOQ, so let's try to cover some approaches that can be useful in real applications.

Using the ternary operator

The Java ternary operator (`?`) is probably the simplest approach for shaping a query at runtime. Check out this sample:

```
public List<ProductRecord> fetchCarsOrNoCars(
       float buyPrice, boolean cars) {

   return ctx.selectFrom(PRODUCT)
    .where((buyPrice > 0f ? PRODUCT.BUY_PRICE.gt(
        BigDecimal.valueOf(buyPrice)) : noCondition())
      .and(cars ? PRODUCT.PRODUCT_LINE.in("Classic Cars",
        "Motorcycles", "Trucks and Buses", "Vintage Cars") :
    PRODUCT.PRODUCT_LINE.in("Plains","Ships", "Trains")))
    .fetch();
}
```

The `PRODUCT.BUY_PRICE.gt(BigDecimal.valueOf(buyPrice))` condition is appended only if the passed `buyPrice` is greater than 0; otherwise, we rely on the handy `noCondition()` method. Next, depending on the `cars` flag, we shape the range of values for `PRODUCT.PRODUCT_LINE.in()`. Via this single jOOQ query, we can shape four different SQL queries at runtime, depending on the `buyPrice` and `cars` values.

Using jOOQ comparators

The jOOQ `Comparator` API is quite handy for toggling comparison operators in conditions while remaining fluent. For instance, let's assume that the client (user, service, and so on) can choose between two categories of employees – a category of all the sales reps and a category of non-sales reps. If the client chooses the first category, then we want to fetch all employees (`EMPLOYEE`) that have a `salary` less than *65,000*. However, if the client chooses the second category, then we want to fetch all employees (`EMPLOYEE`) that have a `salary` greater than or equal to *65,000*. Instead of writing two queries or using any other approach, we can rely on jOOQ's `Comparator.IN` and `Comparator.NOT_IN` (note that, in the case of `NOT_IN`, the projected column(s) should be `NOT NULL`), as follows:

```
List<EmployeeRecord> fetchEmployees(boolean isSaleRep) {

  return ctx.selectFrom(EMPLOYEE)
   .where(EMPLOYEE.SALARY.compare(isSaleRep
      ? Comparator.IN : Comparator.NOT_IN,
        select(EMPLOYEE.SALARY).from(EMPLOYEE)
            .where(EMPLOYEE.SALARY.lt(65000))))
```

```
        .orderBy(EMPLOYEE.SALARY)
        .fetch();
}
```

jOOQ provides a comprehensive list of built-in comparators, including `EQUALS`, `GREATER`, `LIKE`, and `IS_DISTINCT_FROM`. While you can find all of them covered in the jOOQ documentation, here is another example that uses `Comparator.LESS` and `Comparator.GREATER` to express in jOOQ a query that can be translated into four SQL queries, depending on the values of `buyPrice` and `msrp`:

```
public List<ProductRecord> fetchProducts(
                   float buyPrice, float msrp) {

  return ctx.selectFrom(PRODUCT)
    .where(PRODUCT.BUY_PRICE.compare(
      buyPrice < 55f ? Comparator.LESS : Comparator.GREATER,
       select(avg(PRODUCT.MSRP.minus(
        PRODUCT.MSRP.mul(buyPrice / 100f))))
    .from(PRODUCT).where(PRODUCT.MSRP.coerce(Float.class)
     .compare(msrp > 100f ?
         Comparator.LESS : Comparator.GREATER, msrp))))
     .fetch();
}
```

You can check out these examples next to others in *DynamicQuery*.

Using SelectQuery, InsertQuery, UpdateQuery, and DeleteQuery

The goal of the `SelectQuery` (`InsertQuery`, `UpdateQuery`, and `DeleteQuery`) types is to allow the expression of dynamic queries in an imperative style. However, it is recommended to avoid this imperative style and use a more functional style, as you'll see soon in this chapter. So, while you read this section, consider this sentence as a disclaimer.

When the previous approaches can't be used or the query becomes cluttered, it is time to turn your attention to the `SelectQuery` (`InsertQuery`, `UpdateQuery`, and `DeleteQuery`) APIs. These jOOQ APIs are very useful for expressing dynamic queries because they contain dedicated methods for appending different parts of a query effortlessly (for example, conditions, joins, having, and order by).

Using SelectQuery

For instance, let's assume that our application exposes a filter for optionally selecting a price range (`startBuyPrice` and `endBuyPrice`), the product vendor (`productVendor`), and the product scale (`productScale`) for ordering a `PRODUCT`. Based on the client selections, we should execute the proper `SELECT` query, so we start by writing a `SelectQuery`, as shown here:

```
SelectQuery select = ctx.selectFrom(PRODUCT)
  .where(PRODUCT.QUANTITY_IN_STOCK.gt(0))
  .getQuery();
```

So far, this query doesn't involve any of the client selections. Furthermore, we take each client selection and rely on `addConditions()` to enrich them accordingly:

```
if (startBuyPrice != null && endBuyPrice != null) {
   select.addConditions(PRODUCT.BUY_PRICE
         .betweenSymmetric(startBuyPrice, endBuyPrice));
}

if (productVendor != null) {
    select.addConditions(PRODUCT.PRODUCT_VENDOR
          .eq(productVendor));
}

if (productScale != null) {
    select.addConditions(PRODUCT.PRODUCT_SCALE
          .eq(productScale));
}
```

Finally, we execute the query and fetch the results:

```
select.fetch();
```

Done! The same thing can be expressed like this as well:

```
Condition condition = PRODUCT.QUANTITY_IN_STOCK.gt(0);

if (startBuyPrice != null && endBuyPrice != null) {
    condition = condition.and(PRODUCT.BUY_PRICE
       .betweenSymmetric(startBuyPrice, endBuyPrice));
```

```
}

if (productVendor != null) {
    condition = condition.and(PRODUCT.PRODUCT_VENDOR
        .eq(productVendor));
}

if (productScale != null) {
    condition = condition.and(PRODUCT.PRODUCT_SCALE
        .eq(productScale));
}

SelectQuery select = ctx.selectFrom(PRODUCT)
   .where(condition)
   .getQuery();

select.fetch();
```

If you don't have a start condition (a fix condition), then you can start from a dummy `true` condition:

```
Condition condition = trueCondition();
```

Besides `trueCondition()`, we can use `falseCondition()` or `noCondition()`. More details are available here: `https://www.jooq.org/doc/latest/manual/sql-building/conditional-expressions/true-false-no-condition/`.

Next, use `and()`, `or()`, `andNot()`, `andExists()`, and so on to chain the optional conditions as you feel appropriate.

However, conditions are not the only flexible parts of a dynamic query. For instance, let's assume that we have a query that returns the office's cities and countries (`OFFICE.CITY` and `OFFICE.COUNTRY`). However, depending on client selections, this query should also return the employees from these offices (`EMPLOYEE`) and the sales of these employees (`SALE`). This means that our query should be dynamically appended with joins. Via the `SelectQuery` API, this can be done via `addJoin()` methods, exemplified here:

```
public List<Record> appendTwoJoins(
        boolean andEmp, boolean addSale) {
```

```
  SelectQuery select = ctx.select(OFFICE.CITY,
       OFFICE.COUNTRY).from(OFFICE).limit(10).getQuery();

  if (andEmp) {
    select.addSelect(EMPLOYEE.FIRST_NAME, EMPLOYEE.LAST_NAME);
    select.addJoin(EMPLOYEE,
              OFFICE.OFFICE_CODE.eq(EMPLOYEE.OFFICE_CODE));

    if (addSale) {
     select.addSelect(SALE.FISCAL_YEAR,
                SALE.SALE_, SALE.EMPLOYEE_NUMBER);
     select.addJoin(SALE, JoinType.LEFT_OUTER_JOIN,
          EMPLOYEE.EMPLOYEE_NUMBER.eq(SALE.EMPLOYEE_NUMBER));
    }
  }

 return select.fetch();
 }
```

As you can see, `addJoin()` comes in different flavors. Mainly, there is a set of `addJoin()` that implicitly generates an `INNER JOIN` (as `addJoin(TableLike<?> tl, Condition cndtn)`), a set of `addJoin()` that allows us to specify the type of join via the `JoinType` enumeration (as `addJoin(TableLike<?> tl, JoinType jt, Condition... cndtns)`), a set of `addJoinOnKey()` that generates the `ON` predicate based on the given foreign key (as `addJoinOnKey(TableLike<?> tl, JoinType jt, ForeignKey<?,?> fk)`), and a set of `addJoinUsing()` that relies on the `USING` clause (as `addJoinUsing(TableLike<?> table, Collection<? extends Field<?>> fields)`).

Next to the `addFoo()` methods used/mentioned here, we have `addFrom()`, `addHaving()`, `addGroupBy()`, `addLimit()`, `addWindow()`, and so on. You can find all of them and their flavors in the jOOQ documentation.

Sometimes, we need to simply reuse a part of a query a number of times. For instance, the following figure is obtained by `UNION` the (near) same query:

customer_number	clazz_[0, 1]	clazz_[2, 2]	clazz_[3, 5]	clazz_[6, 15]
211	1	0	0	0
450	1	0	0	0
415	1	0	0	0
239	1	0	0	0
471	0	2	0	0
299	0	2	0	0
242	0	0	3	0
398	0	0	4	0
323	0	0	4	0
187	0	0	3	0
424	0	0	3	0
124	0	0	0	9
141	0	0	0	13

Figure 12.7 – Apply UNION to count and classify customer's payment

The query behind this figure applies `UNION` to count and classifies the customers' payments based on the given classes, as shown here:

```
SELECT `classicmodels`.`customer`.`customer_number`,
  count(*) AS `clazz_[0, 1]`, 0 AS `clazz_[2, 2]`,
            0 AS `clazz_[3, 5]`, 0 AS `clazz_[6, 15]`
FROM `classicmodels`.`customer`
JOIN `classicmodels`.`payment` ON
  `classicmodels`.`customer`.`customer_number`
    = `classicmodels`.`payment`.`customer_number`
GROUP BY `classicmodels`.`customer`.`customer_number`
HAVING count(*) BETWEEN 0 AND 1
UNION
...
HAVING count(*) BETWEEN 2 AND 2
UNION
...
HAVING count(*) BETWEEN 3 AND 5
```

```
UNION
SELECT `classicmodels`.`customer`.`customer_number`,
       0, 0, 0, count(*)
FROM `classicmodels`.`customer`
JOIN `classicmodels`.`payment` ON
  `classicmodels`.`customer`.`customer_number`
    = `classicmodels`.`payment`.`customer_number`
GROUP BY `classicmodels`.`customer`.`customer_number`
HAVING count(*) BETWEEN 6 AND 15
```

However, if the number of classes varies (being an input parameter provided by the client), then the number of `UNION` statements also varies, and the `HAVING` clause must be appended dynamically. First, we can isolate the fixed part of our query, like this:

```
private SelectQuery getQuery() {

  return ctx.select(CUSTOMER.CUSTOMER_NUMBER)
    .from(CUSTOMER)
    .join(PAYMENT)
    .on(CUSTOMER.CUSTOMER_NUMBER.eq(PAYMENT.CUSTOMER_NUMBER))
    .groupBy(CUSTOMER.CUSTOMER_NUMBER)
    .getQuery();
}
```

Next, we should `UNION` a `getQuery()` for each given class and generate the specific `HAVING` clause, but not before reading the following important note.

> **Important Note**
>
> Note that it is not possible to use the same instance of a `SelectQuery` on both sides of a set operation such as `s.union(s)`, so you'll need a new `SelectQuery` for each `UNION`. This seems to be a fixable bug, so this note may not be relevant when you read this book.

This can be done as shown in this simple code:

```
public record Clazz(int left, int right) {}

public List<CustomerRecord> classifyCustomerPayments(
```

```
                                    Clazz... clazzes) {

  SelectQuery[] sq = new SelectQuery[clazzes.length];

  for (int i = 0; i < sq.length; i++) {
     sq[i] = getQuery(); // create a query for each UNION
  }

  sq[0].addSelect(count().as("clazz_[" + clazzes[0].left()
    + ", " + clazzes[0].right() + "]"));
  sq[0].addHaving(count().between(clazzes[0].left(),
    clazzes[0].right()));

  for (int i = 1; i < sq.length; i++) {
       sq[0].addSelect(val(0).as("clazz_[" + clazzes[i]
         .left() + ", " + clazzes[i].right() + "]"));
  }

  for (int i = 1; i < sq.length; i++) {
     for (int j = 0; j < i; j++) {
         sq[i].addSelect(val(0));
     }

     sq[i].addSelect(count());

     for (int j = i + 1; j < sq.length; j++) {
         sq[i].addSelect(val(0));
     }

     sq[i].addHaving(count().between(clazzes[i].left(),
       clazzes[i].right()));
     sq[0].union(sq[i]);
  }

  return sq[0].fetch();
}
```

Of course, you can try to write this much cleverly and more compact. A call of this method is shown here:

```
List<CustomerRecord> result = classicModelsRepository
  .classifyCustomerPayments(new Clazz(0, 1), new Clazz(2, 2),
                  new Clazz(3, 5), new Clazz(6, 15));
```

In order to check out these examples, refer to the *DynamicQuery* application.

InsertQuery, UpdateQuery, and DeleteQuery

Dynamic queries representing DML operations are also supported by jOOQ. `InsertQuery`, `UpdateQuery`, and `DeleteQuery` work on the same principle as `SelectQuery` and expose a comprehensive API intended to chain SQL parts into a valid and dynamic SQL query.

Let's see an example of using `InsertQuery` to insert a `PRODUCT` (a classic car), based on data provided by the client, and return the generated identity:

```
public long insertClassicCar(
    String productName, String productVendor,
    String productScale, boolean price) {

  InsertQuery iq = ctx.insertQuery(PRODUCT);

  iq.addValue(PRODUCT.PRODUCT_LINE, "Classic Cars");
  iq.addValue(PRODUCT.CODE, 599302);

  if (productName != null) {
    iq.addValue(PRODUCT.PRODUCT_NAME, productName);
  }

  if (productVendor != null) {
    iq.addValue(PRODUCT.PRODUCT_VENDOR, productVendor);
  }

  if (productScale != null) {
    iq.addValue(PRODUCT.PRODUCT_SCALE, productScale);
  }
```

```
  if (price) {
    iq.addValue(PRODUCT.BUY_PRICE,
           select(avg(PRODUCT.BUY_PRICE)).from(PRODUCT));
  }

  iq.setReturning(PRODUCT.getIdentity());

  iq.execute();

  return iq.getReturnedRecord()
    .getValue(PRODUCT.getIdentity().getField(), Long.class);
}
```

As you'll see in the jOOQ documentation, the `InsertQuery` API supports many more methods, such as `addConditions()`, `onDuplicateKeyIgnore()`, `onConflict()`, `setSelect()`, and `addValueForUpdate()`.

How about a dynamic update or delete? Here is a very intuitive example of a dynamic update:

```
public int updateProduct(float oldPrice, float value) {

  UpdateQuery uq = ctx.updateQuery(PRODUCT);

  uq.addValue(PRODUCT.BUY_PRICE,
     PRODUCT.BUY_PRICE.plus(PRODUCT.BUY_PRICE.mul(value)));
  uq.addConditions(PRODUCT.BUY_PRICE
     .lt(BigDecimal.valueOf(oldPrice)));

  return uq.execute();
}
```

And here's what code for a dynamic delete of sales (`SALE`) looks like:

```
public int deleteSale(int fiscalYear, double sale) {

  DeleteQuery dq = ctx.deleteQuery(SALE);

  Condition condition = SALE.FISCAL_YEAR
```

```
    .compare(fiscalYear <= 2003
       ? Comparator.GREATER : Comparator.LESS, fiscalYear);

  if (sale > 5000d) {
     condition = condition.or(SALE.SALE_.gt(sale));
  }

  dq.addConditions(condition);

  return dq.execute();
}
```

You can check out these examples in *DynamicQuery*. After exploring these APIs, take your time and challenge yourself to write your own dynamic queries. It's really fun and helps you get familiar with this amazingly simple but powerful API.

Writing generic dynamic queries

Sooner or later, you'll realize that what you need is a generic dynamic query. For instance, you may encounter a scenario that sounds like this. You need to select from an arbitrary table a number of arbitrary columns, based on arbitrary conditions. In such a scenario, it will be inefficient to duplicate the code only to vary the name of the table, columns, and conditions. So, most probably, you'll prefer a generic dynamic query, as shown here:

```
public <R extends Record> List<R> select(
    Table<R> table, Collection<SelectField<?>> fields,
                               Condition... conditions) {

   SelectQuery sq = ctx.selectQuery(table);

   sq.addSelect(fields);
   sq.addConditions(conditions);

   return sq.fetch();
}
```

Calling this method can be done as shown in the following examples:

```
List<ProductRecord> rs1 =
  classicModelsRepository.select(PRODUCT,
     List.of(PRODUCT.PRODUCT_LINE, PRODUCT.PRODUCT_NAME,
             PRODUCT.BUY_PRICE, PRODUCT.MSRP),
     PRODUCT.BUY_PRICE.gt(BigDecimal.valueOf(50)),
     PRODUCT.MSRP.gt(BigDecimal.valueOf(80)));

List<Record> rs2 =
  classicModelsRepository.select(table("product"),
     List.of(field("product_line"), field("product_name"),
       field("buy_price"), field("msrp")),
         field("buy_price").gt(50), field("msrp").gt(80));
```

If you want to rely only on the first type of call – that is, calls based on jOOQ-generated code – then you can enforce the type-safety of the generic method by replacing `SelectField<?>` with `TableField<R, ?>`, as follows:

```
public <R extends Record> List<R> select(
    Table<R> table, Collection<TableField<R, ?>> fields,
       Condition... conditions) {

  SelectQuery sq = ctx.selectQuery(table);

  sq.addSelect(fields);
  sq.addConditions(conditions);

  return sq.fetch();
}
```

This time, only the first call (`List<ProductRecord> rs1 = ...`) compiles and works. The same thing applies to DML operations – for instance, inserting in an arbitrary table:

```
public <R extends Record> int insert (
    Table<R> table, Map<TableField<R, ?>, ?> values) {

    InsertQuery iq = ctx.insertQuery(table);
```

```
    iq.addValues(values);

    return iq.execute();
  }
```

Here is a call example of the previous method:

```
int ri = classicModelsRepository.insert(PRODUCT,
  Map.of(PRODUCT.PRODUCT_LINE, "Classic Cars",
         PRODUCT.CODE, 599302,
         PRODUCT.PRODUCT_NAME, "1972 Porsche 914"));
```

Alternatively, for an arbitrary update, we can write the following method:

```
public <R extends Record> int update(Table<R> table,
    Map<TableField<R, ?>, ?> values, Condition... conditions) {

  UpdateQuery uq = ctx.updateQuery(table);

  uq.addValues(values);
  uq.addConditions(conditions);

  return uq.execute();
}
```

Here is a call example of the previous method:

```
int ru = classicModelsRepository.update(SALE,
    Map.of(SALE.TREND, "UP", SALE.HOT, true),
       SALE.TREND.eq("CONSTANT"));
```

Alternatively, for an arbitrary delete, we can write the following method:

```
public <R extends Record> int delete(Table<R> table,
       Condition... conditions) {

  DeleteQuery dq = ctx.deleteQuery(table);

  dq.addConditions(conditions);
```

```
    return dq.execute();
  }
```

Here is a call example of the previous method:

```
int rd1 = classicModelsRepository.delete(SALE,
    SALE.TREND.eq("UP"));

int rd2 = classicModelsRepository.delete(table("sale"),
    field("trend").eq("CONSTANT"));
```

You can see these examples next to others in *GenericDynamicQuery*.

Writing functional dynamic queries

Functional dynamic queries take this topic to the next level. However, let's try to see how and why we should evolve a query from zero to functional implementation. Let's assume that we develop an application for an organization's sales department, and we have to write a query that filters sales (`SALE`) by fiscal year (`SALE.FISCAL_YEAR`). Initially (day 1), we can do it as shown here:

```
// Day 1
public List<SaleRecord>
    filterSaleByFiscalYear(int fiscalYear) {

    return ctx.selectFrom(SALE)
      .where(SALE.FISCAL_YEAR.eq(fiscalYear))
      .fetch();
}
```

While everybody was satisfied with the result, we got a new request for a filter that obtains the sales of a certain trend (`SALE.TREND`). We've done this on day 1, so there is no problem to repeat it on day 2:

```
// Day 2
public List<SaleRecord> filterSaleByTrend(String trend) {

    return ctx.selectFrom(SALE)
      .where(SALE.TREND.eq(trend))
```

```
        .fetch();
}
```

This filter is the same as the filter from day 1, only that it has a different condition/filter. We realize that continuing like this will end up with a lot of similar methods that just repeat the code for different filters, which means a lot of *boilerplate* code. While reflecting on this aspect, we just got an urgent request for a filter of sales by fiscal year and trend. So, our horrible solution on day 3 is shown here:

```
// Day 3
public List<SaleRecord> filterSaleByFiscalYearAndTrend(
        int fiscalYear, String trend) {

    return ctx.selectFrom(SALE)
        .where(SALE.FISCAL_YEAR.eq(fiscalYear)
        .and(SALE.TREND.eq(trend)))
        .fetch();
}
```

After 3 days, we realize that this becomes unacceptable. The code becomes verbose, hard to maintain, and prone to errors.

On day 4, while looking for a solution, we noticed in the jOOQ documentation that the `where()` method also comes with `where(Collection<? extends Condition> clctn)` and `where(Condition... cndtns)`. This means that we can simplify our solution to something like this:

```
// Day 4
public List<SaleRecord>
    filterSaleBy(Collection<Condition> cf) {

    return ctx.selectFrom(SALE)
            .where(cf)
            .fetch();
}
```

This is quite nice because we can pass any set of conditions without modifying the `filterSaleBy()` method. Here is a call example:

```
List<SaleRecord> result =
  classicModelsRepository.filterSaleBy(
    List.of(SALE.FISCAL_YEAR.eq(2004), SALE.TREND.eq("DOWN"),
          SALE.EMPLOYEE_NUMBER.eq(1370L)));
```

However, this is not type-safe. For instance, the error from this call is discovered only at compile time (check out the code in bold):

```
List<SaleRecord> result =
  classicModelsRepository.filterSaleBy(
     List.of(SALE.FISCAL_YEAR.eq(2004), SALE.TREND.eq("DOWN"),
         EMPLOYEE.EMPLOYEE_NUMBER.eq(1370L)));
```

Well, a new day brings a new idea! On day 5, we defined an interface to prevent the type-safety issues from day 4. This is a functional interface, as shown here:

```
// Day 5
@FunctionalInterface
public interface SaleFunction<Sale, Condition> {

    Condition apply(Sale s);
}
```

And `filterSaleBy()` becomes the following:

```
public List<SaleRecord> filterSaleBy(
     SaleFunction<Sale, Condition> sf) {

  return ctx.selectFrom(SALE)
     .where(sf.apply(SALE))
     .fetch();
}
```

Problem solved! This time, we can run this type-safe call:

```
List<SaleRecord> result = classicModelsRepository
  .filterSaleBy(s -> s.SALE_.gt(4000d));
```

On day 6, Mark (our colleague) noticed this code, and he enlightens us that Java 8 already has this functional interface, which is called `java.util.function.Function<T, R>`. So, there is no need to define our `SaleFunction`, since `Function` can do the job, as shown here:

```
// Day 6
public List<SaleRecord> filterSaleBy(
       Function<Sale, Condition> f) {

  return ctx.selectFrom(SALE)
    .where(f.apply(SALE))
    .fetch();
}
```

On day 7, we noticed that calling `filterSaleBy()` works only for a single condition. However, we need to pass multi-conditions as well (as we did earlier when we were using `Collection<Condition>`). This led to the decision of modifying `filterSaleBy()` to accept an array of `Function`. The challenge is represented by applying this array of `Function`, and the solution relies on `Arrays.stream(array)` or `Stream.of(array)`, as shown here (use the one that you find more expressive; as an example, behind the scenes, `Stream.of()` calls `Arrays.stream()`):

```
// Day 7
public List<SaleRecord> filterSaleBy(
        Function<Sale, Condition>... ff) {

  return ctx.selectFrom(SALE)
    .where(Stream.of(ff).map(f -> f.apply(SALE))
                        .collect(toList()))
    .fetch();
}
```

Now, we can write type-safe calls, like this one:

```
List<SaleRecord> result = classicModelsRepository
     .filterSaleBy(s -> s.SALE_.gt(4000d),
                   s -> s.TREND.eq("DOWN"),
                   s -> s.EMPLOYEE_NUMBER.eq(1370L));
```

Cool! Day 8 was an important day because we managed to adjust this code to work for any table and conditions by writing it generically:

```
public <T extends Table<R>, R extends Record> List<R>
    filterBy(T t, Function<T, Condition>... ff) {

  return ctx.selectFrom(t)
  .where(Stream.of(ff).map(f -> f.apply(t)).collect(toList()))
  .fetch();
}
```

Here are some call samples:

```
List<SaleRecord> result1
 = classicModelsRepository.filterBy(SALE,
    s -> s.SALE_.gt(4000d), s -> s.TREND.eq("DOWN"),
    s -> s.EMPLOYEE_NUMBER.eq(1370L));

List<EmployeeRecord> result2 = classicModelsRepository
  .filterBy(EMPLOYEE, e -> e.JOB_TITLE.eq("Sales Rep"),
            e -> e.SALARY.gt(55000));

List<Record> result3 = classicModelsRepository
  .filterBy(table("employee"),
   e -> field("job_title", String.class).eq("Sales Rep"),
   e -> field("salary", Integer.class).gt(55000));
```

On day 9, we start to consider tuning this query. For instance, instead of fetching all fields via `selectFrom()`, we decided to add an argument to receive the fields that should be fetched as a `Collection<TableField<R, ?>>`. Moreover, we decided to defer the creation of such a collection until they are really used, and in order to accomplish this, we wrapped the collection in a `Supplier`, as shown here:

```
public <T extends Table<R>, R extends Record> List<Record>
 filterBy(T t, Supplier<Collection<TableField<R, ?>>>
          select, Function<T, Condition>... ff) {

  return ctx.select(select.get())
    .from(t)
```

```
    .where(Stream.of(ff).map(
       f -> f.apply(t)).collect(toList()))
    .fetch();
}
```

Here is a call sample:

```
List<Record> result = classicModelsRepository.filterBy(SALE,
   () -> List.of(SALE.SALE_ID, SALE.FISCAL_YEAR),
      s -> s.SALE_.gt(4000d), s -> s.TREND.eq("DOWN"),
      s -> s.EMPLOYEE_NUMBER.eq(1370L));
```

Maybe you want to support a call like the following one as well:

```
List<Record> result = classicModelsRepository
   .filterBy(table("sale"),
    () -> List.of(field("sale_id"), field("fiscal_year")),
       s -> field("sale").gt(4000d),
       s -> field("trend").eq("DOWN"),
       s -> field("employee_number").eq(1370L));
```

In such a case, replace `TableField<R, ?>` with `SelectField<?>`, as shown here:

```
public <T extends Table<R>, R extends Record> List<Record>
  filterBy (T t, Supplier<Collection<SelectField<?>>>
            select, Function<T, Condition>... ff) {

 return ctx.select(select.get())
 .from(t)
 .where(Stream.of(ff).map(f -> f.apply(t))
 .collect(toList()))
 .fetch();
}
```

Done! I hope you've found this story and examples useful and inspirational for your own functional generic dynamic queries. Until then, you can see these examples in the bundled code, named *FunctionalDynamicQuery*.

Infinite scrolling and dynamic filters

In the last section of this chapter, let's bring together our two main topics – pagination and dynamic queries. Earlier, in the *Implementing infinite scroll* section, we implemented infinite scrolling for the `ORDERDETAIL` table. Now, let's add some filters for `ORDERDETAIL` that allows a client to choose the price and quantity ordered range, as shown in this figure:

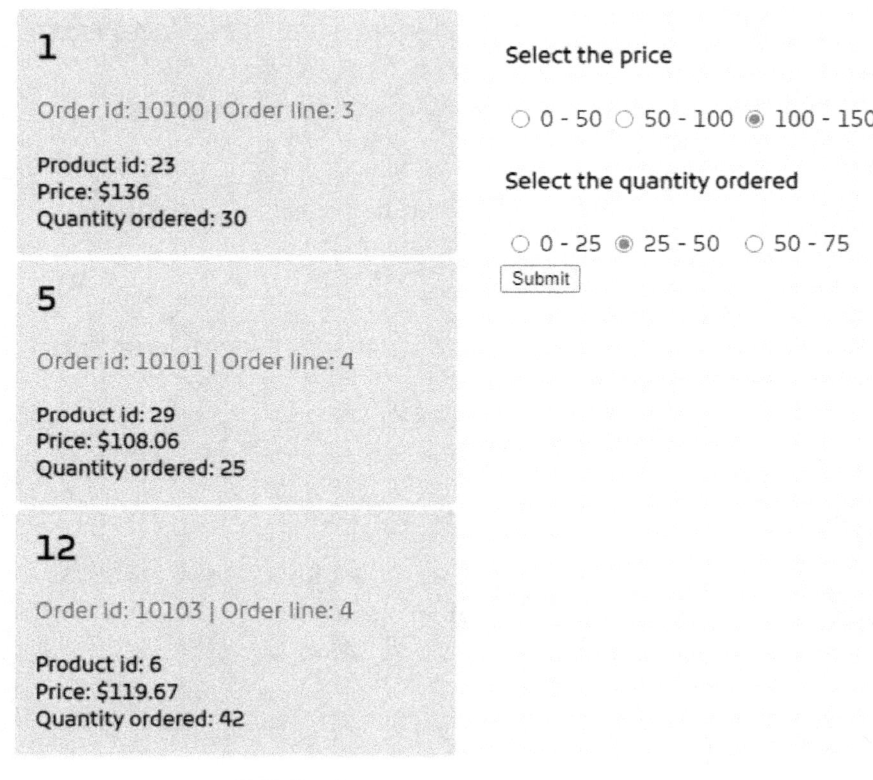

Figure 12.8 – Infinite scrolling and dynamic filters

We can easily implement this behavior by fusing the powers of `SEEK` and `SelectQuery`:

```
public List<Orderdetail> fetchOrderdetailPageAsc(
    long orderdetailId, int size, BigDecimal priceEach,
      Integer quantityOrdered) {

  SelectQuery sq = ctx.selectFrom(ORDERDETAIL)
     .orderBy(ORDERDETAIL.ORDERDETAIL_ID)
```

```
        .seek(orderdetailId)
        .limit(size)
        .getQuery();

    if (priceEach != null) {
     sq.addConditions(ORDERDETAIL.PRICE_EACH.between(
      priceEach.subtract(BigDecimal.valueOf(50)), priceEach));
    }

    if (quantityOrdered != null) {
      sq.addConditions(ORDERDETAIL.QUANTITY_ORDERED.between(
        quantityOrdered - 25, quantityOrdered));
    }

    return sq.fetchInto(Orderdetail.class);
  }
```

The following example URL involves loading the first page of three records that have prices between *50* and *100*, and an order quantity between *50* and *75*:

```
http://localhost:8080/orderdetail/0/3
        ?priceEach=100&quantityOrdered=75
```

You can find the complete example in *SeekInfiniteScrollFilter* for MySQL.

Summary

This was a relatively short chapter about pagination and dynamic queries. As you saw, jOOQ excels on both topics and provides support and APIs that allow us to intuitively and quickly implement the simplest to the most complex scenarios. In the first part of this chapter, we covered offset and keyset pagination (including infinite scrolling, the fancy `DENSE_RANK()`, and the `ROW_NUMBER()` approach). In the second part, we covered dynamic queries, including the ternary operator, the `Comparator` API, the `SelectQuery`, `InsertQuery`, `UpdateQuery`, and `DeleteQuery` APIs, and their respective generic and functional dynamic queries.

In the next chapter, we will talk about exploiting SQL functions.

Part 4: jOOQ and Advanced SQL

In this part, we will cover advanced SQL. This is where jOOQ feels like a fish in the water.

By the end of this part, you will know how to use jOOQ to exploit some of the most powerful SQL features, such as stored procedures, CTEs, window functions, and database views.

This part contains the following chapters:

- *Chapter 13, Exploiting SQL Functions*
- *Chapter 14, Derived Tables, CTEs, and Views*
- *Chapter 15, Calling and Creating Stored Functions and Procedures*
- *Chapter 16, Tackling Aliases and SQL Templating*
- *Chapter 17, Multitenancy in jOOQ*

13
Exploiting SQL Functions

From mathematical and statistical computations to string and date-time manipulations, respectively, to different types of aggregations, rankings, and groupings, SQL built-in functions are quite handy in many scenarios. There are different categories of functions depending on their goal and usage and, as you'll see, jOOQ has accorded major attention to their support. Based on these categories, our agenda for this chapter follows these points:

- Regular functions
- Aggregate functions
- Window functions
- Aggregates as window functions
- Aggregate functions and `ORDER BY`
- Ordered set aggregate functions (`WITHIN GROUP`)
- Grouping, filtering, distinctness, and functions
- Grouping sets

Let's get started!

Technical requirements

The code for this chapter can be found on GitHub at https://github.com/PacktPublishing/jOOQ-Masterclass/tree/master/Chapter13.

Regular functions

Being a SQL user, you've probably worked with a lot of regular or common SQL functions such as functions for dealing with NULL values, numeric functions, string functions, date-time functions, and so on. While the jOOQ manual represents a comprehensive source of information structured as a nomenclature of all the supported SQL built-in functions, we are trying to complete a series of examples designed to get you familiar with the jOOQ syntax in different scenarios. Let's start by talking about SQL functions for dealing with NULL values.

Just in case you need a quick overview about some simple and common NULL stuff, then quickly check out the someNullsStuffGoodToKnow() method available in the bundled code.

SQL functions for dealing with NULLs

SQL provides several functions for handling NULL values in our queries. Next, let's cover COALESCE(), DECODE(), IIF(), NULLIF(), NVL(), and NVL2() functions. Let's start with COALESCE().

COALESCE()

One of the most popular functions for dealing with NULL values is COALESCE(). This function returns the first non-null value from its list of *n* arguments.

For instance, let's assume that for each DEPARTMENT, we want to compute a deduction of 25% from CASH, ACCOUNTS_RECEIVABLE, or INVENTORIES, and a deduction of 25% from ACCRUED_LIABILITIES, ACCOUNTS_PAYABLE, or ST_BORROWING. Since this order is strict, if one of these is a NULL value, we go for the next one, and so on. If all are NULL, then we replace NULL with 0. Relying on the jOOQ coalesce() method, we can write the query as follows:

```
ctx.select(DEPARTMENT.NAME, DEPARTMENT.OFFICE_CODE,
      DEPARTMENT.CASH, ...,
     round(coalesce(DEPARTMENT.CASH,
      DEPARTMENT.ACCOUNTS_RECEIVABLE,
          DEPARTMENT.INVENTORIES,inline(0)).mul(0.25),
              2).as("income_deduction"),
```

```
    round(coalesce(DEPARTMENT.ACCRUED_LIABILITIES,
     DEPARTMENT.ACCOUNTS_PAYABLE, DEPARTMENT.ST_BORROWING,
      inline(0)).mul(0.25), 2).as("expenses_deduction"))
  .from(DEPARTMENT).fetch();
```

Notice the explicit usage of `inline()` for inlining the integer 0. As long as you know that this integer is a constant, there is no need to rely on `val()` for rendering a bind variable (placeholder). Using `inline()` fits pretty well for SQL functions, which typically rely on constant arguments or mathematical formulas having constant terms that can be easily inlined. If you need a quick reminder of `inline()` versus `val()`, then consider a quick revisit of *Chapter 3, jOOQ Core Concepts.*

Besides `coalesce(Field<T> field, Field<?>... fields)` used here, jOOQ provides two other flavors: `coalesce(Field<T> field, T value)` and `coalesce(T value, T... values)`.

Here is another example that relies on the `coalesce()` method to fill the gaps in the `DEPARTMENT.FORECAST_PROFIT` column. Each `FORECAST_PROFIT` value that is `NULL` is filled by the following query:

```
ctx.select(DEPARTMENT.NAME, DEPARTMENT.OFFICE_CODE, …
 coalesce(DEPARTMENT.FORECAST_PROFIT,
     select(
       avg(field(name("t", "forecast_profit"), Double.class)))
      .from(DEPARTMENT.as("t"))
      .where(coalesce(field(name("t", "profit")), 0)
      .gt(coalesce(DEPARTMENT.PROFIT, 0))
       .and(field(name("t", "forecast_profit")).isNotNull())))
        .as("fill_forecast_profit"))
   .from(DEPARTMENT)
   .orderBy(DEPARTMENT.DEPARTMENT_ID).fetch();
```

So, for each row having `FORECAST_PROFIT` equal to `NULL`, we use a custom interpolation formula represented by the average of all the non-null `FORECAST_PROFIT` values where the profit (`PROFIT`) is greater than the profit of the current row.

Next, let's talk about `DECODE()`.

DECODE()

In some dialects (for instance, in Oracle), we have the `DECODE()` function that acts as an if-then-else logic in queries. Having `DECODE(x, a, r1, r2)` is equivalent to the following:

```
IF x = a THEN
    RETURN r1;
ELSE
    RETURN r2;
END IF;
```

Or, since `DECODE` makes `NULL` safe comparisons, it's more like `IF x IS NOT DISTINCT FROM a THEN ....`

Let's attempt to compute a financial index as ((`DEPARTMENT.LOCAL_BUDGET` * 0.25) * 2) / 100. Since `DEPARTMENT.LOCAL_BUDGET` can be `NULL`, we prefer to replace such occurrences with 0. Relying on the jOOQ `decode()` method, we have the following:

```
ctx.select(DEPARTMENT.NAME, DEPARTMENT.OFFICE_CODE,
   DEPARTMENT.LOCAL_BUDGET, decode(DEPARTMENT.LOCAL_BUDGET,
   castNull(Double.class), 0, DEPARTMENT.LOCAL_BUDGET.mul(0.25))
  .mul(2).divide(100).as("financial_index"))
  .from(DEPARTMENT)
  .fetch();
```

The `DECODE()` part can be perceived like this:

```
IF DEPARTMENT.LOCAL_BUDGET = NULL THEN
    RETURN 0;
ELSE
    RETURN DEPARTMENT.LOCAL_BUDGET * 0.25;
END IF;
```

But don't conclude from here that `DECODE()` accepts only this simple logic. Actually, the `DECODE()` syntax is more complex and looks like this:

```
DECODE (x, a1, r1[, a2, r2], ...,[, an, rn] [, d]);
```

In this syntax, the following applies:

- x is compared with the other argument, a_1, a_2, ..., a_n.
- a_1, a_2, ..., or a_n is sequentially compared with the first argument; if any comparison x = a_1, x = a_2, ..., x = a_n returns true, then the DECODE() function terminates by returning the result.
- r_1, r_2, ..., or r_n is the result corresponding to x_i = a_i, i = (1...n).
- d is an expression that should be returned if no match for x_i=a_i, i = (1...n) was found.

Since jOOQ emulates DECODE() using CASE expressions, you can safely use it in all dialects supported by jOOQ, so let's see another example here:

```
ctx.select(DEPARTMENT.NAME, DEPARTMENT.OFFICE_CODE,...,
    decode(DEPARTMENT.NAME,
           "Advertising", "Publicity and promotion",
           "Accounting", "Monetary and business",
           "Logistics", "Facilities and supplies",
           DEPARTMENT.NAME).concat("department")
              .as("description"))
   .from(DEPARTMENT)
   .fetch();
```

So, in this case, if the department name is *Advertising*, *Accounting*, or *Logistics*, then it is replaced with a meaningful description; otherwise, we simply return the current name.

Moreover, DECODE() can be used with ORDER BY, GROUP BY, or next to aggregate functions as well. While more examples can be seen in the bundled code, here is another one of using DECODE() with GROUP BY for counting BUY_PRICE larger/equal/smaller than half of MSRP:

```
ctx.select(field(name("t", "d")), count())
   .from(select(decode(sign(
     PRODUCT.BUY_PRICE.minus(PRODUCT.MSRP.divide(2))),
       1, "Buy price larger than half of MSRP",
       0, "Buy price equal to half of MSRP",
      -1, "Buy price smaller than half of MSRP").as("d"))
       .from(PRODUCT)
       .groupBy(PRODUCT.BUY_PRICE, PRODUCT.MSRP).asTable("t"))
   .groupBy(field(name("t", "d")))
   .fetch();
```

And here is another example of using imbricated `DECODE()`:

```
ctx.select(DEPARTMENT.NAME, DEPARTMENT.OFFICE_CODE,
     DEPARTMENT.LOCAL_BUDGET, DEPARTMENT.PROFIT,
     decode(DEPARTMENT.LOCAL_BUDGET,
     castNull(Double.class), DEPARTMENT.PROFIT,
      decode(sign(DEPARTMENT.PROFIT.minus(
       DEPARTMENT.LOCAL_BUDGET)),
        1, DEPARTMENT.PROFIT.minus(DEPARTMENT.LOCAL_BUDGET),
        0, DEPARTMENT.LOCAL_BUDGET.divide(2).mul(-1),
       -1, DEPARTMENT.LOCAL_BUDGET.mul(-1)))
            .as("profit_balance"))
   .from(DEPARTMENT)
   .fetch();
```

For given `sign(a, b)`, it returns 1 if a > b, 0 if a = b, and -1 if a < b. So, this code can be easily interpreted based on the following output:

```
+-----------+-----------+------------+--------+--------------+
|name       |office_code|local_budget|  profit|profit_balance|
+-----------+-----------+------------+--------+--------------+
|Advertising|1          |      {null}|  {null}|        {null}|
|Sales      |1          |    650000.0|900000.0|      250000.0|
|Finance    |3          |    120000.0| 90000.0|     -120000.0|
|Finance    |7          |    450000.0|  {null}|        {null}|
|Logistics  |8          |      {null}|120000.0|      120000.0|
...
```

Figure 13.1 – Output

More examples are available in the *Functions* bundled code.

IIF()

The `IIF()` function implements the if-then-else logic via three arguments, as follows (this acts as the `NVL2()` function presented later):

```
IIF(boolean_expr, value_for_true_case, value_for_false_case)
```

It evaluates the first argument (`boolean_expr`) and returns the second argument (`value_for_true_case`) and third one (`value_for_false_case`), respectively.

For instance, the following usage of the jOOQ `iif()` function evaluates the `DEPARTMENT.LOCAL_BUDGET.isNull()` expression and outputs the text *NO BUDGET* or *HAS BUDGET*:

```
ctx.select(DEPARTMENT.DEPARTMENT_ID, DEPARTMENT.NAME,
      iif(DEPARTMENT.LOCAL_BUDGET.isNull(),
        "NO BUDGET", "HAS BUDGET").as("budget"))
   .from(DEPARTMENT).fetch();
```

More examples, including imbricated `IIF()` usage, are available in the bundled code.

NULLIF()

The `NULLIF(expr1, expr2)` function returns `NULL` if the arguments are equal. Otherwise, it returns the first argument (`expr1`).

For instance, in legacy databases, it is a common practice to have a mixture of `NULL` and empty strings for missing values. We have intentionally created such a case in the `OFFICE` table for `OFFICE.COUNTRY`.

Since empty strings are not `NULL` values, using `ISNULL()` will not return them even if, for us, `NULL` values and empty strings may have the same mining. Using the jOOQ `nullif()` method is a handy approach for finding all missing data (`NULL` values and empty strings), as follows:

```
ctx.select(OFFICE.OFFICE_CODE, nullif(OFFICE.COUNTRY, ""))
    .from(OFFICE).fetch();

ctx.selectFrom(OFFICE)
    .where(nullif(OFFICE.COUNTRY, "").isNull()).fetch();
```

These examples are available in the *Functions* bundled code.

IFNULL() and ISNULL()

The `IFNULL(expr1, expr2)` and `ISNULL(expr1, expr2)` functions take two arguments and return the first one if it is not `NULL`. Otherwise, they return the second argument. The former is similar to Oracle's `NVL()` function presented later, while the latter is specific to SQL Server. Both of them are emulated by jOOQ via `CASE` expressions for all dialects that don't support them natively.

For instance, the following snippet of code produces 0 for each NULL value of DEPARTMENT.LOCAL_BUDGET via both jOOQ methods, ifnull() and isnull():

```
ctx.select(DEPARTMENT.DEPARTMENT_ID, DEPARTMENT.NAME,
      ifnull(DEPARTMENT.LOCAL_BUDGET, 0).as("budget_if"),
      isnull(DEPARTMENT.LOCAL_BUDGET, 0).as("budget_is"))
   .from(DEPARTMENT)
   .fetch();
```

Here is another example that fetches the customer's postal code or address:

```
ctx.select(
     ifnull(CUSTOMERDETAIL.POSTAL_CODE,
       CUSTOMERDETAIL.ADDRESS_LINE_FIRST).as("address_if"),
     isnull(CUSTOMERDETAIL.POSTAL_CODE,
       CUSTOMERDETAIL.ADDRESS_LINE_FIRST).as("address_is"))
  .from(CUSTOMERDETAIL).fetch();
```

More examples are available in the *Functions* bundled code.

NVL() and NVL2()

Some dialects (for example, Oracle) support two functions named NVL() and NVL2(). Both of them are emulated by jOOQ for all dialects that don't support them natively. The former acts like IFNULL(), while the latter acts like IIF(). So, NVL(expr1, expr2) produces the first argument if it is not NULL; otherwise, it produces the second argument.

For instance, let's use the jOOQ nvl() method for applying the variance formula used in finance to calculate the difference between a forecast and an actual result for DEPARTMENT.FORECAST_PROFIT and DEPARTMENT.PROFIT as ((ACTUAL PROFIT ÷ FORECAST PROFIT) - 1) * 100, as follows:

```
ctx.select(DEPARTMENT.NAME, ...,
       round((nvl(DEPARTMENT.PROFIT, 0d).divide(
       nvl(DEPARTMENT.FORECAST_PROFIT, 10000d)))
       .minus(1d).mul(100), 2).concat("%").as("nvl"))
   .from(DEPARTMENT)
   .fetch();
```

If PROFIT is NULL, then we replace it with 0, and if FORECAST_PROFIT is NULL, then we replace it with a default profit of 10,000. Challenge yourself to write this query via ISNULL() as well.

On the other hand, `NVL2(expr1, expr2, expr3)` evaluates the first argument (`expr1`). If `expr1` is not `NULL`, then it returns the second argument (`expr2`); otherwise, it returns the third argument (`expr3`).

For instance, each `EMPLOYEE` has a salary and an optional `COMMISSION` (a missing commission is `NULL`). Let's fetch salary + commission via jOOQ `nvl2()` and `iif()`, as follows:

```
ctx.select(EMPLOYEE.FIRST_NAME, EMPLOYEE.LAST_NAME,
 iif(EMPLOYEE.COMMISSION.isNull(),EMPLOYEE.SALARY,
  EMPLOYEE.SALARY.plus(EMPLOYEE.COMMISSION))
    .as("iif1"),
 iif(EMPLOYEE.COMMISSION.isNotNull(),
  EMPLOYEE.SALARY.plus(EMPLOYEE.COMMISSION), EMPLOYEE.SALARY)
   .as("iif2"),
 nvl2(EMPLOYEE.COMMISSION,
  EMPLOYEE.SALARY.plus(EMPLOYEE.COMMISSION), EMPLOYEE.SALARY)
   .as("nvl2"))
 .from(EMPLOYEE)
 .fetch();
```

All three columns—`iif1`, `iif2`, and `nvl2`—should contain the same data. Regrettably, `NVL` can perform better than `COALESCE` in some Oracle cases. For more details, consider reading this article: `https://connor-mcdonald.com/2018/02/13/nvl-vs-coalesce/`. You can check out all the examples from this section in the *Functions* bundled code. Next, let's talk about numeric functions.

Numeric functions

jOOQ supports a comprehensive list of numeric functions, including `ABS()`, `SIN()`, `COS()`, `EXP()`, `FLOOR()`, `GREATEST()`, `LEAST()`, `LN()`, `POWER()`, `SIGN()`, `SQRT()`, and much more. Mainly, jOOQ exposes a set of methods that mirrors the names of these SQL functions and supports the proper number and type of arguments.

Since you can find all these functions listed and exemplified in the jOOQ manual, let's try here two examples of combining several of them to accomplish a common goal. For instance, a famous formula for computing the Fibonacci number is the Binet formula (notice that no recursion is required!):

```
Fib(n) = (1.6180339^n - (-0.6180339)^n) / 2.236067977
```

Writing this formula in jOOQ/SQL requires us to use the `power()` numeric function as follows (`n` is the number to compute):

```
ctx.fetchValue(round((power(1.6180339, n).minus(
      power(-0.6180339, n))).divide(2.236067977), 0));
```

How about computing the distance between two points expressed as `(latitude1, longitude1)`, respectively `(latitude2, longitude2)`? Of course, exactly as in the case of the Fibonacci number, such computations are commonly done outside the database (directly in Java) or in a UDF or stored procedure, but trying to solve them in a `SELECT` statement is a good opportunity to quickly practice some numeric functions and get familiar with jOOQ syntax. So, here we go with the required math:

```
a = POWER(SIN((latitude2 - latitude1) / 2.0)), 2)
  + COS(latitude1) * COS(latitude2)
      * POWER (SIN((longitude2 - longitude1) / 2.0), 2);

result = (6371.0 * (2.0 * ATN2(SQRT(a),SQRT(1.0 - a))));
```

This time, we need the jOOQ `power()`, `sin()`, `cos()`, `atn2()`, and `sqrt()` numeric methods, as shown here:

```
double pi180 = Math.PI / 180;

Field<BigDecimal> a = (power(sin(val((latitude2 - latitude1)
 * pi180).divide(2d)), 2d).plus(cos(latitude1 * pi180)
   .mul(cos(latitude2 * pi180)).mul(power(sin(val((
     longitude2 - longitude1) * pi180).divide(2d)), 2d))));

ctx.fetchValue(inline(6371d).mul(inline(2d)
   .mul(atan2(sqrt(a), sqrt(inline(1d).minus(a))))));
```

You can practice these examples in the *Functions* bundled code.

String functions

Exactly as in case of SQL numeric functions, jOOQ supports an impressive set of SQL string functions, including `ASCII()`, `CONCAT()`, `OVERLAY()`, `LOWER()`, `UPPER()`, `LTRIM()`, `RTRIM()`, and so on. You can find each of them exemplified in the jOOQ manual, so here, let's try to use several string functions to obtain an output, as in this screenshot:

HAVE

```
+----------+--------+-------------------+
|first_name|last_name|job_title         |
+----------+--------+-------------------+
|Diane     |Murphy   |President         |
|Mary      |Patterson|VP Sales          |
|Jeff      |Firrelli |VP Marketing      |
|William   |Patterson|Sales Manager (APAC)|
|Gerard    |Bondur   |Sales Manager (EMEA)|
+----------+--------+-------------------+
|...record(s) truncated...
23
```

WANT

```
+-----------------------------------+
|employee                           |
+-----------------------------------+
|DIANE M. (president)...            |
|MARY P. (vp sales)...              |
|JEFF F. (vp marketing)...          |
|WILLIAM P. (sales manager (apac))...|
|GERARD B. (sales manager (emea))... |
+-----------------------------------+
|...record(s) truncated...
23
```

Figure 13.2 – Applying several SQL string functions

Transforming what we have in what we want can be expressed in jOOQ via several methods, including `concat()`, `upper()`, `space()`, `substring()`, `lower()`, and `rpad()`—of course, you can optimize or write the following query in different ways:

```
ctx.select(concat(upper(EMPLOYEE.FIRST_NAME), space(1),
         substring(EMPLOYEE.LAST_NAME, 1, 1).concat(". ("),
           lower(EMPLOYEE.JOB_TITLE),
              rpad(val(")"), 4, '.')).as("employee"))
   .from(EMPLOYEE)
   .fetch();
```

You can check out this example next to several examples of splitting a string by a delimiter in the *Functions* bundled code.

Date-time functions

The last category of functions covered in this section includes date-time functions. Mainly, jOOQ exposes a wide range of date-time functions that can be roughly categorized as functions that operate with `java.sql.Date`, `java.sql.Time`, and `java.sql.Timestamp`, and functions that operate with Java 8 date-time, `java.time.LocalDate`, `java.time.LocalDateTime`, and `java.time.OffsetTime`. jOOQ can't use the `java.time.Duration` or `Period` classes as they work differently from standard SQL intervals, though of course, converters and bindings can be applied.

Moreover, jOOQ comes with a substitute for JDBC missing `java.sql.Interval` data type, named `org.jooq.types.Interval`, having three implementations as `DayToSecond`, `YearToMonth`, and `YearToSecond`.

Here are a few examples that are pretty simple and intuitive. This first example fetches the current date as `java.sql.Date` and `java.time.LocalDate`:

```
Date r = ctx.fetchValue(currentDate());
LocalDate r = ctx.fetchValue(currentLocalDate());
```

This next example converts an ISO *8601* `DATE` string literal into a `java.sql.Date` data type:

```
Date r = ctx.fetchValue(date("2024-01-29"));
```

Adding an interval of 10 days to a `Date` and a `LocalDate` can be done like this:

```
var r = ctx.fetchValue(
 dateAdd(Date.valueOf("2022-02-03"), 10).as("after_10_days"));

var r = ctx.fetchValue(localDateAdd(
 LocalDate.parse("2022-02-03"), 10).as("after_10_days"));
```

Or, adding an interval of 3 months can be done like this:

```
var r = ctx.fetchValue(dateAdd(Date.valueOf("2022-02-03"),
  new YearToMonth(0, 3)).as("after_3_month"));
```

Extracting the day of week (1 = Sunday, 2 = Monday, ..., 7 = Saturday) via the SQL `EXTRACT()` and jOOQ `dayOfWeek()` functions can be done like this:

```
int r = ctx.fetchValue(dayOfWeek(Date.valueOf("2021-05-06")));
int r = ctx.fetchValue(extract(
  Date.valueOf("2021-05-06"), DatePart.DAY_OF_WEEK));
```

You can check out more examples in the *Functions* bundled code. In the next section, let's tackle aggregate functions.

Aggregate functions

The most common aggregate functions (in an arbitrary order) are `AVG()`, `COUNT()`, `MAX()`, `MIN()`, and `SUM()`, including their `DISTINCT` variants. I'm pretty sure that you are very familiar with these aggregates and you've used them in many of your queries. For instance, here are two `SELECT` statements that compute the popular harmonic and geometric means for sales grouped by fiscal year. Here, we use the jOOQ `sum()` and `avg()` functions:

```
// Harmonic mean: n / SUM(1/xi), i=1…n
ctx.select(SALE.FISCAL_YEAR, count().divide(
      sum(inline(1d).divide(SALE.SALE_))).as("harmonic_mean"))
    .from(SALE).groupBy(SALE.FISCAL_YEAR).fetch();
```

And here, we compute the geometric mean:

```
// Geometric mean: EXP(AVG(LN(n)))
ctx.select(SALE.FISCAL_YEAR, exp(avg(ln(SALE.SALE_)))
             .as("geometric_mean"))
    .from(SALE).groupBy(SALE.FISCAL_YEAR).fetch();
```

But as you know (or as you'll find out shortly), there are many other aggregates that have the same goal of performing some calculations across a set of rows and returning a single output row. Again, jOOQ exposes dedicated methods whose names mirror the aggregates' names or represent suggestive shortcuts.

Next, let's see several aggregate functions that are less popular and are commonly used in statistics, finance, science, and other fields. One of them is dedicated to computing Standard Deviation, (`https://en.wikipedia.org/wiki/Standard_deviation`). In jOOQ, we have `stddevSamp()` for Sample and `stddevPop()` for Population. Here is an example of computing SSD, PSD, and emulation of PSD via population variance (introduced next) for sales grouped by fiscal year:

```
ctx.select(SALE.FISCAL_YEAR,
     stddevSamp(SALE.SALE_).as("samp"),    // SSD
     stddevPop(SALE.SALE_).as("pop1"),     // PSD
     sqrt(varPop(SALE.SALE_)).as("pop2")) // PSD emulation
  .from(SALE).groupBy(SALE.FISCAL_YEAR).fetch();
```

Both SSD and PSD are supported in MySQL, PostgreSQL, SQL Server, Oracle, and many other dialects and are useful in different kinds of problems, from finance, statistics, forecasting, and so on. For instance, in statistics, we have the standard score (or so-called z-score) that represents the number of SDs placed above or below the population mean for a certain observation and having the formula $z = (x - \mu) / \sigma$ (z is z-score, x is the observation, μ is the mean, and σ is the SD). You can read further information on this here: `https://en.wikipedia.org/wiki/Standard_score`.

Now, considering that we store the number of sales (`DAILY_ACTIVITY.SALES`) and visitors (`DAILY_ACTIVITY.VISITORS`) in `DAILY_ACTIVITY` and we want to get some information about this data, since there is no direct comparison between sales and visitors, we have to come up with some meaningful representation, and this can be provided by z-scores. By relying on Common Table Expressions (**CTEs**) and SD, we can express in jOOQ the following query (of course, in production, using a stored procedure may be a better choice for such queries):

```
ctx.with("sales_stats").as(
      select(avg(DAILY_ACTIVITY.SALES).as("mean"),
        stddevSamp(DAILY_ACTIVITY.SALES).as("sd"))
      .from(DAILY_ACTIVITY))
   .with("visitors_stats").as(
      select(avg(DAILY_ACTIVITY.VISITORS).as("mean"),
        stddevSamp(DAILY_ACTIVITY.VISITORS).as("sd"))
      .from(DAILY_ACTIVITY))
   .select(DAILY_ACTIVITY.DAY_DATE,
     abs(DAILY_ACTIVITY.SALES
     .minus(field(name("sales_stats", "mean"))))
     .divide(field(name("sales_stats", "sd"), Float.class))
     .as("z_score_sales"),
      abs(DAILY_ACTIVITY.VISITORS
     .minus(field(name("visitors_stats", "mean"))))
     .divide(field(name("visitors_stats", "sd"), Float.class))
        .as("z_score_visitors"))
   .from(table("sales_stats"),
     table("visitors_stats"), DAILY_ACTIVITY).fetch();
```

Among the results produced by this query, we remark the z-score of sales on *2004-01-06*, which is 2.00. In the context of z-score analysis, this output is definitely worth a deeper investigation (typically, z-scores > 1.96 or < -1.96 are considered outliers that should be further investigated). Of course, this is not our goal, so let's jump to another aggregate.

Going further through statistical aggregates, we have variance, which is defined as the average of the squared differences from the mean or the average squared deviations from the mean (`https://en.wikipedia.org/wiki/Variance`). In jOOQ, we have sample variance via `varSamp()` and population variance via `varPop()`, as illustrated in this code example:

```
Field<BigDecimal> x = PRODUCT.BUY_PRICE;

ctx.select(varSamp(x)) // Sample Variance
```

```
    .from(PRODUCT).fetch();

ctx.select(varPop(x)) // Population Variance
    .from(PRODUCT).fetch();
```

Both of them are supported in MySQL, PostgreSQL, SQL Server, Oracle, and many other dialects, but just for fun, you can emulate sample variance via the `COUNT()` and `SUM()` aggregates as has been done in the following code snippet—just another opportunity to practice these aggregates:

```
ctx.select((count().mul(sum(x.mul(x)))
      .minus(sum(x).mul(sum(x)))).divide(count()
      .mul(count().minus(1))).as("VAR_SAMP"))
   .from(PRODUCT).fetch();
```

Next, we have linear regression (or correlation) functions applied for determining regression relationships between the dependent (denoted as Y) and independent (denoted as X) variable expressions (`https://en.wikipedia.org/wiki/Regression_analysis`). In jOOQ, we have `regrSXX()`,`regrSXY()`, `regrSYY()`, `regrAvgX()`, `regrAvgXY()`, `regrCount()`, `regrIntercept()`, `regrR2()`, and `regrSlope()`.

For instance, in the case of `regrSXY(y, x)`, `y` is the dependent variable expression and `x` is the independent variable expression. If `y` is `PRODUCT.BUY_PRICE` and `x` is `PRODUCT.MSRP`, then the linear regression per `PRODUCT_LINE` looks like this:

```
ctx.select(PRODUCT.PRODUCT_LINE,
    (regrSXY(PRODUCT.BUY_PRICE, PRODUCT.MSRP)).as("regr_sxy"))
    .from(PRODUCT).groupBy(PRODUCT.PRODUCT_LINE).fetch();
```

The functions listed earlier (including `regrSXY()`) are supported in all dialects, but they can be easily emulated as well. For instance, `regrSXY()` can be emulated as `(SUM(X*Y)-SUM(X) * SUM(Y)/COUNT(*))`, as illustrated here:

```
ctx.select(PRODUCT.PRODUCT_LINE,
     sum(PRODUCT.BUY_PRICE.mul(PRODUCT.MSRP))
      .minus(sum(PRODUCT.BUY_PRICE).mul(sum(PRODUCT.MSRP)
      .divide(count()))).as("regr_sxy"))
   .from(PRODUCT).groupBy(PRODUCT.PRODUCT_LINE).fetch();
```

In addition, `regrSXY()` can also be emulated as `SUM(1) * COVAR_POP(expr1, expr2)`, where `COVAR_POP()` represents the population covariance and `SUM(1)` is actually `REGR_COUNT(expr1, expr2)`. You can see this example in the bundled code next to many other emulations for `REGR_FOO()` functions and an example of calculating `y = slope * x - intercept` via `regrSlope()` and `regrIntercept()`, linear regression coefficients, but also via `sum()`, `avg()`, and `max()`.

After population covariance (`https://en.wikipedia.org/wiki/Covariance`), `COVAR_POP()`, we have sample covariance, `COVAR_SAMP()`, which can be called like this:

```
ctx.select(PRODUCT.PRODUCT_LINE,
  covarSamp(PRODUCT.BUY_PRICE, PRODUCT.MSRP).as("covar_samp"),
  covarPop(PRODUCT.BUY_PRICE, PRODUCT.MSRP).as("covar_pop"))
    .from(PRODUCT)
    .groupBy(PRODUCT.PRODUCT_LINE)
    .fetch();
```

If your database doesn't support the covariance functions (for instance, MySQL or SQL Server), then you can emulate them via common aggregates—`COVAR_SAMP()` as `(SUM(x*y) - SUM(x) * SUM(y) / COUNT(*)) / (COUNT(*) - 1)`, and `COVAR_POP()` as `(SUM(x*y) - SUM(x) * SUM(y) / COUNT(*)) / COUNT(*)`. You can find examples in the *AggregateFunctions* bundled code.

An interesting function that is not supported by most databases (Exasol is one of the exceptions) but is provided by jOOQ is the synthetic `product()` function. This function represents multiplicative aggregation emulated via `exp(sum(log(arg)))` for positive numbers, and it performs some extra work for zero and negative numbers. For instance, in finance, there is an index named Compounded Month Growth Rate (CMGR) that is computed based on monthly revenue growth, as we have in `SALE.REVENUE_GROWTH`. The formula is `(PRODUCT (1 + SALE.REVENUE_GROWTH))) ^ (1 / COUNT())`, and we've applied it for each year here:

```
ctx.select(SALE.FISCAL_YEAR,
      round((product(one().plus(
         SALE.REVENUE_GROWTH.divide(100)))
            .power(one().divide(count()))).mul(100) ,2)
               .concat("%").as("CMGR"))
   .from(SALE).groupBy(SALE.FISCAL_YEAR).fetch();
```

We also multiply everything by 100 to obtain the result as a percent. You can find this example in the *AggregateFunctions* bundled code, next to other aggregation functions such as `BOOL_AND()`, `EVERY()`, `BOOL_OR()`, and functions for bitwise operations.

When you have to use an aggregate function that is partially supported or not supported by jOOQ, you can rely on the `aggregate()`/`aggregateDistinct()` methods. Of course, your database must support the called aggregate function. For instance, jOOQ doesn't support the Oracle `APPROX_COUNT_DISTINCT()` aggregation function, which represents an alternative to the `COUNT (DISTINCT expr)` function. This is useful for approximating the number of distinct values while processing large amounts of data significantly faster than the traditional `COUNT` function, with negligible deviation from the exact number. Here is a usage of the `(String name, Class<T> type, Field<?>... arguments)` aggregate, which is just one of the provided flavors (check out the documentation for more):

```
ctx.select(ORDERDETAIL.PRODUCT_ID,
     aggregate("approx_count_distinct", Long.class,
       ORDERDETAIL.ORDER_LINE_NUMBER).as("approx_count"))
   .from(ORDERDETAIL)
   .groupBy(ORDERDETAIL.PRODUCT_ID)
   .fetch();
```

You can find this example in the *AggregateFunctions* bundled code for Oracle.

Window functions

Window functions are extremely useful and powerful; therefore, they represent a must-know topic for every developer that interacts with a database via SQL. In a nutshell, the best way to quickly overview window functions is to start from a famous diagram representing a comparison between an aggregation function and a window function that highlights the main difference between them, as represented here:

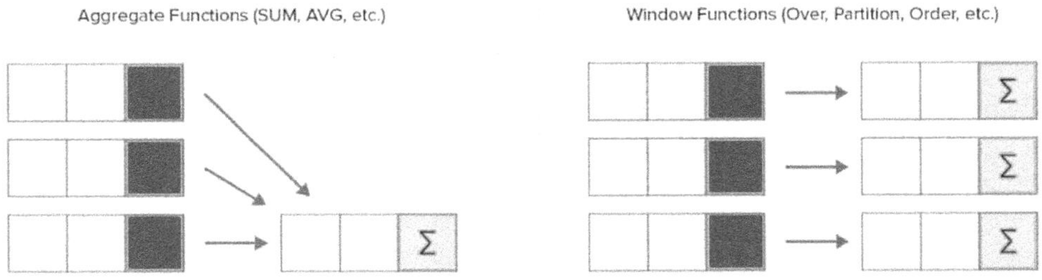

Figure 13.3 – Aggregate functions versus window functions

As you can see, both the aggregate function and the window function calculate something on a set of rows, but a window function doesn't aggregate or group these rows into a single output row. A window function relies on the following syntax:

```
window_function_name (expression) OVER (
    Partition Order Frame
)
```

This syntax can be explained as follows:

Obviously, `window_function_name` represents the window function name, such as `ROW_NUMBER()`, `RANK()`, and so on.

`expression` identifies the column (or target expression) on which this window function will operate.

The `OVER` clause signals that this is a window function, and it consists of three clauses: `Partition`, `Order`, and `Frame`. By adding the `OVER` clause to any aggregate function, you transform it into a window function.

The `Partition` clause is optional, and its goal is to divide the rows into partitions. Next, the window function will operate on each partition. It has the following syntax: `PARTITION BY expr1, expr2, ...`. If `PARTITION BY` is omitted, then the entire result set represents a single partition. To be entirely accurate, if `PARTITION BY` is omitted, then all the data produced by `FROM/WHERE/GROUP BY/HAVING` represents a single partition.

The `Order` clause is also optional, and it handles the order of rows in a partition. Its syntax is `ORDER BY expression [ASC | DESC] [NULLS {FIRST| LAST}] ,...`.

The `Frame` clause demarcates a subset of the current partition. The common syntax is `mode BETWEEN start_of_frame AND end_of_frame [frame_exclusion]`.

`mode` instructs the database about how to treat the input rows. Three possible values indicate the type of relationship between the frame rows and the current row: `ROWS`, `GROUPS`, and `RANGE`.

ROWS

The `ROWS` mode specifies that the offsets of the frame rows and the current row are row numbers (the database sees each input row as an individual unit of work). In this context, `start_of_frame` and `end_of_frame` determine which rows the window frame starts and ends with.

In this context, `start_of_frame` can be `N PRECEDING`, which means that the frame starts at n^th^ rows before the currently evaluated row (in jOOQ, `rowsPreceding(n)`), `UNBOUNDED PRECEDING`, which means that the frame starts at the first row of the current partition (in jOOQ, `rowsUnboundedPreceding()`), and `CURRENT ROW` (jOOQ `rowsCurrentRow()`).

The `end_of_frame` value can be `CURRENT ROW` (previously described), `N FOLLOWING`, which means that the frame ends at the nth row after the currently evaluated row (in jOOQ, `rowsFollowing(n)`), and `UNBOUNDED FOLLOWING`, which means that the frame ends at the last row of the current partition (in jOOQ, `rowsUnboundedFollowing()`).

Check out the following diagram containing some examples:

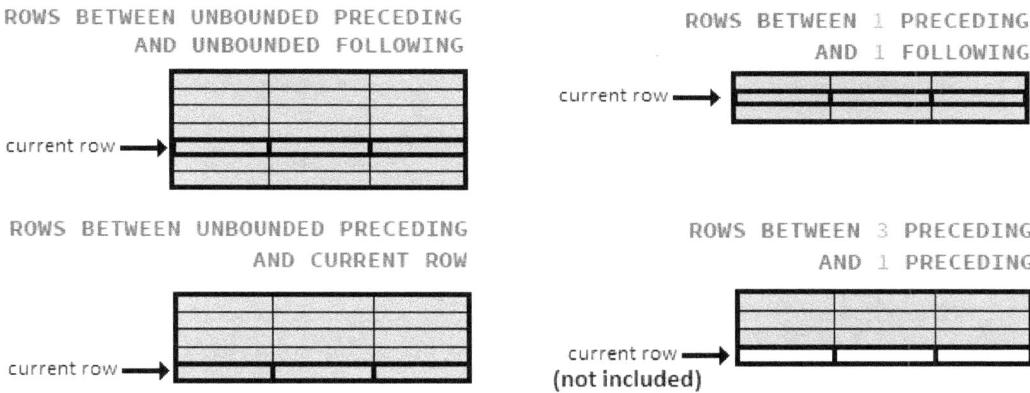

Figure 13.4 – ROWS mode examples

What's in gray represents the included rows.

GROUPS

The `GROUPS` mode instructs the database that the rows with duplicate sorting values should be grouped together. So, `GROUPS` is useful when duplicate values are present.

In this context, `start_of_frame` and `end_of_frame` accept the same values as `ROWS`. But, in the case of `start_of_frame`, `CURRENT_ROW` points to the first row in a group that contains the current row, while in the case of `end_of_frame`, it points to the last row in a group that contains the current row. Moreover, `N PRECEDING/FOLLOWING` refers to groups that should be considered as the number of groups before, respectively, after the current group. On the other hand, `UNBOUNDED PRECEDING/FOLLOWING` has the same meaning as in the case of `ROWS`.

Check out the following diagram containing some examples:

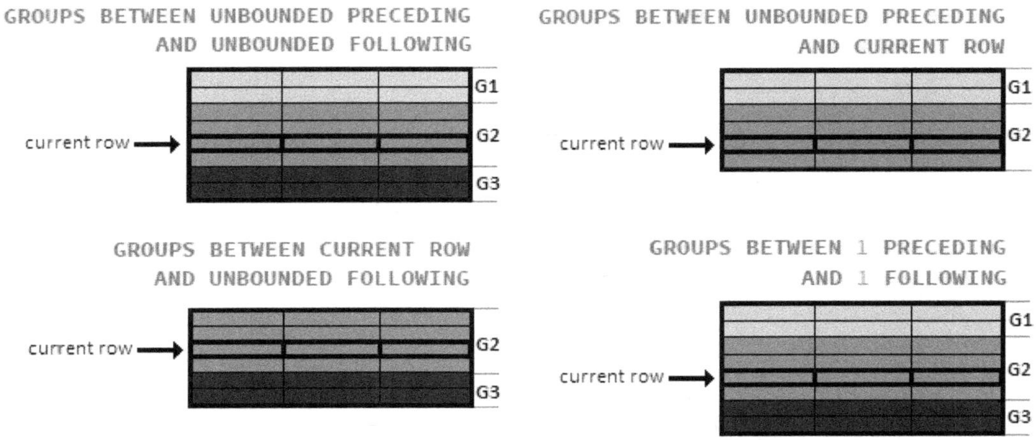

GROUPS BETWEEN UNBOUNDED PRECEDING AND UNBOUNDED FOLLOWING

GROUPS BETWEEN UNBOUNDED PRECEDING AND CURRENT ROW

GROUPS BETWEEN CURRENT ROW AND UNBOUNDED FOLLOWING

GROUPS BETWEEN 1 PRECEDING AND 1 FOLLOWING

current row → G1 G2 G3

Figure 13.5 – GROUPS mode examples

There are three groups (G1, G2, and G3) represented in different shades of gray.

RANGE

The `RANGE` mode doesn't tie rows as `ROWS/GROUPS`. This mode works on a given range of values of the sorting column. This time, for `start_of_frame` and `end_of_frame`, we don't specify the number of rows/groups; instead, we specify the maximum difference of values that the window frame should contain. Both values must be expressed in the same units (or, meaning) as the sorting column is.

In this context, for `start_of_frame`,we have the following: (this time, `N` is a value in the same unit as the sorting column is) `N PRECEDING` (in jOOQ, `rangePreceding(n)`), `UNBOUNDED PRECEDING` (in jOOQ, `rangeUnboundedPreceding()`), and `CURRENT ROW` (in jOOQ, `rangeCurrentRow()`). For `end_of_frame`, we have `CURRENT ROW`, `UNBOUNDED FOLLOWING` (in jOOQ, `rangeUnboundedFollowing()`), `N FOLLOWING` (in jOOQ, `rangeFollowing(n)`).

Check out the following diagram containing some examples:

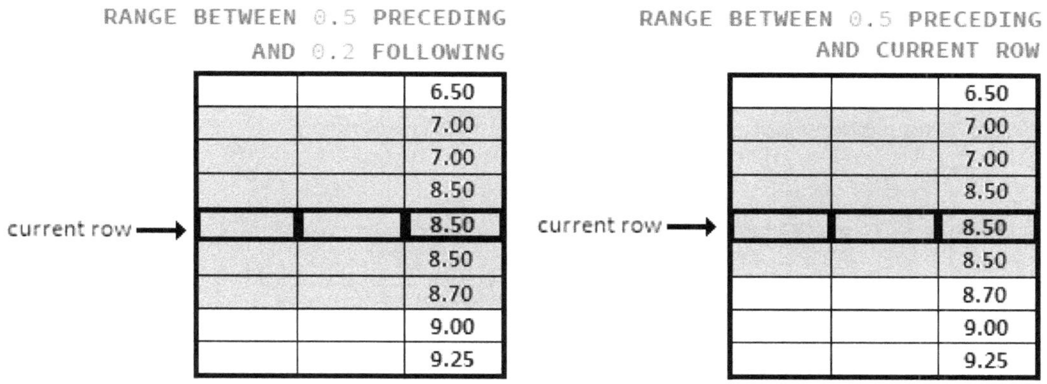

Figure 13.6 – RANGE mode examples

What's in gray represents the included rows.

BETWEEN start_of_frame AND end_of_frame

Especially for the `BETWEEN start_of_frame AND end_of_frame` construction, jOOQ comes with `fooBetweenCurrentRow()`, `fooBetweenFollowing(n)`, `fooBetweenPreceding(n)`, `fooBetweenUnboundedFollowing()`, and `fooBetweenUnboundedPreceding()`. In all these methods, `foo` can be replaced with `rows`, `groups`, or `range`.

In addition, for creating compound frames, jOOQ provides `andCurrentRow()`, `andFollowing(n)`, `andPreceding(n)`, `andUnboundedFollowing()`, and `andUnboundedPreceding()`.

frame_exclusion

Via the `frame_exclusion` optional part, we can exclude certain rows from the window frame. `frame_exclusion` works exactly the same for all three modes. Possible values are listed here:

- `EXCLUDE CURRENT ROW`—Exclude the current row (in jOOQ, `excludeCurrentRow()`).
- `EXCLUDE GROUP`—Exclude the current row but also exclude all peer rows (for instance, exclude all rows having the same value in the sorting column). In jOOQ, we have the `excludeGroup()` method.

- `EXCLUDE TIES`—Exclude all peer rows, but not the current row (in jOOQ, `excludeTies()`).
- `EXCLUDE NO OTHERS`—This is the default, and it means to exclude nothing (in jOOQ, `excludeNoOthers()`).

To better visualize these options, check out the following diagram:

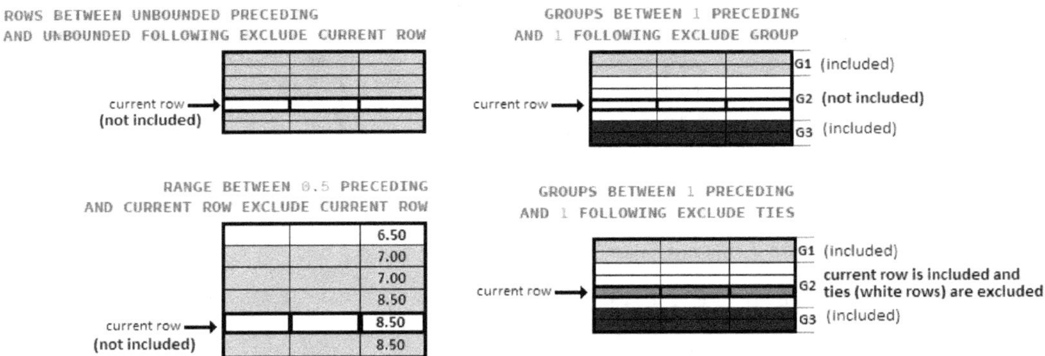

Figure 13.7 – Examples of excluding rows

Speaking about the logical order of operations in SQL, we notice here that window functions are placed between `HAVING` and `SELECT`:

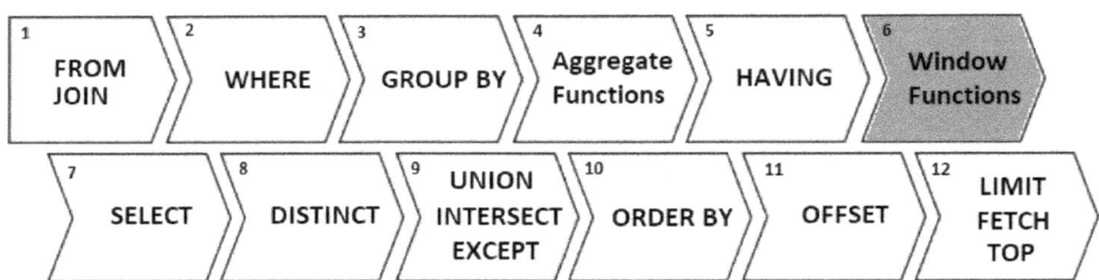

Figure 13.8 – Logical order of operations in SQL

Also, I think is useful to explain that window functions can act upon data produced by all the previous steps *1-5*, and can be declared in all the following steps *7-12* (effectively only in 7 and 10). Before jumping into some window functions examples, let's quickly cover a less-known but quite useful SQL clause.

The QUALIFY clause

Some databases (for instance, Snowflake) support a clause named `QUALIFY`. Via this clause, we can filter (apply a predicate) the results of window functions. Mainly, a `SELECT` ... `QUALIFY` clause is evaluated after window functions are computed, so after **Window Functions** (*Step 6* in *Figure 13.8*) and before **DISTINCT** (*Step 8* in *Figure 13.8*). The syntax of `QUALIFY` is `QUALIFY <predicate>`, and in the following screenshot, you can see how it makes the difference (this query returns every 10th product from the `PRODUCT` table via the `ROW_NUMBER()` window function):

```
ctx.select(field("product_name"))
   .from(select(PRODUCT.PRODUCT_NAME,
          rowNumber().over()
             .orderBy(PRODUCT.PRODUCT_NAME)
             .as("seq")).from(PRODUCT))
   .where(field("seq").mod(10).eq(0))
   .fetch();
```

```
ctx.select(PRODUCT.PRODUCT_NAME)
   .from(PRODUCT)
   .qualify(rowNumber().over()
     .orderBy(PRODUCT.PRODUCT_NAME).mod(10).eq(0))
   .fetch();
```

Figure 13.9 – Logical order of operations in SQL

By using the `QUALIFY` clause, we eliminate the subquery and the code is less verbose. Even if this clause has poor native support among database vendors, jOOQ emulates it for all the supported dialects. Cool, right?! During this chapter, you'll see more examples of using the `QUALIFY` clause.

Working with ROW_NUMBER()

`ROW_NUMBER()` is a ranking window function that assigns a sequential number to each row (it starts from 1). A simple example is shown here:

```
ctx.select(rowNumber().over()
     .orderBy(PRODUCT.PRODUCT_NAME).as("seq_nr"),
     PRODUCT.PRODUCT_NAME, PRODUCT.BUY_PRICE)
   .from(PRODUCT)
   .orderBy(PRODUCT.PRODUCT_NAME)
   .fetch();
```

```
+-------+------------------------+---------+
|seq_nr |product_name            |buy_price|
+-------+------------------------+---------+
|      1|18th century schooner   |    82.34|
|      2|18th Century Vintage    |    60.74|
|      3|1900s Vintage Bi-Plane  |    34.25|
|      4|1900s Vintage Tri-Plane |    36.23|
|      5|1903 Ford Model A       |    68.30|
...
```

Figure 13.10 – Simple example of ROW_NUMBER()

You already saw an example of paginating database views via `ROW_NUMBER()` in *Chapter 12, Pagination and Dynamic Queries*, so you should have no problem understanding the next two examples.

Let's assume that we want to compute the median (`https://en.wikipedia.org/wiki/Median`) of `PRODUCT.QUANTITY_IN_STOCK`. In Oracle and PostgreSQL, this can be done via the built-in `median()` aggregate function, but in MySQL and SQL Server, we have to emulate it somehow, and a good approach consists of using `ROW_NUMBER()`, as follows:

```
Field<Integer> x = PRODUCT.QUANTITY_IN_STOCK.as("x");
Field<Double> y = inline(2.0d).mul(rowNumber().over()
   .orderBy(PRODUCT.QUANTITY_IN_STOCK))
   .minus(count().over()).as("y");

ctx.select(avg(x).as("median")).from(select(x, y)
   .from(PRODUCT))
   .where(y.between(0d, 2d))
   .fetch();
```

That was easy! Next, let's try to solve a different kind of problem, and let's focus on the `ORDER` table. Each order has a `REQUIRED_DATE` and `STATUS` value as `Shipped`, `Cancelled`, and so on. Let's assume that we want to see the clusters (also known as islands) represented by continuous periods of time where the score (`STATUS`, in this case) stayed the same. An output sample can be seen here:

```
+-------------+-----------+-------------+
|cluster_start|cluster_end|cluster_score|
+-------------+-----------+-------------+
|2003-11-17   |2003-11-17 |Cancelled    |
|2003-11-19   |2004-05-13 |Shipped      |
|2004-05-14   |2004-05-14 |Cancelled    |
|2004-05-17   |2004-06-04 |Shipped      |
...
```

Figure 13.11 – Clusters

If we have a requirement to solve this problem via `ROW_NUMBER()` and to express it in jOOQ, then we may come up with this query:

```
Table<?> t = select(
  ORDER.REQUIRED_DATE.as("rdate"), ORDER.STATUS.as("status"),
  (rowNumber().over().orderBy(ORDER.REQUIRED_DATE)
    .minus(rowNumber().over().partitionBy(ORDER.STATUS)
  .orderBy(ORDER.REQUIRED_DATE))).as("cluster_nr"))
  .from(ORDER).asTable("t");
```

```
ctx.select(min(t.field("rdate")).as("cluster_start"),
           max(t.field("rdate")).as("cluster_end"),
           min(t.field("status")).as("cluster_score"))
   .from(t)
   .groupBy(t.field("cluster_nr"))
   .orderBy(1)
   .fetch();
```

You can practice these examples in the *RowNumber* bundled code.

Working with RANK()

`RANK()` is a ranking window function that assigns a rank to each row within the partition of a result set. The rank of a row is computed as *1 + the number of ranks before it*. The columns having the same values get the same ranks; therefore, if multiple rows have the same rank, then the rank of the next row is not consecutive. Think of a competition where two athletes share the first place (or the gold medal) and there is no second place (so, no silver medal). A simple example is provided here:

```
ctx.select(rank().over().orderBy
   (PRODUCT.PRODUCT_LINE).as("rank_nr"),
  PRODUCT.PRODUCT_LINE,
  PRODUCT.PRODUCT_NAME,
  PRODUCT.BUY_PRICE)
  .from(PRODUCT)
  .orderBy(PRODUCT.PRODUCT_LINE)
  .fetch();
```

rank_nr	product_line	product_name	buy_price
1	Classic Cars	1957 Ford Thunderbird	34.21
1	Classic Cars	1957 Corvette Convertible	69.93
1	Classic Cars	1970 Triumph Spitfire	91.92
...			
39	Motorcycles	1969 Harley Davidson	48.81
39	Motorcycles	1996 Moto Guzzi 1100i	68.99
39	Motorcycles	2003 Eagle Drag Bike	91.02
39	Motorcycles	2002 Suzuki XREO	66.27
...			
52	Planes	Corsair F4U (Bird Cage)	29.34
52	Planes	1980s Black Hawk Helicopter	77.27
...			

Figure 13.12 – Simple example of RANK()

Here is another example that ranks `ORDER` by year and months of `ORDER.ORDER_DATE`:

```
ctx.select(ORDER.ORDER_ID, ORDER.CUSTOMER_NUMBER,
      ORDER.ORDER_DATE, rank().over().orderBy(
        year(ORDER.ORDER_DATE), month(ORDER.ORDER_DATE)))
   .from(ORDER).fetch();
```

The `year()` and `month()` shortcuts are provided by jOOQ to avoid the usage of the SQL `EXTRACT()` function. For instance, `year(ORDER.ORDER_DATE)` can be written as `extract(ORDER.ORDER_DATE, DatePart.YEAR)` as well.

How about ranking YEAR is the partition? This can be expressed in jOOQ like this:

```
ctx.select(SALE.EMPLOYEE_NUMBER, SALE.FISCAL_YEAR,
  sum(SALE.SALE_), rank().over().partitionBy(SALE.FISCAL_YEAR)
     .orderBy(sum(SALE.SALE_).desc()).as("sale_rank"))
   .from(SALE)
   .groupBy(SALE.EMPLOYEE_NUMBER, SALE.FISCAL_YEAR)
   .fetch();
```

Finally, let's see an example that ranks the products. Since a partition can be defined via multiple columns, we can easily rank the products by PRODUCT_VENDOR and PRODUCT_SCALE, as shown here:

```
ctx.select(PRODUCT.PRODUCT_NAME, PRODUCT.PRODUCT_VENDOR,
    PRODUCT.PRODUCT_SCALE, rank().over().partitionBy(
      PRODUCT.PRODUCT_VENDOR, PRODUCT.PRODUCT_SCALE)
   .orderBy(PRODUCT.PRODUCT_NAME))
   .from(PRODUCT)
   .fetch();
```

You can practice these examples and more in *Rank*.

Working with DENSE_RANK()

DENSE_RANK() is a window function that assigns a rank to each row within a partition or result set with no gaps in ranking values. A simple example is shown here:

```
ctx.select(SALE.EMPLOYEE_NUMBER, SALE.FISCAL_YEAR, SALE.SALE_,
     denseRank().over().partitionBy(SALE.FISCAL_YEAR)
          .orderBy(SALE.SALE_.desc()).as("sales_rank"))
     .from(SALE)
     .fetch();
```

employee_number	fiscal_year	sale	sales_rank
1370	2000	12434.21	1
1143	2003	25241.42	1
1504	2003	5571.79	2
1370	2003	5282.64	3
1611	2003	4704.91	4
1611	2003	3449.26	5
1166	2003	3213.0	6
1504	2003	1491.38	7
1143	2004	51241.53	1
1611	2004	5657.39	2
1504	2004	5241.43	3
...			

Figure 13.13 – Simple example of DENSE_RANK()

In *Chapter 12, Pagination and Dynamic Queries*, you already saw an example of using `DENSE_RANK()` for paginating `JOIN` statements. Next, let's have another case of ranking employees (`EMPLOYEE`) in offices (`OFFICE`) by their salary (`EMPLOYEE.SALARY`), as follows:

```
ctx.select(EMPLOYEE.FIRST_NAME, EMPLOYEE.LAST_NAME,
  EMPLOYEE.SALARY, OFFICE.CITY, OFFICE.COUNTRY,
  OFFICE.OFFICE_CODE, denseRank().over().partitionBy(
    OFFICE.OFFICE_CODE).orderBy(EMPLOYEE.SALARY.desc())
      .as("salary_rank"))
   .from(EMPLOYEE)
   .innerJoin(OFFICE)
   .on(OFFICE.OFFICE_CODE.eq(EMPLOYEE.OFFICE_CODE)).fetch();
```

An output fragment looks like this (notice that the employees having the same salary get the same rank):

```
+----------+---------+------+-------------+---------+-----------+-----------+
|first_name|last_name|salary|city         |country  |office_code|salary_rank|
+----------+---------+------+-------------+---------+-----------+-----------+
|Diane     |Murphy   |120000|San Francisco|USA      |1          |          1|
|Mary      |Patterson|100000|San Francisco|USA      |1          |          2|
|Jeff      |Firrelli |100000|San Francisco|USA      |1          |          2|
|Anthony   |Bow      | 75000|San Francisco|USA      |1          |          3|
|Leslie    |Jennings | 60000|San Francisco|USA      |1          |          4|
|Leslie    |Thompson | 60000|San Francisco|USA      |1          |          4|
|Julie     |Firrelli | 60000|Boston       |USA      |2          |          1|
|Steve     |Patterson| 55000|Boston       |USA      |2          |          2|
|Foon Yue  |Tseng    | 55000|NYC          |USA      |3          |          1|
|George    |Vanauf   | 55000|NYC          |USA      |3          |          1|
 ...
```

Figure 13.14 – Output

Finally, let's use `DENSE_RANK()` for selecting the highest salary from each office, including duplicates. This time, let's use the `QUALIFY` clause as well. The code is illustrated in the following snippet:

```
select(EMPLOYEE.FIRST_NAME, EMPLOYEE.LAST_NAME,
    EMPLOYEE.SALARY, OFFICE.CITY, OFFICE.COUNTRY,
    OFFICE.OFFICE_CODE)
  .from(EMPLOYEE)
```

```
.innerJoin(OFFICE)
  .on(OFFICE.OFFICE_CODE.eq(EMPLOYEE.OFFICE_CODE))
.qualify(denseRank().over().partitionBy(OFFICE.OFFICE_CODE)
   .orderBy(EMPLOYEE.SALARY.desc()).eq(1))
.fetch();
```

Before going further, here is a nice read: `https://blog.jooq.org/2014/08/12/the-difference-between-row_number-rank-and-dense_rank/`. You can check out these examples in the *DenseRank* bundled code.

Working with PERCENT_RANK()

The `PERCENT_RANK()` window function calculates the percentile rankings `((rank - 1) / (total_rows - 1))` of rows in a result set and returns a value between 0 exclusive and 1 inclusive. The first row in the result set always has the percent rank equal to 0. This function doesn't count `NULL` values and is nondeterministic. Usually, the final result is multiplied by 100 to express as a percentage.

The best way to understand this function is via an example. Let's assume that we want to compute the percentile rank for employees in each office by their salaries. The query expressed in jOOQ will look like this:

```
ctx.select(EMPLOYEE.FIRST_NAME, EMPLOYEE.LAST_NAME,
   EMPLOYEE.SALARY, OFFICE.OFFICE_CODE, OFFICE.CITY,
   OFFICE.COUNTRY, round(percentRank().over()
      .partitionBy(OFFICE.OFFICE_CODE)
      .orderBy(EMPLOYEE.SALARY).mul(100), 2)
      .concat("%").as("PERCENTILE_RANK"))
   .from(EMPLOYEE)
   .innerJoin(OFFICE)
   .on(EMPLOYEE.OFFICE_CODE.eq(OFFICE.OFFICE_CODE))
   .fetch();
```

The following screenshot represents a snippet of the result:

```
+----------+---------+------+-----------+-------------+-------+---------------+
|FIRST_NAME|LAST_NAME|SALARY|OFFICE_CODE|CITY         |COUNTRY|PERCENTILE_RANK|
+----------+---------+------+-----------+-------------+-------+---------------+
|Leslie    |Jennings |  60000|1          |San Francisco|USA    |0%             |
|Leslie    |Thompson |  60000|1          |San Francisco|USA    |0%             |
|Anthony   |Bow      |  75000|1          |San Francisco|USA    |40%            |
|Mary      |Patterson|100000|1          |San Francisco|USA    |60%            |
|Jeff      |Firrelli |100000|1          |San Francisco|USA    |60%            |
|Diane     |Murphy   |120000|1          |San Francisco|USA    |100%           |
|Steve     |Patterson|  55000|2          |Boston       |USA    |0%             |
|Julie     |Firrelli |  60000|2          |Boston       |USA    |100%           |
|Foon Yue  |Tseng    |  55000|3          |NYC          |USA    |0%             |
|George    |Vanauf   |  55000|3          |NYC          |USA    |0%             |
...
```

Figure 13.15 – Percent rank output

So, how do we interpret this output? A percentile rank is commonly defined as the proportion of results (or scores) in a distribution that a certain result (or score) is greater than or equal to (sometimes only greater than counts). For example, if you get a result/score of 90 on a certain test and this result/score was greater than (or equal to) the results/scores of 75% of the participants taking the test, then your percentile rank is 75. You would be in the 75th percentile.

In other words, in office 1, we can say that 40% of employees have salaries lower than *Anthony Bow* (check the third row), so *Anthony Bow* is in the 40th percentile. Also, in office 1, *Diane Murphy* has the highest salary since 100% of employees have salaries lower than her salary (check the sixth row). When the current row is the first in the partition then there is no previous data to consider, therefore the percentile rank is 0. An interesting case is *George Vanauf* (last row) having a percentile rank of 0%. Because his salary (*$55,000*) is equal to the salary of *Foon Yue Tseng*, we can say that nobody has a salary lower than his.

A common use case for `PERCENT_RANK()` is to categorize data into custom groups (also known as custom binning). For example, let's consider that we want to count departments having a low (smaller than the 20th percentile), medium (between the 20th and 80th percentile), and high (greater than 80th percentile) profit. Here's the code we'd use to calculate this:

```
ctx.select(count().filterWhere(field("p").lt(0.2))
     .as("low_profit"),
  count().filterWhere(field("p").between(0.2, 0.8))
     .as("good_profit"),
```

```
 count().filterWhere(field("p").gt(0.8))
    .as("high_profit"))
.from(select(percentRank().over()
        .orderBy(DEPARTMENT.PROFIT).as("p"))
        .from(DEPARTMENT)
        .where(DEPARTMENT.PROFIT.isNotNull()))
.fetch();
```

You can practice these examples—and more—in the *PercentRank* bundled code.

Working with CUME_DIST()

`CUME_DIST()` is a window function that computes the cumulative distribution of a value within a set of values. In other words, `CUME_DIST()` divides the number of rows having values less than or equal to the current row's value by the total number of rows. The returned value is greater than zero and less than or equal to one (0 < `CUME_DIST()` <= 1). The columns having repeated values get the same `CUME_DIST()` value. A simple example is provided here:

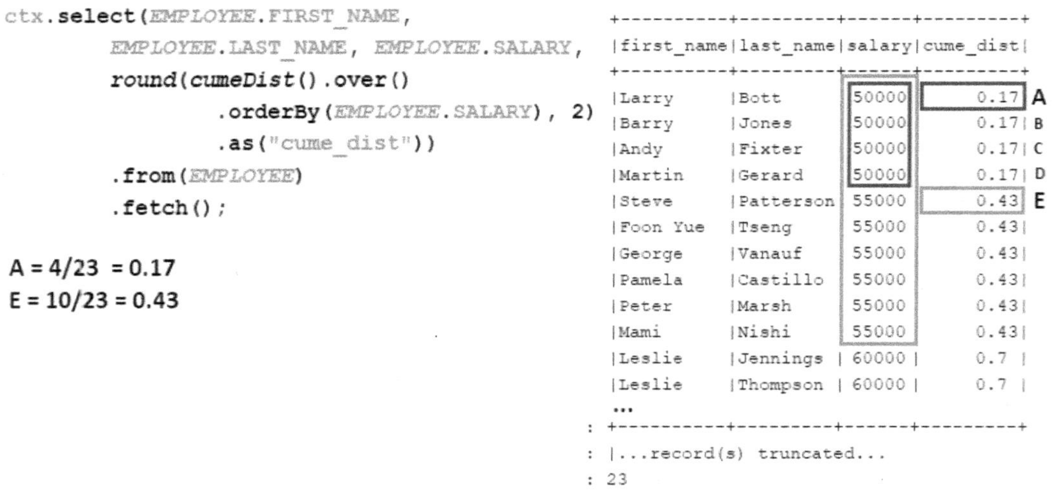

```
ctx.select(EMPLOYEE.FIRST_NAME,
        EMPLOYEE.LAST_NAME, EMPLOYEE.SALARY,
        round(cumeDist().over()
                .orderBy(EMPLOYEE.SALARY), 2)
                .as("cume_dist"))
        .from(EMPLOYEE)
        .fetch();
```

A = 4/23 = 0.17
E = 10/23 = 0.43

```
+----------+---------+------+---------+
|first_name|last_name|salary|cume_dist|
+----------+---------+------+---------+
|Larry     |Bott     |50000 |     0.17| A
|Barry     |Jones    |50000 |     0.17| B
|Andy      |Fixter   |50000 |     0.17| C
|Martin    |Gerard   |50000 |     0.17| D
|Steve     |Patterson| 55000|     0.43| E
|Foon Yue  |Tseng    | 55000|     0.43|
|George    |Vanauf   | 55000|     0.43|
|Pamela    |Castillo | 55000|     0.43|
|Peter     |Marsh    | 55000|     0.43|
|Mami      |Nishi    | 55000|     0.43|
|Leslie    |Jennings | 60000|     0.7 |
|Leslie    |Thompson | 60000|     0.7 |
...
: +----------+---------+------+---------+
: |...record(s) truncated...
: 23
```

Figure 13.16 – Simple example of CUME_DIST()

So, we have a result set of 23 rows. For the first row (denoted as A), `CUME_DIST()` finds the number of rows having a value less than or equal to *50000*. The result is 4. Then, the function divides 4 by the total number of rows, which is 23: 4/23. The result is 0.17 or 17%. The same logic is applied to the next rows.

How about fetching the top 25% of sales in *2003* and *2004*? This can be solved via `CUME_DIST()` and the handy `QUALIFY` clause, as follows:

```
ctx.select(concat(EMPLOYEE.FIRST_NAME, inline(" "),
 EMPLOYEE.LAST_NAME).as("name"), SALE.SALE_, SALE.FISCAL_YEAR)
   .from(EMPLOYEE)
   .join(SALE)
   .on(EMPLOYEE.EMPLOYEE_NUMBER.eq(SALE.EMPLOYEE_NUMBER)
    .and(SALE.FISCAL_YEAR.in(2003, 2004)))
   .qualify(cumeDist().over().partitionBy(SALE.FISCAL_YEAR)
   .orderBy(SALE.SALE_.desc()).lt(BigDecimal.valueOf(0.25)))
   .fetch();
```

You can practice these examples in the *CumeDist* bundled code.

Working with LEAD()/LAG()

`LEAD()` is a window function that looks forward a specified number of rows (offset, by default 1) and accesses that row from the current row. `LAG()` works the same as `LEAD()`, but it looks back. For both, we can optionally specify a default value to be returned when there is no subsequent row (`LEAD()`) or there is no preceding row (`LAG()`) instead of returning `NULL`. A simple example is provided here:

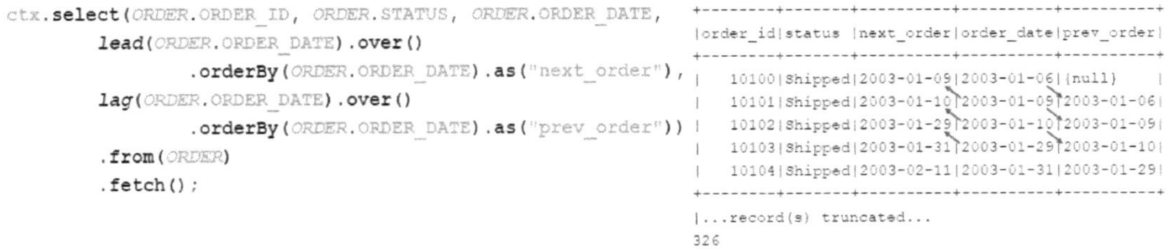

```
ctx.select(ORDER.ORDER_ID, ORDER.STATUS, ORDER.ORDER_DATE,
        lead(ORDER.ORDER_DATE).over()
                .orderBy(ORDER.ORDER_DATE).as("next_order"),
        lag(ORDER.ORDER_DATE).over()
                .orderBy(ORDER.ORDER_DATE).as("prev_order"))
        .from(ORDER)
        .fetch();
```

```
+--------+-------+----------+----------+----------+
|order_id|status |next_order|order_date|prev_order|
+--------+-------+----------+----------+----------+
|   10100|Shipped|2003-01-09|2003-01-06|{null}    |
|   10101|Shipped|2003-01-10|2003-01-09|2003-01-06|
|   10102|Shipped|2003-01-29|2003-01-10|2003-01-09|
|   10103|Shipped|2003-01-31|2003-01-29|2003-01-10|
|   10104|Shipped|2003-02-11|2003-01-31|2003-01-29|
+--------+-------+----------+----------+----------+
|...record(s) truncated...
326
```

Figure 13.17 – Simple example of LEAD() and LAG()

Besides the `lead/lag(Field<T> field)` syntax used in this example, jOOQ also exposes `lead/lag(Field<T> field, int offset)`, `lead/lag(Field<T> field, int offset, Field<T> defaultValue)`, and `lead/lag(Field<T> field, int offset, T defaultValue)`. In this example, `lead/lag(ORDER.ORDER_DATE)` uses an offset of 1, so is the same thing as `lead/lag(ORDER.ORDER_DATE, 1)`.

Here is an example that, for each employee, displays the salary and next salary using the office as a partition of `LEAD()`. When `LEAD()` reaches the end of the partition, we use 0 instead of `NULL`:

```
ctx.select(OFFICE.OFFICE_CODE, OFFICE.CITY, OFFICE.COUNTRY,
 EMPLOYEE.FIRST_NAME, EMPLOYEE.LAST_NAME, EMPLOYEE.SALARY,
 lead(EMPLOYEE.SALARY, 1, 0).over()
    .partitionBy(OFFICE.OFFICE_CODE)
      .orderBy(EMPLOYEE.SALARY).as("next_salary"))
   .from(OFFICE)
   .innerJoin(EMPLOYEE)
   .on(OFFICE.OFFICE_CODE.eq(EMPLOYEE.OFFICE_CODE))
   .fetch();
```

Next, let's tackle an example of calculating the Month-Over-Month (MOM) growth rate. This financial indicator is useful for benchmarking the business, and we already have it in the `SALE.REVENUE_GROWTH` column. But here is the query that can calculate it via the `LAG()` function for the year *2004*:

```
ctx.select(SALE.FISCAL_MONTH,
  inline(100).mul((SALE.SALE_.minus(lag(SALE.SALE_, 1)
    .over().orderBy(SALE.FISCAL_MONTH)))
    .divide(lag(SALE.SALE_, 1).over()
     .orderBy(SALE.FISCAL_MONTH))).concat("%").as("MOM"))
   .from(SALE)
   .where(SALE.FISCAL_YEAR.eq(2004))
   .orderBy(SALE.FISCAL_MONTH)
   .fetch();
```

For more examples, including an example of funneling drop-off metrics and one about time-series analysis, please check out the *LeadLag* bundled code.

Working with NTILE()

`NTILE(n)` is a window function commonly used for distributing the number of rows in the specified `n` number of groups or buckets. Each bucket has a number (starting at 1) that indicates the bucket to which this row belongs. A simple example is provided here:

```
ctx.select(EMPLOYEE.FIRST_NAME,
      EMPLOYEE.LAST_NAME, EMPLOYEE.SALARY,
      ntile(10).over()
            .orderBy(EMPLOYEE.SALARY.desc())
            .as("salary_group"))
      .from(EMPLOYEE)
      .fetch();
```

```
+----------+---------+------+------------+
|first_name|last_name|salary|salary_group|
+----------+---------+------+------------+
|Diane     |Murphy   |120000|           1|
|Mary      |Patterson|100000|           1|
|Jeff      |Firrelli |100000|           1|
|William   |Patterson| 80000|           2|
|Gerard    |Bondur   | 80000|           2|
|Anthony   |Bow      | 75000|           2|
|Gerard    |Hernandez| 65000|           3|
|Leslie    |Jennings | 60000|           3|
|Leslie    |Thompson | 60000|           3|
...
```

Figure 13.18 – Simple example of NTILE()

So, in this example, we've distributed `EMPLOYEE.SALARY` in 10 buckets. `NTILE()` strives to determine how many rows should be in each bucket in order to provide the number of buckets and to keep them approximately equal.

Among its use cases, `NTILE()` is useful for calculating Recency, Frequency, and Monetary (RFM) indices (`https://en.wikipedia.org/wiki/RFM_(market_research)`). In short, the RFM analysis is basically an indexing technique that relies on past purchase behavior to determine different segments of customers.

In our case, the past purchase behavior of each customer (`ORDER.CUSTOMER_NUMBER`) is stored in the `ORDER` table, especially in `ORDER.ORDER_ID`, `ORDER.ORDER_DATE`, and `ORDER.AMOUNT`.

Based on this information, we attempt to divide customers into four equal groups based on the distribution of values for R, F, and M. Four equal groups across RFM variables produce $4^3=64$ potential segments. The result consists of a table having a score between 1 and 4 for each of the quantiles (R, F, and M). The query speaks for itself, as we can see here:

```
ctx.select(field("customer_number"),
    ntile(4).over().orderBy(field("last_order_date"))
     .as("rfm_recency"),
    ntile(4).over().orderBy(field("count_order"))
     .as("rfm_frequency"),
    ntile(4).over().orderBy(field("avg_amount"))
     .as("rfm_monetary")).from(
     select(ORDER.CUSTOMER_NUMBER.as("customer_number"),
      max(ORDER.ORDER_DATE).as("last_order_date"),
      count().as("count_order"),
      avg(ORDER.AMOUNT).as("avg_amount"))
        .from(ORDER)
        .groupBy(ORDER.CUSTOMER_NUMBER))
  .fetch();
```

A sample output is provided here:

```
+---------------+-----------+-------------+------------+
|customer_number|rfm_recency|rfm_frequency|rfm_monetary|
+---------------+-----------+-------------+------------+
|219            |          3|            2|           1|
|103            |          2|            2|           1|
|381            |          3|            4|           1|
|198            |          3|            2|           1|
|473            |          1|            1|           1|
+---------------+-----------+-------------+------------+
|...record(s) truncated...
98
```

Figure 13.19 – RFM sample

By combining the RFM result as R*100+F*10+M, we can obtain an aggregate score. This is available next to more examples in the *Ntile* bundled code.

Working with FIRST_VALUE() and LAST_VALUE()

`FIRST_VALUE(expr)` returns the value of the specified expression (`expr`) with respect to the first row in the window frame.

`NTH_VALUE(expr, offset)` returns the value of the specified expression (`expr`) with respect to the offset row in the window frame.

`LAST_VALUE(expr)` returns the value of the specified expression (`expr`) with respect to the last row in the window frame.

Let's assume that our goal is to obtain the cheapest and most expensive product per product line, as in the following screenshot:

```
+------------+-----------------------------------+---------+-----------------------------------+--------------------------+
|product_line|procuct_name                       |buy_price|cheapest                           |most_expensive            |
+------------+-----------------------------------+---------+-----------------------------------+--------------------------+
Classic Cars|1958 Chevy Corvette Limited Edition|    15.91|1958 Chevy Corvette Limited Edition|1962 LanciaA Delta 16V
Classic Cars|1982 Lamborghini Diablo            |    16.24|1958 Chevy Corvette Limited Edition|1962 LanciaA Delta 16V
...
Classic Cars|1998 Chrysler Plymouth Prowler     |   101.51|1958 Chevy Corvette Limited Edition|1962 LanciaA Delta 16V
Classic Cars|1962 LanciaA Delta 16V             |   103.42|1958 Chevy Corvette Limited Edition|1962 LanciaA Delta 16V
Planes      |Corsair F4U ( Bird Cage)           |    29.34|Corsair F4U ( Bird Cage)           |1980s Black Hawk Helicopter
Planes      |Boeing X-32A JSF                   |    32.77|Corsair F4U ( Bird Cage)           |1980s Black Hawk Helicopter
...
Planes      |America West Airlines B757-200     |    68.80|Corsair F4U ( Bird Cage)           |1980s Black Hawk Helicopter
Planes      |1980s Black Hawk Helicopter     ...|    77.27|Corsair F4U ( Bird Cage)           |1980s Black Hawk Helicopter
|Ships      |Pont Yacht                         |    33.30|Pont Yacht                         |18th century schooner     |
|Ships      |The USS Constitution Ship          |    33.97|Pont Yacht                         |18th century schooner     |
...
```

Figure 13.20 – Cheapest and most expensive product per product line

Accomplishing this task via `FIRST_VALUE()` and `LAST_VALUE()` can be done like this:

```
ctx.select(PRODUCT.PRODUCT_LINE,
  PRODUCT.PRODUCT_NAME, PRODUCT.BUY_PRICE,
  firstValue(PRODUCT.PRODUCT_NAME).over()
   .partitionBy(PRODUCT.PRODUCT_LINE)
     .orderBy(PRODUCT.BUY_PRICE).as("cheapest"),
  lastValue(PRODUCT.PRODUCT_NAME).over()
   .partitionBy(PRODUCT.PRODUCT_LINE)
     .orderBy(PRODUCT.BUY_PRICE)
       .rangeBetweenUnboundedPreceding()
       .andUnboundedFollowing().as("most_expensive"))
  .from(PRODUCT)
  .fetch();
```

If the window frame is not specified, then the default window frame depends on the presence of `ORDER BY`. If `ORDER BY` is present, then the window frame is `RANGE BETWEEN UNBOUNDED PRECEDING AND CURRENT ROW`. If `ORDER BY` is not present, then the window frame is `RANGE BETWEEN UNBOUNDED PRECEDING AND UNBOUNDED FOLLOWING`.

Having this in mind, in our case, `FIRST_VALUE()` can rely on the default window frame to return the first row of the partition, which is the smallest price. On the other hand, `LAST_VALUE()` must explicitly define the window frame as `RANGE BETWEEN UNBOUNDED PRECEDING AND UNBOUNDED FOLLOWING` to return the highest price.

Here is another example of fetching the second most expensive product by product line via `NTH_VALUE()`:

```
ctx.select(PRODUCT.PRODUCT_LINE,
  PRODUCT.PRODUCT_NAME, PRODUCT.BUY_PRICE,
  nthValue(PRODUCT.PRODUCT_NAME, 2).over()
   .partitionBy(PRODUCT.PRODUCT_LINE)
    .orderBy(PRODUCT.BUY_PRICE.desc())
     .rangeBetweenUnboundedPreceding()
     .andUnboundedFollowing().as("second_most_expensive"))
   .from(PRODUCT)
   .fetch();
```

The preceding query orders `BUY_PRICE` in descending order for fetching the second most expensive product by product line. But this is mainly the second row from the bottom, therefore we can rely on the `FROM LAST` clause (in jOOQ, `fromLast()`) to express it, as follows:

```
ctx.select(PRODUCT.PRODUCT_LINE,
   PRODUCT.PRODUCT_NAME, PRODUCT.BUY_PRICE,
   nthValue(PRODUCT.PRODUCT_NAME, 2).fromLast().over()
   .partitionBy(PRODUCT.PRODUCT_LINE)
    .orderBy(PRODUCT.BUY_PRICE)
     .rangeBetweenUnboundedPreceding()
     .andUnboundedFollowing().as("second_most_expensive"))
   .from(PRODUCT)
   .fetch();
```

This query works fine in Oracle, which supports `FROM FIRST` (`fromFirst()`), `FROM LAST` (`fromLast()`), `IGNORE NULLS` (`ignoreNulls()`), and `RESPECT NULLS` (`respectNulls()`).

You can practice these examples in the *FirstLastNth* bundled code.

Working with RATIO_TO_REPORT()

`RATIO_TO_REPORT(expr)` computes the ratio of the specified value to the sum of values in the set. If the given `expr` value is evaluated as `null`, then this function returns `null`. A simple example is provided here:

```
ctx.select(SALE.EMPLOYEE_NUMBER, SALE.FISCAL_YEAR,
       round(SALE.SALE_, 2).as("sale"),
       round(ratioToReport(SALE.SALE_).over(), 2)
            .as("ratio_to_report_sale"))
       .from(SALE)
       .fetch();
```

employee_number	fiscal_year	sale	ratio_to_report_sale
1143	2004	51241.54	0.14
1143	2003	25241.43	0.07
1370	2000	12434.22	0.03
1504	2007	9008.22	0.02
1611	2007	150399.34	0.41
1102	2005	52343.12	0.14
...			

Figure 13.21 – Simple example of RATIO_TO_REPORT()

For instance, for the first row, the ratio is computed as 51241.54 / 369418.38, where 369418.38 is the sum of all sales. After applying the `round()` function, the result is 0.14 or 14%, but if we want to compute the ratio of the current sale per fiscal year, we can do it via `PARTITION BY`, as shown here:

```
ctx.select(SALE.EMPLOYEE_NUMBER, SALE.FISCAL_YEAR, SALE.SALE_,
  round(ratioToReport(SALE.SALE_).over()
          .partitionBy(SALE.FISCAL_YEAR), 2)
          .as("ratio_to_report_sale"))
  .from(SALE).fetch();
```

Let's compute the ratio of the current sum of salaries per employee and express it in percentages, like so:

```
ctx.select(OFFICE.OFFICE_CODE,
     sum(EMPLOYEE.SALARY).as("salaries"),
       ratioToReport(sum(EMPLOYEE.SALARY)).over()
        .mul(100).concat("%").as("ratio_to_report"))
   .from(OFFICE)
   .join(EMPLOYEE)
   .on(OFFICE.OFFICE_CODE.eq(EMPLOYEE.OFFICE_CODE))
   .groupBy(OFFICE.OFFICE_CODE)
   .orderBy(OFFICE.OFFICE_CODE)
   .fetch();
```

You can check out these examples in the *RatioToReport* bundled code.

Aggregates as window functions

Aggregate functions can be used as window functions as well. For instance, let's use the `SUM()` aggregate function as a window function for computing the sum of the successfully transferred amount per customer until each caching date, as illustrated in the following screenshot:

customer_number	caching_date	transfer_amount	status	result
447	2003-09-15T18:22:54	6631.36	SUCCESS	6631.36
447	2004-12-27T16:15:22	12304.13	SUCCESS	18935.49
462	2003-11-08T18:57:25	9977.85	SUCCESS	9977.85
462	2004-07-01T12:31:56	3382.13	SUCCESS	13359.98
462	2004-07-03T11:21:16	2566.17	SUCCESS	15926.15
462	2004-11-27T14:30:22	48355.87	SUCCESS	64282.02
...				

Figure 13.22 – Sum of the transferred amount until each caching date

The jOOQ query can be expressed like this:

```
ctx.select(BANK_TRANSACTION.CUSTOMER_NUMBER,
  BANK_TRANSACTION.CACHING_DATE,
  BANK_TRANSACTION.TRANSFER_AMOUNT, BANK_TRANSACTION.STATUS,
  sum(BANK_TRANSACTION.TRANSFER_AMOUNT).over()
   .partitionBy(BANK_TRANSACTION.CUSTOMER_NUMBER)
```

```
    .orderBy(BANK_TRANSACTION.CACHING_DATE)
    .rowsBetweenUnboundedPreceding().andCurrentRow().as("result"))
  .from(BANK_TRANSACTION)
  .where(BANK_TRANSACTION.STATUS.eq("SUCCESS")).fetch();
```

Or, let's use the AVG() aggregate function as a window function for computing the average of prices for the preceding three ordered products on each order, as illustrated in the following screenshot:

order_id	product_id	order_line_number	quantity_ordered	price_each	avg_prec_3_prices
10103	53	9	41	40.75	40.750000
10103	30	2	22	58.34	49.545000
10103	36	14	35	61.84	53.643333
10103	90	7	45	63.35	61.176667
10103	44	16	46	86.31	70.500000
10103	38	13	25	86.92	78.860000
10103	82	15	25	88.62	87.283333
10103	35	12	27	92.19	89.243333
10103	85	3	31	92.46	91.090000
...					

Figure 13.23 – Average of prices for the preceding three ordered products on each order

The query looks like this:

```
ctx.select(ORDERDETAIL.ORDER_ID, ORDERDETAIL.PRODUCT_ID, ...,
       avg(ORDERDETAIL.PRICE_EACH).over()
        .partitionBy(ORDERDETAIL.ORDER_ID)
        .orderBy(ORDERDETAIL.PRICE_EACH)
        .rowsPreceding(2).as("avg_prec_3_prices"))
   .from(ORDERDETAIL).fetch();
```

How about calculating a running average flavor—in other words, create a report that shows every transaction in *March 2005* for *Visa Electron* cards? Additionally, this report shows the daily average transaction amount relying on a 3-day moving average. The code to accomplish this is shown in the following snippet:

```
ctx.select(
  BANK_TRANSACTION.CACHING_DATE, BANK_TRANSACTION.CARD_TYPE,
  sum(BANK_TRANSACTION.TRANSFER_AMOUNT).as("daily_sum"),
  avg(sum(BANK_TRANSACTION.TRANSFER_AMOUNT)).over()
    .orderBy(BANK_TRANSACTION.CACHING_DATE)
```

```
        .rowsBetweenPreceding(2).andCurrentRow()
           .as("transaction_running_average"))
   .from(BANK_TRANSACTION)
   .where(BANK_TRANSACTION.CACHING_DATE
   .between(LocalDateTime.of(2005, 3, 1, 0, 0, 0),
            LocalDateTime.of(2005, 3, 31, 0, 0, 0))
   .and(BANK_TRANSACTION.CARD_TYPE.eq("VisaElectron")))
   .groupBy(BANK_TRANSACTION.CACHING_DATE,
            BANK_TRANSACTION.CARD_TYPE)
   .orderBy(BANK_TRANSACTION.CACHING_DATE).fetch();
```

As Lukas Eder mentioned: "*What's most mind-blowing about aggregate window functions is that even user-defined aggregate functions can be used as window functions!*"

You can check out more examples (for instance, in the *PostgreSQL* bundled code, you can find queries for *How many other employees have the same salary as me?* and *How many sales are better by 5,000 or less?*) in the *AggregateWindowFunctions* bundled code.

Aggregate functions and ORDER BY

Certain aggregate functions output significantly different results depending on their input order. By default, this ordering is not specified, but it can be controlled via an optional `ORDER BY` clause as an argument. So, in the presence of `ORDER BY` on these aggregate function calls, we can fetch ordered aggregated results. Let's see how we can use such functions in jOOQ and start with a category of functions having their names suffixed with `AGG`, such as `ARRAY_AGG()`, `JSON_ARRAYAGG()`, `XML_AGG()`, `MULTISET_AGG()` (covered in *Chapter 8, Fetching and Mapping*), and so on.

FOO_AGG()

For instance, `ARRAY_AGG()` is a function that aggregates data into an array and, in the presence of `ORDER BY`, it aggregates data into an array conforming to the specified order. Here is an example of using `ARRAY_AGG()` to aggregate `EMPLOYEE.FIRST_NAME` in descending order by `EMPLOYEE.FIRST_NAME` and `LAST_NAME`:

```
ctx.select(arrayAgg(EMPLOYEE.FIRST_NAME).orderBy(
      EMPLOYEE.FIRST_NAME.desc(),
               EMPLOYEE.LAST_NAME.desc()))
   .from(EMPLOYEE).fetch();
```

For PostgreSQL, jOOQ renders this SQL:

```
SELECT ARRAY_AGG(
    "public"."employee"."first_name"
    ORDER BY
      "public"."employee"."first_name" DESC,
      "public"."employee"."last_name" DESC
  ) FROM "public"."employee"
```

The result is an array as [*Yoshimi*, *William*, *Tom*, *Steve*, *Peter*,...], wrapped as `Result<Record1<String[]>>` (extract `String[]` via `get(0).value1()`). Do not confuse `ARRAY_AGG()` with jOOQ's `fetchArray()`. In the case of `ARRAY_AGG()`, the array is built by the database, while in the case of `fetchArray()`, the array is built by jOOQ after fetching the result set.

Another two aggregation functions that accept `ORDER BY` are `JSON_ARRAYAGG()` and `XML_AGG()`. You should be familiar with these functions from *Chapter 8, Fetching and Mapping*, but you can also see several simple examples in the code bundled with this section.

COLLECT()

An interesting method that accepts `ORDER BY` is Oracle's `COLLECT()` method. While `ARRAY_AGG()` represents the standard SQL function for aggregating data into an array, the `COLLECT()` function is specific to Oracle and produces a structurally typed array. Let's assume the following Oracle user-defined type:

```
CREATE TYPE "SALARY_ARR" AS TABLE OF NUMBER(7);
```

The jOOQ Code Generator will produce for this user-defined type the `SalaryArrRecord` class in `jooq.generated.udt.records`. Via this UDT record, we can collect in descending order by salary and ascending order by job title the employees' salaries, as follows:

```
var result = ctx.select(
      collect(EMPLOYEE.SALARY, SalaryArrRecord.class)
  .orderBy(EMPLOYEE.SALARY.asc(),
           EMPLOYEE.JOB_TITLE.desc()))
  .from(EMPLOYEE).fetch();
```

jOOQ fetches `Result<Record1<SalaryArrRecord>>` via the following SQL:

```
SELECT CAST(COLLECT(
    "CLASSICMODELS"."EMPLOYEE"."SALARY"
ORDER BY
    "CLASSICMODELS"."EMPLOYEE"."SALARY" ASC,
    "CLASSICMODELS"."EMPLOYEE"."JOB_TITLE" DESC)
     AS "CLASSICMODELS"."SALARY_ARR")
FROM "CLASSICMODELS"."EMPLOYEE"
```

By calling `get(0).value1().toArray(Integer[]::new)`, you can access the array of salaries. Or, by calling `get(0).value1().get(5)`, you can access the fifth salary. Relying on `fetchOneInto()`/`fetchSingleInto()` is also an option, as illustrated here:

```
SalaryArrRecord result = ctx.select(
  collect(EMPLOYEE.SALARY, SalaryArrRecord.class)
  .orderBy(EMPLOYEE.SALARY.asc(), EMPLOYEE.JOB_TITLE.desc()))
  .from(EMPLOYEE)
  .fetchOneInto(SalaryArrRecord.class);
```

Now, you can access the array of salaries as `result.toArray(Integer[]::new)`, and via `result.get(5)`, you can access the fifth salary.

GROUP_CONCAT()

Another cool aggregate function that accepts an `ORDER BY` clause is the `GROUP_CONCAT()` function (very popular in MySQL), useful to get the aggregated concatenation for a field. jOOQ emulates this function in Oracle, PostgreSQL, SQL Server, and other dialects that don't support it natively.

For instance, let's use `GROUP_CONCAT()` to fetch a string containing employees' names in descending order by salary, as follows:

```
ctx.select(groupConcat(concat(EMPLOYEE.FIRST_NAME,
      inline(" "), EMPLOYEE.LAST_NAME))
   .orderBy(EMPLOYEE.SALARY.desc()).separator(";")
     .as("names_of_employees"))
   .from(EMPLOYEE).fetch();
```

The output will be something like this: *Diane Murphy*; *Mary Patterson*; *Jeff Firrelli*; ….

Oracle's KEEP() clause

Here's a quick one—have you seen in a query an aggregate function like this: `SUM(some_value) KEEP (DENSE_RANK FIRST ORDER BY some_date)`? Or this analytic variant: `SUM(some_value) KEEP (DENSE_RANK LAST ORDER BY some_date) OVER (PARTITION BY some_partition)`?

If you did, then you know that what you saw is Oracle's `KEEP()` clause at work, or—in other words—the SQL `FIRST()` and `LAST()` functions prefixed by the `KEEP()` clause for semantic clarity, and `DENSE_RANK()` for indicating that Oracle should aggregate only on Olympic rank (those rows with the maximum (`LAST()`) or minimum (`FIRST()`) dense rank with respect to a given sorting), respectively suffixed by `ORDER BY()` and, optionally, by `OVER(PARTITION BY())`. Both `LAST()` and `FIRST()` can be treated as aggregates (if you omit the `OVER()` clause) or as analytic functions.

But let's have a scenario based on `CUSTOMER` and `ORDER` tables. Each customer (`CUSTOMER.CUSTOMER_NUMBER`) has one or more order, and let's assume that we want to fetch the `ORDER.ORDER_DATE` value closest to 2004-June-06 (or any other date, including the current date) for each `CUSTOMER` type. This can be easily accomplished in a query, as here:

```
ctx.select(ORDER.CUSTOMER_NUMBER, max(ORDER.ORDER_DATE))
   .from(ORDER)
   .where(ORDER.ORDER_DATE.lt(LocalDate.of(2004, 6, 6)))
   .groupBy(ORDER.CUSTOMER_NUMBER)
   .fetch();
```

How about selecting `ORDER.SHIPPED_DATE` and `ORDER.STATUS` as well? One approach could be to rely on the `ROW_NUMBER()` window function and the `QUALIFY()` clause, as shown here:

```
ctx.select(ORDER.CUSTOMER_NUMBER, ORDER.ORDER_DATE,
     ORDER.SHIPPED_DATE, ORDER.STATUS)
   .from(ORDER)
   .where(ORDER.ORDER_DATE.lt(LocalDate.of(2004, 6, 6)))
   .qualify(rowNumber().over()
     .partitionBy(ORDER.CUSTOMER_NUMBER)
       .orderBy(ORDER.ORDER_DATE.desc()).eq(1))
   .fetch();
```

As you can see in the bundled code, another approach could be to rely on SELECT DISTINCT ON (as *@dmitrygusev* suggested on Twitter) or on an anti-join, but if we write our query for Oracle, then most probably you'll go for the KEEP() clause, as follows:

```
ctx.select(ORDER.CUSTOMER_NUMBER,
           max(ORDER.ORDER_DATE).as("ORDER_DATE"),
           max(ORDER.SHIPPED_DATE).keepDenseRankLastOrderBy(
               ORDER.SHIPPED_DATE).as("SHIPPED_DATE"),
           max(ORDER.STATUS).keepDenseRankLastOrderBy(
               ORDER.SHIPPED_DATE).as("STATUS"))
   .from(ORDER)
   .where(ORDER.ORDER_DATE.lt(LocalDate.of(2004, 6, 6)))
   .groupBy(ORDER.CUSTOMER_NUMBER).fetch();
```

Or, you could do this by exploiting the Oracle's ROWID pseudo-column, as follows:

```
ctx.select(ORDER.CUSTOMER_NUMBER, ORDER.ORDER_DATE,
      ORDER.SHIPPED_DATE, ORDER.STATUS)
   .from(ORDER)
   .where((rowid().in(select(max((rowid()))
      .keepDenseRankLastOrderBy(ORDER.SHIPPED_DATE))
      .from(ORDER)
      .where(ORDER.ORDER_DATE.lt(LocalDate.of(2004, 6, 6)))
      .groupBy(ORDER.CUSTOMER_NUMBER)))).fetch();
```

You can practice these examples in the *AggregateFunctionsOrderBy* bundled code.

Ordered set aggregate functions (WITHIN GROUP)

Ordered set aggregate functions allow operations on a set of rows sorted with ORDER BY via the **mandatory** WITHIN GROUP clause. Commonly, such functions are used for performing computations that depend on a certain row ordering. Here, we can quickly mention *hypothetical set* functions such as RANK(), DENSE_RANK(), PERCENT_RANK(), or CUME_DIST(), and *inverse distribution functions* such as PERCENTILE_CONT(), PERCENTILE_DISC(), or MODE(). A particular case is represented by LISTAGG(), which is covered at the end of this section.

Hypothetical set functions

A hypothetical set function calculates something for a hypothetical value (let's denote it as hv). In this context, `DENSE_RANK()` computes the rank of hv without gaps, while `RANK()` does the same thing but with gaps. `CUME_DIST()` computes the cumulative distribution of hv (the relative rank of a row from 1/n to 1), while `PERCENT_RANK()` computes the percent rank of hv (the relative rank of a row from 0 to 1).

For instance, let's assume that we want to compute the rank without gaps for the hypothetical value (*2004*, *10000*), where *2004* is `SALE.FISCAL_YEAR` and *10000* is `SALE.SALE_`. Next, for the existing data, we want to obtain all ranks without gaps less than the rank of this hypothetical value. For the first part of the problem, we rely on the `DENSE_RANK()` hypothetical set function, while for the second part, on the `DENSE_RANK()` window function, as follows:

```
ctx.select(SALE.EMPLOYEE_NUMBER, SALE.FISCAL_YEAR, SALE.SALE_)
   .from(SALE)
   .qualify(denseRank().over()
    .orderBy(SALE.FISCAL_YEAR.desc(), SALE.SALE_)
    .le(select(denseRank(val(2004), val(10000))
     .withinGroupOrderBy(SALE.FISCAL_YEAR.desc(), SALE.SALE_))
    .from(SALE))).fetch();
```

Now, let's consider another example that uses the `PERCENT_RANK()` hypothetical set function. This time, let's assume that we plan to have a salary of *$61,000* for new sales reps, but before doing that, we want to know the percentage of current sales reps having salaries higher than *$61,000*. This can be done like so:

```
ctx.select(count().as("nr_of_salaries"),
   percentRank(val(61000d)).withinGroupOrderBy(
        EMPLOYEE.SALARY.desc()).mul(100).concat("%")
           .as("salary_percentile_rank"))
   .from(EMPLOYEE)
   .where(EMPLOYEE.JOB_TITLE.eq("Sales Rep")).fetch();
```

Moreover, we want to know the percentage of sales reps' salaries that are higher than *$61,000*. For this, we need the distinct salaries, as shown here:

```
ctx.select(count().as("nr_of_salaries"),
    percentRank(val(61000d)).withinGroupOrderBy(
    field(name("t", "salary")).desc()).mul(100).concat("%")
          .as("salary_percentile_rank"))
   .from(selectDistinct(EMPLOYEE.SALARY.as("salary"))
```

```
    .from(EMPLOYEE)
    .where(EMPLOYEE.JOB_TITLE.eq("Sales Rep"))
    .asTable("t"))
    .fetch();
```

You can practice these examples next to other `RANK()` and `CUME_DIST()` hypothetical set functions in the *OrderedSetAggregateFunctions* bundled code.

Inverse distribution functions

Briefly, the inverse distribution functions compute *percentiles*. There are two distribution models: a discrete model (computed via `PERCENTILE_DISC()`) and a continuous model (computed via `PERCENTILE_CONT()`).

PERCENTILE_DISC() and PERCENTILE_CONT()

But what does it actually mean to compute *percentiles*? Loosely speaking, consider a certain percent, `P` (this percent is a float value between 0 inclusive and 1 inclusive), and an ordering field, `F`. In this context, the percentile computation represents the value below which `P` percent of the `F` values fall.

For instance, let's consider the `SALES` table, and we want to find the 25th percentile sale. In this case, `P` = 0.25, and the ordering field is `SALE.SALE_`. Applying `PERCENTILE_DISC()` and `PERCENTILE_CONT()` results in this query:

```
ctx.select(
     percentileDisc(0.25)
       .withinGroupOrderBy(SALE.SALE_).as("pd - 0.25"),
     percentileCont(0.25)
       .withinGroupOrderBy(SALE.SALE_).as("pc - 0.25"))
    .from(SALE)
    .fetch();
```

In the bundled code, you can see this query extended for the 50th, 75th, and 100th percentile. The resulting value (for instance, 2974.43) represents the sale below which 25% of the sales fall. In this case, `PERCENTILE_DISC()` and `PERCENTILE_CONT()` return the same value (2974.43), but this is not always the case. Remember that `PERCENTILE_DISC()` works on a discrete model, while `PERCENTILE_CONT()` works on a continuous model. In other words, if there is no value (sale) in the sales (also referred to as population) that fall exactly in the specified percentile, `PERCENTILE_CONT()` must interpolate it assuming continuous distribution. Basically, `PERCENTILE_CONT()`

interpolates the value (sale) from the two values (sales) that are immediately after and before the needed one. For instance, if we repeat the previous example for the 11th percentile, then `PERCENTILE_DISC()` returns 1676.14, which is an existent sale, while `PERCENTILE_CONT()` returns 1843.88, which is an interpolated value that doesn't exist in the database.

While Oracle supports `PERCENTILE_DISC()` and `PERCENTILE_CONT()` as ordered set aggregate functions and window function variants, PostgreSQL supports them only as ordered set aggregate functions, SQL Server supports only the window function variants, and MySQL doesn't support them at all. Emulating them is not quite simple, but this great article by Lukas Eder is a must-read in this direction: `https://blog.jooq.org/2019/01/28/how-to-emulate-percentile_disc-in-mysql-and-other-rdbms/`.

Now, let's see an example of using `PERCENTILE_DISC()` as the window function variant and `PERCENTILE_CONT()` as the ordered set aggregate function. This time, the focus is on `EMPLOYEE.SALARY`. First, we want to compute the 50th percentile of salaries per office via `PERCENTILE_DISC()`. Second, we want to keep only those 50th percentiles less than the general 50th percentile calculated via `PERCENTILE_CONT()`. The code is illustrated in the following snippet:

```
ctx.select().from(
  select(OFFICE.OFFICE_CODE, OFFICE.CITY, OFFICE.COUNTRY,
    EMPLOYEE.FIRST_NAME, EMPLOYEE.LAST_NAME, EMPLOYEE.SALARY,
    percentileDisc(0.5).withinGroupOrderBy(EMPLOYEE.SALARY)
     .over().partitionBy(OFFICE.OFFICE_CODE)
     .as("percentile_disc"))
  .from(OFFICE)
  .join(EMPLOYEE)
   .on(OFFICE.OFFICE_CODE.eq(EMPLOYEE.OFFICE_CODE)).asTable("t"))
   .where(field(name("t", "percentile_disc"))
     .le(select(percentileCont(0.5)
       .withinGroupOrderBy(EMPLOYEE.SALARY))
       .from(EMPLOYEE))).fetch();
```

You can practice these examples in the *OrderedSetAggregateFunctions* bundled code.

The MODE() function

Mainly, the `MODE()` function works on a set of values to produce a result (referred to as the *mode*) representing the value that appears with the greatest frequency. The `MODE()` function comes in two flavors, as outlined here:

- `MODE(field)` aggregate function
- `MODE WITHIN GROUP (ORDER BY [order clause])` ordered set aggregate function

If multiple results (modes) are available, then `MODE()` returns only one value. If there is a given ordering, then the first value will be chosen.

The `MODE()` aggregate function is emulated by jOOQ in PostgreSQL and Oracle and is not supported in MySQL and SQL Server. For instance, let's assume that we want to find out in which month of the year we have the most sales, and for this, we may come up with the following query (notice that an explicit `ORDER BY` clause for `MODE()` is not allowed):

```
ctx.select(mode(SALE.FISCAL_MONTH).as("fiscal_month"))
   .from(SALE).fetch();
```

Running this query in PostgreSQL reveals that jOOQ emulates the `MODE()` aggregate function via the ordered set aggregate function, which is supported by PostgreSQL:

```
SELECT MODE() WITHIN GROUP (ORDER BY
 "public"."sale"."fiscal_month") AS "fiscal_month"
FROM "public"."sale"
```

In this case, if multiple modes are available, then the first one is returned with respect to the ascending ordering. On the other hand, for the Oracle case, jOOQ uses the `STATS_MODE()` function, as follows:

```
SELECT
  STATS_MODE("CLASSICMODELS"."SALE"."FISCAL_MONTH")
        "fiscal_month" FROM "CLASSICMODELS"."SALE"
```

In the following case, there is no ordering in the generated SQL, and if multiple modes are available, then only one is returned. On the other hand, the `MODE()` ordered set aggregate function is supported only by PostgreSQL:

```
ctx.select(mode().withinGroupOrderBy(
       SALE.FISCAL_MONTH.desc()).as("fiscal_month"))
   .from(SALE).fetch();
```

If multiple results (modes) are available, then `MODE()` returns only one value representing the highest value (in this particular case, the month closest to December inclusive) since we have used a descending order.

Nevertheless, how to return all modes (if more are available)? Commonly, statisticians refer to a bimodal distribution if two modes are available, to a trimodal distribution if three modes are available, and so on. Emulating `MODE()` to return all modes can be done in several ways. Here is one way (in the bundled code, you can see one more):

```
ctx.select(SALE.FISCAL_MONTH)
   .from(SALE)
   .groupBy(SALE.FISCAL_MONTH)
   .having(count().ge(all(select(count())
     .from(SALE)
     .groupBy(SALE.FISCAL_MONTH))))
   .fetch();
```

But having 1,000 cases where the value of `X` is `'foo'` and 999 cases where the value is `'buzz'`, `MODE()` is `'foo'`. By adding two more instances of `'buzz'`, `MODE()` switches to `'buzz'`. Maybe a good idea would be to allow for some variation in the values via a percent. In other words, the emulation of `MODE()` using a percentage of the total number of occurrences can be done like so (here, 75%):

```
ctx.select(avg(ORDERDETAIL.QUANTITY_ORDERED))
   .from(ORDERDETAIL)
   .groupBy(ORDERDETAIL.QUANTITY_ORDERED)
   .having(count().ge(all(select(count().mul(0.75))
      .from(ORDERDETAIL)
      .groupBy(ORDERDETAIL.QUANTITY_ORDERED))))
   .fetch();
```

You can practice these examples in the *OrderedSetAggregateFunctions* bundled code.

LISTAGG()

The last ordered set aggregate function discussed in this section is `LISTAGG()`. This function is used for aggregating a given list of values into a string delimited via a separator (for instance, it is useful for producing CSV files). The SQL standard imposes the presence of the separator and `WITHIN GROUP` clause. Nevertheless, some databases treat these standards as being optional and apply certain defaults or expose an undefined behavior if the `WITHIN GROUP` clause is omitted. jOOQ provides `listAgg(Field<?> field)` having no explicit separator, and `listAgg(Field<?> field, String separator)`.

The `WITHIN GROUP` clause cannot be omitted. jOOQ emulates this function for dialects that don't support it, such as MySQL (emulates it via `GROUP_CONCAT()`, so a comma is a default separator), PostgreSQL (emulates it via `STRING_AGG()`, so no default separator), and SQL Server (same as in PostgreSQL) via proprietary syntax that offers similar functionality. Oracle supports `LISTAGG()` and there is no default separator.

Here are two simple examples with and without an explicit separator that produces a list of employees names' in ascending order by salary as `Result<Record1<String>>`:

```
ctx.select(listAgg(EMPLOYEE.FIRST_NAME)
   .withinGroupOrderBy(EMPLOYEE.SALARY).as("listagg"))
   .from(EMPLOYEE).fetch();

ctx.select(listAgg(EMPLOYEE.FIRST_NAME, ";")
   .withinGroupOrderBy(EMPLOYEE.SALARY).as("listagg"))
   .from(EMPLOYEE).fetch();
```

Fetching directly, the `String` can be achieved via `fetchOneInto(String.class)`.

`LISTAGG()` can be used in combination with `GROUP BY` and `ORDER BY`, as in the following example that fetches a list of employees per job title:

```
ctx.select(EMPLOYEE.JOB_TITLE,
      listAgg(EMPLOYEE.FIRST_NAME, ",")
   .withinGroupOrderBy(EMPLOYEE.FIRST_NAME).as("employees"))
   .from(EMPLOYEE)
   .groupBy(EMPLOYEE.JOB_TITLE)
   .orderBy(EMPLOYEE.JOB_TITLE).fetch();
```

Moreover, `LISTAGG()` supports a window function variant as well, as shown here:

```
ctx.select(EMPLOYEE.JOB_TITLE,
      listAgg(EMPLOYEE.SALARY, ",")
   .withinGroupOrderBy(EMPLOYEE.SALARY)
      .over().partitionBy(EMPLOYEE.JOB_TITLE))
   .from(EMPLOYEE).fetch();
```

And here is a fun fact from Lukas Eder: "`LISTAGG()` *is not a true ordered set aggregate function. It should use the same* `ORDER BY` *syntax as* `ARRAY_AGG`." See the discussion here: `https://twitter.com/lukaseder/status/1237662156553883648`.

You can practice these examples and more in *OrderedSetAggregateFunctions*.

Grouping, filtering, distinctness, and functions

In this section, *grouping* refers to the usage of `GROUP BY` with functions, *filtering* refers to the usage of the `FILTER` clause with functions, and *distinctness* refers to aggregate functions on distinct values.

Grouping

As you already know, `GROUP BY` is a SQL clause useful for arranging rows in groups via one (or more) column given as an argument. Rows that land in a group have matching values in the given columns/expressions. Typical use cases apply aggregate functions on groups of data produced by `GROUP BY`.

> **Important Note**
>
> Especially when dealing with multiple dialects, it is correct to list all non-aggregated columns from the `SELECT` clause in the `GROUP BY` clause. This way, you avoid potentially indeterminate/random behavior and errors across dialects (some of them will not ask you to do this (for example, MySQL), while others will (for example, Oracle)).

jOOQ supports `GROUP BY` in all dialects, therefore here is an example of fetching offices (`OFFICE`) having fewer than three employees:

```
ctx.select(OFFICE.OFFICE_CODE, OFFICE.CITY,
      nvl(groupConcat(EMPLOYEE.FIRST_NAME), "N/A").as("name"))
   .from(OFFICE)
   .leftJoin(EMPLOYEE)
      .on(OFFICE.OFFICE_CODE.eq(EMPLOYEE.OFFICE_CODE))
   .groupBy(OFFICE.OFFICE_CODE, OFFICE.CITY)
   .having(count().lt(3)).fetch();
```

Here is another example that computes the sum of sales per employee per year, and after that, it computes the average of these sums per employee:

```
ctx.select(field(name("t", "en")),
       avg(field(name("t", "ss"), Double.class))
         .as("sale_avg"))
   .from(ctx.select(SALE.EMPLOYEE_NUMBER,
      SALE.FISCAL_YEAR, sum(SALE.SALE_))
       .from(SALE)
       .groupBy(SALE.EMPLOYEE_NUMBER, SALE.FISCAL_YEAR)
```

```
        .asTable("t", "en", "fy", "ss"))
    .groupBy(field(name("t", "en"))).fetch();
```

You can find more examples of using GROUP BY in *GroupByDistinctFilter*.

Filtering

If we want to refine a query by applying aggregations against a limited set of the values in a column, then we can use CASE expressions, as in this example, which sum the salaries of sales reps and the rest of the employees:

```
ctx.select(EMPLOYEE.SALARY,
  (sum(case_().when(EMPLOYEE.JOB_TITLE.eq("Sales Rep"), 1)
     .else_(0))).as("Sales Rep"),
  (sum(case_().when(EMPLOYEE.JOB_TITLE.ne("Sales Rep"), 1)
     .else_(0))).as("Others"))
    .from(EMPLOYEE).groupBy(EMPLOYEE.SALARY).fetch();
```

As you can see, CASE is flexible but it's a bit tedious. A more straightforward solution is represented by the FILTER clause, exposed by jOOQ via the filterWhere() method, and emulated for every dialect that doesn't support it (usually via CASE expressions). The previous query can be expressed via FILTER, as follows:

```
ctx.select(EMPLOYEE.SALARY,
  (count().filterWhere(EMPLOYEE.JOB_TITLE
      .eq("Sales Rep"))).as("Sales Rep"),
  (count().filterWhere(EMPLOYEE.JOB_TITLE
      .ne("Sales Rep"))).as("Others"))
   .from(EMPLOYEE)
   .groupBy(EMPLOYEE.SALARY).fetch();
```

Or, here is an example of removing NULL values for ARRAY_AGG():

```
ctx.select(arrayAgg(DEPARTMENT.ACCOUNTS_RECEIVABLE)
    .filterWhere(DEPARTMENT.ACCOUNTS_RECEIVABLE.isNotNull()))
    .from(DEPARTMENT).fetch();
```

Another use case for FILTER is related to pivoting rows to columns. For instance, check out this query, which produces the sales per month and per year:

```
ctx.select(SALE.FISCAL_YEAR, SALE.FISCAL_MONTH,
      sum(SALE.SALE_))
```

```
    .from(SALE)
    .groupBy(SALE.FISCAL_YEAR, SALE.FISCAL_MONTH).fetch();
```

The query returns the correct result but in an unexpected form. Its vertical form having one value per row is not quite readable for users. Most probably, a user will be more familiar with a form having one row per year and a dedicated column per month. So, turning the rows of a year into columns should solve the problem, and this can be accomplished in several ways, including the `FILTER` clause, as shown here:

```
ctx.select(SALE.FISCAL_YEAR,
     sum(SALE.SALE_).filterWhere(SALE.FISCAL_MONTH.eq(1))
          .as("Jan_sales"),
     sum(SALE.SALE_).filterWhere(SALE.FISCAL_MONTH.eq(2))
          .as("Feb_sales"),
     ...
     sum(SALE.SALE_).filterWhere(SALE.FISCAL_MONTH.eq(12))
          .as("Dec_sales"))
   .from(SALE).groupBy(SALE.FISCAL_YEAR).fetch();
```

The `FILTER` clause can be considered with aggregate functions used as window functions as well. In such cases, `filterWhere()` comes between the aggregate function and the `OVER()` clause. For instance, the following query sums the salaries of employees per office only for employees that don't get a commission:

```
ctx.select(EMPLOYEE.FIRST_NAME, EMPLOYEE.LAST_NAME,
       EMPLOYEE.SALARY, OFFICE.OFFICE_CODE, OFFICE.CITY,
       OFFICE.COUNTRY, sum(EMPLOYEE.SALARY)
   .filterWhere(EMPLOYEE.COMMISSION.isNull())
    .over().partitionBy(OFFICE.OFFICE_CODE))
   .from(EMPLOYEE)
   .join(OFFICE)
    .on(EMPLOYEE.OFFICE_CODE.eq(OFFICE.OFFICE_CODE)).fetch();
```

Moreover, the `FILTER` clause can be used with ordered set aggregate functions. This way, we can remove rows that don't pass the filter before the aggregation takes place. Here is an example of filtering employees having salaries higher than *$80,000* and collecting the result via `LISTAGG()`:

```
ctx.select(listAgg(EMPLOYEE.FIRST_NAME)
   .withinGroupOrderBy(EMPLOYEE.SALARY)
   .filterWhere(EMPLOYEE.SALARY.gt(80000)).as("listagg"))
   .from(EMPLOYEE).fetch();
```

Since you are here, I am sure that you'll love this article by Lukas Eder about calculating multiple aggregate functions in a single query: `https://blog.jooq.org/2017/04/20/how-to-calculate-multiple-aggregate-functions-in-a-single-query/`.

You can practice the examples and more in the *GroupByDistinctFilter* bundled code.

Distinctness

Most aggregate functions come with a variant for applying them to a distinct set of values. While you can find all of them in the jOOQ documentation, let's quickly list here `countDistinct()`, `sumDistinct()`, `avgDistinct()`, `productDistinct()`, `groupConcatDistinct()`, `arrayAggDistinct()`, and `collectDistinct()`. For completeness' sake, we also have `minDistinct()` and `maxDistinct()`. When a function is not supported by jOOQ, we can still call it via the general `aggregateDistinct()` function.

Here is an example of using `countDistinct()` for fetching employees having sales in at least 3 distinct years:

```
ctx.select(SALE.EMPLOYEE_NUMBER)
   .from(SALE)
   .groupBy(SALE.EMPLOYEE_NUMBER)
   .having(countDistinct(SALE.FISCAL_YEAR).gt(3)).fetch();
```

More examples are available in the *GroupByDistinctFilter* bundled code.

Grouping sets

For those not familiar with *grouping sets*, let's briefly follow a scenario meant to quickly introduce and cover this notion. Consider the following screenshot:

```
ctx.select(SALE.EMPLOYEE_NUMBER, sum(SALE.SALE_))
        .from(SALE)
        .groupBy(SALE.EMPLOYEE_NUMBER)
        .fetch();
+---------------+----------+
|employee_number|       sum|
+---------------+----------+
|           1102| 52343.12 |
|           1504| 25197.18 |
|           1166|  9045.47 |
|           1143| 76482.97 |
|           1370| 34408.68 |
|           1611|171940.95 |
+---------------+----------+
```

```
ctx.select(SALE.FISCAL_YEAR, sum(SALE.SALE_))
        .from(SALE)
        .groupBy(SALE.FISCAL_YEAR)
        .fetch();
+-----------+----------+
|fiscal_year|       sum|
+-----------+----------+
|       2003| 48954.43 |
|       2000| 12434.22 |
|       2005| 58947.5  |
|       2004| 89674.67 |
|       2007|159407.56 |
+-----------+----------+
```

Figure 13.24 – Two queries using a grouping set each

The `groupBy(SALE.EMPLOYEE_NUMBER)` construction from the left-hand side (respectively, `groupBy(SALE.FISCAL_YEAR)` from the right-hand side) is referred to as a *grouping set*. A grouping set can contain none (empty grouping set), one, or more columns. In our case, both grouping sets contain one column.

Getting a unified result set of these two result sets containing the aggregated data of both grouping sets can be done via the `UNION ALL` operator, as illustrated here:

```
ctx.select(SALE.EMPLOYEE_NUMBER,
      castNull(INTEGER).as("fiscal_year"), sum(SALE.SALE_))
   .from(SALE)
   .groupBy(SALE.EMPLOYEE_NUMBER)
   .unionAll(select(castNull(BIGINT).as("employee_number"),
              SALE.FISCAL_YEAR, sum(SALE.SALE_))
      from(SALE)
      groupBy(SALE.FISCAL_YEAR))
   .fetch();
```

```
+---------------+-----------+----------+
|employee_number|fiscal_year|       sum|
+---------------+-----------+----------+
|           1102|     {null}| 52343.12 |
|           1504|     {null}| 25197.18 |
|           1166|     {null}|  9045.47 |
|           1143|     {null}| 76482.97 |
|           1370|     {null}| 34408.68 |
|           1611|     {null}|171940.95 |
|         {null}|       2003| 48954.43 |
|         {null}|       2000| 12434.22 |
|         {null}|       2005| 58947.5  |
|         {null}|       2004| 89674.67 |
|         {null}|       2007|159407.56 |
+---------------+-----------+----------+
```

Figure 13.25 – Union grouping sets

But, as you can see, even for only two grouping sets, this query is quite lengthy. Moreover, it needs to resolve two `SELECT` statements before combining their results into a single result set. Here is where the `GROUPING SETS(column_list)` clause of `GROUP BY` enters the scene. This clause represents a handy shorthand for a series of `UNION`-ed queries, and it can be used in the following example of rewriting the previous query:

```
ctx.select(SALE.EMPLOYEE_NUMBER,
           SALE.FISCAL_YEAR, sum(SALE.SALE_))
   .from(SALE)
   .groupBy(groupingSets(
            SALE.EMPLOYEE_NUMBER, SALE.FISCAL_YEAR))
   .fetch();
```

Cool, right?! Nevertheless, there is an issue that should be considered. `GROUPING SETS()` will generate `NULL` values for each dimension at the subtotal levels. In other words, it is quite hard to distinguish between a real `NULL` value (present in the original data) and a generated `NULL` value. But this job is the responsibility of the `GROUPING()` function, which returns 0 for `NULL` values in the original data and, respectively, 1 for generated `NULL` values that indicate a subtotal.

For instance, if we write a query in the `groupBy(groupingSets(OFFICE.CITY, OFFICE.COUNTRY))` clause, then we will need to distinguish between generated `NULL` values and `NULL` values of `OFFICE.CITY` and, respectively, `OFFICE.COUNTRY`. By using `GROUPING()` to form a condition of a `CASE` expression, we can achieve this, like so:

```
ctx.select(
  case_().when(grouping(OFFICE.CITY).eq(1), "{generated}")
    .else_(OFFICE.CITY).as("city"),
  case_().when(grouping(OFFICE.COUNTRY).eq(1), "{generated}")
    .else_(OFFICE.COUNTRY).as("country"),
  sum(OFFICE.INTERNAL_BUDGET))
   .from(OFFICE)
   .groupBy(groupingSets(OFFICE.CITY, OFFICE.COUNTRY))
   .fetch();
```

In this query, we replaced every generated `NULL` value with the text `{generated}`, while the `NULL` values on the original data will be fetched as `NULL` values. So, we now have a clear picture of `NULL` values' provenience, as illustrated here:

```
+-----------+-------+------+    +-----------+-----------+------+
|city       |country|   sum|    |city       |country    |   sum|
+-----------+-------+------+    +-----------+-----------+------+
|{null}     |{null} |205000|    |{null}     |{generated}|205000|
|Paris      |{null} |113000|    |Paris      |{generated}|113000|
|Sydney     |{null} | 45500|    |Sydney     |{generated}| 45500|
```


Figure 13.26 – No grouping (left-hand side) versus grouping (right-hand side)

Most probably, `{null}` and `{generated}` will not be very attractive for our clients, so we can tune this query a little bit to be more friendly by replacing `{null}` with `"Unspecified"` and `{generated}` with `"-"`, like so:

```
ctx.select(case_().when(grouping(OFFICE.CITY).eq(1), "-")
  .else_(isnull(OFFICE.CITY, "Unspecified")).as("city"),
 case_().when(grouping(OFFICE.COUNTRY).eq(1), "-")
  .else_(isnull(OFFICE.COUNTRY, "Unspecified")).as("country"),
  sum(OFFICE.INTERNAL_BUDGET))
   .from(OFFICE)
   .groupBy(groupingSets(OFFICE.CITY, OFFICE.COUNTRY))
  .fetch();
```

Next to `GROUPING SETS()`, we have `ROLLUP` and `CUBE`. These two extensions of the `GROUP BY` clause are syntactic sugar of `GROUPING SETS()`.

The `ROLLUP` group is a series of grouping sets. For instance, `GROUP BY ROLLUP (x, y, z)` is equivalent to `GROUP BY GROUPING SETS ((x, y, z), (x, y), (x), ())`. `ROLLUP` is typically applied for aggregates of hierarchical data such as sales by year > quarter > month > week, or offices internal budget per territory > state > country > city, as shown here:

```
ctx.select(
   case_().when(grouping(OFFICE.TERRITORY).eq(1), "{generated}")
     .else_(OFFICE.TERRITORY).as("territory"),
   case_().when(grouping(OFFICE.STATE).eq(1), "{generated}")
     .else_(OFFICE.STATE).as("state"),
   case_().when(grouping(OFFICE.COUNTRY).eq(1), "{generated}")
     .else_(OFFICE.COUNTRY).as("country"),
   case_().when(grouping(OFFICE.CITY).eq(1), "{generated}")
     .else_(OFFICE.CITY).as("city"),
   sum(OFFICE.INTERNAL_BUDGET))
   .from(OFFICE)
   .where(OFFICE.COUNTRY.eq("USA"))
   .groupBy(rollup(OFFICE.TERRITORY, OFFICE.STATE,
                   OFFICE.COUNTRY, OFFICE.CITY)).fetch();
```

And the output is shown here:

```
+-----------+-----------+-----------+------------+------+
|territory  |state      |country    |city        |   sum|
+-----------+-----------+-----------+------------+------+
|NA         |CA         |USA        |San Diego   | 65000|
|NA         |CA         |USA        |Los Angeles | 40000|
|NA         |CA         |USA        |San Francisco| 90000|
|NA         |CA         |USA        |{generated} |195000|
|NA         |CA         |{generated}|{generated} |195000|—Subtotal in CA
|NA         |MA         |USA        |Boston      | 85000|
|NA         |MA         |USA        |Springfield | 55000|
|NA         |MA         |USA        |{generated} |140000|
|NA         |MA         |{generated}|{generated} |140000|—Subtotal in MA
|NA         |NY         |USA        |NYC         |100000|
|NA         |NY         |USA        |{generated} |100000|
|NA         |NY         |{generated}|{generated} |100000|—Subtotal in NY
|NA         |{generated}|{generated}|{generated} |435000|—Subtotal in NA
|EMEA       |{null}     |USA        |{null}      | 95000|
|EMEA       |{null}     |USA        |{generated} | 95000|
|EMEA       |{null}     |{generated}|{generated} | 95000|
|EMEA       |{generated}|{generated}|{generated} | 95000|—Subtotal in EMEA
|{generated}|{generated}|{generated}|{generated} |530000|—Grand total
+-----------+-----------+-----------+------------+------+
```

Figure 13.27 – ROLLUP output

As with `ROLLUP`, a `CUBE` group can also be perceived as a series of grouping sets. However, `CUBE` calculates all permutations of the cubed grouping expression along with the grand total. So, for *n* elements, `CUBE` produces 2^n grouping sets. For instance `GROUP BY CUBE (x, y, x)` is equivalent to `GROUP BY GROUPING SETS ((x, y, z), (x, y), (x, z), (y, z), (x), (y), (z), ())`.

Let's apply `CUBE` for computing the sum of the internal budget for offices by state, country, and city. The query is shown here:

```
ctx.select(
 case_().when(grouping(OFFICE.STATE).eq(1), "{generated}")
  .else_(OFFICE.STATE).as("state"),
 case_().when(grouping(OFFICE.COUNTRY).eq(1), "{generated}")
  .else_(OFFICE.COUNTRY).as("country"),
 case_().when(grouping(OFFICE.CITY).eq(1), "{generated}")
  .else_(OFFICE.CITY).as("city"),
 sum(OFFICE.INTERNAL_BUDGET))
  .from(OFFICE)
  .where(OFFICE.COUNTRY.eq("USA"))
```

```
  .groupBy(cube(OFFICE.STATE, OFFICE.COUNTRY, OFFICE.CITY))
  .fetch();
```

Finally, let's talk about the `GROUPING_ID()` function. This function computes the decimal equivalent of the binary value obtained by concatenating the values returned by the `GROUPING()` functions applied to all the columns of the `GROUP BY` clause. Here is an example of using `GROUPING_ID()` via jOOQ `groupingId()`:

```
ctx.select(
 case_().when(grouping(OFFICE.TERRITORY).eq(1), "{generated}")
  .else_(OFFICE.TERRITORY).as("territory"),
 ...
 case_().when(grouping(OFFICE.CITY).eq(1), "{generated}")
  .else_(OFFICE.CITY).as("city"),
 groupingId(OFFICE.TERRITORY, OFFICE.STATE, OFFICE.COUNTRY,
            OFFICE.CITY).as("grouping_id"),
 sum(OFFICE.INTERNAL_BUDGET))
   .from(OFFICE)
   .where(OFFICE.COUNTRY.eq("USA"))
   .groupBy(rollup(OFFICE.TERRITORY, OFFICE.STATE,
                   OFFICE.COUNTRY, OFFICE.CITY))
   .fetch();
```

The following screenshot shows a sample output:

```
+-----------+-----------+-----------+-------------+-----------+------+
|territory  |state      |country    |city         |grouping_id|   sum|
+-----------+-----------+-----------+-------------+-----------+------+
      0           0           0           0
|NA         |CA         |USA        |San Diego    |  0000    0| 65000|
|NA         |CA         |USA        |Los Angeles  |          0| 40000|
 ...
      0           0           1           1
|NA         |CA         |{generated}|{generated}  |  0011    3|195000|
|NA         |MA         |USA        |Boston       |          0| 85000|
 ...
      0           1           1           1
|NA         |{generated}|{generated}|{generated}  |  0111    7|435000|
|EMEA       |{null}     |USA        |{null}       |          0| 95000|
 ...  1           1           1           1
|{generated}|{generated}|{generated}|{generated}  |  1111   15|530000|
+-----------+-----------+-----------+-------------+-----------+------+
```

Figure 13.28 – GROUPING_ID() output

`GROUPING_ID()` can also be used in `HAVING` for creating conditions, as follows:

```
… .having(groupingId(OFFICE.TERRITORY,
    OFFICE.STATE, OFFICE.COUNTRY, OFFICE.CITY).eq(3))…
```

The complete query is available in the *GroupingRollupCube* bundled code.

Summary

Working with SQL functions is such fun! They truly boost the SQL world and allow us to solve so many problems during data manipulation. As you saw in this chapter, jOOQ provides comprehensive support to SQL functions, covering regular and aggregate functions to the mighty window functions, ordered set aggregate functions (`WITHIN GROUP`), and so on. While we're on this topic, allow me to recommend the following article as a great read: `https://blog.jooq.org/how-to-find-the-closest-subset-sum-with-sql/`. In the next chapter, we tackle virtual tables (vtables).

14
Derived Tables, CTEs, and Views

Derived tables, **CTEs**, and views are important players in the SQL context. They're useful to organize and optimize the reuse of long and complex queries – typically, base queries and/or expensive queries (in performance terms), and to improve readability by breaking down the code into separate steps. Mainly, they link a certain query to a name, possibly stored in the schema. In other words, they hold the query text, which can be referenced and executed via the associated name when needed. If results materialize, then the database engine can reuse these cached results, otherwise, they have to be recomputed at each call.

Derived tables, CTEs, and views have specific particularities (including database vendor-specific options), and choosing between them is a decision that strongly depends on the use case, the involved data and queries, the database vendor and optimizer, and so on. As usual, we handle this topic from the jOOQ perspective, so our agenda includes the following:

- Derived tables
- CTEs
- Views

Let's get started!

Technical requirements

The code for this chapter can be found on GitHub at `https://github.com/PacktPublishing/jOOQ-Masterclass/tree/master/Chapter14`.

Derived tables

Have you ever used a nested `SELECT` (a `SELECT` in a table expression)? Of course, you have! Then, you've used a so-called *derived table* having the scope of the statement that creates it. Roughly, a derived table should be treated in the same way as a base table. In other words, it is advisable to give it and its columns meaningful names via the `AS` operator. This way, you can reference the derived table without ambiguity, and you'll respect the fact that most databases don't support unnamed (unaliased) derived tables.

jOOQ allows us to transform any `SELECT` in a derived table via `asTable()`, or its synonym `table()`. Let's have a simple example starting from this `SELECT`:

```
select(inline(1).as("one"));
```

This is not a derived table, but it can become one as follows (these two are synonyms):

```
Table<?> t = select(inline(1).as("one")).asTable();
Table<?> t = table(select(inline(1).as("one")));
```

In jOOQ, we can further refer to this derived table via the local variable `t`. It is convenient to declare `t` as `Table<?>` or to simply use `var`. But, of course, you can explicitly specify the data types as well. Here, `Table<Record1<Integer>>`.

> **Important Note**
>
> The `org.jooq.Table` type can reference a derived table.

Now, the resulting `t` is an unnamed derived table since there is no explicit alias associated with it. Let's see what happens when we select something from `t`:

```
ctx.selectFrom(t).fetch();
```

jOOQ generates the following SQL (we've arbitrarily chosen the PostgreSQL dialect):

```
SELECT "alias_30260683"."one"
  FROM (SELECT 1 AS "one") AS "alias_30260683"
```

jOOQ detected the missing alias for the derived table, therefore it generated one (`alias_30260683`) on our behalf.

Important Note

We earlier iterated that most database vendors require an explicit alias for every derived table. But, as you just saw, jOOQ allows us to omit such aliases, and when we do, jOOQ will generate one on our behalf to guarantee that the generated SQL is syntactically correct. The generated alias is a random number suffixed by `alias_`. This alias should not be referenced explicitly. jOOQ will use it internally to render a correct/valid SQL.

Of course, if we explicitly specify an alias then jOOQ will use it:

```
Table<?> t = select(inline(1).as("one")).asTable("t");
Table<?> t = table(select(inline(1).as("one"))).as("t");
```

The SQL corresponding to PostgreSQL is as follows:

```
SELECT "t"."one" FROM (SELECT 1 AS "one") AS "t"
```

Here is another example using the `values()` constructor:

```
Table<?> t = values(row(1, "John"), row(2, "Mary"),
                    row(3, "Kelly"))
  .as("t", "id", "name"); // or, .asTable("t", "id", "name");
```

Typically, we explicitly specify an alias when we also reference it explicitly, but there is nothing wrong in doing it every time. For instance, jOOQ doesn't require an explicit alias for the following inlined derived table, but there is nothing wrong with adding it:

```
ctx.select()
   .from(EMPLOYEE)
   .crossApply(select(count().as("sales_count")).from(SALE)
     .where(SALE.EMPLOYEE_NUMBER
        .eq(EMPLOYEE.EMPLOYEE_NUMBER)).asTable("t"))
   .fetch();
```

jOOQ relies on the `t` alias instead of generating one.

Extracting/declaring a derived table in a local variable

jOOQ allows us to extract/declare a derived table outside the statement that used it, and, in such a case, its presence and role are better outlined than in the case of nesting it in a table expression.

Extracting/declaring a derived table in a local variable can be useful if we need to refer to the derived table in multiple statements, we need it as part of a dynamic query, or we just want to decongest a complex query.

For instance, consider the following query:

```
ctx.select().from(
    select(ORDERDETAIL.PRODUCT_ID, ORDERDETAIL.PRICE_EACH)
     .from(ORDERDETAIL)
     .where(ORDERDETAIL.QUANTITY_ORDERED.gt(50)))
   .innerJoin(PRODUCT)
   .on(field(name("price_each")).eq(PRODUCT.BUY_PRICE))
   .fetch();
```

The highlighted subquery represents an inlined derived table. jOOQ automatically associates to it an alias and uses that alias to reference the columns `product_id` and `price_each` in the outer `SELECT`. Of course, we can provide an explicit alias as well, but this is not required:

```
ctx.select().from(
    select(ORDERDETAIL.PRODUCT_ID, ORDERDETAIL.PRICE_EACH)
     .from(ORDERDETAIL)
     .where(ORDERDETAIL.QUANTITY_ORDERED.gt(50))
       .asTable("t"))
   .innerJoin(PRODUCT)
   .on(field(name("t", "price_each")).eq(PRODUCT.BUY_PRICE))
   .fetch();
```

This time, jOOQ relies on the `t` alias instead of generating one. Next, let's add this subquery to another query as follows:

```
ctx.select(PRODUCT.PRODUCT_LINE,
           PRODUCT.PRODUCT_NAME, field(name("price_each")))
   .from(select(ORDERDETAIL.PRODUCT_ID,
           ORDERDETAIL.PRICE_EACH).from(ORDERDETAIL)
           .where(ORDERDETAIL.QUANTITY_ORDERED.gt(50)))
   .innerJoin(PRODUCT)
   .on(field(name("product_id")).eq(PRODUCT.PRODUCT_ID))
   .fetch();
```

This query fails at compilation time because the reference to the `product_id` column in `on(field(name("product_id")).eq(PRODUCT.PRODUCT_ID))` is ambiguous.

jOOQ automatically associates a generated alias to the inlined derived table, but it cannot decide whether the product_id column comes from the derived table or from the PRODUCT table. Resolving this issue can be done explicitly by adding and using an alias for the derived table:

```
ctx.select(PRODUCT.PRODUCT_LINE, PRODUCT.PRODUCT_NAME,
           field(name("t", "price_each")))
   .from(select(ORDERDETAIL.PRODUCT_ID,
          ORDERDETAIL.PRICE_EACH).from(ORDERDETAIL)
          .where(ORDERDETAIL.QUANTITY_ORDERED.gt(50))
            .asTable("t"))
   .innerJoin(PRODUCT)
   .on(field(name("t", "product_id"))
     .eq(PRODUCT.PRODUCT_ID))
   .fetch();
```

Now, jOOQ relies on the t alias, and the ambiguity issues have been resolved. Alternatively, we can explicitly associate a unique alias only to the ORDERDETAIL.PRODUCT_ID field as select(ORDERDETAIL.PRODUCT_ID.as("pid")..., and reference it via this alias as field(name("pid"))....

At this point, we have two queries with the same inline derived table. We can avoid code repetition by extracting this derived table in a Java local variable before using it in these two statements. In other words, we declare the derived table in a Java local variable, and we refer to it in the statements via this local variable:

```
Table<?> t = select(
     ORDERDETAIL.PRODUCT_ID, ORDERDETAIL.PRICE_EACH)
  .from(ORDERDETAIL)
  .where(ORDERDETAIL.QUANTITY_ORDERED.gt(50)).asTable("t");
```

So, t is our derived table. Running this snippet of code doesn't have an effect on and doesn't produce any SQL. jOOQ evaluates t only when we reference it in queries, but in order to be evaluated, t must be declared before the queries that use it. This is just Java; we can use a variable only if it was declared upfront. When a query uses t (for instance, via t.field()), jOOQ evaluates t and renders the proper SQL.

For instance, we can use t to rewrite our queries as follows:

```
ctx.select()
   .from(t)
   .innerJoin(PRODUCT)
```

```
    .on(t.field(name("price_each"), BigDecimal.class)
         .eq(PRODUCT.BUY_PRICE))
    .fetch();

ctx.select(PRODUCT.PRODUCT_LINE,
           PRODUCT.PRODUCT_NAME, t.field(name("price_each")))
    .from(t)
    .innerJoin(PRODUCT)
    .on(t.field(name("product_id"), Long.class)
         .eq(PRODUCT.PRODUCT_ID))
    .fetch();
```

But, why this time do we need explicit types in `on(t.field(name("price_each"), BigDecimal.class)` and `.on(t.field(name("product_id"), Long.class)`? The answer is that the fields cannot be dereferenced from `t` in a type-safe way. Therefore it is our job to specify the proper data types. This is a pure Java issue, and has nothing to do with SQL!

But, there is a trick that can help us to keep type safety and reduce verbosity, and that trick consists of using the <T> Field<T> field(Field<T> field) method. The best explanation of this method is given by the jOOQ documentation itself. The following figure is a screenshot from the jOOQ official documentation:

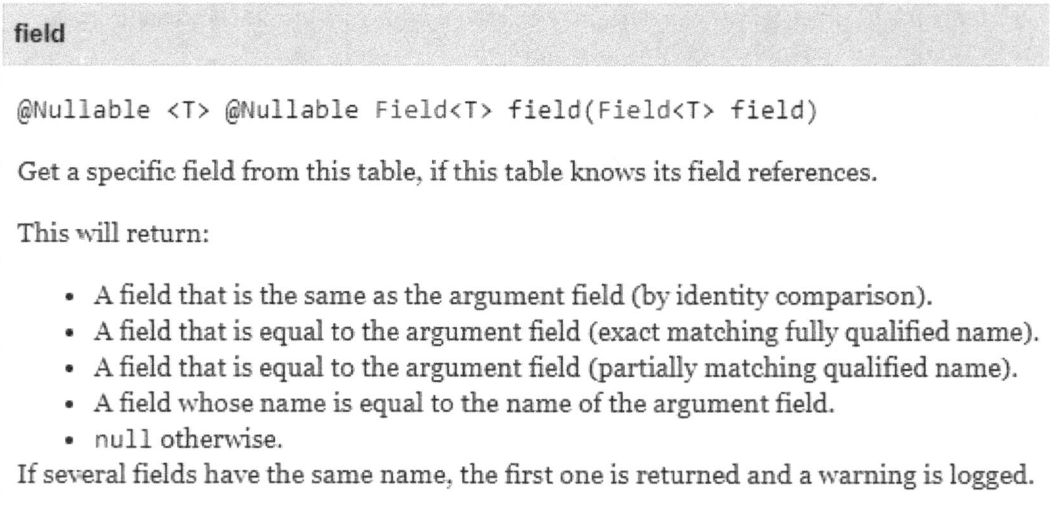

field

@Nullable <T> @Nullable Field<T> field(Field<T> field)

Get a specific field from this table, if this table knows its field references.

This will return:

- A field that is the same as the argument field (by identity comparison).
- A field that is equal to the argument field (exact matching fully qualified name).
- A field that is equal to the argument field (partially matching qualified name).
- A field whose name is equal to the name of the argument field.
- `null` otherwise.

If several fields have the same name, the first one is returned and a warning is logged.

Figure 14.1 – The <T> Field<T> field(Field<T> field) method documentation

The expression `t.field(name("price_each"), …)` indirectly refers to the field `ORDERDETAIL.PRICE_EACH`, and `t.field(name("product_id"), …)` indirectly refers to the field `ORDERDETAIL.PRODUCT_ID`. Therefore, based on the previous figure, we can re-write our queries in a type-safe manner as follows:

```
ctx.select(PRODUCT.PRODUCT_LINE, PRODUCT.PRODUCT_NAME,
           t.field(ORDERDETAIL.PRICE_EACH))
   .from(t)
   .innerJoin(PRODUCT)
   .on(t.field(ORDERDETAIL.PRODUCT_ID)
     .eq(PRODUCT.PRODUCT_ID))
   .fetch();

ctx.select()
   .from(t)
   .innerJoin(PRODUCT)
   .on(t.field(ORDERDETAIL.PRICE_EACH)
     .eq(PRODUCT.BUY_PRICE))
   .fetch();
```

Cool! Now, we can reuse `t` in a "type-safe" manner! However, keep in mind that <T> Field<T> field(Field<T> field) just looks type safe. It's actually as good as an unsafe cast in Java, because the lookup only considers the identifier, not the type. Nor does it coerce the expression. This is why we have the quotes around type-safe.

Here is another example that uses two extracted `Field` in the extracted derived table and the query itself:

```
// fields
Field<BigDecimal> avg = avg(ORDERDETAIL.PRICE_EACH).as("avg");
Field<Long> ord = ORDERDETAIL.ORDER_ID.as("ord");

// derived table
Table<?> t = select(avg, ord).from(ORDERDETAIL)
    .groupBy(ORDERDETAIL.ORDER_ID).asTable("t");

// query
ctx.select(ORDERDETAIL.ORDER_ID, ORDERDETAIL
          .ORDERDETAIL_ID,ORDERDETAIL.PRODUCT_ID,
           ORDERDETAIL.PRICE_EACH)
   .from(ORDERDETAIL, t)
```

```
    .where(ORDERDETAIL.ORDER_ID.eq(ord)
      .and(ORDERDETAIL.PRICE_EACH.lt(avg)))
    .orderBy(ORDERDETAIL.ORDER_ID)
    .fetch();
```

Here, `ord` and `avg` are rendered unqualified (without being prefixed with the derived table alias). But, thanks to `<T> Field<T> field(Field<T> field)`, we can obtain the qualified version:

```
...where(ORDERDETAIL.ORDER_ID.eq(t.field(ord))
   .and(ORDERDETAIL.PRICE_EACH.lt(t.field(avg))))
...
```

Next, let's see an example that uses `fields()` and `asterisk()` to refer to all columns of a derived table extracted in a local variable:

```
Table<?> t = ctx.select(SALE.EMPLOYEE_NUMBER,
                 count(SALE.SALE_).as("sales_count"))
   .from(SALE).groupBy(SALE.EMPLOYEE_NUMBER).asTable("t");

ctx.select(t.fields()).from(t)
   .orderBy(t.field(name("sales_count"))).fetch();

ctx.select(t.asterisk(),
           EMPLOYEE.FIRST_NAME, EMPLOYEE.LAST_NAME)
   .from(EMPLOYEE, t)
   .where(EMPLOYEE.EMPLOYEE_NUMBER.eq(
          t.field(name("employee_number"), Long.class)))
   .orderBy(t.field(name("sales_count"))).fetch();
```

Notice that extracting a subquery is not mandatory for it to be transformed in a `Table`. There are cases when extracting it as a simple `SELECT` is all you need. For instance, when the subquery isn't a derived table, we can do this:

```
ctx.selectFrom(PRODUCT)
   .where(row(PRODUCT.PRODUCT_ID, PRODUCT.BUY_PRICE).in(
     select(ORDERDETAIL.PRODUCT_ID, ORDERDETAIL.PRICE_EACH)
       .from(ORDERDETAIL)
       .where(ORDERDETAIL.QUANTITY_ORDERED.gt(50))))
   .fetch();
```

This subquery (which is not a derived table) can be extracted locally and used like this:

```
// SelectConditionStep<Record2<Long, BigDecimal>>
var s = select(
      ORDERDETAIL.PRODUCT_ID, ORDERDETAIL.PRICE_EACH)
   .from(ORDERDETAIL)
   .where(ORDERDETAIL.QUANTITY_ORDERED.gt(50));

ctx.selectFrom(PRODUCT)
   .where(row(PRODUCT.PRODUCT_ID, PRODUCT.BUY_PRICE).in(s))
   .fetch();
```

There is no need for an alias (jOOQ knows that this is not a derived table and no alias is needed therefore it will not generate one) and no need to transform it into a `Table`. Actually, jOOQ is so flexible that it allows us to do even this:

```
var t = select(ORDERDETAIL.PRODUCT_ID, ORDERDETAIL.PRICE_EACH)
   .from(ORDERDETAIL)
   .where(ORDERDETAIL.QUANTITY_ORDERED.gt(50));

ctx.select(PRODUCT.PRODUCT_LINE, PRODUCT.PRODUCT_NAME,
          t.field(ORDERDETAIL.PRICE_EACH))
   .from(t)
   .innerJoin(PRODUCT)
   .on(t.field(ORDERDETAIL.PRODUCT_ID)
   .eq(PRODUCT.PRODUCT_ID))
   .fetch();

ctx.select()
   .from(t)
   .innerJoin(PRODUCT)
   .on(t.field(ORDERDETAIL.PRICE_EACH)
   .eq(PRODUCT.BUY_PRICE))
   .fetch();
```

Don't worry, jOOQ will not ask you to transform `t` into a `Table`. jOOQ infers that this is a derived table and associates and references a generated alias as expected in the rendered SQL. So, as long as you don't want to associate an explicit alias to the derived table and jOOQ doesn't specifically require a `Table` instance in your query, there is no need to transform the extracted `SELECT` into a `Table` instance. When you need a `Table` instance but not an alias for it, just use the `asTable()` method without arguments:

```
Table<?> t = select(
      ORDERDETAIL.PRODUCT_ID, ORDERDETAIL.PRICE_EACH)
   .from(ORDERDETAIL)
   .where(ORDERDETAIL.QUANTITY_ORDERED.gt(50)).asTable();
```

You can check out these examples along with others in *DerivedTable*.

Exploring Common Table Expressions (CTEs) in jOOQ

CTEs are represented by the SQL-99 `WITH` clause. You already saw several examples of CTE in previous chapters, for instance, in *Chapter 13, Exploiting SQL Functions*, you saw a CTE for computing z-scores.

Roughly, via CTEs, we factor out the code that otherwise should be repeated as derived tables. Typically, a CTE contains a list of derived tables placed in front of a `SELECT` statement in a certain order. The order is important because these derived tables are created conforming to this order and a CTE element can reference only prior CTE elements.

Basically, we distinguish between regular (non-recursive) CTEs and recursive CTEs.

Regular CTEs

A regular CTE associates a name to a temporary result set that has the scope of a statement such as `SELECT`, `INSERT`, `UPDATE`, `DELETE`, or `MERGE` (CTEs for DML statements are very useful vendor-specific extensions). But, a derived table or another type of subquery can have its own CTE as well, such as `SELECT x.a FROM (WITH t (a) AS (SELECT 1) SELECT a FROM t) x`.

The basic syntax of a CTE (for the exact syntax of a certain database vendor, you should consult the documentation) is as follows:

```
WITH CTE_name [(column_name [, ...])]
AS
 (CTE_definition) [, ...]
SQL_statement_using_CTE;
```

In jOOQ, a CTE is represented by the `org.jooq.CommonTableExpression` class and extends the commonly used `org.jooq.Table`, therefore a CTE can be used everywhere a `Table` can be used. The `CTE_name` represents the name used later in the query to refer to the CTE and, in jOOQ, can be specified as the argument of the method `name()` or `with()`.

The `column_name` marks the spot for a list of comma-separated columns that comes after the `CTE_name`. The number of columns specified here, and the number of columns defined in the `CTE_definition` must be equal. In jOOQ, when the `name()` method is used for `CTE_name`, this list can be specified via the `fields()` method. Otherwise, it can be specified as part of the `with()` arguments after the `CTE_name`.

The `AS` keyword is rendered in jOOQ via the `as(Select<?> select)` method. So, the argument of `as()` is the `CTE_definition`. Starting with jOOQ 3.15, the CTE `as(ResultQuery<?>)` method accepts a `ResultQuery<?>` to allow for using `INSERT ... RETURNING` and other `DML ... RETURNING` statements as CTEs in PostgreSQL.

Finally, we have the SQL that uses the CTE and references it via `CTE_name`.

For instance, the following CTE named `cte_sales` computes the sum of sales per employee:

```
CommonTableExpression<Record2<Long, BigDecimal>> t
   = name("cte_sales").fields("employee_nr", "sales")
       .as(select(SALE.EMPLOYEE_NUMBER, sum(SALE.SALE_))
       .from(SALE).groupBy(SALE.EMPLOYEE_NUMBER));
```

This is the CTE declaration that can be referenced via the local variable `t` in any future SQL queries expressed via jOOQ. Running this snippet of code now doesn't execute the `SELECT` and doesn't produce any effect. Once we use `t` in a SQL query, jOOQ will evaluate it to render the expected CTE. That CTE will be executed by the database.

Exactly as in the case of declaring derived tables in local variables, in the case of CTE, the fields cannot be dereferenced from `t` in a type-safe way, therefore it is our job to specify the proper data types in the queries that use the CTE. Again, let me point out that this is a pure Java issue, and has nothing to do with SQL!

Lukas Eder shared this: *Regarding the lack of type safety when dereferencing CTE or derived table fields: This is often an opportunity to rewrite the SQL statement again to something that doesn't use a CTE. On Stack Overflow, I've seen many cases of questions where the person tried to put *everything* in several layers of confusing CTE, when the actual factored-out query could have been *much* easier (for example, if they knew window functions, or the correct logical order of operations, and so on). Just because you can use CTEs, doesn't mean you have to use them *everywhere*.*

So, here is a usage of our CTE, `t`, for fetching the employee having the biggest sales:

```
ctx.with(t)
   .select() // or, .select(t.field("employee_nr"),
             //               t.field("sales"))
   .from(t)
   .where(t.field("sales", Double.class)
      .eq(select(max(t.field("sales" ,Double.class)))
           .from(t))).fetch();
```

By extracting the CTE fields as local variables, we can rewrite our CTE declaration like this:

```
Field<Long> e = SALE.EMPLOYEE_NUMBER;
Field<BigDecimal> s = sum(SALE.SALE_);

CommonTableExpression<Record2<Long, BigDecimal>> t
 = name("cte_sales").fields(e.getName(), s.getName())
    .as(select(e, s).from(SALE).groupBy(e));
```

The SQL that uses this CTE is as follows:

```
ctx.with(t)
   .select() // or, .select(t.field(e.getName()),
             //               t.field(s.getName()))
   .from(t)
   .where(t.field(s.getName(), s.getType())
      .eq(select(max(t.field(s.getName(), s.getType())))
           .from(t))).fetch();
```

And, of course, relying on `<T> Field<T> field(Field<T> field)`, introduced in the previous section, can help us to write a type-safe CTE as follows:

```
ctx.with(t)
   .select() // or, .select(t.field(e), t.field(s))
   .from(t)
   .where(t.field(s)
           .eq(select(max(t.field(s))).from(t))).fetch();
```

As an alternative to the previous explicit CTE, we can write an inline CTE as follows:

```
ctx.with("cte_sales", "employee_nr", "sales")
   .as(select(SALE.EMPLOYEE_NUMBER, sum(SALE.SALE_))
```

```
    .from(SALE)
    .groupBy(SALE.EMPLOYEE_NUMBER))
    .select() // or, field(name("employee_nr")),
              //     field(name("sales"))
    .from(name("cte_sales"))
    .where(field(name("sales"))
       .eq(select(max(field(name("sales"))))
    .from(name("cte_sales")))).fetch();
```

By arbitrarily choosing the PostgreSQL dialect, we have the following rendered SQL for all the previous CTEs:

```
WITH "cte_sales"("employee_nr", "sales") AS
  (SELECT "public"."sale"."employee_number",
          sum("public"."sale"."sale")
   FROM "public"."sale"
   GROUP BY "public"."sale"."employee_number")
SELECT * FROM "cte_sales"
WHERE "sales" = (SELECT max("sales") FROM "cte_sales")
```

You can check these examples in the bundled code named *CteSimple*. So far, our CTE is used only in `SELECT` statements. But, CTE can be used in DML statements such as `INSERT`, `UPDATE`, `DELETE`, and `MERGE` as well.

CTE as SELECT and DML

jOOQ supports using CTE in `INSERT`, `UPDATE`, `DELETE`, and `MERGE`. For instance, the following snippet of code inserts into a brand-new table a random part from the `SALE` table:

```
ctx.createTableIfNotExists("sale_training").as(
      selectFrom(SALE)).withNoData().execute();

ctx.with("training_sale_ids", "sale_id")
    .as(select(SALE.SALE_ID).from(SALE)
        .orderBy(rand()).limit(10))
    .insertInto(table(name("sale_training")))
      .select(select().from(SALE).where(SALE.SALE_ID.notIn(
       select(field(name("sale_id"), Long.class))
         .from(name("training_sale_ids")))))
    .execute();
```

Here is another example that updates the prices of the products (PRODUCT.BUY_PRICE) to the maximum order prices (max(ORDERDETAIL.PRICE_EACH)) via a CTE used in UPDATE:

```
ctx.with("product_cte", "product_id", "max_buy_price")
   .as(select(ORDERDETAIL.PRODUCT_ID,
              max(ORDERDETAIL.PRICE_EACH))
         .from(ORDERDETAIL)
         .groupBy(ORDERDETAIL.PRODUCT_ID))
   .update(PRODUCT)
     .set(PRODUCT.BUY_PRICE, coalesce(field(
     select(field(name("max_buy_price"), BigDecimal.class))
       .from(name("product_cte"))
       .where(PRODUCT.PRODUCT_ID.eq(
          field(name("product_id"), Long.class)))),
          PRODUCT.BUY_PRICE)).execute();
```

You can practice these examples along with others including using CTE in DELETE and MERGE in *CteSelectDml*. Next, let's see how we can express a CTE as DML in PostgreSQL.

A CTE as DML

Starting with jOOQ 3.15, the CTE as(ResultQuery<?>) method accepts a ResultQuery<?> to allow for using INSERT ... RETURNING and other DML ... RETURNING statements as CTE in PostgreSQL. Here is a simple CTE storing the returned SALE_ID:

```
ctx.with("cte", "sale_id")
   .as(insertInto(SALE, SALE.FISCAL_YEAR, SALE.SALE_,
                  SALE.EMPLOYEE_NUMBER, SALE.FISCAL_MONTH,
                  SALE.REVENUE_GROWTH)
   .values(2005, 1250.55, 1504L, 1, 0.0)
   .returningResult(SALE.SALE_ID))
   .selectFrom(name("cte"))
   .fetch();
```

Let's write a CTE that updates the SALE.REVENUE_GROWTH of all employees having a null EMPLOYEE.COMMISSION. All the updated SALE.EMPLOYEE_NUMBER are stored in the CTE and used further to insert in EMPLOYEE_STATUS as follows:

```
ctx.with("cte", "employee_number")
  .as(update(SALE).set(SALE.REVENUE_GROWTH, 0.0)
```

```
    .where(SALE.EMPLOYEE_NUMBER.in(
      select(EMPLOYEE.EMPLOYEE_NUMBER).from(EMPLOYEE)
       .where(EMPLOYEE.COMMISSION.isNull())))
    .returningResult(SALE.EMPLOYEE_NUMBER))
  .insertInto(EMPLOYEE_STATUS, EMPLOYEE_STATUS
              .EMPLOYEE_NUMBER,EMPLOYEE_STATUS.STATUS,
               EMPLOYEE_STATUS.ACQUIRED_DATE)
  .select(select(field(name("employee_number"), Long.class),
     val("REGULAR"), val(LocalDate.now())).from(name("cte")))
  .execute();
```

You can check out more examples in the bundled code named *CteDml* for PostgreSQL. Next, let's see how we can embed plain SQL in a CTE.

CTEs and plain SQL

Using plain SQL in CTE is straightforward as you can see in the following example:

```
CommonTableExpression<Record2<Long, String>>cte = name("cte")
  .fields("pid", "ppl").as(resultQuery(
     // Put any plain SQL statement here
     """
     select "public"."product"."product_id",
            "public"."product"."product_line"
     from "public"."product"
     where "public"."product"."quantity_in_stock" > 0
     """
  ).coerce(field("pid", BIGINT), field("ppl", VARCHAR)));

Result<Record2<Long, String>> result =
   ctx.with(cte).selectFrom(cte).fetch();
```

You can test this example in the bundled code named *CtePlainSql*. Next, let's tackle some common types of CTEs, and let's continue with a query that uses two or more CTEs.

Chaining CTEs

Sometimes, a query must exploit more than one CTE. For instance, let's consider the tables `PRODUCTLINE`, `PRODUCT`, and `ORDERDETAIL`. Our goal is to fetch for each product line some info (for instance, the description), the total number of products, and the total sales. For this, we can write a CTE that joins `PRODUCTLINE` with `PRODUCT` and count the total number of products per product line, and another CTE that joins

PRODUCT with ORDERDETAIL and computes the total sales per product line. Then, both CTEs are used in a SELECT to fetch the final result as in the following inlined CTE:

```
ctx.with("cte_productline_counts")
     .as(select(PRODUCT.PRODUCT_LINE, PRODUCT.CODE,
          count(PRODUCT.PRODUCT_ID).as("product_count"),
          PRODUCTLINE.TEXT_DESCRIPTION.as("description"))
         .from(PRODUCTLINE).join(PRODUCT).onKey()
         .groupBy(PRODUCT.PRODUCT_LINE, PRODUCT.CODE,
                  PRODUCTLINE.TEXT_DESCRIPTION))
     .with("cte_productline_sales")
       .as(select(PRODUCT.PRODUCT_LINE,
             sum(ORDERDETAIL.QUANTITY_ORDERED
               .mul(ORDERDETAIL.PRICE_EACH)).as("sales"))
             .from(PRODUCT).join(ORDERDETAIL).onKey()
       .groupBy(PRODUCT.PRODUCT_LINE))
     .select(field(name("cte_productline_counts",
        "product_line")), field(name("code")),
        field(name("product_count")),
        field(name("description")),
        field(name("sales")))
     .from(name("cte_productline_counts"))
     .join(name("cte_productline_sales"))
     .on(field(name("cte_productline_counts",
              "product_line"))
       .eq(field(name("cte_productline_sales",
                     "product_line"))))
     .orderBy(field(name("cte_productline_counts",
                        "product_line")))
     .fetch();
```

In the bundled code (*CteSimple*), you can see the explicit CTE version as well.

Nested CTEs

CTEs can be nested as well. For instance, here we have a "base" CTE that computes the employees' average salary per office. The next two CTEs fetch from the "base" CTE the minimum and maximum average respectively. Finally, our query cross-joins these CTEs:

```
ctx.with("avg_per_office")
   .as(select(EMPLOYEE.OFFICE_CODE.as("office"),
```

```
        avg(EMPLOYEE.SALARY).as("avg_salary_per_office"))
      .from(EMPLOYEE)
      .groupBy(EMPLOYEE.OFFICE_CODE))
  .with("min_salary_office")
  .as(select(min(field(name("avg_salary_per_office")))
        .as("min_avg_salary_per_office"))
      .from(name("avg_per_office")))
  .with("max_salary_office")
  .as(select(max(field(name("avg_salary_per_office")))
        .as("max_avg_salary_per_office"))
      .from(name("avg_per_office")))
  .select()
      .from(name("avg_per_office"))
      .crossJoin(name("min_salary_office"))
      .crossJoin(name("max_salary_office"))
  .fetch();
```

The potential output is shown in the next figure:

```
+------+---------------------+-------------------------+-------------------------+
|office|avg_salary_per_office|min_avg_salary_per_office|max_avg_salary_per_office|
+------+---------------------+-------------------------+-------------------------+
|6     |   61250.00          |        50000.00         |        85833.33         |
|2     |   57500.00          |        50000.00         |        85833.33         |
|4     |   62000.00          |        50000.00         |        85833.33         |
|7     |   50000.00          |        50000.00         |        85833.33         |
|5     |   57500.00          |        50000.00         |        85833.33         |
|1     |   85833.33          |        50000.00         |        85833.33         |
|3     |   55000.00          |        50000.00         |        85833.33         |
+------+---------------------+-------------------------+-------------------------+
```

Figure 14.2 – Output of nested CTEs example

In the bundled code (*CteSimple*), you can see the explicit CTE version as well.

Some databases (for instance, MySQL and PostgreSQL) allow you to nest CTEs via the `FROM` clause. Here is an example:

```
ctx.with("t2")
   .as(select(avg(field("sum_min_sal", Float.class))
           .as("avg_sum_min_sal")).from(
        with("t1")
           .as(select(min(EMPLOYEE.SALARY).as("min_sal"))
                .from(EMPLOYEE)
                .groupBy(EMPLOYEE.OFFICE_CODE)).select(
```

```
                    sum(field("min_sal", Float.class))
                        .as("sum_min_sal"))
                      .from(name("t1"))
                      .groupBy(field("min_sal"))))
   .select()
   .from(name("t2"))
   .fetch();
```

So, this is a three-step query: first, we compute the minimum salary per office; second, we compute the sum of salaries per minimum salary; and third, we compute the average of these sums.

Materialized CTEs

Do you have an expensive CTE that fetches a relatively small result set and is used two or more times? Then most probably you have a CTE that you may want to materialize. The materialized CTE can then be referenced multiple times by the parent query without recomputing the results.

In jOOQ, materializing a CTE can be done via `asMaterialized()`. Depending on the database, jOOQ will render the proper SQL. For instance, consider the following materialized CTE:

```
ctx.with("cte", "customer_number",
         "order_line_number", "sum_price", "sum_quantity")
   .asMaterialized(
           select(ORDER.CUSTOMER_NUMBER,
                  ORDERDETAIL.ORDER_LINE_NUMBER,
                  sum(ORDERDETAIL.PRICE_EACH),
                  sum(ORDERDETAIL.QUANTITY_ORDERED))
              .from(ORDER)
              .join(ORDERDETAIL)
              .on(ORDER.ORDER_ID.eq(ORDERDETAIL.ORDER_ID))
              .groupBy(ORDERDETAIL.ORDER_LINE_NUMBER,
                       ORDER.CUSTOMER_NUMBER))
   .select(field(name("customer_number")),
      inline("Order Line Number").as("metric"),
      field(name("order_line_number"))).from(name("cte")) // 1
   .unionAll(select(field(name("customer_number")),
      inline("Sum Price").as("metric"),
      field(name("sum_price"))).from(name("cte")))        // 2
   .unionAll(select(field(name("customer_number")),
```

```
      inline("Sum Quantity").as("metric"),
      field(name("sum_quantity"))).from(name("cte")))      // 3
   .fetch();
```

This CTE should be evaluated three times (denoted in code as //1, //2, and //3). Hopefully, thanks to `asMaterialized()`, the result of the CTE should be materialized and reused instead of being recomputed.

Some databases detect that a CTE is used more than once (the `WITH` clause is referenced more than once in the outer query) and automatically try to materialize the result set as an optimization fence. For instance, PostgreSQL will materialize the above CTE even if we don't use `asMaterialized()` and we simply use `as()` because the `WITH` query is called three times.

But, PostgreSQL allows us to control the CTE materialization and change the default behavior. If we want to force inlining the CTE instead of it being materialized, then we add the `NOT MATERIALIZED` hint to the CTE. In jOOQ, this is accomplished via `asNotMaterialized()`:

```
ctx.with("cte", "customer_number",
         "order_line_number", "sum_price", "sum_quantity")
   .asNotMaterialized(select(ORDER.CUSTOMER_NUMBER,
                             ORDERDETAIL.ORDER_LINE_NUMBER, ...
```

On the other hand, in Oracle, we can control materialization via the `/*+ materialize */` and `/*+ inline */` hints. Using jOOQ's `asMaterialized()` renders the `/*+ materialize */` hint, while `asNotMaterialized()` renders the `/*+ inline */` hint. Using jOOQ's `as()` doesn't render any hint, so Oracle's optimizer is free to act as the default.

However, Lukas Eder said: *Note that the Oracle hints aren't documented, so they might change (though all possible Oracle guru blogs document their de facto functionality, so knowing Oracle, they won't break easily). If not explicitly documented, there's never any *guarantee* for any *materialization trick* to keep working.*

Other databases don't support materialization at all or use it only as an internal mechanism of the optimizer (for instance, MySQL). Using jOOQ's `as()`, `asMaterialized()`, and `asNotMaterialized()` renders the same SQL for MySQL, therefore we cannot rely on explicit materialization. In such cases, we can attempt to rewrite our CTE to avoid recalls. For instance, the previous CTE can be optimized to not need materialization in MySQL via `LATERAL`:

```
ctx.with("cte")
  .as(
```

```
   select(ORDER.CUSTOMER_NUMBER,
          ORDERDETAIL.ORDER_LINE_NUMBER,
          sum(ORDERDETAIL.PRICE_EACH).as("sum_price"),
          sum(ORDERDETAIL.QUANTITY_ORDERED)
              .as("sum_quantity"))
    .from(ORDER)
    .join(ORDERDETAIL)
      .on(ORDER.ORDER_ID.eq(ORDERDETAIL.ORDER_ID))
    .groupBy(ORDERDETAIL.ORDER_LINE_NUMBER,
             ORDER.CUSTOMER_NUMBER))
 .select(field(name("customer_number")),
   field(name("t", "metric")), field(name("t", "value")))
    .from(table(name("cte")), lateral(
      select(inline("Order Line Number").as("metric"),
       field(name("order_line_number")).as("value"))
      .unionAll(select(inline("Sum Price").as("metric"),
        field(name("sum_price")).as("value")))
      .unionAll(select(inline("Sum Quantity").as("metric"),
        field(name("sum_quantity")).as("value"))))
 .as('t")).fetch();
```

Here is the alternative for SQL Server (like MySQL, SQL Server doesn't expose any support for explicit materialization; however, there is a proposal for Microsoft to add a dedicated hint similar to what Oracle has) using `CROSS APPLY` and the `VALUES` constructor:

```
ctx.with("cte")
 .as(select(ORDER.CUSTOMER_NUMBER,
            ORDERDETAIL.ORDER_LINE_NUMBER,
            sum(ORDERDETAIL.PRICE_EACH).as("sum_price"),
            sum(ORDERDETAIL.QUANTITY_ORDERED)
                .as("sum_quantity"))
 .from(ORDER)
 .join(ORDERDETAIL)
   .on(ORDER.ORDER_ID.eq(ORDERDETAIL.ORDER_ID))
 .groupBy(ORDERDETAIL.ORDER_LINE_NUMBER,
          ORDER.CUSTOMER_NUMBER))
 .select(field(name("customer_number")),
    field(name("t", "metric")), field(name("t", "value")))
 .from(name("cte")).crossApply(
   values(row("Order Line Number",
```

```
      field(name("cte", "order_line_number"))),
  row("Sum Price", field(name("cte", "sum_price"))),
  row("Sum Quantity", field(name("cte", "sum_quantity"))))
   .as("t", "metric", "value")).fetch();
```

A good cost-based optimizer should always rewrite all SQL statements to the optimal execution plan, so what may work today, might not work tomorrow – such as this `LATERAL/CROSS APPLY` trick. If the optimizer is ever smart enough to detect that `LATERAL/CROSS APPLY` is unnecessary (for example, because of the lack of correlation), then it might be (should be) eliminated.

You can check out all these examples in *CteSimple*. Moreover, in the *CteAggRem* application, you can practice a CTE for calculating the top N items and aggregating (summing) the remainder in a separate row. Basically, while ranking items in the database is a common problem to compute top/bottom N items, another common requirement that is related to this one is to obtain all the other rows (that don't fit in top/bottom N) in a separate row. This is helpful to provide a complete context when presenting data.

In the *CteWMAvg* code, you can check out a statistics problem with the main goal being to highlight recent points. This problem is known as **Weighted Moving Average** (**WMA**). This is part of the moving average family (`https://en.wikipedia.org/wiki/Moving_average`) and, in a nutshell, WMA is a moving average where the previous values (points) range in the sliding window are given different (fractional) weights.

Recursive CTEs

Besides regular CTEs, we have recursive CTEs.

In a nutshell, recursive CTEs reproduce the concept of for-loops in programming. A recursive CTE can handle and explore hierarchical data by referencing themselves. Behind a recursive CTE, there are two main members:

- The *anchor* member – Its goal is to select the starting rows of the involved recursive steps.
- The *recursive* member – Its goal is to generate rows for the CTE. The first iteration step acts against the anchor rows, while the second iteration step acts against the rows previously created in recursions steps. This member occurs after a `UNION ALL` in the CTE definition part. To be more accurate, `UNION ALL` is required by a few dialects, but others are capable of recurring with `UNION` as well, with slightly different semantics.

In jOOQ, recursive CTEs can be expressed via the `withRecursive()` method.

Here's a simple recursive CTE that computes the famous Fibonacci numbers. The *anchor* member is equal to 1, and the *recursive* member applies the Fibonacci formula up to the number 20:

```
ctx.withRecursive("fibonacci", "n", "f", "f1")
   .as(select(inline(1L), inline(0L), inline(1L))
   .unionAll(select(field(name("n"), Long.class).plus(1),
     field(name("f"), Long.class).plus(field(name("f1"))),
     field(name("f"), Long.class))
   .from(name("fibonacci"))
   .where(field(name("n")).lt(20))))
   .select(field(name("n")), field(name("f")).as("f_nbr"))
   .from(name("fibonacci"))
   .fetch();
```

Well, that was easy, wasn't it? Next, let's tackle a famous problem that can be solved via recursive CTE, known as the Travelling Salesman Problem. Consider reading more details here: `https://en.wikipedia.org/wiki/Travelling_salesman_problem`. In a nutshell, we interpret this problem to find the shortest private flight through several cities representing locations of our offices. Basically, in `OFFICE_FLIGHTS`, we have the routes between our offices as `OFFICE_FLIGHTS.DEPART_TOWN`, `OFFICE_FLIGHTS.ARRIVAL_TOWN`, and `OFFICE_FLIGHTS.DISTANCE_KM`. For instance, our CTE will use *Los Angeles* as its anchor city, and then recursively traverse every other city in order to reach *Tokyo*:

```
String from = "Los Angeles";
String to = "Tokyo";

ctx.withRecursive("flights",
      "arrival_town", "steps", "total_distance_km", "path")
 .as(selectDistinct(OFFICE_FLIGHTS.DEPART_TOWN
   .as("arrival_town"), inline(0).as("steps"), inline(0)
   .as("total_distance_km"), cast(from, SQLDataType.VARCHAR)
   .as("path"))
 .from(OFFICE_FLIGHTS)
 .where(OFFICE_FLIGHTS.DEPART_TOWN.eq(from))
 .unionAll(select(field(name("arrivals", "arrival_town"),
           String.class), field(name("flights", "steps"),
           Integer.class).plus(1), field(name("flights",
           "total_distance_km"), Integer.class).plus(
           field(name("arrivals", "distance_km"))),
           concat(field(name("flights", "path")),inline(","),
```

```
              field(name("arrivals", "arrival_town"))))
            .from(OFFICE_FLIGHTS.as("arrivals"),
              table(name("flights")))
            .where(field(name("flights", "arrival_town"))
             .eq(field(name("arrivals", "depart_town")))
              .and(field(name("flights", "path"))
               .notLike(concat(inline("%"),
                field(name("arrivals", "arrival_town")),
                 inline("%")))))))
            .select()
              .from(name("flights"))
              .where(field(name("arrival_town")).eq(to))
              .orderBy(field(name("total_distance_km")))
 .fetch();
```

Some possible output is in the next figure:

```
+------------+-----+-----------------+--------------------------------------------------+
|arrival_town|steps|total_distance_km|path                                              |
+------------+-----+-----------------+--------------------------------------------------+
|Tokyo       |    3|            18983|Los Angeles,Paris,London,Tokyo                    |
|Tokyo       |    5|            19761|Los Angeles,San Diego,Boston,Paris,London,Tokyo   |
|Tokyo       |    5|            19761|Los Angeles,San Diego,Boston,New York,London,Tokyo|
|Tokyo       |    2|            19892|Los Angeles,Sydney,Tokyo                          |
|Tokyo       |    6|            22365|Los Angeles,San Diego,Boston,Springfield,New Yo...|
+------------+-----+-----------------+--------------------------------------------------+
|...record(s) truncated...
40
```

Figure 14.3 – Shortest private flight from Los Angeles to Tokyo, 18,983 km

You can practice these examples in *CteRecursive*.

CTEs and window functions

In this section, let's look at two examples that combine CTE and window functions, and let's start with an example that computes the gaps in IDs. For instance, each `EMPLOYEE` has an associated `EMPLOYEE_NUMBER` and we want to find out how many values are missing from the data values (missing `EMPLOYEE_NUMBER`), and how many existing values are consecutive. This is a job for the `ROW_NUMBER()` window and the following CTE:

```
ctx.with("t", "data_val", "data_seq", "absent_data_grp")
   .as(select(EMPLOYEE.EMPLOYEE_NUMBER,
       rowNumber().over()
         .orderBy(EMPLOYEE.EMPLOYEE_NUMBER))
       EMPLOYEE.EMPLOYEE_NUMBER.minus(
```

```
        rowNumber().over()
            .orderBy(EMPLOYEE.EMPLOYEE_NUMBER)))
        .from(EMPLOYEE))
    .select(field(name("absent_data_grp")), count(),
        min(field(name("data_val"))).as("start_data_val"))
    .from(name("t"))
    .groupBy(field(name("absent_data_grp")))
    .orderBy(field(name("absent_data_grp")))
    .fetch();
```

While you can see this example in the bundled code, let's look at another one that finds the percentile rank of every product line by order values:

```
ctx.with("t", "product_line", "sum_price_each")
    .as(select(PRODUCT.PRODUCT_LINE,
            sum(ORDERDETAIL.PRICE_EACH))
        .from(PRODUCT)
        .join(ORDERDETAIL)
          .on(PRODUCT.PRODUCT_ID.eq(ORDERDETAIL.PRODUCT_ID))
        .groupBy(PRODUCT.PRODUCT_LINE))
    .select(field(name("product_line")),
            field(name("sum_price_each")),
            round(percentRank().over()
            .orderBy(field(name("sum_price_each"))).mul(100), 2)
            .concat("%").as("percentile_rank"))
    .from(name("t"))
    .fetch();
```

You can find these examples along with another one that finds the top three highest-valued orders each year in *CteWf*.

Using CTEs to generate data

CTEs are quite handy for generating data – they act as the source of data for the SQL statement that uses a CTE. For instance, using a CTE and the `VALUES` constructor can be done as follows:

```
ctx.with("dt")
    .as(select()
    .from(values(row(1, "John"), row(2, "Mary"), row(3, "Kelly"))
            .as("t", "id", "name")))
```

```
    .select()
    .from(name("dt"))
    .fetch();
```

Or, using a CTE to unnest an array can be done as follows:

```
ctx.with("dt")
    .as(select().from(unnest(new String[]
        {"John", "Mary", "Kelly"}).as("n")))
    .select()
    .from(name("dt"))
    .fetch();
```

Or, here is an example of unnesting an array to pick up a random value:

```
ctx.with("dt")
    .as(select().from(unnest(new String[]
        {"John", "Mary", "Kelly"}).as("n")))
    .select()
    .from(name("dt"))
    .orderBy(rand())
    .limit(1)
    .fetch();
```

Or, maybe you need a random sample from the database (here, 10 random products):

```
ctx.with("dt")
    .as(selectFrom(PRODUCT).orderBy(rand()).limit(10))
    .select()
    .from(name("dt"))
    .fetch();
```

However, keep in mind that `ORDER BY RAND()` should be avoided for large tables, as `ORDER BY` performs with O(N log N).

If you need more sophisticated sources of data, then probably you'll be interested in generating a series. Here is an example of generating the odd numbers between 1 and 10:

```
ctx.with("dt")
    .as(select().from(generateSeries(1, 10, 2).as("t", "s")))
    .select()
    .from(name("dt"))
    .fetch();
```

Here is an example that associates grades between 1 and 100 with the letters A to F and counts them as well – in other words, custom binning of grades:

```
ctx.with("grades")
   .as(select(round(inline(70).plus(sin(
         field(name("serie", "sample"), Integer.class))
           .mul(30)))).as("grade"))
   .from(generateSeries(1, 100).as("serie", "sample")))
   .select(
     case_().when(field(name("grade")).lt(60), "F")
            .when(field(name("grade")).lt(70), "D")
            .when(field(name("grade")).lt(80), "C")
            .when(field(name("grade")).lt(90), "B")
            .else_("A").as("letter_grade"),count())
   .from(name("grades"))
   .groupBy(field(name("letter_grade")))
   .orderBy(field(name("letter_grade")))
   .fetch();
```

In the bundled code, you can see more binning examples including custom binning of grades via `PERCENT_RANK()`, equal height binning, equal-width binning, the PostgreSQL `width_bucket()` function, and binning with a chart.

After all these snippets, let's tackle the following famous problem: Consider `p` student classes of certain sizes, and `q` rooms of certain sizes, where `q>= p`. Write a CTE for assigning as many classes as possible to rooms of proper size. Let's assume that the given data is in the left-hand side figure and the expected result is in the right-hand side figure:

classes

class_nbr	class_size
c1	80
c2	70
c3	65
c4	55
c5	50
c6	40

rooms

room_nbr	room_size
r1	70
r2	40
r3	50
r4	85
r5	30
r6	65
r7	55

class_nbr	class_size	room_nbr	room_size
c1	80	r4	85
c2	70	r1	70
c3	65	r6	65
c4	55	r7	55
c5	50	r3	50
c6	40	r2	40

Figure 14.4 – Input and expected output

In order to solve this problem, we can generate the input data as follows:

```
ctx.with("classes")
  .as(select()
  .from(values(row("c1", 80), row("c2", 70), row("c3", 65),
               row("c4", 55), row("c5", 50), row("c6", 40))
               .as("t", "class_nbr", "class_size")))
  .with("rooms")
  .as(select()
  .from(values(row("r1", 70), row("r2", 40), row("r3", 50),
    row("r4", 85), row("r5", 30), row("r6", 65), row("r7", 55))
  .as("t", "room_nbr", "room_size")))
...
```

The complete query is quite large to be listed here, but you can find it in the *CteGenData* application next to all the examples from this section.

Dynamic CTEs

Commonly, when we need to dynamically create a CTE, we plan to dynamically shape its name, derived table(s), and the outer query. For instance, the following method allows us to pass these components as arguments and return the result of executing the query:

```
public Result<Record> cte(String cteName, Select select,
          SelectField<?>[] fields, Condition condition,
          GroupField[] groupBy, SortField<?>[] orderBy) {

  var cte = ctx.with(cteName).as(select);

  var cteSelect = fields == null
    ? cte.select() : cte.select(fields)
        .from(table(name(cteName)));

  if (condition != null) {
     cteSelect.where(condition);
  }

  if (groupBy != null) {
     cteSelect.groupBy(groupBy);
  }
```

```
  if (orderBy != null) {
     cteSelect.orderBy(orderBy);
  }

  return cteSelect.fetch();
}
```

Here is a calling sample for solving the problem presented earlier, in the *CTEs and window functions* section:

```
Result<Record> result = cte("t",
 select(EMPLOYEE.EMPLOYEE_NUMBER.as("data_val"),
   rowNumber().over().orderBy(EMPLOYEE.EMPLOYEE_NUMBER)
   .as("data_seq"), EMPLOYEE.EMPLOYEE_NUMBER.minus(
   rowNumber().over().orderBy(EMPLOYEE.EMPLOYEE_NUMBER))
   .as("absent_data_grp"))
   .from(EMPLOYEE),
   new Field[]{field(name("absent_data_grp")), count(),
        min(field(name("data_val"))).as("start_data_val")},
   null, new GroupField[]{field(name("absent_data_grp"))},
   null);
```

Whenever you try to implement a CTE, as here, consider this Lukas Eder note: *This example uses the DSL in a mutable way, which works but is discouraged. A future jOOQ version might turn to an immutable DSL API and this code will stop working. It's unlikely to happen soon, because of the huge backward incompatibility, but the discouragement is real already today :) In IntelliJ, you should already get a warning in this code, because of the API's* `@CheckReturnValue` *annotation usage, at least in jOOQ 3.15.*

On the other hand, if you just need to pass a variable number of CTEs to the outer query, then you can do this:

```
public void CTE(List<CommonTableExpression<?>> CTE) {

  ctx.with(CTE)
     ...
}
```

Or, you can do this:

```
public void CTE(CommonTableExpression<?> cte1,
  CommonTableExpression<?>cte2,
```

```
  CommonTableExpression<?>cte3, ...) {

 ctx.with(cte1, cte2, cte3)
    ...
}
```

You can practice these examples in *CteDynamic*.

Expressing a query via a derived table, a temporary table, and a CTE

Sometimes, we prefer to express a query in several ways to compare their execution plans. For instance, we may have a query and express it via derived tables, temporary tables, and CTE to see which approach fits best. Since jOOQ supports these approaches, let's try to express a query starting from the derived tables approach:

```
ctx.select(EMPLOYEE.FIRST_NAME, EMPLOYEE.LAST_NAME,
   sum(SALE.SALE_), field(select(sum(SALE.SALE_)).from(SALE))
     .divide(field(select(countDistinct(SALE.EMPLOYEE_NUMBER))
    .from(SALE))).as("avg_sales"))
   .from(EMPLOYEE)
   .innerJoin(SALE)
   .on(EMPLOYEE.EMPLOYEE_NUMBER.eq(SALE.EMPLOYEE_NUMBER))
   .groupBy(EMPLOYEE.FIRST_NAME, EMPLOYEE.LAST_NAME)
   .having(sum(SALE.SALE_).gt(field(select(sum(SALE.SALE_))
    .from(SALE))
   .divide(field(select(countDistinct(SALE.EMPLOYEE_NUMBER))
    .from(SALE))))).fetch();
```

So, this query returns all employees with above-average sales. For each employee, we compare their average sales to the total average sales for all employees. Essentially, this query works on the `EMPLOYEE` and `SALE` tables, and we must know the total sales for all employees, the number of employees, and the sum of sales for each employee.

If we extract what we must know in three temporary tables, then we obtain this:

```
ctx.createTemporaryTable("t1").as(
  select(sum(SALE.SALE_).as("sum_all_sales"))
   .from(SALE)).execute();
```

```
ctx.createTemporaryTable("t2").as(
  select(countDistinct(SALE.EMPLOYEE_NUMBER)
    .as("nbr_employee")).from(SALE)).execute();

ctx.createTemporaryTable("t3").as(
  select(EMPLOYEE.FIRST_NAME, EMPLOYEE.LAST_NAME,
       sum(SALE.SALE_).as("employee_sale"))
   .from(EMPLOYEE)
   .innerJoin(SALE)
   .on(EMPLOYEE.EMPLOYEE_NUMBER.eq(SALE.EMPLOYEE_NUMBER))
   .groupBy(EMPLOYEE.FIRST_NAME, EMPLOYEE.LAST_NAME))
   .execute();
```

Having these three temporary tables, we can rewrite our query as follows:

```
ctx.select(field(name("first_name")),field(name("last_name")),
    field(name("employee_sale")), field(name("sum_all_sales"))
 .divide(field(name("nbr_employee"), Integer.class))
   .as("avg_sales"))
 .from(table(name("t1")),table(name("t2")), table(name("t3")))
 .where(field(name("employee_sale")).gt(
   field(name("sum_all_sales")).divide(
     field(name("nbr_employee"), Integer.class))))
 .fetch();
```

Finally, the same query can be expressed via CTE (by replacing `as()` with `asMaterialized()`, you can practice the materialization of this CTE):

```
ctx.with("cte1", "sum_all_sales")
 .as(select(sum(SALE.SALE_)).from(SALE))
 .with("cte2", "nbr_employee")
 .as(select(countDistinct(SALE.EMPLOYEE_NUMBER)).from(SALE))
 .with("cte3", "first_name", "last_name", "employee_sale")
 .as(select(EMPLOYEE.FIRST_NAME, EMPLOYEE.LAST_NAME,
    sum(SALE.SALE_).as("employee_sale"))
     .from(EMPLOYEE)
     .innerJoin(SALE)
     .on(EMPLOYEE.EMPLOYEE_NUMBER.eq(SALE.EMPLOYEE_NUMBER))
     .groupBy(EMPLOYEE.FIRST_NAME, EMPLOYEE.LAST_NAME))
     .select(field(name("first_name")),
```

```
    field(name("last_name")), field(name("employee_sale")),
    field(name("sum_all_sales"))
     .divide(field(name("nbr_employee"), Integer.class))
       .as("avg_sales"))
     .from(table(name("cte1")), table(name("cte2")),
           table(name("cte3")))
     .where(field(name("employee_sale")).gt(
       field(name("sum_all_sales")).divide(
       field(name("nbr_employee"), Integer.class))))
.fetch();
```

Now you just have to run these queries against your database and compare their performances and execution plans. The bundled code contains one more example and is available as *ToCte*.

Handling views in jOOQ

The last section of this chapter is reserved for database views.

A view acts as an actual physical table that can be invoked by name. They fit well for reporting tasks or integration with third-party tools that need a guided query API. By default, the database vendor decides to materialize the results of the view or to rely on other mechanisms to get the same effect. Most vendors (hopefully) don't default to materializing views! Views should behave just like CTE or derived tables and should be transparent to the optimizer. In most cases (in Oracle), we would expect a view to be inlined, even when selected several times, because each time, a different predicate might be pushed down into the view. Actual materialized views are supported only by a few vendors, while the optimizer can decide to materialize the view contents when a view is queried several times. The view's definition is stored in the schema tables so it can be invoked by name wherever a regular/base table could be used. If the view is updatable, then some additional rules come to sustain it.

A view differs by a base, temporary, or derived table in several essential aspects. Base and temporary tables accept constraints, while a view doesn't (in most databases). A view has no presence in the database until it is invoked, whereas a temporary table is persistent. Finally, a derived table has the same scope as the query in which it is created. The view definition cannot contain a reference to itself, since it does not exist yet, but it can contain references to other views.

The basic syntax of a view is as follows (for the exact syntax of a certain database vendor, you should consult the documentation):

```
CREATE VIEW <table name> [(<view column list>)]
   AS <query expression>
    [WITH [<levels clause>] CHECK OPTION]
       <levels clause>::= CASCADED | LOCAL
```

Some RDBMS support constraints on views (for instance, Oracle), though with limitations: `https://docs.oracle.com/en/database/oracle/oracle-database/19/sqlrf/constraint.html`. The documented `WITH CHECK OPTION` is actually a constraint.

Next, let's see some examples of views expressed via jOOQ.

Updatable and read-only views

Views can be either updatable or read-only, but not both. In jOOQ, they can be created via the `createView()` and `createViewIfNotExists()` methods. Dropping a view can be done via `dropView()`, respectively `dropViewIfExists()`. Here is an example of creating a read-only view:

```
ctx.createView("sales_1504_1370")
   .as(select().from(SALE).where(
       SALE.EMPLOYEE_NUMBER.eq(1504L))
   .unionAll(select().from(SALE)
   .where(SALE.EMPLOYEE_NUMBER.eq(1370L))))
   .execute();
```

Roughly, in standard SQL, an updatable view is built on only one table; it cannot contain `GROUP BY`, `HAVING`, `INTERSECT`, `EXCEPT`, `SELECT DISTINCT`, or `UNION` (however, at least in theory, a `UNION` between two disjoint tables, neither of which has duplicate rows in itself, should be updatable), aggregate functions, calculated columns, and any columns excluded from the view must have `DEFAULT` in the base table or be null-able. However, according to the standard SQL T111 optional feature, joins and unions aren't an impediment to updatability per se, so an updatable view doesn't have to be built "on only one table." Also (for the avoidance of any doubt), not all columns of an updatable view have to be updatable, but of course, only updatable columns can be updated.

When the view is modified, the modifications pass through the view to the corresponding underlying base table. In other words, an updatable view has a 1:1 match between its rows and the rows of the underlying base table, therefore the previous view is not updatable. But we can rewrite it without `UNION ALL` to transform it into a valid updatable view:

```
ctx.createView("sales_1504_1370_u")
   .as(select().from(SALE)
   .where(SALE.EMPLOYEE_NUMBER.in(1504L, 1370L)))
   .execute();
```

Some views are "partially" updatable. For instance, views that contain `JOIN` statements like this one:

```
ctx.createView("employees_and_sales", "first_name",
      "last_name", "sale_id", "sale")
   .as(select(EMPLOYEE.FIRST_NAME, EMPLOYEE.LAST_NAME,
              SALE.SALE_ID, SALE.SALE_)
   .from(EMPLOYEE)
   .join(SALE)
   .on(EMPLOYEE.EMPLOYEE_NUMBER.eq(SALE.EMPLOYEE_NUMBER)))
   .execute();
```

While in PostgreSQL this view is not updatable at all, in MySQL, SQL Server, and Oracle, this view is "partially" updatable. In other words, as long as the modifications affect only one of the two involved base tables, the view is updatable, otherwise, it is not. If more base tables are involved in the update, then an error occurs. For instance, in SQL Server, we get an error of *View or function 'employees_and_sales' is not updatable because the modification affects multiple base tables*, while in Oracle, we get *ORA-01776*.

You can check out these examples in *DbViews*.

Types of views (unofficial categorization)

In this section, let's define several common types of views depending on their usage, and let's start with views of the type single-table projection and restriction.

Single-table projection and restriction

Sometimes, for security reasons, we rely on projections/restrictions of a single base table to remove certain rows and/or columns that should not be seen by a particular group of users. For instance, the following view represents a projection of the `BANK_TRANSACTION` base table to restrict/hide the details about the involved banks:

```
ctx.createView("transactions",
               "customer_number", "check_number",
```

```
                "caching_date", "transfer_amount", "status")
    .as(select(BANK_TRANSACTION.CUSTOMER_NUMBER,
               BANK_TRANSACTION.CHECK_NUMBER,
               BANK_TRANSACTION.CACHING_DATE,
               BANK_TRANSACTION.TRANSFER_AMOUNT,
               BANK_TRANSACTION.STATUS)
    .from(BANK_TRANSACTION))
    .execute();
```

Another type of view tackles computed columns.

Calculated columns

Providing summary data is another use case of views. For instance, we prefer to compute the columns in as meaningful a way as possible and expose them to the clients as views. Here is an example of computing the payroll of each employee as salary plus commission:

```
ctx.createView("payroll", "employee_number", "paycheck_amt")
    .as(select(EMPLOYEE.EMPLOYEE_NUMBER, EMPLOYEE.SALARY
      .plus(coalesce(EMPLOYEE.COMMISSION, 0.00)))
    .from(EMPLOYEE))
    .execute();
```

Another type of view tackles translated columns.

Translated columns

Views are also useful for translating codes into texts to increase the readability of the fetched result set. A common case is a suite of `JOIN` statements between several tables via one or more foreign keys. For instance, in the following view, we have a detailed report of customers, orders, and products by translating the `CUSTOMER_NUMBER`, `ORDER_ID`, and `PRODUCT_ID` codes (foreign keys):

```
ctx.createView("customer_orders")
    .as(select(CUSTOMER.CUSTOMER_NAME,
       CUSTOMER.CONTACT_FIRST_NAME, CUSTOMER.CONTACT_LAST_NAME,
       ORDER.SHIPPED_DATE, ORDERDETAIL.QUANTITY_ORDERED,
       ORDERDETAIL.PRICE_EACH, PRODUCT.PRODUCT_NAME,
       PRODUCT.PRODUCT_LINE)
    .from(CUSTOMER)
    .innerJoin(ORDER)
    .on(CUSTOMER.CUSTOMER_NUMBER.eq(ORDER.CUSTOMER_NUMBER))
```

```
   .innerJoin(ORDERDETAIL)
   .on(ORDER.ORDER_ID.eq(ORDERDETAIL.ORDER_ID))
   .innerJoin(PRODUCT)
   .on(ORDERDETAIL.PRODUCT_ID.eq(PRODUCT.PRODUCT_ID)))
   .execute();
```

Next, let's tackle grouped views.

Grouped views

A grouped view relies on a query containing a GROUP BY clause. Commonly, such read-only views contain one or more aggregate functions and they are useful for creating different kinds of reports. Here is an example of creating a grouped view that fetches big sales per employee:

```
ctx.createView("big_sales", "employee_number", "big_sale")
   .as(select(SALE.EMPLOYEE_NUMBER, max(SALE.SALE_))
   .from(SALE)
   .groupBy(SALE.EMPLOYEE_NUMBER))
   .execute();
```

Here is another example that relies on a grouped view to "flatten out" a one-to-many relationship:

```
ctx.createView("employee_sales",
               "employee_number", "sales_count")
   .as(select(SALE.EMPLOYEE_NUMBER, count())
   .from(SALE)
   .groupBy(SALE.EMPLOYEE_NUMBER))
   .execute();

var result = ctx.select(
        EMPLOYEE.FIRST_NAME, EMPLOYEE.LAST_NAME,
        coalesce(field(name("sales_count")), 0)
          .as("sales_count"))
   .from(EMPLOYEE)
   .leftOuterJoin(table(name("employee_sales")))
   .on(EMPLOYEE.EMPLOYEE_NUMBER
    .eq(field(name("employee_sales", "employee_number"),
        Long.class))).fetch();
```

Next, let's tackle UNION-ed views.

UNION-ed views

Using UNION/UNION ALL in views is also a common usage case of views. Here is the previous query of flattening one-to-many relationships rewritten via UNION:

```
ctx.createView("employee_sales_u",
      "employee_number", "sales_count")
   .as(select(SALE.EMPLOYEE_NUMBER, count())
   .from(SALE)
   .groupBy(SALE.EMPLOYEE_NUMBER)
   .union(select(EMPLOYEE.EMPLOYEE_NUMBER, inline(0))
   .from(EMPLOYEE)
   .whereNotExists(select().from(SALE)
   .where(SALE.EMPLOYEE_NUMBER
     .eq(EMPLOYEE.EMPLOYEE_NUMBER))))).execute();

var result = ctx.select(EMPLOYEE.FIRST_NAME,
            EMPLOYEE.LAST_NAME, field(name("sales_count")))
   .from(EMPLOYEE)
   .innerJoin(table(name("employee_sales_u")))
   .on(EMPLOYEE.EMPLOYEE_NUMBER
    .eq(field(name("employee_sales_u", "employee_number"),
       Long.class)))
   .fetch();
```

Finally, let's see an example of nested views.

Nested views

A view can be built on another view. Pay attention to avoid circular references in the query expressions of the views and don't forget that a view must be ultimately built on base tables. Moreover, pay attention if you have different updatable views that reference the same base table at the same time. Using such views in other views may cause ambiguity issues since it is hard to infer what will happen if the highest-level view is modified.

Here is an example of using nested views:

```
ctx.createView("customer_orders_1",
              "customer_number", "orders_count")
   .as(select(ORDER.CUSTOMER_NUMBER, count())
   .from(ORDER)
   .groupBy(ORDER.CUSTOMER_NUMBER)).execute();
```

```
ctx.createView("customer_orders_2", "first_name",
               "last_name", "orders_count")
   .as(select(CUSTOMER.CONTACT_FIRST_NAME,
              CUSTOMER.CONTACT_LAST_NAME,
              coalesce(field(name("orders_count")), 0))
   .from(CUSTOMER)
   .leftOuterJoin(table(name("customer_orders_1")))
   .on(CUSTOMER.CUSTOMER_NUMBER
     .eq(field(name("customer_orders_1",
         "customer_number"), Long.class)))).execute();
```

The first view, `customer_orders_1`, counts the total orders per customer, and the second view, `customer_orders_2`, fetches the name of those customers.

You can see these examples in *DbTypesOfViews*.

Some examples of views

In this section, we rely on views to solve several problems. For instance, the following view is used to compute the cumulative distribution values by the headcount of each office:

```
ctx.createView("office_headcounts",
               "office_code", "headcount")
   .as(select(OFFICE.OFFICE_CODE,
              count(EMPLOYEE.EMPLOYEE_NUMBER))
   .from(OFFICE)
   .innerJoin(EMPLOYEE)
   .on(OFFICE.OFFICE_CODE.eq(EMPLOYEE.OFFICE_CODE))
   .groupBy(OFFICE.OFFICE_CODE))
   .execute();
```

Next, the query that uses this view for computing the cumulative distribution is as follows:

```
ctx.select(field(name("office_code")),
           field(name("headcount")),
           round(cumeDist().over().orderBy(
              field(name("headcount"))).mul(100), 2)
                 .concat("%").as("cume_dist_val"))
   .from(name("office_headcounts"))
   .fetch();
```

Views can be combined with CTE. Here is an example that creates a view on top of the CTE for detecting gaps in IDs – a problem tackled earlier, in the *CTE and window functions* section:

```
ctx.createView("absent_values",
          "data_val", "data_seq", "absent_data_grp")
   .as(with("t", "data_val", "data_seq", "absent_data_grp")
     .as(select(EMPLOYEE.EMPLOYEE_NUMBER,
      rowNumber().over().orderBy(EMPLOYEE.EMPLOYEE_NUMBER),
      EMPLOYEE.EMPLOYEE_NUMBER.minus(rowNumber().over()
           .orderBy(EMPLOYEE.EMPLOYEE_NUMBER)))
           .from(EMPLOYEE))
     .select(field(name("absent_data_grp")), count(),
       min(field(name("data_val"))).as("start_data_val"))
        .from(name("t"))
        .groupBy(field(name("absent_data_grp"))))
   .execute();
```

The query is straightforward:

```
ctx.select().from(name("absent_values")).fetch();
ctx.selectFrom(name("absent_values")).fetch();
```

Finally, let's see an example that attempts to optimize shipping costs in the future based on historical data from *2003*. Let's assume that we are shipping orders with a specialized company that can provide us, on demand, the list of trucks with their available periods per year as follows:

```
Table truck = select().from(values(
 row("Truck1",LocalDate.of(2003,1,1),LocalDate.of(2003,1,12)),
 row("Truck2",LocalDate.of(2003,1,8),LocalDate.of(2003,1,27)),
 ...
)).asTable("truck", "truck_id", "free_from", "free_to");
```

Booking trucks in advance for certain periods takes advantage of certain discounts, therefore, based on the orders from 2003, we can analyze some queries that can tell us whether this action can optimize shipping costs.

We start with a view named `order_truck`, which tells us which trucks are available for each order:

```
ctx.createView("order_truck", "truck_id", "order_id")
   .as(select(field(name("truck_id")), ORDER.ORDER_ID)
```

```
   .from(truck, ORDER)
   .where(not(field(name("free_to")).lt(ORDER.ORDER_DATE)
   .or(field(name("free_from")).gt(ORDER.REQUIRED_DATE)))))
   .execute();
```

Based on this view, we can run several queries that provide important information. For instance, how many orders can be shipped by each truck?

```
ctx.select(field(name("truck_id")), count().as("order_count"))
   .from(name("order_truck"))
   .groupBy(field(name("truck_id")))
   .fetch();
```

Or, how many trucks can ship the same order?

```
ctx.select(field(name("order_id")), count()
   .as("truck_count"))
   .from(name("order_truck"))
   .groupBy(field(name("order_id")))
   .fetch();
```

Moreover, based on this view, we can create another view named `order_truck_all` that can tell us the earliest and latest points in both intervals:

```
ctx.createView("order_truck_all", "truck_id",
               "order_id", "entry_date", "exit_date")
   .as(select(field(name("t", "truck_id")),
              field(name("t", "order_id")),
              ORDER.ORDER_DATE, ORDER.REQUIRED_DATE)
   .from(table(name("order_truck")).as("t"), ORDER)
   .where(ORDER.ORDER_ID.eq(field(name("t", "order_id"),
     Long.class)))
   .union(select(field(name("t", "truck_id")),
                 field(name("t", "order_id")),
                 truck.field(name("free_from")),
                 truck.field(name("free_to")))
          .from(table(name("order_truck")).as("t"), truck)
          .where(truck.field(name("truck_id"))
            .eq(field(name("t", "truck_id"))))))
   .execute();
```

Getting the exact points in both intervals can be determined based on the previous view as follows:

```
ctx.createView("order_truck_exact", "truck_id",
               "order_id", "entry_date", "exit_date")
   .as(select(field(name("truck_id")),
              field(name("order_id")),
              max(field(name("entry_date"))),
              min(field(name("exit_date"))))
         .from(name("order_truck_all"))
         .groupBy(field(name("truck_id")),
                  field(name("order_id"))))
   .execute();
```

Depending on how deeply we want to analyze the data, we can continue adding more queries and views, but I think you've got the idea. You can check out these examples in the bundled code, named *DbViewsEx*.

For those that expected to cover table-valued functions here (also called "parameterized views") as well, please consider the next chapter.

On the other hand, in this chapter, you saw at work several jOOQ methods useful to trigger DDL statements, such as `createView()`, `createTemporaryTable()`, and so on. Actually, jOOQ provides a comprehensive API for programmatically generating DDL that is covered by examples in the bundled code named *DynamicSchema*. Take your time to practice those examples and get familiar with them.

Summary

In this chapter, you've learned how to express derived tables, CTEs, and views in jOOQ. Since these are powerful SQL tools, it is very important to be familiar with them, therefore, besides the examples from this chapter, it is advisable to challenge yourself and try to solve more problems via jOOQ's DSL.

In the next chapter, we will tackle stored functions/procedures.

15

Calling and Creating Stored Functions and Procedures

SQL is a declarative language, but it also has procedural features such as stored functions/procedures, triggers, and cursors, which means SQL is considered to be a **Fourth-Generation Programming Language** (**4GL**). In this chapter, we will see how to call and create stored functions/procedures, or in other words, how to call and create Persistent Stored Modules (SQL/PSM) for MySQL, PostgreSQL, SQL Server, and Oracle.

Just in case you need a quick reminder about the key differences between the stored procedures and functions, check out the following head-to-head table (some of these differences are entirely or partially true depending on the database):

Procedure	Function
Used typically to execute a certain process	Used typically to perform some calculation
Relies on OUT parameters to return results	Relies on RETURN to return the result
It is not mandatory to return a result	It is mandatory to return a result
The returned data type will not be specified	The returned data type is mandatory to be specified
Cannot be called in SELECT statement	Can be called in SELECT statement

Figure 15.1 – Key differences between procedures and functions

As you can infer from the previous comparison, the main difference is that procedures (may) produce a side effect, whereas functions are (generally) expected not to.

So, our agenda includes the following topics:

- Calling stored functions/procedures from jOOQ
- Stored procedures
- Creating stored functions/procedures via jOOQ

Right before getting started, let's have some insight from Lukas Eder who shared that: "*This may come up later, but it might be worth mentioning early: there are some users who use jOOQ *only* for its stored procedure code generation capabilities. When you have a lot of stored procedures, it's almost impossible to bind to them without code generation, and jOOQ works very well out of the box, kind of like when you have a WSDL file (or something comparable), and you generate all the stubs with Axis or Metro, and so on.*"

OK, now let's get started!

Technical requirements

The code for this chapter can be found on GitHub at `https://github.com/PacktPublishing/jOOQ-Masterclass/tree/master/Chapter15`.

Calling stored functions/procedures from jOOQ

Once you start dealing with stored functions/procedures across different database vendors, you'll be hit by the lack of standardization. For instance, the vendor-specific syntax for expressing functions/procedures, the wide variety of vendor-specific functions/procedure types, and different ways of supporting and treating the output parameters are just a few non-standard aspects of stored functions/procedures.

Calling stored functions/procedures via plain JDBC code is not easy either, especially if advanced data types are involved (for instance, arrays or UDTs). But, as you already know, using the jOOQ DSL saves us from interacting directly with the JDBC API, so it saves us from making cutting-edge decisions regarding JDBC workarounds.

The jOOQ DSL represents stored functions/procedures via the `org.jooq.Routine` API, so there is a common API for both. Whenever the jOOQ generator detects a stored function/procedure it generates (among other things) a dedicated class in the proper package (in our case, `jooq.generated.routines`) that reflects its name (for instance, by default, a stored function named `get_emps_in_office()` results in a class named `GetEmpsInOffice`) and extends the jOOQ `AbstractRoutine` class. The generated class exposes the API needed to call this stored function/procedure via jOOQ DSL. Moreover, as you'll see soon, calling a stored function/procedure can also be done in an anonymous procedural block via `DSLContext.begin()` and directly via `DSLContext.call()`. But, enough theory, next let's tackle some different kinds of stored functions/procedures from the jOOQ perspective, and let's start with stored functions.

Stored functions

Stored functions return a result (for instance, the result of a computation). They can be called in SQL statements and, usually, they don't support output (`OUT`) parameters. However, in Oracle and PostgreSQL, stored functions may have output parameters that can be interpreted as returned results. Moreover, until version 11, PostgreSQL supports only stored functions that combine the features of stored functions and procedures. On the other hand, PostgreSQL 11 and beyond, Oracle, MySQL, and SQL Server distinguish between stored functions and procedures.

Next, let's see how we can call from jOOQ different kinds of stored functions expressed in one of these four dialects, and let's start by calling some scalar functions.

Scalar functions

A stored function that takes none, one, or more parameters, and returns a single value is commonly referred to as a scalar function. As a common practice, scalar functions encapsulate complex calculations that appear in many queries. Instead of expressing the calculation in every query, you can write a scalar function that encapsulates this calculation and uses it in each query. Roughly, the syntax of a scalar function is a variation of this skeleton:

```
CREATE FUNCTION name (parameters)
RETURNS data_type AS
BEGIN
    statements/computations
    RETURN value
END
```

For instance, a simple scalar function expressed in MySQL may look as follows:

```
DELIMITER $$
CREATE FUNCTION `sale_price`(
  `quantity` INT, `list_price` REAL, `fraction_of_price` REAL)
```

```
  RETURNS REAL
  DETERMINISTIC
  BEGIN
    RETURN (`list_price` -
      (`list_price` * `fraction_of_price`)) * `quantity`;
  END $$
DELIMITER ;
```

For this scalar function, the jOOQ Code Generator produces a dedicated class named `jooq.generated.routines.SalePrice`. Among its methods, this class exposes setters that allow us to provide the input parameters of the function. In our example, we will have `setQuantity(Integer value)`, `setQuantity(Field<Integer> field)`, `setListPrice(Double value)`, `setListPrice(Field<Double> field)`, `setFractionOfPrice(Double value)`, and `setFractionOfPrice(Field<Double> field)`. The function can be executed via jOOQ's well-known `execute()` methods. If the function already has a `Configuration` attached, then you can rely on `execute()` without parameters, otherwise use the `execute(Configuration c)` method as follows:

```
SalePrice sp = new SalePrice();
sp.setQuantity(25);
sp.setListPrice(15.5);
sp.setFractionOfPrice(0.75);

sp.execute(ctx.configuration());
```

Getting the returned scalar result can be done via the `getReturnValue()` method. In this case, you can use it like this:

```
double result = sp.getReturnValue();
```

As you just saw, each setter has a flavor that gets a `Field` as an argument. This means that we can write something like this (check the first two setters):

```
sp.setQuantity(field(select(PRODUCT.QUANTITY_IN_STOCK)
  .from(PRODUCT).where(PRODUCT.PRODUCT_ID.eq(1L))));
sp.setListPrice(field(select(PRODUCT.MSRP.coerce(Double.class))
  .from(PRODUCT).where(PRODUCT.PRODUCT_ID.eq(1L))));
sp.setFractionOfPrice(0.75);

sp.execute(ctx.configuration());

double result = sp.getReturnValue();
```

The previous examples are useful if the routine has more than 254 parameters (which isn't allowed in Java), if there are default parameters, which users don't want to set, or if the parameters need to be set dynamically. Otherwise, most probably, you'll prefer to use the static convenience API.

Writing these two examples in a more convenient/compact way can be done via the `Routines.salePrice()` static method. The `jooq.generated.Routines` class provides convenient static methods for accessing all stored functions/procedures that jOOQ has found in your database. In this case, the following two examples compact the previous examples (of course, you can shorten this example further by importing the `jooq.generated.Routines.salePrice` static):

```
double sp = Routines.salePrice(
    ctx.configuration(), 25, 15.5, 0.75);

Field<Float> sp = Routines.salePrice(
  field(select(PRODUCT.QUANTITY_IN_STOCK)
         .from(PRODUCT).where(PRODUCT.PRODUCT_ID.eq(1L))),
  field(select(PRODUCT.MSRP.coerce(Double.class))
         .from(PRODUCT).where(PRODUCT.PRODUCT_ID.eq(1L))),
  val(0.75));

double sp = ctx.fetchValue(salePrice(
 field(select(PRODUCT.QUANTITY_IN_STOCK)
  .from(PRODUCT).where(PRODUCT.PRODUCT_ID.eq(1L))),
 field(select(PRODUCT.MSRP.coerce(Double.class))
  .from(PRODUCT).where(PRODUCT.PRODUCT_ID.eq(1L))),
 val(0.75)));
```

Scalar functions can be used in queries as well. Here is an example via another flavor of `Routines.salePrice()`:

```
ctx.select(ORDERDETAIL.ORDER_ID,
    sum(salePrice(ORDERDETAIL.QUANTITY_ORDERED,
        ORDERDETAIL.PRICE_EACH.coerce(Double.class),
        val(0.75))).as("sum_sale_price"))
   .from(ORDERDETAIL)
   .groupBy(ORDERDETAIL.ORDER_ID)
   .orderBy(field(name("sum_sale_price")).desc())
   .fetch();
```

For MySQL, jOOQ renders the following plain SQL:

```
SELECT `classicmodels`.`orderdetail`.`order_id`,
   sum(`classicmodels`.`sale_price`(
       `classicmodels`.`orderdetail`.`quantity_ordered`,
       `classicmodels`.`orderdetail`.`price_each`, 7.5E-1))
         AS `sum_sale_price`
   FROM `classicmodels`.`orderdetail`
   GROUP BY `classicmodels`.`orderdetail`.`order_id`
   ORDER BY `sum_sale_price` DESC
```

You can practice this example in *ScalarFunction*.

In this context, here is another function that was written for PostgreSQL that updates a `PRODUCT.MSRP` and returns it via `UPDATE ... RETURNING` (do not confuse it with the function `RETURN` statement!):

```
CREATE OR REPLACE FUNCTION "update_msrp" (
    "id" BIGINT, "debit" INTEGER) RETURNS REAL AS $$
    UPDATE "public"."product"
        SET "msrp" = "public"."product"."msrp" - "debit"
        WHERE "public"."product"."product_id" = "id"
        RETURNING "public"."product"."msrp";
$$ LANGUAGE SQL;
```

jOOQ can call such a function as you saw in the previous example. For instance, here it is called in a `SELECT`:

```
ctx.select(PRODUCT.PRODUCT_ID, PRODUCT.PRODUCT_NAME,
           PRODUCT.MSRP.as("obsolete_msrp"),
           updateMsrp(PRODUCT.PRODUCT_ID, inline(50)))
   .from(PRODUCT).fetch();
```

And the rendered SQL is as follows:

```
SELECT "public"."product"."product_id",
       "public"."product"."product_name",
       "public"."product"."msrp" AS "obsolete_msrp",
               "public"."update_msrp"("id" := "public"
                   ."product"."product_id", "debit" := 50)
FROM "public"."product"
```

You can practice this example in *UpdateFunction* for PostgreSQL.

If your scalar functions are under Oracle, then you can take advantage of a nice Oracle feature known as *scalar subquery caching*. Basically, this feature renders the calls of stored functions wrapped in SQL statements as a scalar subquery. This feature avoids switching between PL/SQL and SQL contexts and this may result in better performance.

Consider the following scalar function:

```
CREATE OR REPLACE NONEDITIONABLE FUNCTION
     "card_commission"("card_type" IN VARCHAR2)
RETURN NUMBER IS
 "commission" NUMBER := 0;
 BEGIN
   RETURN CASE "card_type"
     WHEN 'VisaElectron' THEN .15
     WHEN 'Mastercard' THEN .22
     ELSE .25
   END;
END;
```

Calling this scalar function in a query can be done here:

```
ctx.select(cardCommission(BANK_TRANSACTION.CARD_TYPE))
   .from(BANK_TRANSACTION)
   .fetch();

List<BigDecimal> commission = ctx.fetchValues(
  select(cardCommission(BANK_TRANSACTION.CARD_TYPE))
    .from(BANK_TRANSACTION));
```

But keep in mind Lukas Eder's note: *"Just in case, scalar subquery caching isn't documented in Oracle, as far as I know. The context switch isn't avoided entirely, but it happens only once per scalar subquery input value, and thus per function argument value, and per query. So, instead of having 1 switch per row, we now have 1 switch per function input value."*

Each time we execute this code, jOOQ renders a query that requires switching between PL/SQL and SQL contexts in order to execute the scalar function:

```
SELECT "CLASSICMODELS"."card_commission"(
       "CLASSICMODELS"."BANK_TRANSACTION"."CARD_TYPE")
FROM "CLASSICMODELS"."BANK_TRANSACTION"
```

But jOOQ has the `withRenderScalarSubqueriesForStoredFunctions()` flag-setting that is by default set to `false`. Once we set it to `true`, jOOQ turns on Oracle's *scalar subquery caching* feature. In the following example, we turn this feature on only for the current SQL:

```
ctx.configuration().derive(new Settings()
   .withRenderScalarSubqueriesForStoredFunctions(true))
   .dsl()
   .select(cardCommission(BANK_TRANSACTION.CARD_TYPE))
   .from(BANK_TRANSACTION)
   .fetch();
```

This time, the call of the `cardCommission()` stored function is rendered as a scalar subquery:

```
SELECT
  (SELECT "CLASSICMODELS"."card_commission"(
    "CLASSICMODELS"."BANK_TRANSACTION"."CARD_TYPE")
      FROM DUAL)
FROM "CLASSICMODELS"."BANK_TRANSACTION"
```

You can practice this example in *ScalarSubqueryCaching* for Oracle.

Functions returning arrays

PostgreSQL is a convenient way to write a function that returns an array. For instance, here is a function returning `DEPARTMENT.TOPIC`, which is declared in our schema as an array of type `TEXT[]`:

```
CREATE OR REPLACE FUNCTION "department_topic_arr"
   (IN "id" BIGINT)
RETURNS TEXT[]
AS $$
  SELECT "public"."department"."topic"
      FROM "public"."department" WHERE
           "public"."department"."department_id" = "id"
$$ LANGUAGE SQL;
```

Calling this function in SELECT via the dedicated method, departmentTopicArr(), generated by jOOQ in Routines can be done by unnesting the returned array as in the following examples:

```
ctx.select().from(unnest(departmentTopicArr(2L))
   .as("t")).fetch();

ctx.fetch(unnest(departmentTopicArr(2L)).as("t"));
```

Next, let's take a function having an anonymous parameter (no explicit name) that builds and returns an array:

```
CREATE OR REPLACE FUNCTION "employee_office_arr"(VARCHAR(10))
  RETURNS BIGINT[]
AS $$
  SELECT ARRAY(SELECT "public"."employee"."employee_number"
    FROM "public"."employee" WHERE "public"."employee"
         ."office_code" = $1)
$$ LANGUAGE sql;
```

This time, let's instantiate EmployeeOfficeArr, and let's pass the required parameter via a setter:

```
EmployeeOfficeArr eoa = new EmployeeOfficeArr();
eoa.set__1("1");

eoa.execute(ctx.configuration());

Long[] result = eoa.getReturnValue();
```

Since the function's parameter doesn't have a name, jOOQ has used its default implementation and generated a set__1() setter. If you had two no-name parameters, then jOOQ would generate set__1() and set__2(), and so on. In other words, jOOQ generates setters based on parameter positions starting from 1.

On the other hand, using the generated Routines.employeeOfficeArr() in a SELECT query can be done as follows:

```
ctx.select(field(name("t", "en")), sum(SALE.SALE_))
   .from(SALE)
   .rightJoin(unnest(employeeOfficeArr("1")).as("t", "en"))
```

```
    .on(field(name("t", "en")).eq(SALE.EMPLOYEE_NUMBER))
  .groupBy(field(name("t", "en"))).fetch();
```

For PostgreSQL, jOOQ renders this SQL:

```
SELECT "t"."en",sum("public"."sale"."sale")
FROM "public"."sale"
RIGHT OUTER JOIN unnest("public"."employee_office_arr"('1'))
AS "t" ("en")
ON "t"."en" = "public"."sale"."employee_number"
GROUP BY "t"."en"
```

You can practice these examples in *ArrayFunction* for PostgreSQL.

Functions with output parameters

As we said earlier, PostgreSQL and Oracle allow output parameters in functions. Here is an example in PostgreSQL:

```
CREATE OR REPLACE FUNCTION "get_salary_stat"(
    OUT "min_sal" INT, OUT "max_sal" INT, OUT "avg_sal" NUMERIC)
  LANGUAGE plpgsql
  AS $$
  BEGIN
    SELECT MIN("public"."employee"."salary"),
           MAX("public"."employee"."salary"),
           AVG("public"."employee"."salary")::NUMERIC(7,2)
      INTO "min_sal", "max_sal", "avg_sal"
      FROM "public"."employee";
  END;
$$;
```

This function doesn't have a `RETURN`, but it has three `OUT` parameters that help us to obtain the results of execution. For each such parameter, jOOQ generates a getter, so we can call it via the generated `jooq.generated.routines.GetSalaryStat` like this:

```
GetSalaryStat salStat = new GetSalaryStat();
salStat.execute(ctx.configuration());

Integer minSal = salStat.getMinSal();
Integer maxSal = salStat.getMaxSal();
BigDecimal avgSal = salStat.getAvgSal();
```

This code (more precisely the `execute()` call) leads to the following `SELECT` (or `CALL`, in Oracle):

```
SELECT "min_sal", "max_sal", "avg_sal"
FROM "public"."get_salary_stat"()
```

Here's the same result via `Routines.getSalaryStat(Configuration c)`:

```
GetSalaryStat salStat = getSalaryStat(ctx.configuration());
// call the getters
```

Here are two more examples:

```
Integer minSal = ctx.fetchValue(val(getSalaryStat(
      ctx.configuration()).getMinSal()));

ctx.select(EMPLOYEE.FIRST_NAME, EMPLOYEE.LAST_NAME,
           EMPLOYEE.SALARY)
   .from(EMPLOYEE)
   .where(EMPLOYEE.SALARY.coerce(BigDecimal.class)
      .gt(getSalaryStat(ctx.configuration()).getAvgSal()))
   .fetch();
```

But pay attention that both of these examples lead to two `SELECT` statements (or, a `CALL` and a `SELECT` statement in Oracle) since functions with output parameters cannot be called from plain SQL. In other words, jOOQ calls the routine and fetches the `OUT` parameters as you can see next:

```
SELECT "min_sal","max_sal","avg_sal"
FROM "public"."get_salary_stat"()
```

And, afterward, it executes the `SELECT` that uses the results extracted from the output parameters. The following `SELECT` fits our second example (65652.17 is the average salary computed via `getSalaryStat(Configuration c)`):

```
SELECT "public"."employee"."first_name",
       "public"."employee"."last_name",
       "public"."employee"."salary"
FROM "public"."employee"
WHERE "public"."employee"."salary" > 65652.17
```

Lukas Eder shared: *"In Oracle, functions with OUT parameters aren't "SQL callable," though... In PostgreSQL, they're just "syntax sugar" (or un-sugar, depending on taste) for a function returning a record."*

You can practice these examples in *InOutFunction* for PostgreSQL (here, you can find an `IN OUT` example as well).

So, using `OUT` (or `IN OUT`) parameters in functions is not such a great idea and must be avoided. As Oracle mentioned, besides preventing a function from being used in SQL queries (more details here: `https://oracle-base.com/articles/12c/with-clause-enhancements-12cr1`), the presence of output parameters in a function prevents a function from being marked as a `DETERMINISTIC` function or used as a result-cached function. Unlike PostgreSQL, Oracle functions having output parameters must have an explicit `RETURN`. Next to the jOOQ getters dedicated to output parameters, you can call `getReturnValue()` to obtain the result returned explicitly via the `RETURN` statement. When this book was written, functions with `OUT` parameters couldn't be called in jOOQ from plain SQL. Follow this feature request here: `https://github.com/jOOQ/jOOQ/issues/3426`.

You can practice an example in *InOutFunction* for Oracle (here, you can find an `IN OUT` example too).

Polymorphic functions

Some databases support the so-called polymorphic functions that accept and return polymorphic types. For instance, PostgreSQL supports the following polymorphic types: `anyelement`, `anyarray`, `anynonarray`, `anyenum`, and `anyrange`. Here is an example that builds an array from two passed arbitrary data type elements:

```
CREATE FUNCTION "make_array"(anyelement, anyelement)
RETURNS anyarray
AS $$
  SELECT ARRAY[$1, $2];
$$ LANGUAGE SQL;
```

Calling this function from jOOQ can be done as follows (notice the positional setters at work again):

```
MakeArray ma = new MakeArray();
ma.set__1(1);
ma.set__2(2);

ma.execute(ctx.configuration());
```

The returned result is `org.postgresql.jdbc.PgArray`:

```
PgArray arr = (PgArray) ma.getReturnValue();
```

In the bundled code you can see further processing of this array. For now, let's call `make_array()` from `SELECT` to build an array of integers and an array of strings:

```
ctx.select(makeArray(1, 2).as("ia"),
           makeArray("a", "b").as("ta")).fetch();
```

How about a function that combines polymorphic types and output parameters? Here is one:

```
CREATE FUNCTION "dup"(IN "f1" anyelement,
   OUT "f2" anyelement, OUT "f3" anyarray)
AS 'select $1, array[$1,$1]' LANGUAGE sql;
```

Call it via jOOQ:

```
Dup dup = new Dup();
dup.setF1(10);

dup.execute(ctx.configuration());
// call here getF2() and/or getF3()
```

Or use it in a `SELECT` (remember from the previous section that this leads to two statements against the database):

```
ctx.select(val(dup(ctx.configuration(), 10).getF2())).fetch();
ctx.fetchValue(val(dup(ctx.configuration(), 10).getF2()));
```

You can practice these examples in *PolymorphicFunction* for PostgreSQL.

Functions returning explicit cursors

A function can return an explicit cursor (an explicit pointer that points to a result of a query) as well. Most probably, you're familiar with PostgreSQL `REFCURSOR` and Oracle `SYS_REFCURSOR`. Here is an example for Oracle:

```
CREATE OR REPLACE NONEDITIONABLE FUNCTION
   "GET_CUSTOMER" ("cl" IN NUMBER)
  RETURN SYS_REFCURSOR
     AS  "cur"  SYS_REFCURSOR;
```

```
BEGIN
    OPEN "cur" FOR
         SELECT *
         FROM "CUSTOMER"
                     WHERE "CUSTOMER"."CREDIT_LIMIT" > "cl"
         ORDER BY "CUSTOMER"."CUSTOMER_NAME";

    RETURN "cur";
END;
```

Calling this function via the generated `jooq.generated.routines.GetCustomer` can be done as follows:

```
GetCustomer customers = new GetCustomer();
customers.setCl(120000);

customers.execute(ctx.configuration());

Result<Record> result = customers.getReturnValue();
```

The result is mapped by jOOQ to `Result<Record>`, so a list of records is fitted entirely into memory. For better accommodation of large datasets, jOOQ has some pending feature requests to stream the cursors, so you can check the progress of issues #4503 and #4472 on the jOOQ repository. Alternatively, you can wrap the returned cursor in a table-valued function and fetch the results via a `SELECT` statement using jOOQ's `ResultQuery.fetchLazy()` method, as you saw in *Chapter 8*, *Fetching and Mapping*.

Now, going further, you can loop the `Result<Record>` and process each record, but to access a certain column (for instance, the customer name) of a certain row (for instance, the first row), then you can use `getValue()` or `get()` as here:

```
String name = (String) result.getValue(0, "CUSTOMER_NAME");
String name = result.get(0).get("CUSTOMER_NAME", String.class);
```

To get multiple names (or other columns) rely on `getValues()`, which comes in many flavors that you can find in the official documentation.

Obtaining the same `Result<Record>` can be done more compactly via the generated static `Routines.getCustomer()`:

```
Result<Record> result = getCustomer(
   ctx.configuration(), 120000);
```

If you need a `Table` instead of this `Result<Record>`, then simply rely on `org.jooq.impl.DSL.table` as in the following two examples:

```
Table<?> t = table(result);
Table<CustomerRecord> t = table(result.into(CUSTOMER));
```

Next, you can use `t` in queries like any regular `Table`:

```
ctx.select(CUSTOMERDETAIL.ADDRESS_LINE_FIRST,
           CUSTOMERDETAIL.POSTAL_CODE,
           t.field(name("CUSTOMER_NAME")))
   .from(t)
   .join(CUSTOMERDETAIL)
    .on(CUSTOMERDETAIL.CUSTOMER_NUMBER.eq(
        t.field(name("CUSTOMER_NUMBER"), Long.class)))
   .fetch();
```

On the other hand, another flavor of `Routines.getCustomer()` returns the result wrapped in a `Field` as `Field<Result<Record>>`. This allows us to use this result as a `Field`. For instance, here is a `SELECT`:

```
ctx.select(getCustomer(field(
      select(avg(CUSTOMER.CREDIT_LIMIT))
         .from(CUSTOMER)))).fetch();
```

You can practice these examples in *CursorFunction* for Oracle.

How about a function that returns multiple cursors? Here is a sample that was written for PostgreSQL:

```
CREATE OR REPLACE FUNCTION "get_offices_multiple"()
  RETURNS SETOF REFCURSOR
AS $$
 DECLARE
  "ref1" REFCURSOR;
  "ref2" REFCURSOR;
 BEGIN
 OPEN "ref1" FOR
  SELECT "public"."office"."city", "public"."office"."country"
  FROM "public"."office"
  WHERE "public"."office"."internal_budget" < 100000;
```

```
  RETURN NEXT "ref1";

 OPEN "ref2" FOR
  SELECT "public"."office"."city", "public"."office"."country"
  FROM "public"."office"
  WHERE "public"."office"."internal_budget" > 100000;
 RETURN NEXT "ref2";
END;
$$ LANGUAGE plpgsql;
```

In this case, each cursor produces a `Result<Record>` wrapped in a generated `Record` class. Here, we have two cursors, therefore two `Result<Record>` wrapped in two instances of the generated `GetOfficesMultipleRecord`. When we call `Routines.getOfficesMultiple(Configuration c)` we get a `Result<GetOfficesMultipleRecord>` that can be expanded as follows:

```
Result<GetOfficesMultipleRecord> results =
        getOfficesMultiple(ctx.configuration());

for (GetOfficesMultipleRecord result : results) {
  Result<Record> records = result.getGetOfficesMultiple();
  System.out.println("-------------------------");
  for (Record r : records) {
    System.out.println(r.get("city") + ", " + r.get("country"));
  }
}
```

You can practice these examples in *CursorFunction* for PostgreSQL.

Table-valued functions

Apart from database views, one of the underrated features of SQL is table-valued functions. This is not the first time in this book that we've discussed this feature, but this time, let's add a few more details. So, a table-valued function is a function that returns a set of data as a table data type. The returned table can be used just like a regular table.

Table-valued functions are not supported in MySQL, but they are supported in PostgreSQL, Oracle, and SQL Server. Next is a snippet of code from a PostgreSQL table-valued function (notice the `RETURNS TABLE` syntax, which indicates that the `SELECT` query from the function returns the data as a table to whatever calls the function):

```
CREATE OR REPLACE FUNCTION "product_of_product_line"(
  IN "p_line_in" VARCHAR)
```

```
RETURNS TABLE("p_id" BIGINT, "p_name" VARCHAR,
              "p_line" VARCHAR) LANGUAGE plpgsql
AS $$
BEGIN
  RETURN QUERY
  SELECT ...;
  END;
$$;
```

By default, the jOOQ Code Generator will generate for this function a class named `ProductOfProductLine` in the `jooq.generated.tables` package, not in the `jooq.generated.routines` package. The explanation is simple; jOOQ (like most databases) treats table-valued functions as ordinary tables that can be used in the `FROM` clause of `SELECT` like any other table. An exception from this practice is Oracle, where it is quite common to treat them as standalone routines – in this context, jOOQ has a flag setting that allows us to indicate whether table-valued functions should be treated as ordinary tables (generated in `jooq.generated.tables`) or as plain routines (generated in `jooq.generated.routines`). This is detailed in *Chapter 6, Tackling Different Kinds of JOIN Statements.*

Calling a table-valued function (with arguments) can be done via the `call()` method:

```
ProductOfProductLine popl = new ProductOfProductLine();
Table<ProductOfProductLineRecord> t = popl.call("Trains");

Result<ProductOfProductLineRecord> r =
   ctx.fetch(popl.call("Trains"));

Result<ProductOfProductLineRecord> r =
   ctx.selectFrom(popl.call("Trains")).fetch();
```

In queries, we may prefer to use the `PRODUCT_OF_PRODUCT_LINE` static field that was generated by the jOOQ generator in `ProductOfProductLine`. Both of the following examples produce the same SQL:

```
ctx.selectFrom(PRODUCT_OF_PRODUCT_LINE.call("Trains"))
   .fetch();

ctx.selectFrom(PRODUCT_OF_PRODUCT_LINE(val("Trains")))
   .fetch();
```

Here are two more examples of calling this table-valued function in the `FROM` clause:

```
ctx.selectFrom(PRODUCT_OF_PRODUCT_LINE.call("Trains"))
   .where(PRODUCT_OF_PRODUCT_LINE.P_NAME.like("1962%"))
   .fetch();

ctx.select(PRODUCT_OF_PRODUCT_LINE.P_ID,
           PRODUCT_OF_PRODUCT_LINE.P_NAME)
   .from(PRODUCT_OF_PRODUCT_LINE.call("Classic Cars"))
   .where(PRODUCT_OF_PRODUCT_LINE.P_ID.gt(100L))
   .fetch();
```

Since a table-valued function returns a table, we should be able to use them in joins. But the regular `JOIN` feature doesn't allow us to join a table-valued function, so we need another approach. Here is where `CROSS APPLY` and `OUTER APPLY` (or `LATERAL`) enter into the scene. In *Chapter 6, Tackling Different Kinds of JOIN Statements*, you saw an example of using `CROSS/OUTER APPLY` to solve the popular task of joining two tables based on the results of a TOP-N query. So, `CROSS/OUTER APPLY` allows us to combine in a query the results returned by a table-valued function with the results of other tables or, in short, to join table-valued functions to other tables.

For instance, let's use `CROSS/OUTER APPLY` (you can think of it as a `Stream.flatMap()` in Java) to join the `PRODUCTLINE` table to our table-valued function. Let's say that we have added a new `PRODUCTLINE` without products named *Helicopters* and let's see how `CROSS APPLY` works:

```
ctx.select(PRODUCTLINE.PRODUCT_LINE, PRODUCTLINE
   .TEXT_DESCRIPTION, PRODUCT_OF_PRODUCT_LINE.asterisk())
   .from(PRODUCTLINE)
   .crossApply(PRODUCT_OF_PRODUCT_LINE(
               PRODUCTLINE.PRODUCT_LINE))
   .fetch();
```

Since the *Helicopters* product line has no products, `CROSS APPLY` will not fetch it because `CROSS APPLY` acts as `CROSS JOIN LATERAL`. How about `OUTER APPLY`?

```
ctx.select(PRODUCTLINE.PRODUCT_LINE, PRODUCTLINE
     .TEXT_DESCRIPTION, PRODUCT_OF_PRODUCT_LINE.asterisk())
   .from(PRODUCTLINE)
   .outerApply(PRODUCT_OF_PRODUCT_LINE(
               PRODUCTLINE.PRODUCT_LINE))
   .fetch();
```

On the other hand, `OUTER APPLY` acts as `LEFT OUTER JOIN LATERAL`, so the *Helicopters* product line is returned as well.

Lukas Eder shared an opinion here: *"In fact, for historic reasons, APPLY or LATERAL is optional in at least Db2, Oracle, and PostgreSQL, under some conditions. SQL Server had APPLY for a long time, but the others introduced LATERAL only relatively recently. I personally don't understand the value of making LATERAL explicit. It's always clear what an implicit LATERAL means..."*

You can check out these examples in the bundled code, *TableValuedFunction* for PostgreSQL, Oracle, and SQL Server.

Oracle's package

Oracle allows us to group the functions/procedures that are commonly logically related into a package. A package has two parts: the first part contains the public items and is known as the *package specification*, while the second part, known as the *package body*, provides the code of the cursors or subprograms declared in the package specification. If no cursors/subprograms were declared in the package specification, then the package body can be skipped. If you are not an Oracle fan, then the following syntax should help you to digest this topic a little bit easier:

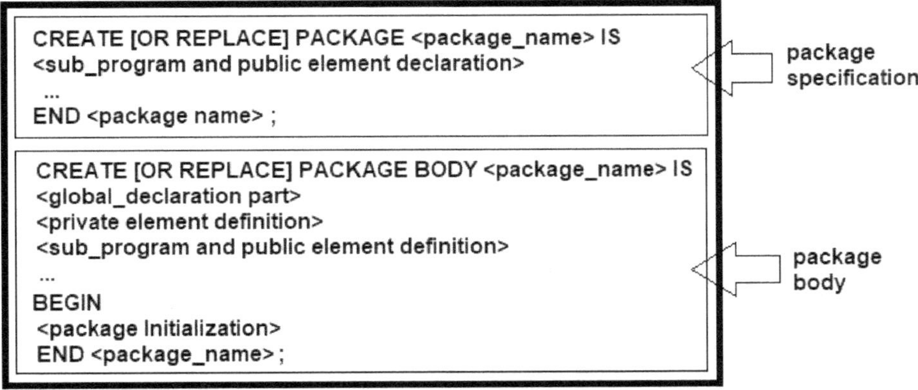

Figure 15.2 – Syntax of Oracle package

Lukas Eder shared an analogy that's useful to better understand this topic: *"If it helps, package specifications are like interfaces, and bodies are like singleton instances, kind of. In a way, like this:* `https://twitter.com/lukaseder/status/1443855980693962755`."

Packages sustain modularity, facilitate clear design, and increase code maintainability by hiding the implementation details in the package body. Moreover, packages are loaded into memory as a whole at the first invocation of a function/procedure so any subsequent invocations of functions/procedures from this package require no disk I/O. While more information on using Oracle packages is available in the official documentation, let's have an example:

```
CREATE OR REPLACE PACKAGE "DEPARTMENT_PKG" AS
 TYPE "BGT" IS RECORD ("LOCAL_BUDGET" FLOAT, "CASH" FLOAT);

  FUNCTION "GET_BGT"("p_profit" IN FLOAT)
    RETURN "BGT";

  FUNCTION "GET_MAX_CASH"
    RETURN FLOAT;
END "DEPARTMENT_PKG";
/
CREATE OR REPLACE PACKAGE BODY "DEPARTMENT_PKG"
-- check bundled code for this skipped part
END"DEPARTMENT_PKG";
/
```

A spicy tip from Lukas Eder: *"The '/' is a SQL*Plus 'spool' token (also supported by SQL Developer), and not an actual PL/SQL syntax element. For example, it doesn't work in DBeaver.*

And, another tip regarding quoted identifiers: *Might be a good reminder that, to be better interoperable with non-jOOQ code, perhaps not using quoted identifiers is better. It will be a PITA if all callers will have to always quote that identifier if they're not using jOOQ :)"*

So, here we have a package named `DEPARTMENT_PKG` containing a user-defined type named `BGT` and two functions, `GET_BGT()` and `GET_MAX_CASH()`. At the source code generation stage, jOOQ will reflect this package and its content via Java sub-packages as follows:

- `jooq.generated.packages` – Contains the `DepartmentPkg` class representing the package and exposing the `DEPARTMENT_PKG` static that can be used to call the functions as `DEPARTMENT_PKG.getMaxCash()` and `DEPARTMENT_PKG.getBgt()` and get the result as a `Field`.
- `jooq.generated.packages.department_pkg` – Contains the `GetBgt` and `GetMaxCash` classes representing the two functions from the package. Moreover, it contains the `UDTs` class containing the static `BGT` for `CLASSICMODELS.DEPARTMENT_PKG.BGT` (`jooq.generated.packages.department_pkg.udt.Bgt.BGT`).

- `jooq.generated.packages.department_pkg.udt` – Contains the class `Bgt` mapping the `BGT` UDT type as an extension of `UDTImpl`.
- `jooq.generated.packages.department_pkg.udt.records` – Contains the `BgtRecord` class representing the `BGT` UDT type as an extension of `UDTRecordImpl`.

Calling these two functions (`GET_BGT()` and `GET_MAX_CASH()`) can be done by instantiating the `GetMaxCash` class, respectively the `GetBgt` class and call `execute()` as follows:

```
GetMaxCash gmc = new GetMaxCash();
gmc.execute(ctx.configuration());
double resultGmc = gmc.getReturnValue();

GetBgt bgt = new GetBgt();
bgt.setPProfit(50000.0);
bgt.execute(ctx.configuration());
BgtRecord resultBgt = bgt.getReturnValue();
```

We can also compact these examples via the statics `DepartmentPkg.getBgt()` and `DepartmentPkg.getMaxCash()` as follows:

```
double resultGmc = getMaxCash(ctx.configuration());
BgtRecord resultBgt = getBgt(ctx.configuration(), 50000.0);
```

Calling these functions from queries is also possible. For instance, here are two trivial examples of calling them in `SELECT`:

```
ctx.select(getMaxCash()).fetch();
double mc = ctx.fetchValue(getMaxCash());

The rendered SQL:
SELECT "CLASSICMODELS"."department_pkg"."get_max_cash"()
FROM DUAL
```

Here is another example that uses both functions in the same query:

```
ctx.select(OFFICE.OFFICE_CODE, OFFICE.CITY, OFFICE.COUNTRY,
           DEPARTMENT.NAME, DEPARTMENT.LOCAL_BUDGET)
   .from(OFFICE)
   .join(DEPARTMENT)
   .on(OFFICE.OFFICE_CODE.eq(DEPARTMENT.OFFICE_CODE)
```

```
    .and(DEPARTMENT.LOCAL_BUDGET
      .in(getBgt(ctx.configuration(),
         getMaxCash(ctx.configuration()))
           .getLocalBudget())))
  .fetch();
```

Check out these examples next to a few others not presented here in *Package* for Oracle.

Oracle's member function/procedure

Mainly, an Oracle PL/SQL object type contains attributes and members (or methods). Attributes or fields have data types and they are used to store data, while members are subprograms (functions/procedures) that are defined in the object type and manipulate the attributes for implementing certain functionalities. In this way, Oracle UDTs are a fully-fledged attempt at implementing an Object Relational Database Management System (ORDBMS). PostgreSQL didn't go quite as far as Oracle.

If you are not an Oracle fan, then the following syntax should shed some light on this topic:

```
CREATE [OR REPLACE] TYPE <object_type_name> AS OBJECT
(
 <attribute_l> <datatype>,
 <attribute_2> <datatype>,
 ...
 MEMBER [PROCEDURE|FUNCTION] <member_name_1>,
 MEMBER [PROCEDURE|FUNCTION] <member_name_2>,
 ...
);

CREATE [OR REPLACE] TYPE BODY<object_type_name> AS

 MEMBER [PROCEDURE|FUNCTION] <member_name_1> IS
  <declarative section>
 BEGIN
  <execution part>
 END <member_name_1>;
 ...
END;
```

Figure 15.3 – Syntax of Oracle members

Based on this syntax, let's have an example. Our `MANAGER` table has a field named `MANAGER_EVALUATION` that is of type `EVALUATION_CRITERIA` (in jOOQ, of type `TableField<ManagerRecord, EvaluationCriteriaRecord>`) defined as follows:

```
CREATE OR REPLACE TYPE "EVALUATION_CRITERIA" AS OBJECT (
  "communication_ability" NUMBER(7),
```

```
  "ethics" NUMBER(7),
  "performance" NUMBER(7),
  "employee_input" NUMBER(7),

  MEMBER FUNCTION "IMPROVE"("k" NUMBER)
      RETURN "EVALUATION_CRITERIA",
  MAP MEMBER FUNCTION "SCORE" RETURN NUMBER
);

CREATE OR REPLACE TYPE BODY "EVALUATION_CRITERIA" AS
-- check bundled code for this skipped part
END;
```

Here, we have an object type containing four attributes (`communication_ability`, `ethics`, `performance`, and `employee_input`), two member functions (`IMPROVE()` and `SCORE()`), and no member procedures.

The jOOQ Code Generator produces the following artifacts:

- `jooq.generated.udt` – In this package, we have the UDT type named `EvaluationCriteria`, representing an extension of jOOQ's `UDTImpl<EvaluationCriteriaRecord>`. It contains the statics `EVALUATION_CRITERIA`, `COMMUNICATION_ABILITY`, `ETHICS`, `PERFORMANCE`, and `EMPLOYEE_INPUT` to refer to these UDT attributes and several flavors of the `improve()` and `score()` member functions returning the plain result or wrapped in a `Field`.
- `jooq.generated.udt.records` – Contains the `EvaluationCriteriaRecord` representing the UDT record as an extension of the jOOQ's `UDTRecordImpl<EvaluationCriteriaRecord>`. The `EvaluationCriteriaRecord` contains getters/setters for the object type attributes and contains the methods `improve()` and `score()` as well. These methods encapsulate the code needed to call the actual `improve()` and `score()` member functions.
- `jooq.generated.udt.evaluation_criteria` – Contains the `Improve` and `Score` classes (routines), so a class for each member function. Both `EvaluationCriteria` and `EvaluationCriteriaRecord` use these classes internally.

So, we can distinguish between calling the member functions starting from an empty record or from an existing record (for instance, a record fetched from the database). The conventional approach for starting from an empty record relies on `DSLContext.newRecord()`:

```
EvaluationCriteriaRecord ecr =
   ctx.newRecord(EVALUATION_CRITERIA);
```

The created record is already attached. Alternatively, we can instantiate `EvaluationCriteriaRecord` or use `EVALUATION_CRITERIA.newRecord()`, but the resulting record is not attached to a configuration (database), so you'll have to explicitly attach it by calling `attach()`.

Next, we set the values of the attributes and call the member functions. Here, we call the `score()` method, which returns a `BigDecimal`:

```
ecr.setCommunicationAbility(58);
ecr.setEthics(30);
ecr.setPerformance(26);
ecr.setEmployeeInput(59);

BigDecimal result = ecr.score();
```

On the other hand, the `improve()` methods increase the evaluation attributes by the given value and return a new `EvaluationCriteriaRecord`:

```
EvaluationCriteriaRecord newEcr = ecr.improve(10);
```

The `newEcr` record has `communication_ability` at *68* instead of *58*, `ethics` at *40* instead of *30*, and `performance` at *36* instead of *26*. Only `employee_input` remains unchanged.

We can use `ecr/newEcr` in queries as well. Here is an example of using `newEcr` that originates from an empty record next to `MANAGER.MANAGER_EVALUATION`, which is fetched from the database (remember that `MANAGER.MANAGER_EVALUATION` is a `TableField<ManagerRecord, EvaluationCriteriaRecord>`):

```
ctx.select(MANAGER.MANAGER_ID, MANAGER.MANAGER_NAME)
   .from(MANAGER)
   .where(score(MANAGER.MANAGER_EVALUATION)
     .lt(newEcr.score()))
   .fetch();
```

Here's another example that combines the calls of `improve()` and `score()`:

```
ctx.select(MANAGER.MANAGER_ID, MANAGER.MANAGER_NAME)
   .from(MANAGER)
   .where(score(improve(
      MANAGER.MANAGER_EVALUATION, inline(10)))
        .gt(BigDecimal.valueOf(57)))
   .fetch();
```

Check out these examples next to others omitted here in *MemberFunction* for Oracle.

User-defined aggregate stored functions

Oracle and SQL Server allow us to define aggregate stored functions (if you are not familiar with this topic, then please search these on Google: *Oracle User-Defined Aggregate Functions Interface* for Oracle and *SQL Server User-Defined Aggregate Functions* for SQL Server).

Moreover, the code of such functions is quite big to be listed here so please check the bundled code. For Oracle, check the application *UserDefinedAggFunction*, which calls a user-defined aggregate function named `secondMax()` that finds the second maximum value:

```
ctx.select(secondMax(ORDERDETAIL.QUANTITY_ORDERED),
                     ORDERDETAIL.PRODUCT_ID)
   .from(ORDERDETAIL)
   .groupBy(ORDERDETAIL.PRODUCT_ID)
   .having(secondMax(ORDERDETAIL.QUANTITY_ORDERED)
     .gt(BigDecimal.valueOf(55)))
   .fetch();
```

And for SQL Server, check the application also named *UserDefinedAggFunction*. Here, we call an aggregate stored function named `concatenate()` that simply concatenates the given strings:

```
ctx.select(concatenate(EMPLOYEE.FIRST_NAME))
   .from(EMPLOYEE)
   .where(EMPLOYEE.FIRST_NAME.like("M%"))
   .fetch();
```

In order to work, please pay attention that you need to place the `StringUtilities.dll` DLL file (available in the bundled code) in the path specified in the function code.

Stored procedures

jOOQ allows you to call stored procedures via the same `Routines` API. Next, let's see several examples of calling different kinds of stored procedures.

Stored procedures and output parameters

For instance, let's consider the following stored procedure expressed in Oracle and having an `OUT` parameter:

```
CREATE OR REPLACE NONEDITIONABLE PROCEDURE
   "GET_AVG_PRICE_BY_PRODUCT_LINE" (
    "pl" IN VARCHAR2, "average" OUT DECIMAL) AS
BEGIN
   SELECT AVG("PRODUCT"."BUY_PRICE")
      INTO "average"
      FROM "PRODUCT"
      WHERE "PRODUCT"."PRODUCT_LINE" = "pl";
END;
```

jOOQ generates the Java version of this stored procedure as a class named `GetAvgPriceByProductLine` in the `jooq.generated.routines` package. The methods of this class allow us to prepare the parameters (each input parameter has associated a setter, while each output parameter has associated a getter) and call our stored procedure as follows:

```
GetAvgPriceByProductLine avg = new GetAvgPriceByProductLine();
avg.setPl("Classic Cars");

avg.execute(ctx.configuration());

BigInteger result = avg.getAverage();
```

We can express this more compactly via the generated `jooq.generated.Routines.getAvgPriceByProductLine()` static as follows:

```
BigInteger result = getAvgPriceByProductLine(
   ctx.configuration(), "Classic Cars");
```

Calling a stored procedure in a jOOQ query can be done via the `getAvgPriceByProductLine(Configuration configuration, String pl)` flavor as in the following example:

```
ctx.select(PRODUCT.PRODUCT_ID, PRODUCT.PRODUCT_NAME,
           PRODUCT.BUY_PRICE)
```

```
    .from(PRODUCT)
    .where(PRODUCT.BUY_PRICE.coerce(BigInteger.class)
      .gt(getAvgPriceByProductLine(
          ctx.configuration(), "Classic Cars"))
      .and(PRODUCT.PRODUCT_LINE.eq("Classic Cars")))
    .fetch();
```

jOOQ first renders the call of the stored procedure and fetches the `OUT` parameter value:

```
call "CLASSICMODELS"."get_avg_price_by_product_line" (
  'Classic Cars', ?)
```

Next, the fetched value (*64*, in this example) is used to render the `SELECT`:

```
SELECT "CLASSICMODELS"."PRODUCT"."PRODUCT_ID",
       "CLASSICMODELS"."PRODUCT"."PRODUCT_NAME",
       "CLASSICMODELS"."PRODUCT"."BUY_PRICE"
FROM "CLASSICMODELS"."PRODUCT"
WHERE ("CLASSICMODELS"."PRODUCT"."BUY_PRICE" > 64
  AND "CLASSICMODELS"."PRODUCT"."PRODUCT_LINE"
    = 'Classic Cars')
```

Next, let's call a stored procedure without output parameters that fetches a single result set.

Stored procedures fetching a single result set

Here is a MySQL example of a stored procedure that doesn't contain output parameters and fetches a single result set via `SELECT`:

```
DELIMITER $$
CREATE PROCEDURE `get_product`(IN `pid` BIGINT)
  BEGIN
   SELECT * FROM `product` WHERE `product`.`product_id` = `pid`;
  END $$
DELIMITER
```

Calling this stored procedure from jOOQ can be done via the generated `GetProduct` class:

```
GetProduct gp = new GetProduct();
gp.setPid(1L);
```

```
gp.execute(ctx.configuration());

Result<Record> result = gp.getResults().get(0);
```

The result is obtained via gp.getResults(). Since there is a single result (the result set produced by the SELECT), we have to call get(0). If there were two results involved (for instance, if we had two SELECT statements in the stored procedure), then we would call get(0) to get the first result and get(1) to get the second result. Or, in the case of even more results, simply loop the results. Notice that, in the case of stored procedures, the getReturnValue() method returns void since stored procedures don't return results as a stored function (they don't contain a RETURN statement). In fact, SQL Server's procedures can return an error code, which is an int. You can see that in SQL Server generated code for procedures.

Calling the previous stored procedure via Routines.getProduct() returns void as well:

```
getProduct(ctx.configuration(), 1L);
```

The results obtained via getResults() are of type Result<Record>. This can be easily transformed into a regular Table as follows:

```
Table<?> t = table(gp.getResults().get(0));
Table<ProductRecord> t = table(gp.getResults()
  .get(0).into(PRODUCT));
```

Next, let's take the get_product() stored procedure and let's express it in Oracle by adding an OUT parameter of type SYS_REFCURSOR.

Stored procedures with a single cursor

The following stored procedure fetches the same result set as the previous get_product(), but it returns it via a cursor:

```
CREATE OR REPLACE NONEDITIONABLE PROCEDURE "GET_PRODUCT"(
    'pid" IN NUMBER, "cursor_result" OUT SYS_REFCURSOR) AS
BEGIN
  OPEN "cursor_result" FOR
    SELECT * FROM "PRODUCT"
    WHERE "PRODUCT"."PRODUCT_ID" = "pid";
END;
```

This time, in `GetProduct`, jOOQ generates a getter for the `OUT` parameter named `getCursorResult()`, which allows us to fetch the result as a `Result<Record>`:

```
GetProduct gp = new GetProduct();
gp.setPid(1L);

gp.execute(ctx.configuration());

Result<Record> result = gp.getCursorResult();
```

Or you can fetch it more compactly via `Routines.getProduct(Configuration configuration, Number pid)`:

```
Result<Record> result = getProduct(ctx.configuration(), 1L);
```

As usual, this `Result<Record>` can be easily transformed into a regular `Table`:

```
Table<?> t = table(gp.getResults().get(0));
Table<?> t = table(getProduct(ctx.configuration(), 1L));

Table<ProductRecord> t =
    table(gp.getCursorResult().into(PRODUCT));
Table<ProductRecord> t =
    table(getProduct(ctx.configuration(), 1L)
        .into(PRODUCT));
```

Next, you can use this `Table` in queries.

Stored procedures fetching multiple result sets

Going further, let's tackle a stored procedure returning multiple result sets. Here, it's expressed in MySQL:

```
DELIMITER $$
CREATE PROCEDURE `get_emps_in_office`(
                           IN `in_office_code` VARCHAR(10))
  BEGIN
    SELECT `office`.`city`, `office`.`country`,
               `office`.`internal_budget`
      FROM `office`
      WHERE `office`.`office_code`=`in_office_code`;
```

```
    SELECT `employee`.`employee_number`,
           `employee`.`first_name`, `employee`.`last_name`
     FROM `employee`
     WHERE `employee`.`office_code`=`in_office_code`;
  END $$
DELIMITER ;
```

As you already know, jOOQ generates the `GetEmpsInOffice` class and the result sets are available via `getResults()`:

```
GetEmpsInOffice geio = new GetEmpsInOffice();
geio.setInOfficeCode("1");

geio.execute(ctx.configuration());

Results results = geio.getResults();

for (Result<Record> result : results) {
  System.out.println("Result set:\n");
  for (Record record : result) {
    System.out.println(record);
  }
}
```

`Routines.getEmpsInOffice(Configuration c, String inOfficeCode)` returns `void`.

Stored procedures with multiple cursors

Next, let's take the `get_emps_in_office()` stored procedure, and let's express it in Oracle by adding two `OUT` parameters of type `SYS_REFCURSOR`:

```
CREATE OR REPLACE NONEDITIONABLE PROCEDURE
  "GET_EMPS_IN_OFFICE"("in_office_code" IN VARCHAR,
    "cursor_office" OUT SYS_REFCURSOR,
    "cursor_employee" OUT SYS_REFCURSOR) AS
BEGIN
  OPEN "cursor_office" FOR
    SELECT "OFFICE"."CITY", "OFFICE"."COUNTRY",
        "OFFICE"."INTERNAL_BUDGET"
```

```
    FROM "OFFICE"
    WHERE "OFFICE"."OFFICE_CODE" = "in_office_code";

  OPEN "cursor_employee" FOR
    SELECT "EMPLOYEE"."EMPLOYEE_NUMBER",
           "EMPLOYEE"."FIRST_NAME", "EMPLOYEE"."LAST_NAME"
    FROM "EMPLOYEE"
    WHERE "EMPLOYEE"."OFFICE_CODE" = "in_office_code";
END;
```

This time, besides `getResults()`, which you are already familiar with, we can take advantage of the getters produced by jOOQ for the `OUT` parameters as follows:

```
GetEmpsInOffice geio = new GetEmpsInOffice();
geio.setInOfficeCode("1");

geio.execute(ctx.configuration());

Result<Record> co = geio.getCursorOffice();
Result<Record> ce = geio.getCursorEmployee();
```

Also, relying on `Routines.getEmpsInOffice(Configuration c, String inOfficeCode)` is quite convenient:

```
GetEmpsInOffice results =
   getEmpsInOffice(ctx.configuration(), "1");
```

Next, you can rely on `results.getCursorInfo()` respectively on `results.getCursorEmployee()` or by looping the results as follows:

```
for (Result<Record> result : results.getResults()) {
  ...
}
```

Next, loop each `Result<Record>` as `for (Record record : result)` ....

Not sure it's worth mentioning, but at least Oracle also knows typed `REF CURSORS` (instead of just `SYS_REFCURSOR`), which jOOQ will support as well soon (when you're reading this book, these features should be available): `https://github.com/jOOQ/jOOQ/issues/11708`.

Calling stored procedures via the CALL statement

Finally, let's tackle the API of calling a stored procedure via the `CALL` statement in an anonymous procedural block and via `CALL` directly. Consider the following stored procedure expressed in Oracle (the complete code is available in the bundled code):

```
CREATE OR REPLACE NONEDITIONABLE PROCEDURE
    "REFRESH_TOP3_PRODUCT"("p_line_in" IN VARCHAR2) AS
BEGIN
  DELETE FROM "TOP3PRODUCT";
  INSERT INTO ...
      FETCH NEXT 3 ROWS ONLY;
END;
```

Calling this stored procedure via the `CALL` statement in an anonymous procedural block can be done via `DSLContext.begin()` as follows:

```
ctx.begin(call(name("REFRESH_TOP3_PRODUCT"))
   .args(val("Trains")))
   .execute();
```

Or, call it directly via `DSLContext.call()`:

```
ctx.call(name("REFRESH_TOP3_PRODUCT"))
   .args(val("Trains"))
   .execute();
```

You can practice all these examples in *CallProcedure*.

jOOQ and creating stored functions/procedures

Starting with version 3.15, jOOQ began to add an API for creating stored functions, procedures, and triggers. Among others, we have support for `CREATE FUNCTION`, `CREATE OR REPLACE FUNCTION`, `CREATE PROCEDURE`, `CREATE OR REPLACE PROCEDURE`, `DROP FUNCTION`, and `DROP PROCEDURE`.

Creating stored functions

For instance, creating a scalar function for MySQL can be done as follows:

```
Parameter<Integer> quantity = in("quantity", INTEGER);
Parameter<Double> listPrice = in("list_price", DOUBLE);
Parameter<Double> fractionOfPrice =
```

```
  in("fraction_of_price", DOUBLE);

ctx.createOrReplaceFunction("sale_price_jooq")
   .parameters(quantity, listPrice, fractionOfPrice)
   .returns(DECIMAL(10, 2))
   .deterministic()
   .as(return_(listPrice.minus(listPrice
     .mul(fractionOfPrice)).mul(quantity)))
   .execute();
```

Here, we create a scalar function having three input parameters created via the intuitive `Parameter` API by specifying their name and types. For MySQL, jOOQ renders the following code:

```
DROP FUNCTION IF EXISTS `sale_price_jooq`;

CREATE FUNCTION `sale_price_jooq`(`quantity` INT,
         `list_price` DOUBLE, `fraction_of_price` DOUBLE)
RETURNS DECIMAL(10, 2)
DETERMINISTIC
BEGIN
RETURN ((`list_price` - (`list_price` *
         `fraction_of_price`)) * `quantity`);
END;
```

Notice that in order to work, you should be aware of the following note.

> **Important Note**
> In MySQL, we can execute statement batches if we turn on the `allowMultiQueries` flag, which defaults to `false`; otherwise, we get an error. The previously generated SQL chains two statements, therefore this flag needs to be turned on – we can do it via the JDBC URL as `jdbc:mysql:.../classicmodels?allowMultiQueries=true`. Alternatively, in this particular case, we can rely on the `dropFunctionIfExists()`,`createFunction()` combo instead of `createOrReplaceFunction()`. I strongly advise you to take a couple of minutes to read this article by Lukas Eder, which explains in detail the implication of this flag in the jOOQ context: `https://blog.jooq.org/mysqls-allowmultiqueries-flag-with-jdbc-and-jooq/`.

You already know how to call this function from jOOQ via the generated code. This means that you have to run this code to create the stored function in the database, and afterward, run the jOOQ Code Generator to obtain the expected jOOQ artifacts. On the other hand, you can call it via plain SQL via the DSL's `function()` as in this example:

```
float result = ctx.select(function(name("sale_price_jooq"),
  DECIMAL(10, 2), inline(10), inline(20.45), inline(0.33)))
  .fetchOneInto(Float.class);
```

You can practice this example in *CreateFunction*.

How about creating the following PostgreSQL function having output parameters?

```
CREATE OR REPLACE FUNCTION "swap_jooq"(
       INOUT "x" INT, INOUT "y" INT)
  RETURNS RECORD LANGUAGE PLPGSQL AS $$
BEGIN
  SELECT "x", "y" INTO "y", "x";
END; $$
```

From jOOQ, this function can be created like this:

```
Parameter<Integer> x = inOut("x", INTEGER);
Parameter<Integer> y = inOut("y", INTEGER);

ctx.createOrReplaceFunction("swap_jooq")
   .parameters(x, y)
   .returns(RECORD)
   .as(begin(select(x, y).into(y, x)))
   .execute();
```

Calling this function can be done via plain SQL as in this example:

```
Record1<Record> result = ctx.select(
  function(name("swap_jooq"),
    RECORD, inline(1), inline(2))).fetchOne();
```

You can practice this example next to another one using `OUT` parameters in *CreateFunction* for PostgreSQL. For more examples that allow you to explore this API in detail, please consider the bundled code and the jOOQ manual.

Creating stored procedures

Let's begin with a stored procedure expressed in the SQL Server dialect:

```
CREATE OR ALTER PROCEDURE [update_msrp_jooq]
   @product_id BIGINT, @debit INT AS
BEGIN
  UPDATE [classicmodels].[dbo].[product]
  SET [classicmodels].[dbo].[product].[msrp] =
      ([classicmodels].[dbo].[product].[msrp] - @debit)
WHERE [classicmodels].[dbo].[product].[product_id] =
      @product_id;
END;
```

This stored procedure, which has two input parameters and updates the PRODUCT.MSRP field, can be created through the jOOQ API as follows:

```
Parameter<Long> id = in("id", BIGINT);
Parameter<Integer> debit = in("debit", INTEGER);

ctx.createOrReplaceProcedure("update_msrp_jooq")
   .parameters(id, debit)
   .as(update(PRODUCT)
        .set(PRODUCT.MSRP, PRODUCT.MSRP.minus(debit))
        .where(PRODUCT.PRODUCT_ID.eq(id)))
   .execute();
```

You already know how to call this procedure from jOOQ via the generated code, so this time, let's call it via the CALL statement:

```
// CALL statement in an anonymous procedural block
var result = ctx.begin(call(name("update_msrp_jooq"))
       .args(inline(1L), inline(100)))
       .execute();

// CALL statement directly
var result = ctx.call(name("update_msrp_jooq"))
       .args(inline(1L), inline(100))
       .execute();
```

The returned result represents the number of rows affected by this UPDATE. This example is available in *CreateProcedure* for SQL Server and PostgreSQL.

Next, let's pick up an example expressed in Oracle dialect:

```
CREATE OR REPLACE NONEDITIONABLE PROCEDURE
   "get_avg_price_by_product_line_jooq" (
      "pl" IN VARCHAR2,"average" OUT DECIMAL) AS
BEGIN
   SELECT AVG("CLASSICMODELS"."PRODUCT"."BUY_PRICE")
   INTO "average" FROM "CLASSICMODELS"."PRODUCT"
   WHERE "CLASSICMODELS"."PRODUCT"."PRODUCT_LINE" = "pl";
END;
```

This time, the jOOQ code for creating this procedure is as follows:

```
Parameter<String> pl = in("pl", VARCHAR);
Parameter<BigDecimal> average = out("average", DECIMAL);

ctx.createOrReplaceProcedure(
    "get_avg_price_by_product_line_jooq")
   .parameters(pl, average)
   .as(select(avg(PRODUCT.BUY_PRICE)).into(average)
   .from(PRODUCT)
   .where(PRODUCT.PRODUCT_LINE.eq(pl)))
   .execute();
```

Since you are already familiar with this stored procedure from the previous section, *Stored procedures and output parameters*, you should have no problem calling it. The example is available in *CreateProcedure* for Oracle.

Finally, let's tackle a MySQL stored procedure that fetches a result set via `SELECT`:

```
CREATE PROCEDURE `get_office_gt_budget_jooq`(`budget` INT)
  BEGIN
    SELECT `classicmodels`.`office`.`city`,
           `classicmodels`.`office`.`country`,
           `classicmodels`.`office`.`state`
    FROM `classicmodels`.`office`
    WHERE `classicmodels`.`office`.`internal_budget` > `budget`;
  END;
```

The jOOQ code that creates the stored procedure is as follows:

```
Parameter<Integer> budget = in("budget", INTEGER);

ctx.createOrReplaceProcedure("get_office_gt_budget_jooq")
    .parameters(budget)
    .as(select(OFFICE.CITY, OFFICE.COUNTRY, OFFICE.STATE)
    .from(OFFICE)
    .where(OFFICE.INTERNAL_BUDGET.gt(budget)))
    .execute();
```

In order for this code to work, we need to turn on the `allowMultiQueries` flag as explained in the previous section, *Creating stored functions*.

You can find this example next to another one that fetches two result sets in the application named *CreateProcedure* for MySQL.

Summary

In this chapter, you've learned how to call and create some typical stored functions and procedures. Since these are powerful SQL tools, jOOQ strives to provide a comprehensive API to cover the tons of possibilities to express these artifacts in different dialects. Most probably, by the time you read this book, jOOQ will have already enriched this API even further and more examples will be available in the bundled code.

In the next chapter, we tackle aliases and SQL templating.

16
Tackling Aliases and SQL Templating

This chapter covers two important topics that sustain your road to becoming a jOOQ power user: aliases and SQL templating.

The first part of this chapter tackles several practices for aliasing tables and columns via the jOOQ DSL. The goal of this part is to make you comfortable when you need to express your SQL aliases via jOOQ and to provide you with a comprehensive list of examples that cover the most common use cases.

The second part of this chapter is all about SQL templating or how to express SQL when the jOOQ DSL cannot help us. There will be rare cases when you'll have to write plain SQL or combine DSL and plain SQL to obtain some corner cases or vendor-specific features.

In this chapter, we will cover the following main topics:

- Expressing SQL aliases in jOOQ
- SQL templating

Let's get started!

Technical requirements

The code for this chapter can be found on GitHub at `https://github.com/PacktPublishing/jOOQ-Masterclass/tree/master/Chapter16`.

Expressing SQL aliases in jOOQ

SQL aliasing is a simple task. After all, it's just about giving some nicknames to your columns and tables and referring to them via these nicknames instead of using their real names. But as simple as this may seem, this is a rather controversial topic. Some of the open questions you may come across will sound like this: Should I only use aliases when it's mandatory (for instance, when I reference the same table twice)? Should I use meaningful names or will single letters work just fine (`p`, `q`, `t1`, `t2`, and so on)? Do they increase readability and decrease typing time? Most probably, the correct answer is that it depends... on the context, on the query, on who is writing the query (a developer, a DBA, a generator), and so on!

As you'll see shortly, using aliasing via a DSL requires us to respect several rules and to be prepared for some verbosity since the host language (here, Java) comes with several shortcomings that a DSL must address as elegantly as possible. Aliasing sits next to derived tables, arithmetic expressions, and casting as one of the main challenges for a DSL, so let's see exactly what we should know about it.

The examples from the following sections are available via *SimpleAliases* and *AliasesSamples*.

Expressing simple aliased tables and columns

Independent of how you like to use SQL aliases, when you want to express them in jOOQ, you must be aware of several methods, including `as()` and `asTable()`, which come in many flavors, such as `as(String alias)`, `as(Name alias)`, `as(Name as, Name... fieldAliases)`, `asTable()`, `asTable(String alias)`, `asTable(Name alias)`, `asTable(Table<?> alias)`, and so on. Commonly, we must deal with aliased tables and fields. Here is a quick sample of using aliased tables in jOOQ:

```
ctx.select(field(name("t", "first_name")),
           field(name("t", "last_name")))
   .from(EMPLOYEE.as("t"))
   .fetch();

ctx.select(field(name("t", "product_id")),
           field(name("t", "product_name")),
```

```
            field(selectCount()
                    .from(PRODUCT)
                    .where(PRODUCT.PRODUCT_ID.eq(
                       field(name("t", "product_id"),
                         Long.class)))).as("count"))
   .from(PRODUCT.as("t"))
   .fetch();
```

The following is an example of using some aliased fields (used here to take full control of the column names that are generated in your SQL):

```
ctx.select(EMPLOYEE.FIRST_NAME.as("fn"),
           EMPLOYEE.LAST_NAME.as("ln"))
   .from(EMPLOYEE).fetch();

ctx.select(concat(EMPLOYEE.FIRST_NAME,
           inline(" "), EMPLOYEE.LAST_NAME).as("name"),
           EMPLOYEE.EMAIL.as("contact"),
           EMPLOYEE.REPORTS_TO.as("boss_id"))
   .from(EMPLOYEE).fetch();
```

Next, we'll look at some more complex examples of using aliases.

Aliases and JOINs

One of the common cases where we see SQL aliases at work is in `JOIN` statements. Instead of repeating the table names, people prefer to associate aliases with the joined tables and refer to them via these aliases. For instance, in the following screenshot, we have a `JOIN` between two MySQL tables (`OFFICE` and `DEPARTMENT`) expressed without aliases (top) and with aliases (bottom):

```
SELECT `classicmodels`.`office`.`city`,
       `classicmodels`.`department`.`name`,
       `classicmodels`.`department`.`profit`
FROM `classicmodels`.`office`
JOIN `classicmodels`.`department` ON `classicmodels`.`office`.`office_code` =
            `classicmodels`.`department`.`office_code`
```

```
SELECT `t1`.`city`,
       `t2`.`name`,
       `t2`.`profit`
FROM `classicmodels`.`office` AS `t1`
JOIN `classicmodels`.`department` AS `t2`
  ON `t1`.`office_code` = `t2`.`office_code`
```

Figure 16.1 – JOIN with and without aliases

If we express the first SQL in jOOQ (without using aliases) then we obtain this – in jOOQ, whenever you can omit the usage of aliases, do it! This way, you have better a chance to obtain clean expressions, as shown here:

```
ctx.select(OFFICE.CITY, DEPARTMENT.NAME, DEPARTMENT.PROFIT)
   .from(OFFICE)
   .join(DEPARTMENT)
   .on(OFFICE.OFFICE_CODE.eq(DEPARTMENT.OFFICE_CODE))
   .fetch();
```

This is a clean and readable jOOQ snippet of code. Since jOOQ generates the SQL on our behalf, we don't feel the need to add some aliases to increase readability or reduce the typing time.

Nevertheless, next, let's add the proper aliases to obtain the second SQL. As our first attempt, we may have written this:

```
ctx.select(field("t1.city"),
           field("t2.name"), field("t2.profit"))
   .from(OFFICE.as("t1"))
   .join(DEPARTMENT.as("t2"))
   .on(field("t1.office_code").eq(field("t2.office_code")))
   .fetch();
```

So, we have associated the `t1` alias with the `OFFICE` table via `OFFICE.as("t1")`, and the `t2` alias with the `DEPARTMENT` table via `DEPARTMENT.as("t2")`. Furthermore, we used our aliases via the `field()` method as `t1` and `t2`, respectively. Besides losing some readability in the jOOQ code, have you spotted other issues in this code compared to the jOOQ code without aliases? Sure you did – it's not type-safe and it renders unquoted identifiers.

When we say `field("t1.city")`, jOOQ renders `t1.city`, not `` `t1`.`city` `` (in MySQL). However, it is advisable to strive for qualified and quoted identifiers to avoid name conflicts and potential errors (for instance, using a keyword such as `ORDER` as an unquoted table name leads to errors). Generally speaking, quoted identifiers allow us to use reserved names as object names (for instance, `ORDER`), use special characters in object names (whitespaces and so on), and instructs (most databases) us to treat case-insensitive identifiers as case-sensitive ones (for example, `"address"` and `"ADDRESS"` are different identifiers, whereas `address` and `ADDRESS` are not).

However, jOOQ can render qualified and quoted identifiers if we rely on explicitly using `DSL.name()`, which is a very handy `static` method that comes in several flavors and it is useful for constructing SQL-injection-safe, syntax-safe SQL identifiers for use in plain SQL. It is commonly used in the `table()` and `field()` methods – for example, `name(table_name, field_name)` – but you can check out all the flavors in the documentation. The following table represents what jOOQ renders for different usages of the `name()` method and different databases:

	MySQL	PostgreSQL	SQL Server	Oracle
"ORDER"	ORDER	ORDER	ORDER	ORDER
name("ORDER")	\`ORDER\`	"ORDER"	[ORDER]	"ORDER"
name("ORDER", "ORDER_ID")	\`ORDER\`.\`ORDER_ID\`	"ORDER"."ORDER_ID"	[ORDER].[ORDER_ID]	"ORDER"."ORDER_ID"

Figure 16.2 – Using jOOQ name()

When an identifier occurs several times, it can be extracted in a local variable as a `Name` and reused in queries as needed, like so:

```
Name orderId = name("ORDER", "ORDER_ID");
Field orderId = field(name("ORDER", "ORDER_ID"));
Table t = table(name("ORDER"));
```

When jOOQ evaluates `name("ORDER", "ORDER_ID")` (for MySQL), it renders `` `ORDER`.`ORDER_ID` ``. Of course, `ORDER_ID` doesn't necessarily need the back ticks – only `ORDER` does. Playing with quotations for identifiers can be done via `quotedName()` and `unquotedName()` of the `DSL` class, like so:

```
// `ORDER`.ORDER_ID
Name orderId = name(quotedName("ORDER"),
                    unquotedName("ORDER_ID"));
```

Moreover, jOOQ allows us to control (globally or at the query level) how identifiers are quoted via the `RenderQuotedNames` setting and cases via the `RenderNameCase` setting. For instance, we can instruct jOOQ to quote all the identifiers in the upper part of the current query, as follows:

```
For MySQL, jOOQ render this SQL:
select `T`.`FIRST_NAME` as `FN`, `T`.`LAST_NAME` as `LN`
from `CLASSICMODELS`.`EMPLOYEE` as `T`
```

```
ctx.configuration().derive(new Settings()
    .withRenderQuotedNames(RenderQuotedNames.ALWAYS)
    .withRenderNameCase(RenderNameCase.UPPER))
    .dsl()
    .select(field(name("t", "first_name")).as("fn"),
            field(name("t", "last_name")).as("ln"))
    .from(EMPLOYEE.as("t"))
    .fetch();
```

While you can find more details about these settings in the documentation (`https://www.jooq.org/doc/latest/manual/sql-building/dsl-context/custom-settings/settings-name-style/`), keep in mind that they only affect identifiers that are expressed via Java-based schemas or `name()`. In other words, they have no effect on `field("identifier")` and `table("identifier")`. These are rendered exactly as you provide them.

jOOQ doesn't force us in any way to use quoting, qualifications, and cases consistently in the same query or across multiple queries (since jOOQ renders by default). However, juggling these aspects may lead to issues, from inconsistent results to SQL errors. This happens because, in some databases (for instance, SQL Server), identifiers are always case insensitive. This means that quoting only helps to allow special characters or escape keywords in identifiers. In other databases (for instance, Oracle), identifiers are only case insensitive if they are unquoted, while quoted identifiers are case sensitive. However, there are also databases (for instance, Sybase ASE) where identifiers are always case sensitive, regardless of them being quoted. Again, quoting only helps to allow special characters or escape keywords in identifiers. And, let's not forget the dialects (for instance, MySQL) that mix the preceding rules, depending on the operating system, object type, configuration, and other events.

So, pay attention to how you decide to handle quoting, qualification, and case sensitivity aspects. The best/safest way is to express queries via the Java-based schema, use aliases only when they are mandatory, and always use `name()` if you have to refer to identifiers as plain strings. From that point on, let jOOQ do the rest.

That being said, if we apply `name()` to our query, we obtain the following code:

```
ctx.select(field(name("t1", "city")),
        field(name("t2", "name")), field(name("t2", "profit")))
   .from(OFFICE.as(name("t1")))
```

```
    .join(DEPARTMENT.as(name("t2")))
    .on(field(name("t1", "office_code"))
         .eq(field(name("t2", "office_code"))))
    .fetch();
```

This time, the rendered identifiers correspond to our expectations, but this snippet of jOOQ code is still not type-safe. To transform this non-type-safe query into a type-safe one, we must extract the aliases and define them in local variables before using them, as follows (notice that there is no reason to explicitly use `name()`):

```
Office t1 = OFFICE.as("t1");
Department t2 = DEPARTMENT.as("t2");

ctx.select(t1.CITY, t2.NAME, t2.PROFIT)
    .from(t1)
    .join(t2)
    .on(t1.OFFICE_CODE.eq(t2.OFFICE_CODE))
    .fetch();
```

Alternatively, you may prefer a minimalist aliasing approach, as follows:

```
Department t = DEPARTMENT.as("t");
// or, Department t = DEPARTMENT;

ctx.select(OFFICE.CITY, t.NAME, t.PROFIT)
    .from(OFFICE)
    .join(t)
    .on(OFFICE.OFFICE_CODE.eq(t.OFFICE_CODE))
    .fetch();
```

Calling the `as()` method on the generated tables (here, on `OFFICE` and `DEPARTMENT`) returns an object of the same type as the table (`jooq.generated.tables.Office` and `jooq.generated.tables.Department`). The resulting object can be used to dereference fields from the aliased table in a type-safe way. So, thanks to `as()`, `Office`, and `Department`, we are type-safe again while using the desired table aliases. And, of course, the identifiers are implicitly rendered, quoted, and qualified.

> **Important Note**
>
> As a rule of thumb, in jOOQ, strive to extract and declare aliases in local variables before using them in queries. Do this especially if your aliases refer to the generated tables, are aliases that should be reused across multiple queries, you wish to increase the readability of the jOOQ expression and/or avoid typos, and so on. Of course, if your jOOQ expression simply associates some aliases to columns (to take full control of the column names that are generated in your SQL), then extracting them as local variables won't produce a significant improvement.

Let's look at a table that's been aliased as follows:

```
Table<Record1<String>> t3 =
  ctx.select(t1.CITY).from(t1).asTable("t3");
```

In this case, we can refer to fields in a non-type-safe manner via `field(Name name)`, as shown in the following example:

```
ctx.select(t3.field(name("city")),
      CUSTOMERDETAIL.CUSTOMER_NUMBER)
   .from(t3)
   .join(CUSTOMERDETAIL)
   .on(t3.field(name("city"), String.class)
     .eq(CUSTOMERDETAIL.CITY))
   .fetch();
```

The same `field()` method can be applied to any type-unsafe aliased table as `Table<?>` to return `Field<?>`.

In this case, we can make these fields look type-safe via the `<T> Field<T> field(Field<T> field)` method as well (introduced in *Chapter 14, Derived Tables, CTEs, and Views*). The `t3.field(name("city"))` expression indirectly refers to the `t1.CITY` field, so we can rewrite our queries in a type-safe manner, as follows:

```
ctx.select(t3.field(t1.CITY), CUSTOMERDETAIL.CUSTOMER_NUMBER)
   .from(t3)
   .join(CUSTOMERDETAIL)
   .on(t3.field(t1.CITY).eq(CUSTOMERDETAIL.CITY))
   .fetch();
```

However, remember that `Table.field(Field<T>):Field<T>` just looks type-safe. It's as good as an unsafe cast in Java because the lookup only considers the identifier, not the type. Nor does it coerce the expression.

So far, so good! You can practice these examples in *AliasesSamples*. Now, let's take some time to cover several fundamental aspects of jOOQ aliases and practice some simple but essential exercises.

Aliases and GROUP BY/ORDER BY

Let's consider the following SQL expressed in SQL Server:

```
SELECT [classicmodels].[dbo].[product].[product_line] [pl]
FROM [classicmodels].[dbo].[product]
GROUP BY [classicmodels].[dbo].[product].[product_line]
ORDER BY [pl]
```

This query uses an alias named `pl` for the `PRODUCT_LINE` column. Attempting to express this query via jOOQ, based on what we learned earlier, may result in something like this:

```
Field<String> pl = PRODUCT.PRODUCT_LINE.as("pl");

ctx.select(pl)
   .from(PRODUCT)
   .groupBy(pl)
   .orderBy(pl)
   .fetch();
```

But this isn't correct! The problem here is related to our expectations. We expect `PRODUCT.PRODUCT_LINE.as("pl")` to produce `[pl]` in `ORDER BY`, `[classicmodels].[dbo].[product].[product_line]` in `GROUP BY`, and `[classicmodels].[dbo].[product].[product_line] [pl]` in `SELECT`. In other words, we expect that the three usages of the local `pl` variable will magically render the output that makes more sense for us. Well, this isn't true!

Think about the jOOQ DSL more as an expression tree. So, we can store both `PRODUCT.PRODUCT_LINE` and `PRODUCT.PRODUCT_LINE.as("pl")` in separate local variables, and explicitly reuse the one that makes the most sense:

```
Field<String> pl1 = PRODUCT.PRODUCT_LINE.as("pl");
Field<String> pl2 = PRODUCT.PRODUCT_LINE;
```

```
ctx.select(pl1)
    .from(PRODUCT)
    .groupBy(pl2)
    .orderBy(pl1)
    .fetch();
```

This time, it is correct!

> **Important Note**
>
> Reusing the `x.as("y")` expression in a query and thinking that it "magically" produces `x` or `y`, whatever makes more sense, is a really bad understanding of jOOQ aliases. Thinking that `x.as("y")` generates `x` in `GROUP BY`, `y` in `ORDER BY`, and `x.as("y")` in `SELECT` is dangerous logic that will give you headaches. The aliased expression, `x.as("y")`, produces `y` "everywhere" outside of `SELECT`, and it produces the alias declaration in `SELECT` (but only immediately in `SELECT`). It "never" produces only `x`.

You can practice these examples in *AliasesSamples*.

Aliases and bad assumptions

Next, let's consider the examples shown in the following screenshot:

```
ctx.select(OFFICE.CITY.as("t"))
        .from(OFFICE)
        .fetch();
```

(A)

```
ctx.select(OFFICE.CITY)
        .from(OFFICE.as("t"))
        .fetch();
```

(B)

Figure 16.3 – Aliases use case

What can you say about (A) and (B)? If you said that (A) is correct while (B) is wrong, then you are right. Congratulations! Speaking about (B), since we assigned an alias to the `[office]` table, the `[office].[city]` column becomes unknown. The rendered SQL highlights the following aspect:

```
SELECT [classicmodels].[dbo].[office].[city]
FROM [classicmodels].[dbo].[office] [t]
```

So, one simple and straightforward solution is to simply remove the alias. Now, let's examine a few bad choices. First, let's explore this one:

```
// SELECT t FROM [classicmodels].[dbo].[office] [t]
ctx.select(field("t", "city"))
   .from(OFFICE.as("t"))
   .fetch();
```

This construction is based on the assumption that jOOQ exposes a method, `field(String table_name, String field_name)`, but there is no such method! Then why does the preceding code compile? Because DSL exposes a `field(String sql, Object... bindings)` that is used for SQL templating, it's being used in the wrong context. Pay attention to such silly mistakes! Who didn't feel lucky and tried to use an API without reading the documentation?!

Now, how about this one?

```
// SELECT [t].[office_code], [t].[city], ..., [t].[location]
// FROM [classicmodels].[dbo].[office] [t]
ctx.select(table("t").field("city"))
   .from(OFFICE.as("t"))
   .fetch();
```

This is just another example built on wrong assumptions. While jOOQ exposes a `table(String sql)`, which is useful for returning a table that wraps the given plain SQL, this example assumes the existence of a `table(String alias)` that returns a table that wraps an alias and is aware of its fields.

Going further, let's try this approach:

```
// SELECT [city] FROM [classicmodels].[dbo].[office] [t]
ctx.select(field(name("city")))
   .from(OFFICE.as("t"))
   .fetch();
```

This approach works just fine but you must be aware that the unqualified `[city]` is prone to ambiguities. For instance, let's say that we enrich this query, as follows:

```
ctx.select(field(name("city")))
   .from(OFFICE.as("t1"), CUSTOMERDETAIL.as("t2"))
   .fetch();
```

This leads to an ambiguous column, `[city]`, because it's unclear if we are referring to `OFFICE.CITY` or `CUSTOMERDETAIL.CITY`. In this case, table aliases can help us express this clearly:

```
ctx.select(field(name("t1", "city")).as("city_office"),
           field(name("t2", "city")).as("city_customer"))
   .from(OFFICE.as("t1"), CUSTOMERDETAIL.as("t2"))
   .fetch();
```

It's much better to declare aliases before using them:

```
Office t1 = OFFICE.as("t1");
Customerdetail t2 = CUSTOMERDETAIL.as("t2");

ctx.select(t1.CITY, t2.CITY)
   .from(t1, t2)
   .fetch();

Field<String> c1 = t1.CITY.as("city_office");
Field<String> c2 = t2.CITY.as("city_customer");

ctx.select(c1, c2)
   .from(t1, t2)
   .fetch();
```

Now, let's look at another case and start with two wrong approaches. So, what's wrong here?

```
ctx.select()
   .from(OFFICE
   .leftOuterJoin(DEPARTMENT)
     .on(OFFICE.OFFICE_CODE.eq(DEPARTMENT.OFFICE_CODE))
   .innerJoin(EMPLOYEE)
     .on(EMPLOYEE.OFFICE_CODE.eq(
        field(name("office_code"), String.class)))
   .fetch();
```

After joining `OFFICE` and `DEPARTMENT`, the result contains two columns named `office_code` – one from `OFFICE` and another from `DEPARTMENT`. Joining this result with `EMPLOYEE` reveals that the `office_code` column in the `ON` clause is ambiguous. To remove this ambiguity, we can use aliased tables:

```
ctx.select()
   .from(OFFICE.as("o")
```

```
    .leftOuterJoin(DEPARTMENT.as("d"))
      .on(field(name("o","office_code"))
        .eq(field(name("d","office_code"))))
    .innerJoin(EMPLOYEE)
      .on(EMPLOYEE.OFFICE_CODE.eq(OFFICE.OFFICE_CODE))
    .fetch();
```

Is this correct? This time, we have associated aliases with our `OFFICE.as("o")` and `DEPARTMENT.as("d")` tables. While joining `OFFICE` with `DEPARTMENT`, we correctly used the aliases, but when we joined the result to `EMPLOYEE`, we didn't use the `OFFICE` alias – we used the un-aliased `OFFICE.OFFICE_CODE`. This is rendered in MySQL as `` `classicmodels`.`office`.`office_code` `` and it represents an unknown column in the `ON` clause. So, the correct expression is as follows:

```
ctx.select()
    .from(OFFICE.as("o")
    .leftOuterJoin(DEPARTMENT.as("d"))
      .on(field(name("o","office_code"))
        .eq(field(name("d","office_code"))))
    .innerJoin(EMPLOYEE)
      .on(EMPLOYEE.OFFICE_CODE
        .eq(field(name("o","office_code"), String.class)))
    .fetch();
```

Can we write this more compact and type-safe? Sure we can – via local variables:

```
Office o = OFFICE.as("o");
Department d = DEPARTMENT.as("d");

ctx.select()
    .from(o.leftOuterJoin(d)
    .on(o.OFFICE_CODE.eq(d.OFFICE_CODE)))
    .innerJoin(EMPLOYEE)
      .on(EMPLOYEE.OFFICE_CODE.eq(o.OFFICE_CODE))
    .fetch();
```

Again, local variables help us express aliases and obtain elegant code.

Aliases and typos

Next, let's look at another way to extract aliases in local variables. Check out the following code:

```
ctx.select(field("s1.msrp"), field("s2.msrp"))
   .from(PRODUCT.as("s1"), PRODUCT.as("s2"))
   .where(field("s1.msrp").lt(field("s2.msrp"))
      .and(field("s1.product_line").eq("s2.product_line")))
   .groupBy(field("s1.msrp"), field("s2.msrp"))
   .having(count().eq(selectCount().from(PRODUCT.as("s3"))
   .where(field("s3.msrp").eq(field("s1.msrp"))))
      .and(count().eq(selectCount().from(PRODUCT.as("s4"))
      .where(field("s4.msrp").eq(field("s2.msrp"))))))
   .fetch();
```

There is a mistake (a typo) in this expression. Can you spot it? (It isn't easy!) If not, you'll end up with valid SQL that returns inaccurate results. The typo snuck into the `.and(field("s1.product_line").eq("s2.product_line")))` part of the code, which should be `.and(field("s1.product_line").eq(field("s2.product_line"))))`. But if we extract the aliases in local variables, then the code eliminates the risk of a typo and increases the readability of the expression (notice that `s1`, `s2`, `s3`, and `s4` are not equal objects and that they cannot be used interchangeably):

```
Product s1 = PRODUCT.as("s1");
Product s2 = PRODUCT.as("s2");
Product s3 = PRODUCT.as("s3");
Product s4 = PRODUCT.as("s4");

ctx.select(s1.MSRP, s2.MSRP)
   .from(s1, s2)
   .where(s1.MSRP.lt(s2.MSRP)
   .and(s1.PRODUCT_LINE.eq(s2.PRODUCT_LINE)))
   .groupBy(s1.MSRP, s2.MSRP)
   .having(count().eq(selectCount().from(s3)
     .where(s3.MSRP.eq(s1.MSRP)))
     .and(count().eq(selectCount().from(s4)
     .where(s4.MSRP.eq(s2.MSRP)))))
   .fetch();
```

You can practice these examples in *AliasesSamples*.

Aliases and derived tables

Let's look at another example that starts with the following snippet of code:

```
ctx.select().from(
 select(CUSTOMER.CUSTOMER_NUMBER,
        CUSTOMER.CUSTOMER_NAME, field("t.invoice_amount"))
 .from(CUSTOMER)
 .join(select(PAYMENT.CUSTOMER_NUMBER,
              PAYMENT.INVOICE_AMOUNT)
        .from(PAYMENT).asTable("t"))
  .on(field("t.customer_number")
    .eq(CUSTOMER.CUSTOMER_NUMBER)))
 .fetch();
```

So, what's wrong here?! Let's inspect the generated SQL (this is for MySQL):

```
SELECT `alias_84938429`.`customer_number`,
       `alias_84938429`.`customer_name`,
       `alias_84938429`.t.invoice_amount
FROM
  (SELECT `classicmodels`.`customer`.`customer_number`,
          `classicmodels`.`customer`.`customer_name`,
          t.invoice_amount
   FROM `classicmodels`.`customer`
   JOIN
     (SELECT `classicmodels`.`payment`.`customer_number`,
             `classicmodels`.`payment`.`invoice_amount`
      FROM `classicmodels`.`payment`) AS `t` ON
       t.customer_number =
        `classicmodels`.`customer`.`customer_number`)
      AS `alias_84938429
```

As you can see, jOOQ has automatically associated an alias with the derived table (`alias_84938429`) that was obtained from `JOIN` and used this alias to reference `customer_number`, `customer_name`, and `invoice_amount`. While `customer_number` and `customer_name` are correctly qualified and quoted, `invoice_amount` has been incorrectly rendered as `t.invoice_amount`. The problem is in `field("t.invoice_amount")`, which instructs jOOQ that the column name is `t.invoice_amount`, not `invoice_amount`, so the resulting `` `alias_84938429`.t.invoice_amount `` is an unknown column.

There are a few solutions to this problem, and one of them consists of using `name()` for proper quoting and qualifying:

```
ctx.select().from(select(CUSTOMER.CUSTOMER_NUMBER,
CUSTOMER.CUSTOMER_NAME, field(name("t", "invoice_amount")))
   .from(CUSTOMER)
   .join(
     select(PAYMENT.CUSTOMER_NUMBER,
            PAYMENT.INVOICE_AMOUNT)
      .from(PAYMENT).asTable("t"))
   .on(field(name("t", "customer_number"))
     .eq(CUSTOMER.CUSTOMER_NUMBER)))
   .fetch();
```

This time, jOOQ renders `` `alias_10104609`.`invoice_amount` ``. In the bundled code, you can see four more solutions to this problem.

To understand this context, let's check out the following example:

```
ctx.select()
   .from(select(EMPLOYEE.EMPLOYEE_NUMBER.as("en"),
                EMPLOYEE.SALARY.as("sal"))
             .from(EMPLOYEE)
             .where(EMPLOYEE.MONTHLY_BONUS.isNull()))
             .innerJoin(SALE)
             .on(field(name("en"))
             .eq(SALE.EMPLOYEE_NUMBER))
   .fetch();
```

Here, we have explicitly associated column aliases with the inner `SELECT`, but we did not associate an alias with the derived table produced by `JOIN`. These aliases are further used to reference the columns outside this `SELECT` (in the outer `SELECT`). Notice that we let jOOQ qualify these aliases to the generated alias for the divided table accordingly:

```
SELECT `alias_41049514`.`en`,
       `alias_41049514`.`sal`,
       `classicmodels`.`sale`.`sale_id`,
       ...
FROM
  (SELECT `classicmodels`.`employee`.`employee_number`
      AS `en`, `classicmodels`.`employee`.`salary` AS `sal`
   FROM `classicmodels`.`employee`
```

```
    WHERE `classicmodels`.`employee`.`monthly_bonus` IS NULL
          ) AS `alias_41049514`
JOIN `classicmodels`.`sale` ON `en` =
     `classicmodels`.`sale`.`employee_number`
```

If we want to control the alias of the derived table as well, then we can do the following:

```
ctx.select(SALE.SALE_, SALE.FISCAL_YEAR,
           field(name("t", "sal")))
   .from(select(EMPLOYEE.EMPLOYEE_NUMBER.as("en"),
                EMPLOYEE.SALARY.as("sal"))
         .from(EMPLOYEE)
         .where(EMPLOYEE.MONTHLY_BONUS.isNull())
                .asTable("t"))
         .innerJoin(SALE)
          .on(field(name("t", "en"))
          .eq(SALE.EMPLOYEE_NUMBER))
   .fetch();
```

This time, the rendered SQL uses our table alias:

```
SELECT `classicmodels`.`sale`.`sale`,
       `classicmodels`.`sale`.`fiscal_year`,
       `t`.`sal`
FROM
  (SELECT `classicmodels`.`employee`.`employee_number`
    AS `en`, `classicmodels`.`employee`.`salary` AS `sal`
   FROM `classicmodels`.`employee`
   WHERE `classicmodels`.`employee`.`monthly_bonus`
      IS NULL) AS `t`
JOIN `classicmodels`.`sale` ON `t`.`en` =
     `classicmodels`.`sale`.`employee_number`
```

Finally, here is a more verbose example of using aliases:

```
ctx.select(field(name("t2", "s")).as("c1"),
           field(name("t2", "y")).as("c2"),
           field(name("t2", "i")).as("c3"))
   .from(select(SALE.SALE_.as("s"), SALE.FISCAL_YEAR.as("y"),
                field(name("t1", "emp_sal")).as("i"))
          .from(select(EMPLOYEE.EMPLOYEE_NUMBER.as("emp_nr"),
```

```
                EMPLOYEE.SALARY.as("emp_sal"))
                  .from(EMPLOYEE)
                  .where(EMPLOYEE.MONTHLY_BONUS.isNull())
                  .asTable("t1"))
    .innerJoin(SALE)
    .on(field(name("t1","emp_nr"))
      .eq(SALE.EMPLOYEE_NUMBER)).asTable("t2"))
    .fetch();
```

Take your time to analyze this expression and the generated SQL:

```
SELECT `t2`.`s` AS `c1`,
       `t2`.`y` AS `c2`,
       `t2`.`i` AS `c3`
FROM
  (SELECT `classicmodels`.`sale`.`sale` AS `s`,
          `classicmodels`.`sale`.`fiscal_year` AS `y`,
          `t1`.`emp_sal` AS `i`
   FROM
     (SELECT `classicmodels`.`employee`.`employee_number`
                AS `emp_nr`,
             `classicmodels`.`employee`.`salary`
                AS `emp_sal`
      FROM `classicmodels`.`employee`
      WHERE `classicmodels`.`employee`.`monthly_bonus`
        IS NULL) AS `t1`
   JOIN `classicmodels`.`sale` ON `t1`.`emp_nr` =
        `classicmodels`.`sale`.`employee_number`) AS `t2`
```

Now, let's look at a few more examples of using aliases.

Derived column list

When column names are not known in advance (but the table's degree is!), we can use the so-called *derived column list*. You saw many examples of using this feature with unnested tables, so here are two more for the `VALUES()` table constructor and a regular table:

```
ctx.select().from(values(row("A", "John", 4333, false))
  .as("T", "A", "B", "C", "D")).fetch();
```

The following code is for a regular table:

```
ctx.select(min(field(name("t", "rdate"))).as("cluster_start"),
         max(field(name("t", "rdate"))).as("cluster_end"),
         min(field(name("t", "status"))).as("cluster_status"))
    .from(select(ORDER.REQUIRED_DATE, ORDER.STATUS,
           rowNumber().over().orderBy(ORDER.REQUIRED_DATE)
           .minus(rowNumber().over().partitionBy(ORDER.STATUS)
            .orderBy(ORDER.REQUIRED_DATE)))
            .from(ORDER)
            .asTable("t", "rdate", "status", "cluster"))
    .groupBy(field(name("t", "cluster")))
    .orderBy(1)
    .fetch();
```

If you are not familiar with these kinds of aliases, take your time to inspect the rendered SQL and read some documentation.

Aliases and the CASE expression

Aliases can be used with `CASE` expressions as well. Here is an example:

```
ctx.select(EMPLOYEE.SALARY,
  count(case_().when(EMPLOYEE.SALARY
   .gt(0).and(EMPLOYEE.SALARY.lt(50000)), 1)).as("< 50000"),
  count(case_().when(EMPLOYEE.SALARY.gt(50000)
   .and(EMPLOYEE.SALARY.lt(100000)), 1)).as("50000 - 100000"),
  count(case_().when(EMPLOYEE.SALARY
   .gt(100000), 1)).as("> 100000"))
 .from(EMPLOYEE)
 .groupBy(EMPLOYEE.SALARY)
 .fetch();
```

They can also be used in `FILTER WHERE` expressions:

```
ctx.select(EMPLOYEE.SALARY,
  count().filterWhere(EMPLOYEE.SALARY
   .gt(0).and(EMPLOYEE.SALARY.lt(50000))).as("< 50000"),
  count().filterWhere(EMPLOYEE.SALARY.gt(50000)
   .and(EMPLOYEE.SALARY.lt(100000))).as("50000 - 100000"),
  count().filterWhere(EMPLOYEE.SALARY
```

```
    .gt(100000)).as("> 100000"))
  .from(EMPLOYEE)
  .groupBy(EMPLOYEE.SALARY)
  .fetch();
```

As you can see, using aliases in `CASE/FILTER` expressions is quite handy since it allows us to express the meaning of each case better.

Aliases and IS NOT NULL

Aliases can be used with `IS NOT NULL` (and companions) if we wrap our `Condition` in `field()` to obtain a `Field<Boolean>`:

```
ctx.select(EMPLOYEE.FIRST_NAME,
           EMPLOYEE.LAST_NAME, EMPLOYEE.COMMISSION,
           field(EMPLOYEE.COMMISSION.isNotNull()).as("C"))
   .from(EMPLOYEE)
   .fetch();
```

Finally, let's take a quick look at aliases and CTEs.

Aliases and CTEs

In *Chapter 14, Derived Tables, CTEs, and Views*, we looked at tons of examples of using aliases in CTEs and derived tables, so please consider that chapter if you wish to become familiar with this topic. Next, let's talk about SQL templating.

SQL templating

When we talk about SQL templating or the Plain SQL Templating Language, we're talking about covering those cases where the DSL cannot help us express our SQL. The jOOQ DSL strives to cover SQL as much as possible by constantly adding more and more features, but it is normal to still find some corner case syntax or vendor-specific features that won't be covered by the DSL. In such cases, jOOQ allows us to express SQL as plain SQL strings with bind values or query parts via the Plain SQL API.

The Plain SQL API materializes in a set of overloaded methods that can be used where the DSL doesn't help. Here are some examples:

```
field/table(String sql)
field(String sql, Class<T> type)
field(String sql, Class<T> type, Object... bindings)
```

```
field(String sql, Class<T> type, QueryPart... parts)
field/table(String sql, Object... bindings)
field(String sql, DataType<T> type)
field(String sql, DataType<T> type, Object... bindings)
field(String sql, DataType<T> type, QueryPart... parts)
field/table(String sql, QueryPart... parts)

from/where/join …(String string)
from/where/join …(String string, Object... os)
from/where/join …(String string, QueryPart... qps)
```

So, we can pass SQL as follows:

- Plain SQL strings
- Plain SQL strings and bindings (?)
- Plain SQL strings and `QueryPart` ({0}, {1}, …)

Binding and the query part argument overloads use the so-called Plain SQL Templating Language.

Here are several examples of using plain SQL with bind values (these examples are available in *SQLTemplating*):

```
ctx.fetch("""
          SELECT first_name, last_name
          FROM employee WHERE salary > ? AND job_title = ?
          """, 5000, "Sales Rep");

ctx.resultQuery("""
           SELECT first_name, last_name
           FROM employee WHERE salary > ? AND job_title = ?
           """, 5000, "Sales Rep")
   .fetch();

ctx.query("""
          UPDATE product SET product.quantity_in_stock = ?
          WHERE product.product_id = ?
          """, 0, 2)
   .execute();
```

```
ctx.queries(query(""), query(""), query(""))
           .executeBatch();
```

Now, let's look at some examples to help you become familiar with the technique of mixing plain SQL with SQL expressed via DSL. Let's consider the following MySQL query:

```
SELECT `classicmodels`.`office`.`office_code`,
       ...
       `classicmodels`.`customerdetail`.`customer_number`,
       ...
FROM `classicmodels`.`office`
JOIN `classicmodels`.`customerdetail`
ON `classicmodels`.`office`.`postal_code` =
   `classicmodels`.`customerdetail`.`postal_code`
WHERE not((
  `classicmodels`.`office`.`city`,
  `classicmodels`.`office`.`country`)
   <=> (`classicmodels`.`customerdetail`.`city`,
         `classicmodels`.`customerdetail`.`country`))
```

If you are a jOOQ novice and you're trying to express this query via the jOOQ DSL, then you'll probably encounter some issues in the highlighted code. Can we express that part via the DSL? The answer is yes, but if we cannot find the proper solution, then we can embed it as plain SQL as well. Here is the code:

```
ctx.select()
   .from(OFFICE)
   .innerJoin(CUSTOMERDETAIL)
   .on(OFFICE.POSTAL_CODE.eq(CUSTOMERDETAIL.POSTAL_CODE))
   .where("""
           not(
                (
                  `classicmodels`.`office`.`city`,
                  `classicmodels`.`office`.`country`
                ) <=> (
                  `classicmodels`.`customerdetail`.`city`,
                  `classicmodels`.`customerdetail`.`country`
                )
              )
         """)
   .fetch();
```

Done! Of course, once you become more familiar with the jOOQ DSL, you'll be able to express this query 100% via the DSL, and let jOOQ emulate it accordingly (much better!):

```
ctx.select()
   .from(OFFICE)
   .innerJoin(CUSTOMERDETAIL)
   .on(OFFICE.POSTAL_CODE.eq(CUSTOMERDETAIL.POSTAL_CODE))
   .where(row(OFFICE.CITY, OFFICE.COUNTRY)
   .isDistinctFrom(row(
      CUSTOMERDETAIL.CITY, CUSTOMERDETAIL.COUNTRY)))
   .fetch();
```

But sometimes, you'll need SQL templating. For instance, MySQL defines a function, `CONCAT_WS(separator, exp1, exp2, exp3,...)`, that adds two or more expressions together with the given separator. This function doesn't have a jOOQ correspondent, so we can use it via SQL templating (here, plain SQL and query parts), as follows:

```
ctx.select(PRODUCT.PRODUCT_NAME,
           field("CONCAT_WS({0}, {1}, {2})",
                 String.class, val("-"),
                 PRODUCT.BUY_PRICE, PRODUCT.MSRP))
   .from(PRODUCT)
   .fetch();
```

Since the number of parts to concatenate can vary, it will be more practical to rely on the convenient `DSL.list(QueryPart...)`, which allows us to define a comma-separated list of query parts in a single template argument:

```
ctx.select(PRODUCT.PRODUCT_NAME,
           field("CONCAT_WS({0}, {1})",
                 String.class, val("-"),
                 list(PRODUCT.BUY_PRICE, PRODUCT.MSRP)))
   .from(PRODUCT)
   .fetch();
```

This time, the template argument, `{1}`, has been replaced with the list of strings that should be concatenated. Now, you can simply pass that list.

The jOOQ DSL also doesn't support MySQL variables (`@variable`). For instance, how would you express the following MySQL query, which uses the `@type` and `@num` variables?

```
SELECT `classicmodels`.`employee`.`job_title`,
   `classicmodels`.`employee`.`salary`,
   @num := if(@type = `classicmodels`.`employee`.
              `job_title`, @num + 1, 1) AS `rn`,
   @type := `classicmodels`.`employee`.`job_title` AS `dummy`
FROM `classicmodels`.`employee`
ORDER BY `classicmodels`.`employee`.`job_title`,
         `classicmodels`.`employee`.`salary`
```

Here's SQL templating to the rescue:

```
ctx.select(EMPLOYEE.JOB_TITLE, EMPLOYEE.SALARY,
        field("@num := if(@type = {0}, @num + 1, 1)",
        EMPLOYEE.JOB_TITLE).as("rn"),
        field("@type := {0}", EMPLOYEE.JOB_TITLE).as("dummy"))
   .from(EMPLOYEE)
   .orderBy(EMPLOYEE.JOB_TITLE, EMPLOYEE.SALARY)
   .fetch();
```

You can practice these examples, along with others, in *SQLTemplating*.

SQL templating is also useful when we need to work with certain data types, such as the PostgreSQL `HSTORE` data type. We know that jOOQ allows us to define converters and bindings, especially for dealing with such types. In *Chapter 7*, *Types, Converters, and Bindings*, we wrote an `org.jooq.Converter` and an `org.jooq.Binding` for the `HSTORE` data type. Moreover, the `jooq-postgres-extensions` module supports `HSTORE` as well.

However, using SQL templating can represent a quick solution as well – for instance, you may only need to write a few queries and you don't have time to write a converter/binding. We can insert this into our `HSTORE` (`PRODUCT.SPECS`) via SQL templating, as follows:

```
ctx.insertInto(PRODUCT, PRODUCT.PRODUCT_NAME,
        PRODUCT.PRODUCT_LINE, PRODUCT.CODE, PRODUCT.SPECS)
   .values("2002 Masserati Levante", "Classic Cars",
          599302L, field("?::hstore", String.class,
          HStoreConverter.toString(Map.of("Length (in)",
           "197", "Width (in)", "77.5", "Height (in)",
```

```
            "66.1", "Engine", "Twin Turbo Premium Unleaded
             V-6"))))
    .execute();
```

We can select everything from a HSTORE like so:

```
List<Map<String, String>> specs =
 ctx.select(PRODUCT.SPECS.coerce(String.class))
    .from(PRODUCT)
    .where(PRODUCT.PRODUCT_NAME.eq("2002 Masserati Levante"))
    .fetch(rs -> {
      return HStoreConverter.fromString(
        rs.getValue(PRODUCT.SPECS).toString());
    });
```

Notice that both examples rely on org.postgresql.util.HStoreConverter.

Other operations that are performed against an HSTORE rely on vendor-specific operators. Using such operators is a perfect job for SQL templating. For instance, getting an HSTORE entry by its key can be done by respecting the PostgreSQL syntax, as shown here:

```
ctx.select(PRODUCT.PRODUCT_ID, PRODUCT.PRODUCT_NAME,
           field("{0} -> {1}", String.class, PRODUCT.SPECS,
           val("Engine")).as("engine"))
   .from(PRODUCT)
   .where(field("{0} -> {1}", String.class, PRODUCT.SPECS,
           val("Length (in)")).eq("197"))
   .fetch();
```

Alternatively, we can delete entries by key, as shown here:

```
ctx.update(PRODUCT)
   .set(PRODUCT.SPECS, (field("delete({0}, {1})",
     Record.class, PRODUCT.SPECS, val("Engine"))))
   .execute();
```

We can also convert an HSTORE into JSON:

```
ctx.select(PRODUCT.PRODUCT_NAME,
         field("hstore_to_json ({0}) json", PRODUCT.SPECS))
   .from(PRODUCT)
   .fetch();
```

More examples are available in the bundled code – *SQLTemplating* for PostgreSQL. If you need these operators more often, then you should retrieve their SQL templating code in static utility methods and simply call those methods. For example, a get-by-key method can be expressed as follows:

```
public static Field<String> getByKey(
         Field<Map<String, String>> hstore, String key) {
  return field("{0} -> {1}", String.class, hstore, val(key));
}
```

We can also define CTE via SQL templating. Here is an example of defining a CTE via `ResultQuery` and SQL templating:

```
Result<Record1<BigDecimal>> msrps = ctx.resultQuery(
  "with \"updatedMsrp\" as ({0}) {1}",
    update(PRODUCT).set(PRODUCT.MSRP,
      PRODUCT.MSRP.plus(PRODUCT.MSRP.mul(0.25)))
      .returning(PRODUCT.MSRP),
    select().from(name("updatedMsrp")))
   .coerce(PRODUCT.MSRP)
   .fetch();
```

This code still uses `ResultQuery` and SQL templating, but this time, the plain SQL looks as follows:

```
Result<Record1<BigDecimal>> msrps = ctx.resultQuery(
    "with \"updatedMsrp\" as ({0}) {1}",
     resultQuery("""
                 update
                     "public"."product"
                 set
                     "msrp" = (
                       "public"."product"."msrp" + (
                       "public"."product"."msrp" * 0.25
                     )
                 ) returning "public"."product"."msrp"
                 """),
     resultQuery("""
                 select *
                 from "updatedMsrp"
```

```
                """))
    .coerce(PRODUCT.MSRP)
    .fetch();
```

More examples are available in *SQLTemplating* for PostgreSQL.

How about calling some SQL Server functions? Let's try to call a function that returns an integer that measures the difference between the `SOUNDEX()` values of two different character expressions. Yes – the `DIFFERENCE()` function:

```
ctx.select(field("DIFFERENCE({0}, {1})",
           SQLDataType.INTEGER, "Juice", "Jucy"))
   .fetch();
```

How about calling the `FORMAT()` function?

```
ctx.select(field("FORMAT({0}, {1})",
      123456789, "##-##-#####"))
   .fetch();
```

Now, let's try the following SQL Server batch, which uses SQL Server local variables:

```
DECLARE @var1 VARCHAR(70)

select @var1=(select
  [classicmodels].[dbo].[product].[product_name]
from [classicmodels].[dbo].[product]
where [classicmodels].[dbo].[product].[product_id] = 1)

update [classicmodels].[dbo].[product]
set [classicmodels].[dbo].[product].[quantity_in_stock] = 0
where [classicmodels].[dbo].[product].[product_name] = @var1
```

Again, combining SQL and SQL templating comes to the rescue:

```
ctx.batch(
  query("DECLARE @var1 VARCHAR(70)"),
  select(field("@var1=({0})", select(PRODUCT.PRODUCT_NAME)
   .from(PRODUCT).where(PRODUCT.PRODUCT_ID.eq(1L)))),
  update(PRODUCT).set(PRODUCT.QUANTITY_IN_STOCK, 0)
   .where(PRODUCT.PRODUCT_NAME
      .eq(field("@var1", String.class)))
).execute();
```

You can practice these examples in *SQLTemplating* for SQL Server.

So far, we've looked at examples that are specific to MySQL, PostgreSQL, and SQL Server. Finally, let's add one for Oracle. For instance, if you plan to update/delete records that are referenced by a `SELECT FOR UPDATE` statement, you can use a `WHERE CURRENT OF` statement. The following example uses SQL templating to build such a SQL sample:

```
String sql = ctx.resultQuery("{0} WHERE CURRENT OF cur",
   deleteFrom(PRODUCT)).getSQL();
```

The SQL is as follows:

```
delete from "CLASSICMODELS"."PRODUCT" WHERE CURRENT OF cur
```

You can practice these examples in *SQLTemplating* for Oracle.

Moreover, especially for those corner cases that require complex SQL clauses, jOOQ exposes a set of classes that are very well exemplified in the official documentation: `https://www.jooq.org/doc/latest/manual/sql-building/queryparts/custom-queryparts/`.

Summary

This was a short but comprehensive chapter about jOOQ aliases and SQL templating. In jOOQ, most of the time, you can have a peaceful life without being a power user of these features, but when they come into play, it is nice to understand their basics and exploit them.

In the next chapter, we'll tackle multitenancy.

17 Multitenancy in jOOQ

Sometimes, our applications need to operate in a multitenant environment, that is, in an environment that operates on multiple tenants (different databases, different tables, or generally speaking, different instances that are logically isolated, but physically integrated). In this chapter, we will cover some common use cases of integrating jOOQ in a multitenant environment based on the following agenda:

- Connecting to a separate database per role/login via the `RenderMapping` API
- Connecting to a separate database per role/login via a connection switch
- Generating code for two schemas of the same vendor
- Generating code for two schemas of different vendors

Let's get started!

Technical requirements

The code for this chapter can be found on GitHub at `https://github.com/PacktPublishing/jOOQ-Masterclass/tree/master/Chapter17`.

Connecting to a separate database per role/login via the RenderMapping API

Connecting to a separate database per role/login is a classical use case of multitenancy. Commonly, you have a pillar database (let's call it the `development` database) and several other databases with the same schema (let's call them the `stage` database and the `test` database). All three databases belong to the same vendor (here, MySQL) and have the same schema, but they hold data for different roles, accounts, organizations, partners, and so on of the application.

For simplicity, the `development` database has a single table named `product`. This database is used for generating jOOQ artifacts, but we want to execute the queries depending on the current role (currently logged in user) against the `stage` or `test` databases.

The key to such implementation relies on juggling with the jOOQ `RenderMapping` API. jOOQ allows us to specify at runtime an input schema (for instance, `development`) and an output schema (for instance, `stage`), and, in queries, it will render the output schema. The climax of the code relies on these settings, as you can see here (the authentication is specific to the Spring Security API):

```
Authentication auth = SecurityContextHolder
  .getContext().getAuthentication();
String authority = auth.getAuthorities().iterator()
  .next().getAuthority();
String database = authority.substring(5).toLowerCase();

ctx.configuration().derive(new Settings()
    .withRenderMapping(new RenderMapping()
      .withSchemata(
        new MappedSchema().withInput("development")
                          .withOutput(database)))).dsl()
   .insertInto(PRODUCT, PRODUCT.PRODUCT_NAME,
               PRODUCT.QUANTITY_IN_STOCK)
   .values("Product", 100)
   .execute();
```

Depending on the role of the currently authenticated user, jOOQ renders the expected database name (for instance, `` `stage`.`product` `` or `` `test`.`product` ``). Basically, each user has a role (for instance, `ROLE_STAGE` or `ROLE_TEST`; for simplicity, a user has a single role), and we extract the output database name by removing `ROLE_` and lowercase the remaining text; by convention, the extracted text represents the name of the database as well. Of course, you can use the username, organization name, or whatever convention makes sense in your case.

You can test this example in the application named *MT* for MySQL.

The `withInput()` method takes the complete name of the input schema. If you want to match the name of the input schema against a regular expression, then instead of `withInput()`, use `withInputExpression(Pattern.compile("reg_exp"))` (for instance, `("development_(.*)")`).

If you are in a database that supports catalogs (for instance, SQL Server), then simply use `MappedCatalog()` and `withCatalogs()`, as in the following example:

```
String catalog = …;
Settings settings = new Settings()
    .withRenderMapping(new RenderMapping()
    .withCatalogs(new MappedCatalog()
    .withInput("development")
    .withOutput(catalog))
    .withSchemata(…); // optional, if you need schema as well
```

If you don't need a runtime schema and instead need to hardwire mappings at code generation time (jOOQ will always render at runtime, conforming to these settings), then, for Maven, use the following:

```
<database>
  <schemata>
    <schema>
     <inputSchema>…</inputSchema>
     <outputSchema>…</outputSchema>
    </schema>
  </schemata>
</database>
```

For Gradle, use the following:

```
database {
  schemata {
    schema {
     inputSchema = '…'
     outputSchema = '…'
   }
```

```
    }
}
```

Use the following for programmatic:

```
new org.jooq.meta.jaxb.Configuration()
  .withGenerator(new Generator()
    .withDatabase(new Database()
      .withSchemata(
        new SchemaMappingType()
          .withInputSchema("...")
          .withOutputSchema("...")
      )
    )
  )
```

You can see such an example in *MTM* for MySQL. As you'll see, all accounts/roles act against the database that was hardwired at code generation time (the `stage` database).

If you are using a database that supports catalogs (for instance, SQL Server), then simply rely on `<catalogs>`, `<catalog>`, `<inputCatalog>`, and `<outputCatalog>`. For Maven, use the following:

```
<database>
  <catalogs>
    <catalog>
      <inputCatalog>...</inputCatalog>
      <outputCatalog>...</outputCatalog>

      <!-- Optionally, if you need schema mapping -->
      <schemata>
      </schemata>
    </catalog>
  </catalogs>
</database>
```

For Gradle, use the following:

```
database {
  catalogs {
   catalog {
    inputCatalog = '...'
```

```
    outputCatalog = '...'

    // Optionally, if you need schema mapping
    schemata {}
    }
  }
}
```

For programmatic, use the following:

```
new org.jooq.meta.jaxb.Configuration()
  .withGenerator(new Generator()
    .withDatabase(new Database()
      .withCatalogs(
        new CatalogMappingType()
          .withInputCatalog("...")
          .withOutputCatalog("...")

          // Optionally, if you need schema mapping
          .withSchemata()
      )
    )
  )
```

So far, the `development` database has a single table named `product`. This table has the same name in the `stage` and `test` databases but let's assume that we decide to call it `product_dev` in the `development` database, `product_stage` in the `stage` database, and `product_test` in the `test` database. In this case, even if jOOQ renders the database name per role correctly, it doesn't render the table's names correctly. Fortunately, jOOQ allows us to configure this aspect via `withTables()` and `MappedTable()`, as follows:

```
ctx.configuration().derive(new Settings()
    .withRenderMapping(new RenderMapping()
      .withSchemata(
        new MappedSchema().withInput("development")
                          .withOutput(database)
      .withTables(
        new MappedTable().withInput("product_dev")
          .withOutput("product_" + database))))).dsl()
   .insertInto(PRODUCT_DEV, PRODUCT_DEV.PRODUCT_NAME,
               PRODUCT_DEV.QUANTITY_IN_STOCK)
```

```
    .values("Product", 100)
    .execute();
```

You can check out this example in the application named *MTT* for MySQL.

Connecting to a separate database per role/login via a connection switch

Another quick solution for connecting to a separate database per role/login consists of switching to the proper connection at runtime. In order to accomplish this task, we have to suppress the jOOQ default behavior of rendering the schema/catalog name. This way, we don't risk connecting to database `A` but get database `B` rendered in front of our tables, and so on. In other words, we need unqualified names.

jOOQ allows us to turn off rendering the schema/catalog name via the `withRenderSchema(false)` and `withRenderCatalog(false)` settings. The following example connects to the database having the same name as the role of the logged in user and suppresses rendering the schema/catalog names:

```
Authentication auth = SecurityContextHolder
     .getContext().getAuthentication();

if (auth != null && auth.isAuthenticated()) {

   String authority = auth.getAuthorities()
     .iterator().next().getAuthority();
   String database = authority.substring(5).toLowerCase();

   DSL.using(
    "jdbc:mysql://localhost:3306/" + database,
        "root", "root")
      .configuration().derive(new Settings()
         .withRenderCatalog(Boolean.FALSE)
         .withRenderSchema(Boolean.FALSE))
         .dsl()
      .insertInto(PRODUCT, PRODUCT.PRODUCT_NAME,
           PRODUCT.QUANTITY_IN_STOCK)
      .values("Product", 100)
      .execute();
}
```

You can check out this example in the application named *MTC* for MySQL.

Alternatively, we can instruct jOOQ to remove any schema references from the generated code via the `outputSchemaToDefault` flag. For Maven, use the following:

```
<outputSchemaToDefault>true</outputSchemaToDefault>
```

For Gradle, use the following:

```
outputSchemaToDefault = true
```

Since there are no more schema references in the generated code, the generated classes can run on all your schemas:

```
String database = …;

DSL.using(
  "jdbc:mysql://localhost:3306/" + database,
      "root", "root")
  .insertInto(PRODUCT, PRODUCT.PRODUCT_NAME,
              PRODUCT.QUANTITY_IN_STOCK)
  .values("Product", 100)
  .execute();
```

You can test this example in the application named *MTCO* for MySQL.

Generating code for two schemas of the same vendor

Consider two schemas of the same vendor named `db1` and `db2`. In the first schema (`db1`), we have a table named `productline`, and in the second schema (`db2`), we have a table named `product`. Our goal is to generate the jOOQ artifacts (to run the jOOQ Code Generator) for these two schemas of the same vendor (here, MySQL) and to execute queries against one or another, and even join these two tables.

Basically, as long as we don't specify any input schema, jOOQ generates code for all the schemas it can find. But since we want to instruct jOOQ to work only on the `db1` and `db2` schemas, we can do it as follows (here, for Maven):

```
<database>
 <schemata>
   <schema>
```

```
    <inputSchema>db1</inputSchema>
   </schema>
   <schema>
    <inputSchema>db2</inputSchema>
   </schema>
 </schemata>
</database>
```

I am sure you have enough experience now to intuit how to write this for Gradle or programmatic, so I'll skip those examples.

Once we run the jOOQ Code Generator, we are ready to execute queries, as follows:

```
ctx.select().from(DB1.PRODUCTLINE).fetch();
ctx.select().from(DB2.PRODUCT).fetch();
```

Or, here is a join between `PRODUCTLINE` and `PRODUCT`:

```
ctx.select(DB1.PRODUCTLINE.PRODUCT_LINE,
   DB2.PRODUCT.PRODUCT_ID, DB2.PRODUCT.PRODUCT_NAME,
   DB2.PRODUCT.QUANTITY_IN_STOCK)
   .from(DB1.PRODUCTLINE)
   .join(DB2.PRODUCT)
   .on(DB1.PRODUCTLINE.PRODUCT_LINE
     .eq(DB2.PRODUCT.PRODUCT_LINE))
   .fetch();
```

`DB1` and `DB2` were statically imported, as follows:

```
import static jooq.generated.db1.Db1.DB1;
import static jooq.generated.db2.Db2.DB2;
```

The complete example is available in the application named *MTJ* for MySQL.

Generating code for two schemas of different vendors

Consider two schemas of different vendors – for instance, our `classicmodels` schema for MySQL and PostgreSQL. Our goal is to generate the jOOQ artifacts for both schemas and execute queries against one or another.

Considering a Maven-based application, we can accomplish this task by using two `<execution>` entries, for the `flyway-maven-plugin` plugin and the `jooq-codegen-maven` plugin. Here is the skeleton code for `jooq-codegen-maven` (the complete code is available in the bundled code):

```
<plugin>
  <groupId>org.jooq</groupId>
  <artifactId>jooq-codegen-maven</artifactId>
  <executions>
    <execution>
      <id>generate-mysql</id>
      <phase>generate-sources</phase>
      <goals>
        <goal>generate</goal>
      </goals>
      <configuration xmlns="... jooq-codegen-3.16.0.xsd">
        ... <!-- MySQL schema configuration -->
      </configuration>
    </execution>
    <execution>
      <id>generate-postgresql</id>
      <phase>generate-sources</phase>
      <goals>
        <goal>generate</goal>
      </goals>
      <configuration xmlns="...jooq-codegen-3.16.0.xsd">
        ... <!-- PostgreSQL schema configuration -->
      </configuration>
    </execution>
  </executions>
</plugin>
```

Next, jOOQ generates artifacts for both vendors and we can switch between connections and tables, as follows:

```
DSL.using(
 "jdbc:mysql://localhost:3306/classicmodels",
    "root", "root")
   .select().from(mysql.jooq.generated.tables.Product.PRODUCT)
   .fetch();
```

```
DSL.using(
 "jdbc:postgresql://localhost:5432/classicmodels",
        "postgres", "root")
   .select().from(
          postgresql.jooq.generated.tables.Product.PRODUCT)
   .fetch();
```

Or, considering that we have already programmatically configured our `DataSource` objects, we can configure two `DSLContext` as well (the complete code is available in the bundled code):

```
@Bean(name="mysqlDSLContext")
public DSLContext mysqlDSLContext(@Qualifier("configMySql")
         DataSourceProperties properties) {

  return DSL.using(
    properties.getUrl(), properties.getUsername(),
    properties.getPassword());
}

@Bean(name="postgresqlDSLContext")
public DSLContext postgresqlDSLContext(
    @Qualifier("configPostgreSql")
       DataSourceProperties properties) {

  return DSL.using(
    properties.getUrl(), properties.getUsername(),
    properties.getPassword());
}
```

You can also inject these two `DSLContext` and use the one you want:

```
private final DSLContext mysqlCtx;
private final DSLContext postgresqlCtx;

public ClassicModelsRepository(
 @Qualifier("mysqlDSLContext") DSLContext mysqlCtx,
 @Qualifier("postgresqlDSLContext") DSLContext postgresqlCtx){
     this.mysqlCtx = mysqlCtx;
     this.postgresqlCtx = postgresqlCtx;
}
```

```
...
mysqlCtx.select().from(
  mysql.jooq.generated.tables.Product.PRODUCT).fetch();

postgresqlCtx.select().from(
  postgresql.jooq.generated.tables.Product.PRODUCT).fetch();
```

The complete code is named *MT2DB*. If you want to generate the artifacts for only one vendor depending on the active profile, then you'll love the *MP* application.

Summary

Multitenancy is not a regular task but it is good to know that jOOQ is quite versatile and allows us to configure multiple databases/schemas in seconds. Moreover, as you just saw, the jOOQ + Spring Boot combo is a perfect match for accomplishing multitenancy tasks.

In the next chapter, we talk about jOOQ SPI.

Part 5: Fine-tuning jOOQ, Logging, and Testing

In this part, we cover tuning jOOQ via configurations and settings. Moreover, we discuss logging jOOQ output and testing.

By the end of this part, you will know how to fine-tune jOOQ, how to log jOOQ output, and how to write tests.

This part contains the following chapters:

- *Chapter 18, jOOQ SPI (Providers and Listeners)*
- *Chapter 19, Logging and Testing*

18 jOOQ SPI (Providers and Listeners)

jOOQ provides a lot of hooks that allow us to alter its default behavior at different levels. Among these hooks, we have lightweight settings and configurations, and the heavy-duty, extremely stable **Service Provider Interface** (**SPI**) made of generators, providers, listeners, parsers, and so on. So, like any robust and mature technology, jOOQ comes with an impressive SPI dedicated to those corner cases where the core technology cannot help.

In this chapter, we scratch the surface of each of these hooks in order to expose the usage steps and some examples that will help you to understand how to develop your own implementations. Our agenda includes the following:

- jOOQ settings
- jOOQ configuration
- jOOQ providers
- jOOQ listeners
- Altering the jOOQ code generation process

Let's get started!

Technical requirements

The code for this chapter can be found on GitHub at `https://github.com/PacktPublishing/jOOQ-Masterclass/tree/master/Chapter18`.

jOOQ settings

jOOQ comes with a comprehensive list of settings (`org.jooq.conf.Settings`) that attempts to cover the most popular use cases related to rendering the SQL code. These settings are available declaratively (via `jooq-settings.xml` in the classpath) or programmatically via methods such as `setFooSetting()` or `withFooSetting()`, which can be chained in a fluent style. To take effect, `Settings` must be part of `org.jooq.Configuration`, and this can be done in multiple ways, as you can read in the jOOQ manual at `https://www.jooq.org/doc/latest/manual/sql-building/dsl-context/custom-settings/`. But most probably, in a Spring Boot application, you'll prefer one of the following approaches:

Pass global `Settings` to the default `Configuration` via `jooq-settings.xml` in the classpath (the `DSLContext` prepared by Spring Boot will take advantage of these settings):

```
<?xml version="1.0" encoding="UTF-8"?>

<settings>
  <renderCatalog>false</renderCatalog>
  <renderSchema>false</renderSchema>
  <!-- more settings added here -->
</settings>
```

Pass global `Settings` to the default `Configuration` via an `@Bean` (the `DSLContext` prepared by Spring Boot will take advantage of these settings):

```
@org.springframework.context.annotation.Configuration
public class JooqConfig {

  @Bean
  public Settings jooqSettings() {
    return new Settings()
      .withRenderSchema(Boolean.FALSE) // this is a setting
      ... // more settings added here
  }
  ...
}
```

At some point, set a new global `Settings` that will be applied from this point onward (this is a global `Settings` because we use `Configuration#set()`):

```
ctx.configuration().set(new Settings()
   .withMaxRows(5)
   ... // more settings added here
   ).dsl()
   . // some query
```

Append new global settings to the current global `Settings`:

```
ctx.configuration().settings()
   .withRenderKeywordCase(RenderKeywordCase.UPPER);
ctx. // some query
```

You can practice these examples in *GlobalSettings* for MySQL.

At some point, set a new local `Settings` that will be applied only to the current query (this is a local `Settings` because we use `Configuration#derive()`):

```
ctx.configuration().derive(new Settings()
   .withMaxRows(5)
   ... // more settings added here
   ).dsl()
   . // some query
```

Or, setting a global/local setting and appends to it more local settings:

```
ctx.configuration().settings()
   .withRenderMapping(new RenderMapping()
      .withSchemata(
         new MappedSchema()
            .withInput("classicmodels")
            .withOutput("classicmodels_test")));

// 'derivedCtx' inherits settings of 'ctx'
DSLContext derivedCtx = ctx.configuration().derive(
    ctx.settings() // using here new Settings() will NOT
                   // inherit 'ctx' settings
       .withRenderKeywordCase(RenderKeywordCase.UPPER)).dsl();
```

You can practice this example in *LocalSettings* for MySQL. It is highly recommended to reserve some time and at least to briefly scroll the entire list of jOOQ-supported settings at `https://www.jooq.org/javadoc/latest/org.jooq/org/jooq/conf/Settings.html`. Next, let's talk about jOOQ `Configuration`.

jOOQ Configuration

`org.jooq.Configuration` represents the spine of `DSLContext`. `DSLContext` needs the precious information provided by `Configuration` for query rendering and execution. While `Configuration` takes advantage of `Settings` (as you just saw), it also has a lot more other configurations that can be specified as in the examples from this section.

By default, Spring Boot gives us a `DSLContext` built on the default `Configuration` (the `Configuration` accessible via `ctx.configuration()`), and as you know, while providing custom settings and configurations, we can alter this `Configuration` globally via `set()` or locally by creating a derived one via `derive()`.

But, in some scenarios, for instance, when you build custom providers or listeners, you'll prefer to build the `Configuration` to be aware of your artifacts right from the start instead of extracting it from `DSLContext`. In other words, when `DSLContext` is built, it should use the ready-to-go `Configuration`.

Before Spring Boot 2.5.0, this step required a little bit of effort, as you can see here:

```
@org.springframework.context.annotation.Configuration
public class JooqConfig {

  @Bean
  @ConditionalOnMissingBean(org.jooq.Configuration.class)
  public DefaultConfiguration jooqConfiguration(
       JooqProperties properties, DataSource ds,
       ConnectionProvider cp, TransactionProvider tp) {

    final DefaultConfiguration defaultConfig =
      new DefaultConfiguration();

    defaultConfig
     .set(cp)                                  // must have
     .set(properties.determineSqlDialect(ds))  // must have
     .set(tp) // for using SpringTransactionProvider
     .set(new Settings().withRenderKeywordCase(
```

```
        RenderKeywordCase.UPPER)); // optional
      // more configs ...

    return defaultConfig;
}
```

This is a `Configuration` created from scratch (actually from the jOOQ built-in `DefaultConfiguration`) that will be used by Spring Boot to create the `DSLContext`. At a minimum, we need to specify a `ConnectionProvider` and the SQL dialect. Optionally, if we want to use `SpringTransactionProvider` as the default provider for jOOQ transactions, then we need to set it as in this code. After this minimum configuration, you can continue adding your settings, providers, listeners, and so on. You can practice this example in *Before250Config* for MySQL.

Starting with version 2.5.0, Spring Boot facilitates easier customization of jOOQ's `DefaultConfiguration` via a bean that implements a functional interface named `DefaultConfigurationCustomizer`. This acts as a callback and can be used as in the following example:

```
@org.springframework.context.annotation.Configuration
public class JooqConfig
     implements DefaultConfigurationCustomizer {

  @Override
  public void customize(DefaultConfiguration configuration) {

     configuration.set(new Settings()
       .withRenderKeywordCase(RenderKeywordCase.UPPER));
       ... // more configs
    }
}
```

This is more practical because we can add only what we need. You can check out this example in *After250Config* for MySQL. Next, let's talk about jOOQ providers.

jOOQ providers

The jOOQ SPI exposes a suite of providers such as `TransactionProvider`, `RecordMapperProvider`, `ConverterProvider`, and so on. Their overall goal is simple—to provide some feature that is not provided by the jOOQ default providers. For instance, let's check out `TransactionProvider`.

TransactionProvider

For instance, we know that jOOQ transactions are backed in Spring Boot by a transaction provider named `SpringTransactionProvider` (the Spring Boot built-in implementation of jOOQ's `TransactionProvider`) that exposes by default a read-write transaction with no name (`null`), having the propagation set to `PROPAGATION_NESTED` and the isolation level to the default isolation level of the underlying database, `ISOLATION_DEFAULT`.

Now, let's assume that we implement a module of our application that serves only reports via jOOQ transactions (so we don't use `@Transactional`). In such a module, we don't want to allow writing, we want to run each query in a separate/new transaction with a timeout of 1 second, and we want to avoid the **dirty reads** phenomenon (a transaction reads the uncommitted modifications of another concurrent transaction that rolls back in the end). In other words, we need to provide a read-only transaction having the propagation set to `PROPAGATION_REQUIRES_NEW`, the isolation level set to `ISOLATION_READ_COMMITTED`, and the timeout set to 1 second.

To obtain such a transaction, we can implement a `TransactionProvider` and override the `begin()` method as in the following code:

```
public class MyTransactionProvider
        implements TransactionProvider {

  private final PlatformTransactionManager transactionManager;

  public MyTransactionProvider(
       PlatformTransactionManager transactionManager) {
   this.transactionManager = transactionManager;
 }

 @Override
 public void begin(TransactionContext context) {

  DefaultTransactionDefinition definition =
   new DefaultTransactionDefinition(
    TransactionDefinition.PROPAGATION_REQUIRES_NEW);
  definition.setIsolationLevel(
   TransactionDefinition.ISOLATION_READ_COMMITTED);
  definition.setName("TRANSACTION_" + Math.round(1000));
```

```
    definition.setReadOnly(true);
    definition.setTimeout(1);

    TransactionStatus status =
     this.transactionManager.getTransaction(definition);
    context.transaction(new SpringTransaction(status));
  }
  ...
}
```

Once we have the transaction provider, we have to configure it in jOOQ. Assuming that we are using Spring Boot 2.5.0+, and based on the previous section, this can be done as follows:

```
@org.springframework.context.annotation.Configuration
public class JooqConfig
       implements DefaultConfigurationCustomizer {

  private final PlatformTransactionManager txManager;

  public JooqConfig(PlatformTransactionManager txManager) {
   this.txManager = txManager;
  }

  @Override
  public void customize(DefaultConfiguration configuration) {

   configuration.set(newMyTransactionProvider(txManager));
  }
}
```

You can practice this example in *A250MyTransactionProvider* for MySQL. When you run the application, you'll notice at the console that the created transaction has these coordinates: *Creating new transaction with name [TRANSACTION_1000]: PROPAGATION_REQUIRES_NEW, ISOLATION_READ_COMMITTED, timeout_1, readOnly.*

If you are using a Spring Boot version prior to 2.5.0, then check out the application named *B250MyTransactionProvider* for MySQL.

And, of course, you can configure the provider via `DSLContext` as well:

```
ctx.configuration().set(
   new MyTransactionProvider(txManager)).dsl() ...;
```

Or, you can use the following:

```
ctx.configuration().derive(
  new MyTransactionProvider(txManager)).dsl() ...;
```

Now, let's consider another scenario solved via `ConverterProvider`.

ConverterProvider

We have to project some JSON functions and map them hierarchically. We already know that this is no issue in a Spring Boot + jOOQ combo since jOOQ can fetch the JSON and can call Jackson (the default in Spring Boot) to map it accordingly. But, we don't want to use Jackson; we want to use Flexjson (`http://flexjson.sourceforge.net/`). jOOQ is not aware of this library (jOOQ can detect only the presence of Jackson and Gson), so we need to provide a converter such as `org.jooq.ConverterProvider` that uses Flexjson to accomplish this task. Take your time to check the source in *{A,B}250ConverterProvider* for MySQL. Finally, let's focus on this scenario solved via `RecordMapperProvider`.

RecordMapperProvider

We have a ton of legacy POJOs implemented via the `Builder` pattern and we decide to write a bunch of jOOQ `RecordMapper` for mapping queries to these POJOs. In order to streamline the process of using these `RecordMapper`, we also decide to write a `RecordMapperProvider`. Basically, this will be responsible for using the proper `RecordMapper` without our explicit intervention. Are you curious about how to do it? Then check out the *{A,B}250RecordMapperProvider* and *RecordMapperProvider* applications for MySQL. Mainly, these applications are the same, but they use different approaches to configure `RecordMapperProvider`.

With `ConverterProvider` and `RecordMapperProvider`, I think it's important to mention that these *replace* out-of-the-box behavior, they don't *enhance* it. So, custom providers have to make sure to fall back to the default implementations if they can't handle a conversion/mapping.

jOOQ listeners

jOOQ comes with a significant number of listeners that are quite versatile and useful in hooking us into jOOQ life cycle management for solving a wide range of tasks. Let's "arbitrarily" pick up the mighty `ExecuteListener`.

ExecuteListener

For instance, one of the listeners that you'll love is `org.jooq.ExecuteListener`. This listener comes with a bunch of methods that can hook in the life cycle of a `Query`, `Routine`, or `ResultSet` to alter the default rendering, preparing, binding, executing, and fetching stage. The most convenient approach to implement your own listener is to extend the jOOQ default implementation, `DefaultExecuteListener`. This way, you can override only the methods that you want and you keep up with the SPI evolution (however, by the time you read this book, it is possible that this default listener will have been removed, and all methods are now default methods on the interface). Consider applying this technique to any other jOOQ listener, since jOOQ provides a default implementation for all (mainly, for `FooListener`, there is a `DefaultFooListener`).

For now, let's write an `ExecuteListener` that alters the rendered SQL that is about to be executed. Basically, all we want is to alter every MySQL `SELECT` by adding the `/*+ MAX_EXECUTION_TIME(n) */` hint, which allows us to specify a query timeout in milliseconds. The jOOQ DSL allows for adding MySQL/Oracle-style hints. :) Use `ctx.select(...).hint("/*+ ... */").from(...)`. But only `ExecuteListener` can patch multiple queries without modifying the queries themselves. So, `ExecuteListener` exposes callbacks such as `renderStart(ExecuteContext)` and `renderEnd(ExecuteContext)`, which are called before rendering SQL from `QueryPart` and after rendering SQL from `QueryPart`, respectively. Once we are in control, we can rely on `ExecuteContext`, which gives us access to the underlying connection (`ExecuteContext.connection()`), query (`ExecuteContext.query()`), rendered SQL (`ExecuteContext.sql()`), and so on. In this specific case, we are interested in accessing the rendered SQL and modifying it, so we override `renderEnd(ExecuteContext)` and call `ExecuteContext.sql()` as follows:

```
public class MyExecuteListener extends
    DefaultExecuteListener{

  private static final Logger logger =
    Logger.getLogger(MyExecuteListener.class.getName());
```

```
  @Override
  public void renderEnd(ExecuteContext ecx) {

    if (ecx.configuration().data()
        .containsKey("timeout_hint_select") &&
                ecx.query() instanceof Select) {

      String sql = ecx.sql();

      if (sql != null) {
        ecx.sql(sql.replace(
         "select",
         "select " + ecx.configuration().data()
            .get("timeout_hint_select")
      ));

      logger.info(() -> {
        return "Executing modified query : " + ecx.sql();
      });
    }
   }
  }
}
```

The code from inside the decisional block is quite simple: we just capture the rendered SQL (the SQL that is about to be executed shortly) and modify it accordingly by adding the MySQL hint. But, what is `...data().containsKey("timeout_hint_select")`? Mainly, `Configuration` comes with three methods that work together to pass custom data through `Configuration`. These methods are `data(Object key, Object value)`, which allows us to set some custom data; `data(Object key)`, which allows us to get some custom data based on a key; and `data()`, which returns the entire `Map` of custom data. So, in our code, we check whether the custom data of the current `Configuration` contains a key named `timeout_hint_select` (this is a name we have chosen). If such a key exists, it means that we want to add the MySQL hint (which was set as the value corresponding to this key) to the current `SELECT`, otherwise, we take no action. This piece of custom information was set as follows:

```
Configuration derived = ctx.configuration().derive();
derived.data("timeout_hint_select",
             "/*+ MAX_EXECUTION_TIME(5) */");
```

Once this custom data is set, we can execute a `SELECT` that will be enriched with the MySQL hint by our custom `ExecuteListener`:

```
derived.dsl().select(...).fetch();
```

You can practice this example in *A250ExecuteListener* for MySQL. If you are using a Spring Boot version prior to 2.5.0, then go for *B250ExecuteListener* for MySQL. There is also an application named *ExecuteListener* for MySQL that does the same thing but it "inlines" `ExecuteListener` via `CallbackExecuteListener` (this represents `ExecuteListener` – useful if you prefer functional composition):

```
ctx.configuration().derive(new CallbackExecuteListener()
                       .onRenderEnd(ecx -> {
   ...}))
   .dsl()
   .select(...).fetch();
```

Most listeners have a functional composition approach as well that can be used as in the previous snippet of code. Next, let's talk about a listener named `ParseListener`.

jOOQ SQL parser and ParseListener

`ParseListener` (SQL Parser Listener) was introduced in jOOQ 3.15, but before discussing it, we should discuss the SQL `Parser` (`org.jooq.Parser`).

SQL Parser

jOOQ comes with a powerful and mature `Parser` API that is capable of parsing an arbitrary SQL string (or a fragment of it) into different jOOQ API elements. For instance, we have `Parser.parseQuery(String sql)`, which returns the `org.jooq.Query` type containing a single query that corresponds to the passed `sql`.

One of the main functionalities of the `Parser` API is that it can act as a translator between two dialects. In other words, we have SQL in dialect *X*, and we can programmatically pass it through the SQL `Parser` to obtain the SQL translated/emulated for dialect *Y*. For instance, consider a Spring Data JPA application that contains a significant number of native queries written for the MySQL dialect like this one:

```
@Query(value = "SELECT c.customer_name as customerName, "
  + "d.address_line_first as addressLineFirst,
     d.address_line_second as addressLineSecond "
  + "FROM customer c JOIN customerdetail d "
  + "ON c.customer_number = d.customer_number "
```

```
  + "WHERE (NOT d.address_line_first <=>
    d.address_line_second)", nativeQuery=true)
    List<SimpleCustomer> fetchCustomerNotSameAddress();
```

The idea is that management took the decision to switch to PostgreSQL, so you should migrate all these queries to the PostgreSQL dialect and you should do it with insignificant downtime. Even if you are familiar with the differences between these two dialects and you don't have a problem expressing both, you are still under time pressure. This is a scenario where jOOQ can save you because all you have to do is to pass to the jOOQ `Parser` your native queries and jOOQ will translate/emulate them for PostgreSQL. Assuming that you are using Spring Data JPA backed by Hibernate, then all you need to do is to add a Hibernate interceptor that exposes the SQL string that is about to execute:

```
@Configuration
public class SqlInspector implements StatementInspector {

  @Override
  public String inspect(String sql) {

    Query query = DSL.using(SQLDialect.POSTGRES)
      .parser()
      .parseQuery(sql);

    if (query != null) {
        return query.getSQL();
    }

    return null; // interpreted as the default SQL string
  }
}
```

Done in 5 minutes! How cool is that?! Obviously, your colleagues will ask you what sorcery this was, so you have a good opportunity to introduce them to jOOQ. :)

If you check out the console output, you'll see that Hibernate reports the following SQL string to be executed against the PostgreSQL database:

```
SELECT c.customer_name AS customername,
       d.address_line_first AS addresslinefirst,
       d.address_line_second AS addresslinesecond
FROM customer AS c
JOIN customerdetail AS d
```

```
  ON c.customer_number = d.customer_number
WHERE NOT (d.address_line_first IS NOT DISTINCT
           FROM d.address_line_second)
```

Of course, you can change the dialect and obtain the SQL for any of the jOOQ-supported dialects. Now, you have time to copy the jOOQ output and replace your native queries accordingly since the application continues to run as usual. At the end, simply decouple this interceptor. You can practice this application in *JPAParser*.

Besides `parseQuery()`, we have `parseName(String sql)`, which parses the given `sql` into `org.jooq.Name`; `parseField(String sql)`, which parses the given `sql` into `org.jooq.Field`; `parseCondition(String sql)`, which parses the given `sql` into `org.jooq.Condition`; and so on. Please check out the jOOQ documentation to see all the methods and their flavors.

But jOOQ can do even more via the so-called *parsing connection* feature (available for R2DBC as well). Basically, this means that the SQL string is passed through the jOOQ `Parser` and the output SQL can become the source of a `java.sql.PreparedStatement` or `java.sql.Statement`, which can be executed via these JDBC APIs (`executeQuery(String sql)`). This happens as long as the SQL string comes through a JDBC connection (`java.sql.Connection`) that is obtained as in this example:

There's no way from syntax alone to decide which input semantics it could be:

```
try (Connection c = DSL.using(url, user, pass)
      .configuration()
      .set(new Settings()
          .withParseDialect(SQLDialect.MYSQL))
      .dsl()
      .parsingConnection();   // this does the trick

      PreparedStatement ps = c.prepareStatement(sql);
    ) {
     ...
}
```

The `sql` passed to the `PreparedStatement` represents any SQL string. For instance, it can be produced by `JdbcTemplate`, the Criteria API, `EntityManager`, and so on. Gathering the SQL string from the Criteria API and `EntityManager` can be a little bit tricky (since it requires a Hibernate `AbstractProducedQuery` action) but you can find the complete solution in *JPAParsingConnection* for MySQL.

Besides the `Parser` API, jOOQ also exposes a translator between dialects via the `Parser` CLI (`https://www.jooq.org/doc/latest/manual/sql-building/sql-parser/sql-parser-cli/`) and via this website: . Now, we can talk about `ParseListener`.

SQL Parser Listener

It is quite easy to intuit that the SQL Parser Listener (`org.jooq.ParseListener` introduced in jOOQ 3.15) is responsible for providing hooks that allow altering the default behavior of the jOOQ parser.

For instance, let's consider the following `SELECT`, which uses the SQL `CONCAT_WS(separator, str1, str2, ...)` function:

```
SELECT concat_ws('|', city, address_line_first,
  address_line_second, country, territory) AS address
FROM office
```

This variadic function that ignores `NULL` values and uses a string separator/delimiter to separate all arguments concatenated in the resulting string is natively supported by MySQL, PostgreSQL, and SQL Server but is not supported by Oracle. Moreover, jOOQ (at least until version 3.16.4) doesn't support it either. One way to use it in our queries is via plain SQL as follows:

```
ctx.resultQuery("SELECT concat_ws('|', city,
  address_line_first, address_line_second, country, territory)
AS address FROM office").fetch();
```

But, if we try to execute this query against Oracle, it will not work since Oracle doesn't support it and jOOQ doesn't emulate it in Oracle syntax. A solution consists of implementing our own `ParseListener` that can emulate the `CONCAT_WS()` effect. For instance, the following `ParseListener` accomplishes this via the `NVL2()` function (please read all comments in the code in order to get you familiar with this API):

```
public class MyParseListener extends DefaultParseListener {

 @Override
 public Field parseField(ParseContext pcx) {

  if (pcx.parseFunctionNameIf("CONCAT_WS")) {

   pcx.parse('(');
```

```
    String separator = pcx.parseStringLiteral();
    pcx.parse(',');

    // extract the variadic list of fields
    List<Field<?>> fields = pcx.parseList(",",
        c -> c.parseField());

    pcx.parse(')'); // the function CONCAT_WS() was parsed
    ...
```

After parsing, we prepare the Oracle emulation:

```
    ...
    // prepare the Oracle emulation
    return CustomField.of("", SQLDataType.VARCHAR, f -> {
     switch (f.family()) {
      case ORACLE -> {
       Field result = inline("");
       for (Field<?> field : fields) {
        result = result.concat(DSL.nvl2(field,
                   inline(separator).concat(
                     field.coerce(String.class)), field));
       }

       f.visit(result); // visit this QueryPart
      }

      // case other dialect ...
      }
    });
   }

  // pass control to jOOQ
  return null;
 }
}
```

To keep the code simple and short, we have considered some assumptions. Mainly, the separator and string literals should be enclosed in single quotes, the separator itself is a single character, and it should be at least one argument after the separator.

This time, when we do this:

```
String sql = ctx.configuration().derive(SQLDialect.ORACLE)
  .dsl()
  .render(ctx.parser().parseQuery("""
   SELECT concat_ws('|', city, address_line_first,
     address_line_second, country, territory) AS address
   FROM office"""));

ctx.resultQuery(sql).fetch();
```

Our parser (followed by the jOOQ parser) produces this SQL compatible with Oracle syntax:

```
SELECT ((((('' || nvl2(CITY, ('|' || CITY), CITY)) ||
  nvl2(ADDRESS_LINE_FIRST, ('|' || ADDRESS_LINE_FIRST),
       ADDRESS_LINE_FIRST)) ||
  nvl2(ADDRESS_LINE_SECOND, ('|' || ADDRESS_LINE_SECOND),
       ADDRESS_LINE_SECOND)) ||
  nvl2(COUNTRY, ('|' || COUNTRY), COUNTRY)) ||
  nvl2(TERRITORY, ('|' || TERRITORY), TERRITORY)) ADDRESS
FROM OFFICE
```

You can practice this example in *A250ParseListener* for Oracle (for Spring Boot 2.5.0+), and in *B250ParseListener* for Oracle (for Spring Boot prior 2.5.0). Besides parsing fields (`Field`), `ParseListener` can also parse tables (`org.jooq.Table` via `parseTable()`) and conditions (`org.jooq.Condition` via `parseCondition()`).

If you prefer functional composition, then check out `CallbackParseListener`. Next, let's quickly cover other jOOQ listeners.

RecordListener

Via the jOOQ `RecordListener` implementations, we can add custom behavior during `UpdatableRecord` events such as insert, update, delete, store, and refresh (if you are not familiar with `UpdatableRecord`, then consider *Chapter 3, jOOQ Core Concepts*).

For each *event* listen by `RecordListener` we have an `eventStart()` and `eventEnd()` method. `eventStart()` is a callback invoked before the *event* takes place, while the `eventEnd()` callback is invoked after the *event* has happened.

For instance, let's consider that every time an `EmployeeRecord` is inserted, we have an algorithm that generates the primary key, `EMPLOYEE_NUMBER`. Next, the `EXTENSION` field is always of type *xEmployee_number* (for instance, if `EMPLOYEE_NUMBER` is *9887* then `EXTENSION` is *x9887*). Since we don't want to let people do this task manually, we can easily automate this process via `RecordListener` as follows:

```
public class MyRecordListener extends DefaultRecordListener {

 @Override
 public void insertStart(RecordContext rcx) {

  if (rcx.record() instanceof EmployeeRecord employee) {

   // call the secret algorithm that produces the PK
   long secretNumber = (long) (10000 * Math.random());

   employee.setEmployeeNumber(secretNumber);
   employee.setExtension("x" + secretNumber);
  }
 }
}
```

Probably worth mentioning, `RecordListener` doesn't apply to ordinary DML statements (let alone plain SQL templates). People often think they can add some security stuff in there, which is then bypassed. It really only works on `TableRecord`/`UpdatableRecord` types. Starting from jOOQ 3.16, a lot of tasks that are currently solved with `RecordListener` are probably better solved with `VisitListener`, which will become *much* more powerful once the new query object model is in place (`https://blog.jooq.org/traversing-jooq-expression-trees-with-the-new-traverser-api/`). In jOOQ 3.16, it won't be ready for this task yet, but it might be in jOOQ 3.17.

You can practice this application in *{A,B}250RecordListener1* for MySQL. Moreover, you can find the *{A,B}250RecordListener2* application for MySQL, which extends this one by overriding `insertEnd()` to automatically insert a row in `EMPLOYEE_STATUS` based on the inserted `EmployeeRecord`:

```
@Override
public void insertEnd(RecordContext rcx) {
```

```
  if (rcx.record() instanceof EmployeeRecord employee) {

   EmployeeStatusRecord status =
      rcx.dsl().newRecord(EMPLOYEE_STATUS);

   status.setEmployeeNumber(employee.getEmployeeNumber());
   status.setStatus("REGULAR");
   status.setAcquiredDate(LocalDate.now());

   status.insert();
  }
}
```

If you prefer functional composition, then check out `CallbackRecordListener`.

DiagnosticsListener

`DiagnosticsListener` is available from jOOQ 3.11 and it fits perfectly in scenarios where you want to detect inefficiencies in your database interaction. This listener can act at different levels, such as jOOQ, JDBC, and SQL levels.

Mainly, this listener exposes a suite of callbacks (one callback per problem it detects). For instance, we have `repeatedStatements()` for detecting N+1 problems, `tooManyColumnsFetched()` for detecting whether `ResultSet` fetches more columns than necessary, `tooManyRowsFetched()` for detecting whether `ResultSet` fetches more rows than necessary, and so on (you can find all of them in the documentation).

Let's assume a Spring Data JPA application that runs the following classical N+1 scenario (the `Productline` and `Product` entities are involved in a lazy bidirectional `@OneToMany` relationship):

```
@Transactional(readOnly = true)
public void fetchProductlinesAndProducts() {

  List<Productline> productlines
   = productlineRepository.findAll();

  for (Productline : productlines) {

   List<Product> products = productline.getProducts();
```

```
    System.out.println("Productline: "
     + productline.getProductLine()
     + " Products: " + products);
  }
}
```

So, there is a `SELECT` triggered for fetching the product lines, and for each product line, there is a `SELECT` for fetching its products. Obviously, in performance terms, this is not efficient, and jOOQ can signal this via a custom `DiagnosticsListener` as shown next:

```
public class MyDiagnosticsListener
         extends DefaultDiagnosticsListener {

 private static final Logger = ...;

 @Override
 public void repeatedStatements(DiagnosticsContext dcx) {

  log.warning(() ->
   "These queries are prone to be a N+1 case: \n"
     + dcx.repeatedStatements());
 }
}
```

Now, the previous N+1 case will be logged, so you have been warned!

jOOQ can diagnose over a `java.sql.Connection` (`diagnosticsConnection()`) or a `javax.sql.DataSource` (`diagnosticsDataSource()` wraps a `java.sql.Connection` in a `DataSource`). Exactly as in the case of a *parsing connection*, this JDBC connection proxies the underlying connection, therefore you have to pass your SQL through this proxy. In a Spring Data JPA application, you can quickly improvise a diagnose profile that relies on a `SingleConnectionDataSource`, as you can see in *JPADiagnosticsListener* for MySQL. The same case is available in *SDJDBCDiagnosticsListener* for MySQL, which wraps a Spring Data JDBC application. Also, the jOOQ manual has some cool JDBC examples that you should check (`https://www.jooq.org/doc/latest/manual/sql-execution/diagnostics/`).

TransactionListener

As its name suggests, `TransactionListener` provides hooks for interfering with transaction events such as begin, commit, and rollback. For each such *event*, there is an `eventBegin()`, called before the *event*, and an `eventEnd()`, called after the *event*. Moreover, for functional composition purposes, there is `CallbackTransactionListener`.

Let's consider a scenario that requires us to back up the data after each update of `EmployeeRecord`. By "back up," we understand that we need to save an `INSERT` containing the data before this update in the file corresponding to the employee to be updated.

`TransactionListener` doesn't expose information about the underlying SQL, therefore we cannot determine whether `EmployeeRecord` is updated or not from inside of this listener. But, we can do it from `RecordListener` and the `updateStart()` callback. When an `UPDATE` occurs, `updateStart()` is called and we can inspect the record type. If it is an `EmployeeRecord`, we can store its original (`original()`) state via `data()` as follows:

```
@Override
public void updateStart(RecordContext rcx) {

  if (rcx.record() instanceof EmployeeRecord) {
   EmployeeRecord employee =
    (EmployeeRecord) rcx.record().original();

  rcx.configuration().data("employee", employee);
  }
}
```

Now, you may think that, at the update end (`updateEnd()`), we can write the `EmployeeRecord` original state in the proper file. But a transaction can be rolled back, and in such a case, we should roll back the entry from the file as well. Obviously, this is cumbersome. It will be much easier to alter the file only after the transaction commits, so when we are sure that the update succeeded. Here is where `TransactionListener` and `commitEnd()` become useful:

```
public class MyTransactionListener
      extends DefaultTransactionListener {

 @Override
```

```
 public void commitEnd(TransactionContext tcx) {

  EmployeeRecord employee =
    (EmployeeRecord) tcx.configuration().data("employee");

  if (employee != null) {
    // write to file corresponding to this employee
  }
 }
}
```

Cool, right!? You just saw how to combine two listeners to accomplish a common task. Check the source in *{A,B}250RecordTransactionListener* for MySQL.

VisitListener

The last listener that we'll briefly cover is probably the most complex one, `VisitListener`. Mainly, `VisitListener` is a listener that allows us to manipulate the jOOQ **Abstract Syntax Tree** (**AST**), which contains query parts (`QueryPart`) and clauses (`Clause`). So, we can visit `QueryPart` (via `visitStart()` and `visitEnd()`) and `Clause` (via `clauseStart()` and `clauseEnd()`).

A very simple example could be like this: we want to create some views via the jOOQ DSL (`ctx.createOrReplaceView("product_view").as(...).execute()`) and we also want to add them to the `WITH CHECK OPTION` clause. Since the jOOQ DSL doesn't support this clause, we can do it via `VisitListener` as follows:

```
public class MyVisitListener extends DefaultVisitListener {

 @Override
 public void clauseEnd(VisitContext vcx) {

  if (vcx.clause().equals(CREATE_VIEW_AS)) {

    vcx.context().formatSeparator()
       .sql("WITH CHECK OPTION");
  }
 }
}
```

While you can practice this trivial example in *{A,B}250VisitListener* for MySQL, I strongly recommend you read these two awesome articles from the jOOQ blog as well: `https://blog.jooq.org/implementing-client-side-row-level-security-with-jooq/` and `https://blog.jooq.org/jooq-internals-pushing-up-sql-fragments/`. You'll have the chance to learn a lot about the `VisitListener` API. You never know when you'll need it! For instance, you may want to implement your *soft deletes* mechanism, add a condition for each query, and so on. In such scenarios, `VisitListener` is exactly what are you looking for! Moreover, when this book was written, jOOQ started to add a new player, called **Query Object Model** (**QOM**), as a public API. This API facilitates an easy, intuitive, and powerful traversal of the jOOQ AST. You don't want to miss this article: `https://blog.jooq.org/traversing-jooq-expression-trees-with-the-new-traverser-api/`.

Next, let's talk about altering the jOOQ code generation process.

Altering the jOOQ code generation process

We already know that jOOQ comes with three Code Generators (for Java, Scala, and Kotlin). For Java, we use `org.jooq.codegen.JavaGenerator`, which can be shaped/customized declaratively (or, programmatically) via a comprehensive set of configurations grouped under `<configuration>` (Maven), `configurations` (Gradle), or `org.jooq.meta.jaxb.Configuration`. But, sometimes, we need more control, or in other words, we need a custom generator implementation.

Implementing a custom generator

Imagine a scenario where we need a query method and it would be very handy if it was provided by the built-in jOOQ DAO. Obviously, the jOOQ goal is to maintain a thin DAO layer that avoids a large number of methods caused by different types of query combinations (don't expect to see in the default DAO a query method such as `fetchByField1AndField2()` since trying to cover all combinations of fields (even for only two fields) leads to a heavy DAO layer that most probably will not be fully exploited).

But, we can enrich the generated DAOs via a custom generator. An important aspect is the fact that a custom generator requires a separate project (or module) that will work as a dependency for the project that is going to use it. This is needed because the generator must run at compilation time, so the way to achieve this is by adding it as a dependency. Since we use a multi-module Spring Boot application, we can easily achieve this by adding the custom generator as a separate module of the project. This is very handy since most Spring Boot production apps are developed in multi-module style.

Speaking about the effective implementation of a custom generator, we have to extend the Java generator, `org.jooq.codegen.JavaGenerator`, and override the default-empty method, `generateDaoClassFooter(TableDefinition table, JavaWriter out)`. The stub code is listed next:

```
public class CustomJavaGenerator extends JavaGenerator {

   @Override
   protected void generateDaoClassFooter(
         TableDefinition table, JavaWriter out) {
      ...
   }
}
```

Based on this stub code, let's generate additional DAO query methods.

Adding a query method to all DAOs

Let's assume that we want to add a query method to all the generated DAOs, for instance, a method that limits the number of fetched POJOs (records), such as `List<POJO> findLimitedTo(Integer value)`, where `value` represents the number of POJOs to fetch in the `List`. Check out the code:

```
@Override
protected void generateDaoClassFooter(
             TableDefinition table, JavaWriter out) {

   final String pType =
    getStrategy().getFullJavaClassName(table, Mode.POJO);

   // add a method common to all DAOs
   out.tab(1).javadoc("Fetch the number of records
                 limited by <code>value</code>");
   out.tab(1).println("public %s<%s> findLimitedTo(
         %s value) {", List.class, pType, Integer.class);
   out.tab(2).println("return ctx().selectFrom(%s)",
               getStrategy().getFullJavaIdentifier(table));
   out.tab(3).println(".limit(value)");
   out.tab(3).println(".fetch(mapper());");
   out.tab(1).println("}");
}
```

Let's quickly see what is happening here:

- In line 5, we ask jOOQ to give the name of the generated POJO that corresponds to the current `table` and that is used in our query method to return a `List<POJO>`. For instance, for the `ORDER` table, `getFullJavaClassName()` returns `jooq.generated.tables.pojos.Order`.
- In line 9, we generate some Javadoc.
- In lines 11-17, we generate the method signature and its body. The `getFullJavaIdentifier()` used at line 14 gives us the fully qualified name of the current table (for example, `jooq.generated.tables.Order.ORDER`).
- The `ctx()` method used on line 13 and `mapper()` used in line 16 are defined in the `org.jooq.impl.DAOImpl` class. Each generated DAO extends `DAOImpl`, and therefore has access to these methods.

Based on this code, the jOOQ generator adds at the end of each generated DAO a method as follows (this method is added in `OrderRepository`):

```
/**
 * Fetch the number of records limited by <code>value</code>
 */
public List<jooq.generated.tables.pojos.Order>
                    findLimitedTo(Integer value) {

   return ctx().selectFrom(jooq.generated.tables.Order.ORDER)
               .limit(value)
               .fetch(mapper());
}
```

How about adding methods only in certain DAOs?

Adding a query method in certain DAOs

Let's add a query method named `findOrderByStatusAndOrderDate()` only in the `OrderRepository` DAO. A simple and quick solution consists of checking the table name via the `TableDefinition` argument of the `generateDaoClassFooter()` method. For instance, the following code adds the `findOrderByStatusAndOrderDate()` method only in the DAO that corresponds to the `ORDER` table:

```
@Override
protected void generateDaoClassFooter(
```

```
            TableDefinition table, JavaWriter out) {

   final String pType
      = getStrategy().getFullJavaClassName(table, Mode.POJO);

   // add a method specific to Order DAO
   if (table.getName().equals("order")) {
      out.println("public %s<%s>
         findOrderByStatusAndOrderDate(
            %s statusVal, %s orderDateVal) {",
               List.class, pType,
                    String.class, LocalDate.class);
      ...
   }
}
```

This code generates `findOrderByStatusAndOrderDate()` only in `jooq.generated.tables.daos.OrderRepository`:

```
/**
 * Fetch orders having status <code>statusVal</code>
 *  and order date after <code>orderDateVal</code>
 */
public List<jooq.generated.tables.pojos.Order>
      findOrderByStatusAndOrderDate(String statusVal,
         LocalDate orderDateVal) {

   return ctx().selectFrom(jooq.generated.tables.Order.ORDER)
      .where(jooq.generated.tables.Order.ORDER.STATUS
         .eq(statusVal))
         .and(jooq.generated.tables.Order.ORDER.ORDER_DATE
            .ge(orderDateVal))
      .fetch(mapper());
}
```

Besides `table.getName()`, you can enforce the previous condition for more control via `table.getCatalog()`, `table.getQualifiedName()`,`table.getSchema()`, and so on.

The complete example is available in *AddDAOMethods* for MySQL and Oracle.

As a bonus, if you need to enrich the jOOQ-generated DAOs with the corresponding interfaces, then you need a custom generator as in the application named *InterfacesDao* for MySQL and Oracle. If you check out this code, you'll see a so-called *custom generator strategy*. Next, let's detail this aspect.

Writing a custom generator strategy

You already know how to use `<strategy>` (Maven), `strategy {}` (Gradle), or `withStrategy()` (programmatic) to inject custom behavior for naming classes, methods, members, and so on during the jOOQ code generation process. For instance, we have used this technique for renaming our DAO classes in Spring Data JPA style.

But, overriding naming schemes during code generation can be accomplished via a custom generator strategy as well. For instance, this is useful when we want to generate certain method names as in our scenario that starts from the following query:

```
ctx.select(concat(EMPLOYEE.FIRST_NAME, inline(" "),
        EMPLOYEE.LAST_NAME).as("employee"),
        concat(EMPLOYEE.employee().FIRST_NAME, inline(" "),
               EMPLOYEE.employee().LAST_NAME).as("reports_to"))
   .from(EMPLOYEE)
   .where(EMPLOYEE.JOB_TITLE.eq(
          EMPLOYEE.employee().JOB_TITLE))
   .fetch();
```

This is a self-join that relies on the `employee()` navigation method. Conforming to the default generator strategy, writing a self-join is done via a navigation method having the same name as the table itself (for the `EMPLOYEE` table, we have the `employee()` method).

But, if you find `EMPLOYEE.employee()` a little bit confusing, and you prefer something more meaningful, such as `EMPLOYEE.reportsTo()` (or something else), then you need a custom generator strategy. This can be accomplished by extending the jOOQ `DefaultGeneratorStrategy` and overriding the proper methods described in the jOOQ manual: `https://www.jooq.org/doc/latest/manual/code-generation/codegen-generatorstrategy/`.

So, in our case, we need to override `getJavaMethodName()` as follows:

```
public class MyGeneratorStrategy
        extends DefaultGeneratorStrategy {

  @Override
  public String getJavaMethodName(
```

```
        Definition, Mode mode) {

    if (definition.getQualifiedName()
          .equals("classicmodels.employee")
             && mode.equals(Mode.DEFAULT)) {
        return "reportsTo";
    }

    return super.getJavaMethodName(definition, mode);
  }
}
```

Finally, we have to set this custom generator strategy as follows (here, for Maven, but you can easily intuit how to do it for Gradle or programmatically):

```
<generator>
 <strategy>
  <name>
   com.classicmodels.strategy.MyGeneratorStrategy
  </name>
 </strategy>
</generator>
```

Done! Now, after code generation, you can re-write the previous query as follows (notice the `reportsTo()` method instead of `employee()`):

```
ctx.select(concat(EMPLOYEE.FIRST_NAME, inline(" "),
         EMPLOYEE.LAST_NAME).as("employee"),
         concat(EMPLOYEE.reportsTo().FIRST_NAME, inline(" "),
            EMPLOYEE.reportsTo().LAST_NAME).as("reports_tc"))
   .from(EMPLOYEE)
   .where(EMPLOYEE.JOB_TITLE.eq(gma
          EMPLOYEE.reportsTo().JOB_TITLE))
   .fetch();
```

The jOOQ Java default generator strategy follows the *Pascal* naming strategy, which is the most popular in the Java language. But, besides the *Pascal* naming strategy, jOOQ also comes with a `KeepNamesGeneratorStrategy` custom generator strategy that simply holds names in place. Moreover, you may like to study `JPrefixGeneratorStrategy`, respectively the `JVMArgsGeneratorStrategy`. These are just some examples (they are not part of the jOOQ Code Generator) that can be found on GitHub at `https://github.com/jOOQ/jOOQ/tree/main/jOOQ-codegen/src/main/java/org/jooq/codegen/example`.

Summary

In this chapter, we have briefly covered the jOOQ SPI. Obviously, the tasks solved via an SPI are not daily tasks and require overall solid knowledge about the underlying technology. But, since you have read earlier chapters in this book, you should have no problems assimilating the knowledge in this chapter as well. But, of course, using this SPI to solve real problems requires more study of the documentation and more practice.

In the next chapter, we tackle logging and testing jOOQ applications.

19
Logging and Testing

In this chapter, we will cover logging and testing from the jOOQ perspective. Relying on the fact that these are common-sense notions, I won't explain what logging and testing are, nor will I highlight their obvious importance. That being said, let's jump directly into the agenda of this chapter:

- jOOQ logging
- jOOQ testing

Let's get started!

Technical requirements

The code for this chapter can be found on GitHub at `https://github.com/PacktPublishing/jOOQ-Masterclass/tree/master/Chapter19`.

jOOQ logging

By default, you'll see the jOOQ logs at the `DEBUG` level during code generation and during queries/routine execution. For instance, during a regular `SELECT` execution, jOOQ logs the query SQL string (with and without the bind values), the first 5 records from the fetched result set as a nice formatted table, and the size of the result is set as shown in the following figure:

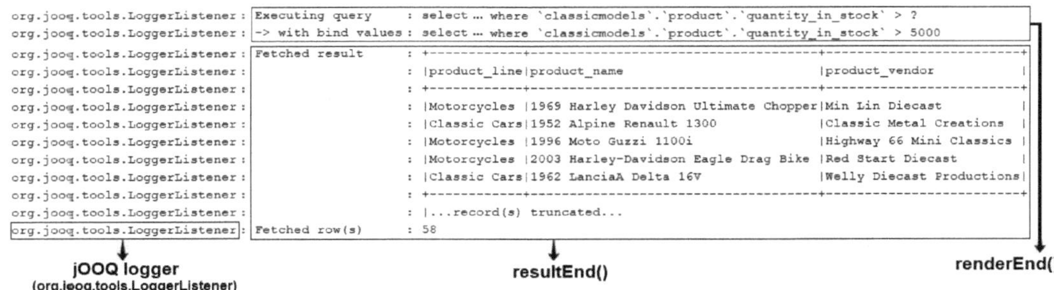

Figure 19.1 – A default jOOQ log for a SELECT execution

This figure reveals a few important aspects of jOOQ logging. First of all, the jOOQ logger is named `org.jooq.tools.LoggerListener` and represents an implementation of the `ExecuteListener` SPI presented in *Chapter 18, jOOQ SPI (Providers and Listeners)*. Under the hood, `LoggerListener` uses an internal abstraction (`org.jooq.tools.JooqLogger`) that attempts to interact with any of the famous loggers, sl4j, log4j, or the Java Logging API (`java.util.logging`). So, if your application uses any of these loggers, jOOQ hooks into it and uses it.

As you can see in this figure, jOOQ logs the query SQL string when the `renderEnd()` callback is invoked, and the fetched result set when the `resultEnd()` callback is invoked. Nevertheless, the jOOQ methods that rely on lazy (sequential) access to the underlying JDBC `ResultSet` (so, methods that uses `Iterator` of the `Cursor`– for instance, `ResultQuery.fetchStream()` and `ResultQuery.collect()`) don't pass through `resultStart()` and `resultEnd()`. In such cases, only the first five records from `ResultSet` are buffered by jOOQ and are available for logging in `fetchEnd()` via `ExecuteContext.data("org.jooq.tools.LoggerListener.BUFFER")`. The rest of the records are either lost or skipped.

If we execute a routine or the query is a DML, then other callbacks are involved as well. Are you curious to find out more?! Then you'll enjoy studying the `LoggerListener` source code by yourself.

jOOQ logging in Spring Boot – default zero-configuration logging

In Spring Boot 2.x, without providing any explicit logging configurations, we see logs printed in the console at the `INFO` level. This is happening because the Spring Boot default logging functionality uses the popular Logback logging framework.

Mainly, the Spring Boot logger is determined by the `spring-boot-starter-logging` artifact that (based on the provided configuration or auto-configuration) activates any of the supported logging providers (`java.util.logging`, log4j2, and Logback). This artifact can be imported explicitly or transitively (for instance, as a dependency of `spring-boot-starter-web`).

In this context, having a Spring Boot application with no explicit logging configurations will not log jOOQ messages. However, we can take advantage of jOOQ logging if we simply enable the `DEBUG` level (or `TRACE` for more verbose logging). For instance, we can do it in the `application.properties` as follows:

```
// set DEBUG level globally
logging.level.root=DEBUG

// or, set DEBUG level only for jOOQ
logging.level.org.jooq.tools.LoggerListener=DEBUG
```

You can practice this example in *SimpleLogging* for MySQL.

jOOQ logging with Logback/log4j2

If you already have Logback configured (for instance, via `logback-spring.xml`), then you'll need to add the jOOQ logger, as follows:

```
...
<!-- SQL execution logging is logged to the
     LoggerListener logger at DEBUG level -->
<logger name="org.jooq.tools.LoggerListener"
        level="debug" additivity="false">
  <appender-ref ref="ConsoleAppender"/>
</logger>

<!-- Other jOOQ related debug log output -->
<logger name="org.jooq" level="debug" additivity="false">
  <appender-ref ref="ConsoleAppender"/>
</logger>
...
```

You can practice this example in *Logback* for MySQL. If you prefer log4j2, then consider the *Log4j2* application for MySQL. The jOOQ logger is configured in `log4j2.xml`.

Turn off jOOQ logging

Turning on/off jOOQ logging can be done via the `set/withExecuteLogging()` setting. For instance, the following query will not be logged:

```
ctx.configuration().derive(
       new Settings().withExecuteLogging(Boolean.FALSE))
   .dsl()
   .select(PRODUCT.PRODUCT_NAME, PRODUCT.PRODUCT_VENDOR)
   .from(PRODUCT).fetch();
```

You can practice this example in *TurnOffLogging* for MySQL. Note that this setting doesn't affect the jOOQ Code Generator logging. That logging is configured with `<logging>LEVEL</logging>` (Maven), `logging = 'LEVEL'` (Gradle), or `.withLogging(Logging.LEVEL)` (programmatically). `LEVEL` can be any of `TRACE`, `DEBUG`, `INFO`, `WARN`, `ERROR`, and `FATAL`. Here is the Maven approach for setting the `WARN` level – log everything that is bigger or equal to the `WARN` level:

```
<configuration xmlns="...">
  <logging>WARN</logging>
</configuration>
```

You can practice this example in *GenCodeLogging* for MySQL.

In the second part of this section, let's tackle a suite of examples that should help you to get familiar with different techniques of customizing jOOQ logging. Based on these examples, you should be capable of solving your scenarios. Since these are just examples, they won't cover all possible cases, which is worth remembering in your real scenarios.

Customizing result set logging

By default, jOOQ truncates the logged result set to five records. However, we can easily log the entire result set via `format(int size)`, as shown here:

```
private static final Logger log =
        LoggerFactory.getLogger(...);

var result = ctx.select(...)...
                  .fetch();

log.debug("Result set:\n" + result.format
    (result.size()));
```

How about logging the whole result set for every query (excluding queries that rely on lazy, sequential access to an underlying JDBC `ResultSet`)? Moreover, let's assume that we plan to log the row number, as shown in the following figure:

```
+--------+----------------+----------+------------------------------------+
|order_id|quantity_ordered|product_id|product_name                        |
+--------+----------------+----------+------------------------------------+
|   10107|              30|         1|1969 Harley Davidson Ultimate Chopper| # 1
|   10121|              34|         1|1969 Harley Davidson Ultimate Chopper| # 2
|   10134|              41|         1|1969 Harley Davidson Ultimate Chopper| # 3
    ...
|   10386|              43|       110|Pont Yacht                          | # 2994
|   10397|              34|       110|Pont Yacht                          | # 2995
|   10414|              47|       110|Pont Yacht                          | # 2996
+--------+----------------+----------+------------------------------------+
 Total fetched row(s) : 2996
```

Figure 19.2 – Customizing result set logging

One approach for accomplishing this consists of writing a custom logger as `ExecuteListener` and overriding the `resultEnd()` method:

```
public class MyLoggerListener extends DefaultExecuteListener {

 private static final JooqLogger log =
    JooqLogger.getLogger(LoggerListener.class);

 @Override
 public void resultEnd(ExecuteContext ctx) {

  Result<?> result = ctx.result();

  if (result != null) {

    logMultiline("Total Fetched result",
                 result.format(), Level.FINE, result.size());
    log.debug("Total fetched row(s)", result.size());
  }
 }

 // inspired from jOOQ source code
```

```
  private void logMultiline(String comment,
            String message, Level level, int size) {
    // check the bundled code
  }
}
```

You can practice this example in *LogAllRS* for MySQL.

Customizing binding parameters logging

If we switch the logging level to `TRACE` (`logging.level.root=TRACE`), then we get more verbose jOOQ logging. For instance, the bind parameters are logged as a separate list, as shown in this example:

```
Binding variable 1 : 5000 (integer /* java.lang.Integer */)
Binding variable 2 : 223113 (bigint /* java.lang.Long */)
...
```

Challenge yourself to customize this list to look different and to be logged at `DEBUG` level. You can find some inspiration in *LogBind* for MySQL, which logs bindings like this:

```
... : [1] as integer /* java.lang.Integer */ - [5000]
... : [vintageCars] as bigint /* java.lang.Long */ - [223113]
...
```

How about log bindings as a nice formatted table? I'm looking forward to seeing your code.

Customizing logging invocation order

Let's assume that we plan to enrich jOOQ logging to log a chart, as shown here:

```
Executing query : select `classicmodels`.`orderdetail`.`order_id`, ...
Start Chart     :
                : 1,312.58|xxxxxxxxx xxx xxx x xxx xxxxxxxxxxx  xx x        xx x  x   xx  x x xxx
                : 1,137.76|xxxxxxxxx xxx xxx x xxx xxxxxxxxxxx  xx x x   x  xx x  x   xx  x xxxxx
                :   962.94|xxxxxxxxx xxx xxx x xxx xxxxxxxxxxx  xx x x   xx xx xx x   xxx xxxxxxx
                :   788.12|xxxxxxxxxxxxx xxxxx xxxxxxxxxxxxxxxxxxx x x x xx xx xxxx  xxxxxxxxxxxxx
                :   613.30|xxxxxxxxxxxxxxxxxxxxxxxxxxxxxxxxxxxxxxxxx x x xxxxx xxxxxxxxxxxxxxxxxxx
                :   438.48|xxxxxxxxxxxxxxxxxxxxxxxxxxxxxxxxxxxxxxxxxxxxxxxxxxxxxxxxxxxxxxxxxxxxxxx
                : --------+-----------------------------------------------------------------------
                :         | 1  5  10  16  22  28  34  40  48  54  60  66  72  78  84  91  97  103
End Chart       :
Fetched result  : +--------+----------+----------+----------+------------+--------------------
                : |order_id|product_id|price_each|product_id|product_line|product_name
                : +--------+----------+----------+----------+------------+--------------------
                : |   10107|         1|     81.35|         1|Motorcycles |1969 Harley Davidson
                : |   10121|         1|     86.13|         1|Motorcycles |1969 Harley Davidson   ...
                : |   10134|         1|     90.92|         1|Motorcycles |1969 Harley Davidson
                : |   10145|         1|     76.56|         1|Motorcycles |1969 Harley Davidson
                : |   10159|         1|     81.35|         1|Motorcycles |1969 Harley Davidson
                : +--------+----------+----------+----------+------------+--------------------
                : |...record(s) truncated...
Fetched row(s)  : 2996
```

Figure 19.3 – Logging a chart

This chart is logged only for `SELECT` statements that contain `PRODUCT.PRODUCT_ID` (represented on the X axis of the chart – category) and `PRODUCT.BUY_PRICE` (represented on the Y axis of the chart – value). Moreover, we don't take into account the queries that rely on lazy sequential access to the underlying JDBC `ResultSet`, such as `ctx.selectFrom(PRODUCT).collect(Collectors.toList());`. In such cases, jOOQ buffers for logging only the first five records, so, in most of the cases, the chart will be irrelevant.

The first step consists of writing a custom `ExecuteListener` (our own logger) and overriding the `resultEnd()` method – called after fetching a set of records from `ResultSet`. In this method, we search for `PRODUCT.PRODUCT_ID` and `PRODUCT.BUY_PRICE`, and if we find them, then we use the jOOQ `ChartFormat` API, as shown here:

```
@Override
public void resultEnd(ExecuteContext ecx) {

 if (ecx.query() != null && ecx.query() instanceof Select) {

  Result<?> result = ecx.result();
  if (result != null && !result.isEmpty()) {

   final int x = result.indexOf(PRODUCT.PRODUCT_ID);
```

```
    final int y = result.indexOf(PRODUCT.BUY_PRICE);

    if (x != -1 && y != -1) {

     ChartFormat cf = new ChartFormat()
      .category(x)
      .values(y)
      .shades('x');

     String[] chart = result.formatChart(cf).split("\n");

     log.debug("Start Chart", "");
     for (int i = 0; i < chart.length; i++) {
      log.debug("", chart[i]);
     }
     log.debug("End Chart", "");

    } else {
      log.debug("Chart", "The chart cannot be
                          constructed (missing data)");
    }
  }
 }
}
```

There is one more thing that we need. At this moment, our `resultEnd()` is invoked after the jOOQ's `LoggerListener.resultEnd()` is invoked, which means that our chart is logged after the result set. However, if you look at the previous figure, you can see that our chart is logged before the result set. This can be accomplished by reversing the order of invocation for the `fooEnd()` methods:

```
configuration.settings()
 .withExecuteListenerEndInvocationOrder(
    InvocationOrder.REVERSE);
```

So, by default, as long as the jOOQ logger is enabled, our loggers (the overridden `fooStart()` and `fooEnd()` methods) are invoked after their counterparts from the default logger (`LoggingLogger`). But, we can reverse the default order via two settings: `withExecuteListenerStartInvocationOrder()` for `fooStart()` methods and `withExecuteListenerEndInvocationOrder()` for `fooEnd()` methods. In our case, after reversion, our `resultEnd()` is called before `LoggingLogger`.

`resultEnd()`, and this is how we slipped our chart in the proper place. You can practice this example in *ReverseLog* for MySQL.

Wrapping jOOQ logging into custom text

Let's assume that we plan to wrap each query/routine default logging into some custom text, as shown in the following figure:

```
jOOQ LOG START          : ==========================START=========================
jOOQ LOG START          : Dialect:MYSQL | SQL Type: BATCH
jOOQ LOG START          : ========================================================
Executing batch query   : insert into `classicmodels`.`sale` (`fiscal_year`, `employee_number`, ...
Executing batch query   : insert into `classicmodels`.`sale` (`fiscal_year`, `employee_number`, ...
Executing batch query   : insert into `classicmodels`.`sale` (`fiscal_year`, `employee_number`, ...
Executing batch query   : update `classicmodels`.`sale` set `classicmodels`.`sale`.`fiscal_year`...
jOOQ LOG END            : ===========================END==========================
```

Figure 19.4 – Wrapping jOOQ logging into custom text

Before checking a potential solution in *WrapLog* for MySQL, consider challenging yourself to solve it.

Filtering jOOQ logging

Sometimes, we want to be very selective with what's being logged. For instance, let's assume that only the SQL strings for the `INSERT` and `DELETE` statements should be logged. So, after we turn off the jOOQ default logger, we set up our logger, which should be capable of isolating the `INSERT` and `DELETE` statements from the rest of the queries. A simple approach consists of applying a simple check, such as (`query instanceof Insert || query instanceof Delete`), where `query` is given by `ExecuteContext.query()`. However, this will not work in the case of plain SQL or batches containing the `INSERT` and `DELETE` statements. Specifically for such cases, we can apply a regular expression, such as "`^(?i:(INSERT|DELETE).*)$`", to the SQL string(s) returned via `ExecuteContext` passed in `renderEnd()`. While you can find these words materialized in code lines in *FilterLog* for MySQL, let's focus on another scenario.

Let's assume that we plan to log only regular `SELECT`, `INSERT`, `UPDATE`, and `DELETE` statements that contain a suite of given tables (plain SQL, batches, and routines are not logged at all). For instance, we can conveniently pass the desired tables via `data()` as follows:

```
ctx.data().put(EMPLOYEE.getQualifiedName(), "");
ctx.data().put(SALE.getQualifiedName(), "");
```

So, if a query refers to the EMPLOYEE and SALE tables, then, and only then, should it be logged. This time, relying on regular expressions can be a little bit sophisticated and risky. It would be more proper to rely on a VisitListener that allows us to inspect the AST and extract the referred tables of the current query with a robust approach. Every QueryPart passes through VisitListener, so we can inspect its type and collect it accordingly:

```
private static class TablesExtractor
      extends DefaultVisitListener {

 @Override
 public void visitEnd(VisitContext vcx) {

  if (vcx.renderContext() != null) {
    if (vcx.queryPart() instanceof Table) {
     Table<?> t = (Table<?>) vcx.queryPart();

     vcx.configuration().data()
            .putIfAbsent(t.getQualifiedName(), "");
    }
  }
 }
}
```

When VisitListener finishes its execution, we have already traversed all QueryPart, and we've collected all the tables involved in the current query, so we can compare these tables with the given tables and decide whether or not to log the current query. Note that our VisitListener has been declared as private static class because we use it internally in our ExecuteListener (our logger), which orchestrates the logging process. More precisely, at the proper moment, we append this VisitListener to a configuration derived from the configuration of ExecuteContext, passed to our ExecuteListener. So, this VisitListener is not appended to the configuration of DSLContext that is used to execute the query.

The relevant part of our logger (ExecuteListener) is listed here:

```
public class MyLoggerListener extends DefaultExecuteListener {
 ...

 @Override
```

```
public void renderEnd(ExecuteContext ecx) {

  if (ecx.query() != null &&
      !ecx.configuration().data().isEmpty()) {
   ...
   Configuration configuration = ecx.configuration()
           .deriveAppending(new TablesExtractor());
   ...
   if (configuration.data().keySet().containsAll(tables)) {
    ...
   }
   ...
}
```

Check out the highlighted code. The `deriveAppending()` method creates a derived `Configuration` from this one (by "this one", we understand `Configuration` of the current `ExecuteContext`, which was automatically derived from the `Configuration` of `DSLContext`), with appended visit listeners. Practically, this `VisitListener` is inserted into `Configuration` through `VisitListenerProvider`, which is responsible for creating a new listener instance for every rendering life cycle.

However, what's the point of this? In short, it is all about performance and scopes (`org.jooq.Scope`). `VisitListener` is intensively called; therefore, it can have some impact on rendering performance. So, in order to minimize its usage, we ensure that it is used only in the proper conditions from our logger. In addition, `VisitListener` should store the list of tables that are being rendered in some place accessible to our logger. Since we choose to rely on the `data()` map, we have to ensure that the logger and `VisitListener` have access to it. By appending `VisitListener` to the logger via `deriveAppending()`, we append its `Scope` as well, so the `data()` map is accessible from both. This way, we can share custom data between the logger and `VisitContext` for the entire lifetime of the scope.

You can practice this example in *FilterVisitLog* for MySQL. Well, that's all about logging. Next, let's talk about testing.

jOOQ testing

Accomplishing jOOQ testing can be done in several ways, but we can immediately highlight that the less appealing option relies on mocking the jOOQ API, while the best option relies on writing integration tests against the production database (or at least against an in-memory database). Let's start with the option that fits well only in simple cases, mocking the jOOQ API.

Mocking the jOOQ API

While mocking the JDBC API can be really difficult, jOOQ solves this chore and exposes a simple mock API via `org.jooq.tools.jdbc`. The climax of this API is represented by the `MockConnection` (for mocking a database connection) and `MockDataProvider` (for mocking query executions). Assuming that jUnit 5 is used, we can mock a connection like this:

```
public class ClassicmodelsTest {

 public static DSLContext ctx;

 @BeforeAll
 public static void setup() {

  // Initialise your data provider
  MockDataProvider provider = new ClassicmodelsMockProvider();
  MockConnection connection = new MockConnection(provider);

  // Pass the mock connection to a jOOQ DSLContext
  ClassicmodelsTest.ctx = DSL.using(
                  connection, SQLDialect.MYSQL);

  // Optionally, you may want to disable jOOQ logging
  ClassicmodelsTest.ctx.configuration().settings()
          .withExecuteLogging(Boolean.FALSE);
 }

 // add tests here
}
```

Before writing tests, we have to prepare `ClassicmodelsMockProvider` as an implementation of `MockDataProvider` that overrides the `execute()` method. This method returns an array of `MockResult` (each `MockResult` represents a mock result). A possible implementation may look as follows:

```
public class ClassicmodelsMockProvider
                  implements MockDataProvider {

 private static final String ACCEPTED_SQL =
```

```
    "(SELECT|UPDATE|INSERT|DELETE).*";
 ...
 @Override
 public MockResult[] execute(MockExecuteContext mex)
   throws SQLException {

  // The DSLContext can be used to create
  // org.jooq.Result and org.jooq.Record objects
  DSLContext ctx = DSL.using(SQLDialect.MYSQL);

  // So, here we can have maximum 3 results
  MockResult[] mock = new MockResult[3];

  // The execute context contains SQL string(s),
  // bind values, and other meta-data
  String sql = mex.sql();

  // Exceptions are propagated through the JDBC and jOOQ APIs
  if (!sql.toUpperCase().matches(ACCEPTED_SQL)) {
   throw new SQLException("Statement not supported: " + sql);
  }

  // From this point forward, you decide, whether any given
  // statement returns results, and how many
  ...
}
```

Now, we are ready to go! First, we can write a test. Here is an example:

```
@Test
public void sampleTest() {

 Result<Record2<Long, String>> result =
   ctx.select(PRODUCT.PRODUCT_ID, PRODUCT.PRODUCT_NAME)
      .from(PRODUCT)
      .where(PRODUCT.PRODUCT_ID.eq(1L))
      .fetch();
```

```
 assertThat(result, hasSize(equalTo(1)));
 assertThat(result.getValue(0, PRODUCT.PRODUCT_ID),
   is(equalTo(1L)));
 assertThat(result.getValue(0, PRODUCT.PRODUCT_NAME),
   is(equalTo("2002 Suzuki XREO")));
}
```

The code that mocks this behavior is added in `ClassicmodelsMockProvider`:

```
private static final String SELECT_ONE_RESULT_ONE_RECORD =
 "select ... where `classicmodels`.`product`.`product_id`=?";
...
} else if (sql.equals(SELECT_ONE_RESULT_ONE_RECORD)) {

  Result<Record2<Long, String>> result
    = ctx.newResult(PRODUCT.PRODUCT_ID, PRODUCT.PRODUCT_NAME);
  result.add(
    ctx.newRecord(PRODUCT.PRODUCT_ID, PRODUCT.PRODUCT_NAME)
       .values(1L, "2002 Suzuki XREO"));

   mock[0] = new MockResult(-1, result);
}
```

The first argument of the `MockResult` constructor represents the number of affected rows, and `-1` represents that the row count not being applicable. In the bundled code (*Mock* for MySQL), you can see more examples, including testing batching, fetching many results, and deciding the result based on the bindings. However, do not forget that jOOQ testing is equivalent to testing the database interaction, so mocking is proper only for simple cases. Do not use it for transactions, locking, or testing your entire database!

If you don't believe me, then follow Lukas Eder's statement: *"The fact that mocking only fits well in a few cases can't be stressed enough. People will still attempt to use this SPI, because it looks so easy to do, not thinking about the fact that they're about to implement a full-fledged DBMS in the poorest of ways. I've had numerous users to whom I've explained this 3-4x: 'You're about to implement a full-fledged DBMS" and they keep asking me: "Why doesn't jOOQ 'just' execute this query when I mock it?" – "Well jOOQ *isn't* a DBMS, but it allows you to pretend you can write one, using the mocking SPI." And they keep asking again and again. Hard to imagine what's tricky about this, but as much as it helps with SEO (people want to solve this problem, then discover jOOQ), I regret leading some developers down this path... It's excellent though to test some converter and mapping integrations within jOOQ."*

Writing integration tests

A quick approach for writing integration tests for jOOQ relies on simply creating `DSLContext` for the production database. Here is an example:

```
public class ClassicmodelsIT {

 private static DSLContext ctx;

 @BeforeAll
 public static void setup() {

  ctx = DSL.using("jdbc:mysql://localhost:3306/
    classicmodels" + "?allowMultiQueries=true", "root", "root");
 }

 @Test
 ...
}
```

However, this approach (exemplified in *SimpleTest* for MySQL) fits well for simple scenarios that don't require dealing with transaction management (begin, commit, and rollback). For instance, if you just need to test your `SELECT` statements, then most probably this approach is all you need.

Using SpringBoot @JooqTest

On the other hand, it's a common scenario to run each integration test in a separate transaction that rolls back in the end, and to achieve this while testing jOOQ in Spring Boot, you can use the `@JooqTest` annotation, as shown here:

```
@JooqTest
@ActiveProfiles("test") // profile is optional
public class ClassicmodelsIT {

 @Autowired
 private DSLContext ctx;

 // optional, if you need more control of Spring transactions
 @Autowired
 private TransactionTemplate template;
```

```
 @Test
 ...
}
```

This time, Spring Boot automatically creates `DSLContext` for the current profile (of course, using explicit profiles is optional, but I added it here, since it is a common practice in Spring Boot applications) and automatically wraps each test in a separate Spring transaction that is rolled back at the end. In this context, if you prefer to use jOOQ transactions for certain tests, then don't forget to disable Spring transaction by annotating those test methods with `@Transactional(propagation=Propagation.NEVER)`. The same is true for the usage of `TransactionTemplate`. You can practice this example in *JooqTest* for MySQL, which contains several tests, including jOOQ optimistic locking via `TransactionTemplate` and via jOOQ transactions.

By using Spring Boot profiles, you can easily configure a separate database for tests that is (or not) identical to the production database. In *JooqTestDb*, you have the MySQL `classicmodels` database for production and the MySQL `classicmodels_test` database for testing (both of them have the same schema and data and are managed by Flyway).

Moreover, if you prefer an in-memory database that is destroyed at the end of testing, then in *JooqTestInMem* for MySQL, you have the on-disk MySQL `classicmodels` database for production and the in-memory H2 `classicmodels_mem_test` database for testing (both of them have the same schema and data and are managed by Flyway). In these two applications, after you inject the `DSLContext` prepared by Spring Boot, you have to point jOOQ to the test schema – for instance, for the in-memory database, as shown here:

```
@JooqTest
@ActiveProfiles("test")
@TestInstance(Lifecycle.PER_CLASS)
public class ClassicmodelsIT {

 @Autowired
 private DSLContext ctx;

 // optional, if you need more control of Spring transactions
 @Autowired
 private TransactionTemplate template;

 @BeforeAll
```

```
 public void setup() {

  ctx.settings()
  // .withExecuteLogging(Boolean.FALSE) // optional
      .withRenderNameCase(RenderNameCase.UPPER)
      .withRenderMapping(new RenderMapping()
      .withSchemata(
        new MappedSchema().withInput("classicmodels")
                          .withOutput("PUBLIC")));
 }

 @Test
 ...
}
```

You should be familiar with this technique from *Chapter 17, Multitenancy in jOOQ*.

Using Testcontainers

Testcontainers (`https://www.testcontainers.org/`) is a Java library that allows us to perform JUnit tests in lightweight Docker containers, created and destroyed automatically for the most common databases. So, in order to use Testcontainers, you have to install Docker.

Once you've installed Docker and provided the expected dependencies in your Spring Boot application, you can start a container and run some tests. Here, I've done it for MySQL:

```
@JooqTest
@Testcontainers
@ActiveProfiles("test")
public class ClassicmodelsIT {

 private static DSLContext ctx;

 // optional, if you need more control of Spring transactions
 @Autowired
 private TransactionTemplate template;

 @Container
 private static final MySQLContainer sqlContainer =
   new MySQLContainer<>("mysql:8.0")
```

```
      .withDatabaseName("classicmodels")
      .withStartupTimeoutSeconds(1800)
      .withCommand("--authentication-
                      policy=mysql_native_password");

  @BeforeAll
  public static void setup() throws SQLException {

   // load into the database the schema and data
   Flyway flyway = Flyway.configure()
     .dataSource(sqlContainer.getJdbcUrl(),
      sqlContainer.getUsername(), sqlContainer.getPassword())
     .baselineOnMigrate(true)
     .load();
   flyway.migrate();

   // obtain a connection to MySQL
   Connection conn = sqlContainer.createConnection("");

   // intialize jOOQ DSLContext
   ctx = DSL.using(conn, SQLDialect.MYSQL);
  }

  // this is optional since is done automatically anyway
  @AfterAll
  public static void tearDown() {
   if (sqlContainer != null) {
    if (sqlContainer.isRunning()) {
      sqlContainer.stop();
    }
   }
  }

  @Test
  ...
}
```

Note that we've populated the test database via Flyway, but this is not mandatory. You can use any other dedicated utility, such as Commons DbUtils. For instance, you can do it via `org.testcontainers.ext.ScriptUtils`, like this:

```
...
var containerDelegate =
  new JdbcDatabaseDelegate(sqlContainer, "");
ScriptUtils.runInitScript(containerDelegate,
  "integration/migration/V1.1__CreateTest.sql");
ScriptUtils.runInitScript(containerDelegate,
  "integration/migration/afterMigrate.sql");
...
```

That's it! Now, you can spin out a throwaway container for testing database interaction. Most probably, this is the most preferable approach for testing jOOQ applications in production. You can practice this example in *Testcontainers* for MySQL.

Testing R2DBC

Finally, if you are using jOOQ R2DBC, then writing tests is quite straightforward.

In the bundled code, you can find three examples for MySQL, as follows:

- The *TestR2DBC* example: `ConnectionFactory` is created via `ConnectionFactories.get()` and `DSLContext` via `ctx = DSL.using(connectionFactory)`. The tests are executed against a production database.
- The *TestR2DBCDb* example: `ConnectionFactory` is automatically created by Spring Boot and `DSLContext` is created as `@Bean`. The tests are executed against a MySQL test database (`classicmodels_test`), similar to the production one (`classicmodels`).
- The *TestR2DBCInMem* example: `ConnectionFactory` is automatically created by Spring Boot and `DSLContext` is created as `@Bean`. The tests are executed against an H2 in-memory test database (`classicmodels_mem_test`).

Summary

As you just saw, jOOQ has solid support for logging and testing, proving yet again that it is a mature technology ready to meet the most demanding expectations of a production environment. With a high rate of productivity and a small learning curve, jOOQ is the first choice that I use and recommend for projects. I strongly encourage you to do the same!

Index

A

B

C

D

E

F

G

H

I

J

K

L

M

N

O

R

S

T

U

Packt.com

Subscribe to our online digital library for full access to over 7,000 books and videos, as well as industry leading tools to help you plan your personal development and advance your career. For more information, please visit our website.

Why subscribe?

- Spend less time learning and more time coding with practical eBooks and Videos from over 4,000 industry professionals
- Improve your learning with Skill Plans built especially for you
- Get a free eBook or video every month
- Fully searchable for easy access to vital information
- Copy and paste, print, and bookmark content

Did you know that Packt offers eBook versions of every book published, with PDF and ePub files available? You can upgrade to the eBook version at packt.com and as a print book customer, you are entitled to a discount on the eBook copy. Get in touch with us at customercare@packtpub.com for more details.

At www.packt.com, you can also read a collection of free technical articles, sign up for a range of free newsletters, and receive exclusive discounts and offers on Packt books and eBooks.

Other Books You May Enjoy

If you enjoyed this book, you may be interested in these other books by Packt:

Designing Hexagonal Architecture with Java

Davi Vieira

ISBN: 9781801816489

- Find out how to assemble business rules algorithms using the specification design pattern
- Combine domain-driven design techniques with hexagonal principles to create powerful domain models
- Employ adapters to make the system support different protocols such as REST, gRPC, and WebSocket
- Create a module and package structure based on hexagonal principles
- Use Java modules to enforce dependency inversion and ensure isolation between software components
- Implement Quarkus DI to manage the life cycle of input and output ports

Practical Cloud-Native Java Development with MicroProfile

Emily Jiang, Andrew McCright, John Alcorn, David Chan, Alasdair Nottingham

ISBN: 9781801078801

- Understand best practices for applying the 12-Factor methodology while building cloud-native applications
- Create client-server architecture using MicroProfile Rest Client and JAX-RS
- Configure your cloud-native application using MicroProfile Config
- Secure your cloud-native application with MicroProfile JWT
- Become well-versed with running your cloud-native applications in Open Liberty
- Grasp MicroProfile Open Tracing and learn how to use Jaeger to view trace spans
- Deploy Docker containers to Kubernetes and understand how to use ConfigMap and Secrets from Kubernetes

Packt is searching for authors like you

If you're interested in becoming an author for Packt, please visit `authors.packtpub.com` and apply today. We have worked with thousands of developers and tech professionals, just like you, to help them share their insight with the global tech community. You can make a general application, apply for a specific hot topic that we are recruiting an author for, or submit your own idea.

Share Your Thoughts

Now you've finished *jOOQ Masterclass*, we'd love to hear your thoughts! Scan the QR code below to go straight to the Amazon review page for this book and share your feedback or leave a review on the site that you purchased it from.

https://packt.link/r/1800566891

Your review is important to us and the tech community and will help us make sure we're delivering excellent quality content.